Advanced Strength and Applied Stress Analysis

Advanced Strength and Applied Stress Analysis

Richard G. Budynas

Professor of Mechanical Engineering
Rochester Institute of Technology

McGraw-Hill Book Company

New York   St. Louis   San Francisco   Auckland   Bogotá   Düsseldorf
Johannesburg   London   Madrid   Mexico   Montreal   New Delhi   Panama
Paris   São Paulo   Singapore   Sydney   Tokyo   Toronto

Advanced Strength and Applied Stress Analysis

567890 FGRFGR 83210

This book was set in Times. The editors were B. J. Clark and J. W. Maisel;
the cover was designed by Rafael Hernandez; the production supervisor was Leroy A. Young.
The drawings were done by Oxford Illustrators Limited.
Fairfield Graphics was printer and binder.

Library of Congress Cataloging in Publication Data

Budynas, Richard G
    Advanced strength and applied stress analysis.

    1. Strength of materials.   2. Strains and stresses.
I.   Title.
TA405.B82   620.1'12   76-55693
ISBN 0-07-008828-4

To my sons
David and Chris

CONTENTS

Appendixes

When dealing with a design or redesign problem a mechanical design engineer knows that the design must (1) function according to some prescribed requirements and (2) have an acceptable level of structural integrity, i.e., each element within the system must have the strength to carry out its designated function. Stress analysis attempts to ensure that each element of a given system will not fail to meet the structural requirements during the course of the system's performance. This book, in turn, attempts to expand the background of mechanical designers or analysts with respect to understanding the internal behavior of mechanical elements under the action of applied loads.

It is assumed that the reader has previously completed a course in elementary strength of materials, which introduces students to the fundamental aspects of stress analysis and enables them to understand internal forces and moments and their relation to the internal stresses and strains within simple elements under the influence of simple loading configurations and the resulting deflections or deformations of the overall element. Many assumptions are made in a first study that students neither fully understand nor retain. Moreover, simplifications are proffered that tend to restrict students' abilities to expand their analyses to more complicated situations and to a true understanding of the "real" nature of the simple models introduced in the first, elementary study. Thus, the first course in the theory of strength of materials should be considered only as an initial framework for the study of stress analysis.

Practical stress-analysis problems include both simple ones which agree with idealistic models having simple classical, closed-form solutions and those which are of a complex nature not amenable to classical methods. For analysts to be capable of handling this wide range of problems, they must first be aware of the basic phenomenon they are dealing with and the fundamental governing equations related to this phenomenon. They must also acquire a sort of file or collection of known classical problems, complete with solutions and analytical methods. Finally, analysts should acquire a skill and expertise in the area of experimental stress analysis and numerical methods.

With a scope that attempts to include these elements, no single book can ever attain completeness. However, in this book the fundamental aspects of each specific category in terms of the mathematical level of that of a junior to senior undergraduate student in engineering are covered. And with mathematical complexity kept to a minimum, certain fundamental advanced topics traditionally considered graduate-level material are presented with little difficulty, so that the transition to advanced graduate-level topics in the future will be made more easily and effectively.

Without a sound physical understanding of such basic phenomena as stress, strain, etc., it is quite difficult to master the subject matter and to utilize the mathematical formulations in practical situations. It is similarly difficult to understand the bases behind certain assumptions and simplifications made in order to arrive at reasonable approximations of actual behavior. But once such an understanding is achieved, the study of more advanced formulations and techniques used in applied stress analysis is more meaningful. Chapter 1 of this book deals with this aspect. It, as well as Chapter 2, is a review as well as an exposure to the mathematical notation used in the book.

Chapters 2 to 6 deal with problems that can be solved by various classical methods providing closed-form solutions to theoretical models. Chapter 2 is a review of elementary strength of materials. Chapter 3 contains various topics that are natural extensions of the elementary problems and are normally covered in an advanced strength of materials course. Chapter 4 is devoted to energy techniques used in stress analysis, because this method of analysis is extremely important for dealing with complex systems containing a large number of standard elements. Energy techniques also provide approximate solutions to problems difficult to solve by standard classical approaches. Chapter 5 presents analytical approaches used in design and failure analysis and approximation techniques used in practice. Chapter 6 contains a brief introduction to the mathematical theory of elasticity.

For the types of problems for which the classical analytical methods are either impractical or virtually impossible, two introductory chapters on experimental and numerical methods are included: Chapter 7 is an introduction to experimental stress analysis, and Chapter 8 is an introduction to the finite-element technique.

The author has thus attempted to provide a modern and concise introduction to the entire scope of the area of stress analysis, using consistent notation, well-defined coordinate systems, and both USCS and SI units.

Richard G. Budynas

A	area, area of cross section, light-wave vector amplitude, analyzer filter axis
A_m	area bounded by perimeter centerline of thin tube
a	dimension, varying amplitude of light-wave vector
b	dimension, beam width
b_e	equivalent width of composite-beam section
C	material calibration constant for transmission photoelasticity
C_1	correction factor for photoelastic coatings
c	distance from neutral axis to extreme beam fiber, speed of light, speed of strain pulse
D	weight density
d	diameter, distance
d_x, d_y, d_z	direction numbers of a plane surface
E	modulus of elasticity, output voltage
$\mathbf{E}$	modulus-of-elasticity scale factor
e	eccentricity, distance from centroidal axis to neutral axis of curved beam
F	force, concentrated load
$\mathbf{F}$	force scale factor
F_e	equivalent concentrated force
$\bar{F}_x, \bar{F}_y, \bar{F}_z$	body forces per unit volume
f	material calibration constant for photoelastic coatings, finite-element nodal force
f_s	strength uncertainty factor
f_σ	stress uncertainty factor
f	coefficient of friction
G	modulus of elasticity in shear
h	dimension, depth of beam
h_p	depth of plastic region
I	moment of inertia of cross-sectional area
I_m, I_n	principal moments of inertia of cross-sectional area

I_1, I_2, I_3	stress invariants
J	polar moment of inertia of cross-sectional area
K_t	stress concentration factor, strain-gage transverse-sensitivity factor
k	stress-optic coefficient
k_s	spring constant
L	length
L	length scale factor
l	length
l_g	length of strain gage
l_p	length of strain pulse
M	net internal bending moment, reaction moment
M	moment scale factor
M_e	maximum elastic bending moment
M_f	equivalent bending moment for plastic deflections
M_p	maximum plastic bending moment
m	mass
m, n	principal axes of inertia
N	number of cycles, photoelastic fringe order
N_θ	photoelastic fringe order at angle of incidence θ
n	design factor, index of refraction, angular speed, r/min
n_x, n_y, n_z	directional cosines
P	force, concentrated load, product of inertia of cross-sectional area, polarizer filter axis
P	pressure scale factor
P_L	limit force
p	pressure, press-fit interference pressure
Q	first moment of partial area of beam section, dummy force
q	distributed load intensity
R	reaction force, radius, electrical resistance
R_g	strain-gage nominal resistance
r	radius
r, θ, z	cylindrical coordinates
r_g	radius of gyration of cross-sectional area $= \sqrt{I/A}$
S	strength
S_A	strain-gage factor for uniaxial strain field
S_g	strain-gage factor for uniaxial stress field
S_m	length of perimeter centerline of thin tube
s	position, curvilinear coordinate
T	torsional moment
T	temperature
t	thickness, time
t_p	pulse time
t_x, t_y, t_z	directional cosines of net shear stress
U	strain energy
u	strain energy per unit volume
u, v, w	displacements in x, y, z directions, respectively
u_r, u_θ, u_z	displacements in r, θ, z directions, respectively
V	net internal shear force, input voltage
W	weight, work

W_c	complementary work
w	force per unit length, work per unit volume, width
x, y, z	rectangular coordinates
x_p	position of initiation of plastic behavior in beam
y_c	displacement of beam centroidal axis
Z	elastic section modulus $= I/c$
Z_p	plastic section modulus

Greek

α	angle, linear coefficient of thermal expansion, location of neutral plane for unsymmetrical bending
β	angle
γ	shear strain
Δ	change of a designated function, position phase change of light wave
δ	deflection, press-fit radial interference
δ	displacement scale factor
δ_{ij}	Kronecker delta
ϵ	normal strain
ϵ	strain scale factor
ϵ_a	axial strain
ϵ_e	elastic strain limit
ϵ_f	fracture strain
ϵ_t	brittle-coating threshold strain
Θ	angular deflection
θ	angle, angular deflection
θ'	angular deflection per unit length
λ	wavelength, Lamé constant
μ	Lamé constant
ν	Poisson's ratio
ρ	radius of curvature, mass density
σ	normal stress
σ	stress scale factor
σ_e	elastic stress limit
τ	shear stress
Φ	complementary strain energy, Airy stress function
Φ_t	torsional stress function
ϕ	helix angle of twist
ω	angular speed, rad/s
∇^2	Laplacian operator

Advanced Strength and Applied Stress Analysis

BASIC CONCEPTS OF FORCE, STRESS, STRAIN, AND DISPLACEMENT

1-0 INTRODUCTION

The concepts of force and force distributions are introduced in fundamental courses in statics, and the concepts of stress, strain, and elastic displacements are presented in a first course on strength of materials. The importance of the basic ideas developed in these courses cannot be overemphasized. To understand more advanced formulations, experimental methods, and the implementation of design improvements, a comprehensive knowledge of the fundamentals of statics and strength of materials is essential.

This chapter reviews the fundamental properties of state in the traditional ordering; i.e., beginning with external forces, the discussion then leads into the definition of internal force distributions (stress). Once this definition is made, strain is related to stress, and finally displacement is related to strain. However, in terms of physical significance and in approaching real engineering problems the process is reversed, and normally the stress or strain formulations are first based on an analysis of deformation or deflections. For example, the linear bending-stress distribution for straight beams is based on the first assumption that plane surfaces initially perpendicular to the bending axis remain plane as the beam undergoes bending. Once this assumption is made, the equations for the equilibrium of forces are utilized, and the bending equation is then completed. Thus in a physical sense applied forces generate elastic deflections which can be analytically related to strains and subsequently to internal stresses.

1

The material in this chapter tends to emphasize the physical understanding of the important state properties and should be a review for the reader, who is urged to reexamine his own course notes and textbook on strength of materials upon completion of this chapter.

1-1 FORCE DIAGRAMS

Free-body diagrams are a necessary tool in the solution of any stress-analysis problem, and any reader who is weak on this point should attempt to correct this deficiency early, as drawing free-body diagrams is generally the first and most important step of an analysis.

Generally, an element under analysis is supported or attached to another element. In order to analyze the element, the element is completely isolated from the supports; then all the applied loads *on* the element are shown on the element. Finally, every type of internal force and moment which possibly can be transmitted through the supports is shown on the isolated element. The values of the support forces and moments are then obtained using the appropriate equations of motion based on the dynamic state of the element within the overall structure. For example, any element within a structure is always in a specific state of dynamic equilibrium. Hence, when the element is isolated from the structure, the applied and support loads acting on the element must be such that the isolated element is in the same state of dynamic equilibrium as that specified within the overall structure. Several examples of element isolation and support analysis are shown in Fig. 1-1, where the structures given are in a state of static equilibrium. Naturally, these examples are not meant to be comprehensive and are only given to illustrate the method of element isolation. At this point, it is recommended that the reader review basic statics.

In the method of element isolation, it is necessary to examine each support point and determine exactly the types of forces and moments the support can transmit. Note that if the 500-lb force shown in Fig. 1-1a were not present, the support force N would be zero and it would obviously be unnecessary to show N on the diagram. However, in many cases it is a safe procedure to show first that the support is capable of providing a possible reaction. Then, by using the dynamic-equilibrium equations, the value of the reaction is determined. For example, an error might arise in problems like those shown in Fig. 1-1c and d if all the possible reactions were not shown.

1-2 FORCE DISTRIBUTIONS

One is generally quite confident when dealing with concentrated forces and moments like those shown in Sec. 1-1, but when one is dealing with distributed forces over lengths or areas, the force distributions become more abstract to the student and the physical understanding becomes clouded. This is unfortu-

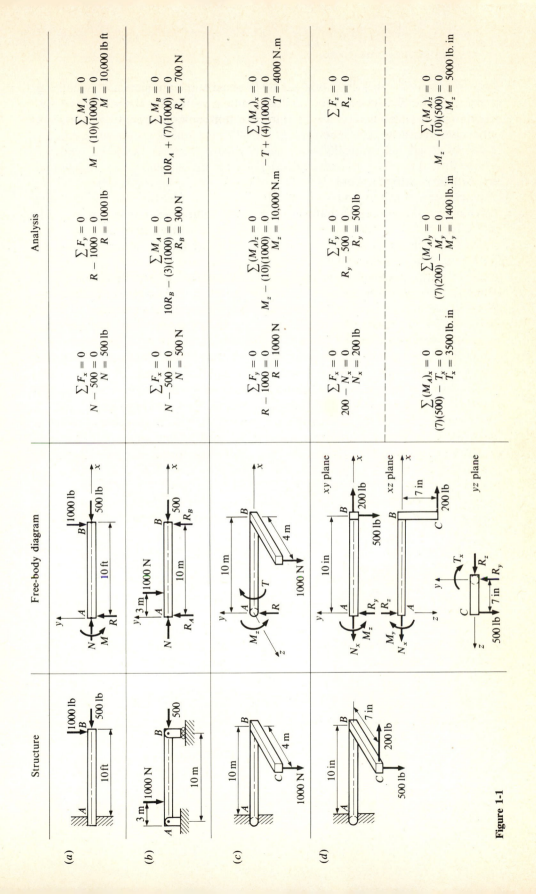

Figure 1-1

nate, since *real* forces in nature are distributed forces, whereas concentrated forces are merely the invention and abstraction of the analyst to simplify the solution of the problem. As an illustration, a uniform weight is placed at the end of a cantilever beam, as shown in Fig. 1-2a. The load distributions shown in Fig. 1-2b to d are all statically equivalent. Figure 1-2b is idealized, showing concentrated forces and moments which are the *net* effects of the applied load and wall reactions. Figure 1-2c is less idealized, force distributions having replaced the concentrated forces and moments. However, the distributions have been *assumed* to be uniform. A more realistic model would incorporate the effects of *deformations* of the weight, beam, and wall, as shown in Fig. 1-2d. One could take further steps to model the exact structure more accurately, but the point to be made is that the model chosen depends on the results desired. If the analyst is concerned with the stresses in section c-c, any one of the three models shown would essentially provide the same results because the bearing stresses dissipate quickly the farther the section is from the wall and load zone. Thus, for section c-c, it makes sense to use the simplest model, Fig. 1-2b. However, within the support or loading zone, each model would give different results. Thus, the analyst should be careful when investigating the stresses in these areas.

Note also that as a structure deforms or deflects, the net reactions will change. This means that in order to perform a precise calculation of the reactions, the final deformed geometry must be known before the equilibrium equations are used. When deformations are small, however, the deformed structure is almost identical to the undeformed geometry and the errors in evaluating forces and moments are negligible. The majority of structural

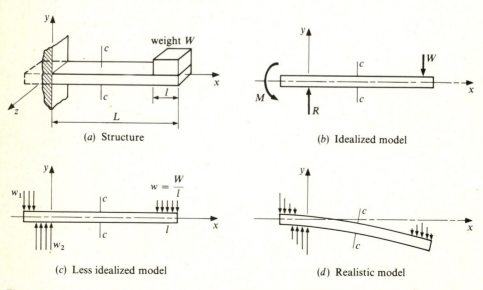

(a) Structure

(b) Idealized model

(c) Less idealized model

(d) Realistic model

Figure 1-2

problems fall in this category, but in a small class of problems the coupling between deflections and the corresponding reaction forces is quite large and the deformations affect the force analysis considerably. These problems, which include elastic stability problems (buckling), contact stresses, and combined bending and axial loading of beams, are considered later in the text.

If the load distribution is known or assumed, a method which simplifies the analysis is the technique of determining an "equivalent" concentrated force. Consider the beam shown in Fig. 1-3a, which is carrying a distributed load $w(x)$, in units of force per unit length, which varies along the length. The actual wall reaction at point A will look as shown in Fig. 1-3b, but what is normally wanted first is the net force and moment exerted on the beam by the wall. Thus, the force R and moment M shown in Fig. 1-3c are the net result of the force distribution w_1 and w_2. For the beam to be in equilibrium R must be equal and opposite to the total force exerted by $w(x)$, and M must be equal and opposite to the total moment exerted by $w(x)$. To simplify the determination of the support reactions, an equivalent force F_e is determined, where F_e is the total force exerted by $w(x)$. The equivalent force must be applied at a particular point x_e such that the force exerts the same net moment as $w(x)$ applies (see Fig. 1-3d). The method of determining F_e and x_e is quite simple and is normally developed in a first course on statics, but because of its importance, it will be repeated here.

Examine a small portion of the load acting at $x = x_1$ over an infinitesimal span of dx_1. The load distribution can be considered constant over this very small distance equal to $w(x_1)$, as shown in Fig. 1-4. The force exerted on dx_1 is then $w(x_1)\,dx_1$. The total force due to the distribution $w(x)$ for every value of x

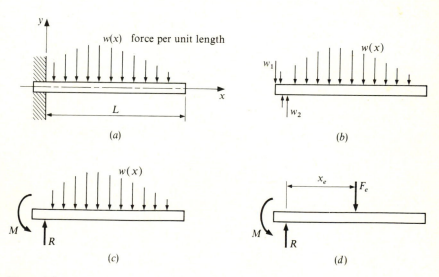

(a)

(b)

(c)

(d)

Figure 1-3

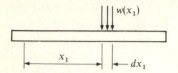

Figure 1-4

from $x_1 = 0$ to $x_1 = L$ is the sum, or integral, of each $w(x_1)\, dx_1$. Thus

$$F_e = \int_0^L w(x_1)\, dx_1 \tag{1-1}$$

and the total moment exerted at $x = 0$ is

$$M_e = \int_0^L x_1 w(x_1)\, dx_1 \tag{1-2}$$

However, $M_e = F_e x_e$; thus

$$x_e = \frac{M_e}{F_e} = \frac{\displaystyle\int_0^L x w(x)\, dx}{\displaystyle\int_0^L w(x)\, dx} \tag{1-3}$$

It can be seen from Eq. (1-1) that the equivalent force is simply the "area" of the force distribution over its length of application, and the position of application of the equivalent force is at the centroid of this area [Eq. (1-3)]. For simple load distributions like those shown in Fig. 1-5, the integrations in Eqs.

Load distribution   Equivalent concentrated force

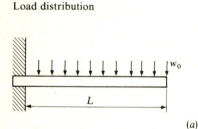
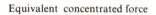
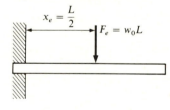

(a)

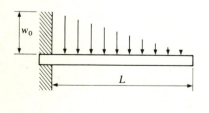
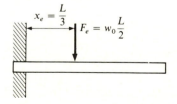

(b)

Figure 1-5

Load distribution

Equivalent concentrated force

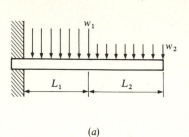

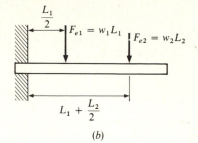

(a)

(b)

Figure 1-6

(1-1) and (1-3) are unnecessary, as the areas and centroids are known. When the distribution is discontinuous, like that in Fig. 1-6, equivalent forces for each zone where the loading is continuous can be found, as shown in Fig. 1-6b

It should be obvious that the equivalent force of the total load distribution is used only to find the support reactions and cannot be used for analyzing internal forces or deflections throughout the basic element. To analyze the internal forces within the element, it is first necessary to isolate that portion of the element which exposes the surface to be examined. The equivalent-force approach can then be reapplied to whatever load distribution remains on that portion of the element. For example, consider the uniformly loaded cantilever beam shown in Fig. 1-7a. The reader should be able to prove that the wall supplies a total force on the beam of w_0L upward and a total moment reaction of $0.5w_0L^2$ counterclockwise. If one wants to find the net internal force and moment occurring at a surface located at $x = x_1$, the equivalent force found in solving the wall reactions *cannot* be used. First, make a "break" at $x = x_1$ and isolate the right-hand portion of the beam (see Fig. 1-7b, showing the actual load distribution that exists on that portion). In order to determine the internal reactions R_1 and M_1 it is necessary to determine the equivalent force of the load distribution acting on the portion of the element under consideration. Thus, from Fig. 1-7b,

$$F_e = w_0(L - x_1) \qquad \text{and} \qquad x_e = \frac{L - x_1}{2}$$

From this the reader should be able to obtain R_1 and M_1:†

$$R_1 = w_0(L - x_1) \qquad \text{and} \qquad M_1 = \frac{w_0}{2}(L - x_1)^2$$

It can be seen that the internal force and moment in the beam depends on the value of x_1, which is illustrated in Fig. 1-7d.

† The notation used here for internal shear force and bending moments does not necessarily agree with the standard conventions used in strength of materials. This is discussed in Chap. 2.

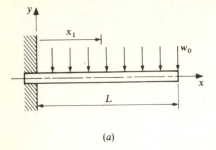

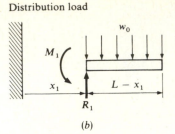

(a)

Distribution load

(b)

Equivalent concentrated force

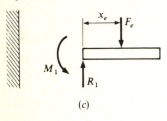

(c)

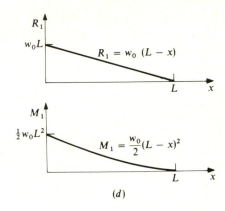

(d)

Figure 1-7

The force distribution in the previous analysis was given in units of force per unit length. When dealing with beams in bending, it is a common practice to express the distribution in this manner, but to be more specific, force distributions should be expressed in pressure units of force per unit area. It is often necessary to integrate the force distributions over a surface in order to obtain the net force and moment acting on that surface.

Example 1-1 Figure 1-8a illustrates the cross section of a rectangular beam in bending. From elementary strength of materials, the internal force distribution corresponding to bending about the z axis is given by $\sigma = -M_z y / I_z$ in units of force per unit area in the x direction, where M_z is the bending moment about the z axis (force times length) and I_z is the area moment of inertia about the z axis (given by $I_z = \frac{1}{12} b h^3$). Using area integration, prove that the net force in the x direction is zero and that the net moment about the z axis from this distribution is M_z.

SOLUTION Since the stress does not vary with respect to the z direction, isolate a dA area where $dA = b\, dy$, as shown in Fig. 1-8b. If σ were positive in the x direction, as shown in Fig. 1-8b, the force on the dA area in the x direction would be $\sigma\, dA$. The total force is found by integration across the entire area:

$$F_x = \int \sigma\, dA = \int_{-h/2}^{h/2} -\frac{M_z y}{I_z}(b\, dy) = -\frac{M_z b}{2I_z}\left[\left(\frac{h}{2}\right)^2 - \left(\frac{-h}{2}\right)^2\right] = 0$$

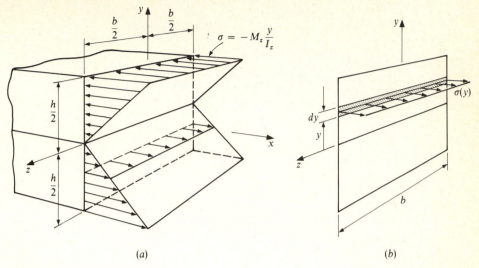

Figure 1-8

The net moment about the z axis due to the force on the dA area is $-y\sigma\, dA$. Integration across the entire area yields the net moment about the z axis.

$$M_z = \int - y\sigma\, dA = \int_{-h/2}^{h/2} -y\frac{-M_z y}{I_z}(b\, dy) = \frac{M_z}{I_z}\frac{bh^3}{12}$$

However, $I_z = bh^3/12$, and the moment about the z axis reduces to $M_z = M_z$.

1-3 STRESS

Stress is simply an internally distributed force within a body. To obtain a physical understanding of this idea, consider being submerged in water at a particular depth. The force one feels at this depth is a pressure (or compressive *stress*) and is not a finite number of "concentrated" forces. Other types of force distributions (stress) can occur in a liquid or solid. Tensile (pulling) and shear (rubbing or sliding) force distributions can also exist. As an example of a compressive force distribution within a solid, a uniform weight is placed on a short block, as shown in Fig. 1-9a. To determine the force distribution acting at section c-c, the first step is to make a break at section c-c and isolate the top portion of the element. The weight can be replaced by its equivalent force F_e, where $F_e = W$. If the portion left is to be in equilibrium, there must be an internal force F at section c-c. If the weight of the block is considered negligible, $F = F_e = W$ (see Fig. 1-9b). However, the material cannot exert a concentrated force such as F. Like all real forces in nature, the internal force F will actually be distributed across the surface in the form of a pressure p, as

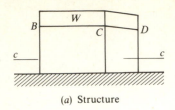

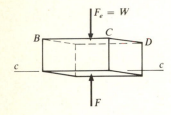

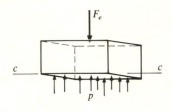

(a) Structure

(b) Force on section c-c   (c) Pressure distribution at section c-c

Figure 1-9

shown in Fig. 1-9c. This pressure is pushing against the surface; hence it is a compressive distributed force, or simply a compressive stress.

The total force F is the equivalent concentrated force of the force distribution p. Since the pressure is a distributed force, it can be examined at a particular point over an infinitesimal span of area, say dA_1. Consider the pressure at this point to have a value of p_1. The force due to the pressure p_1 acting over an area of dA_1 is simply $p_1\,dA_1$. The total force acting on the cross section then is the sum of $p\,dA_1$ across the entire surface, i.e., simply the integral of $p_1\,dA$. Thus

$$F = \int_A p_1\,dA_1 \tag{1-4}$$

If the pressure or stress is uniform or constant across the area, $p_1 = p$, where p is a constant and the integral in Eq. (1-4) becomes simply pA, where A is the total cross-sectional area. Thus

$$p = \frac{F}{A} \qquad \text{force per unit area} \tag{1-5}$$

This should be quite familiar to the reader, but note that the main assumption is that the stress is constant across the area. Compressive stresses are normally designated as negative quantities to differentiate them from tensile stresses where tensile stresses are considered positive. Thus the stress in the above example is

$$\sigma = -\frac{F}{A} \qquad \text{force per unit area} \tag{1-6}$$

where instead of p to designate pressure σ is used to designate *stress*.

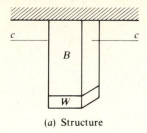

(a) Structure

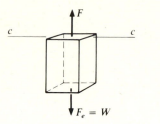

(b) Force on section c-c

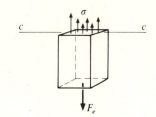

(c) Stress distribution on section c-c

Figure 1-10

An example of pure tensile stress is shown in Fig. 1-10, where a weight W is welded to bar B. If the stress is desired at section c-c, a break is made there; as before, the equivalent applied force and internal reaction is shown (Fig. 1-10b). If the weight of bar B is negligible, then for equilibrium, $F = F_e = W$. If the force distribution of F across section c-c (the stress) is uniform, as in the previous example,

$$\sigma = + \frac{F}{A} \qquad \text{force per unit area} \tag{1-7}$$

where σ in this example is positive, or a tensile stress, since the internal force is pulling at the surface.

An example of pure shear stress is illustrated in Fig. 1-11, where a tensile force P is transmitted through a yoke-and-tongue assembly. Isolating the tongue element requires breaks of the pin above and below the tongue. If no friction exists between the yoke and tongue, then for equilibrium, internal tangential forces at the exposed pin surfaces are present, as shown in Fig. 1-11b. Thanks to symmetry, these forces are equal and for equilibrium are $F = P/2$. These forces are tangential to the surfaces, and, as before, the force manifests itself as a force distribution, or stress. This stress, since it is tangential to the surface (rubbing), is called a *shear stress* τ. As before,

$$F = \int \tau \, dA$$

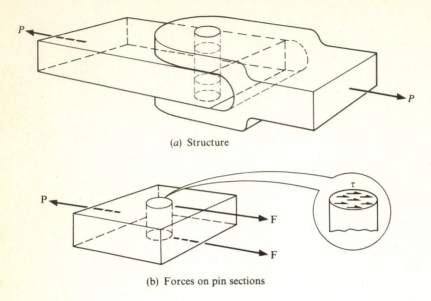

(a) Structure

(b) Forces on pin sections

Figure 1-11

and if the shear stress is assumed constant across the surface,†

$$\tau = \frac{F}{A} \tag{1-8}$$

The sign convention for shear stress will be defined later.

The three examples of uniform stress help clarify the concept of stress. However, stress should not be thought of as necessarily being constant across a finite surface. Stress is a *point function*. That is, stress is actually a force distribution, and in general it can be considered constant only over an infinitesimal area at a specific point located on a given surface.

Consider a general solid body with a number of equivalent concentrated forces acting on it, as shown in Fig. 1-12. P_i are applied forces and R_j are possible support forces. To determine the state of stress at point Q in the body, it is necessary to expose a surface containing point Q. This is done by making a planar slice, or break, through the body intersecting Q. The orientation of this slice is arbitrary, but in analysis this slice is generally made in a convenient plane where the state of stress can be determined easily or where certain geometric relations can be utilized. The first slice, illustrated in Fig. 1-13, is arbitrarily designated as the yz plane. The external forces on the portion of the body remaining are shown, as well as the internal pressure or stress distribution

---

† If the clearances between the elements are not close, the pin tends to bend. The bending stresses cause the transverse shear stress to become nonuniform. Transverse shear stresses in rods in bending are discussed in Sec. 2-3.

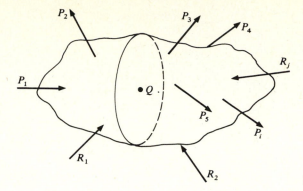

Figure 1-12

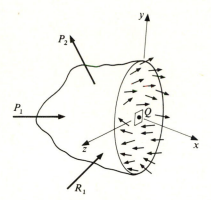

Figure 1-13

across the exposed surface. In the general case, the stress distribution will not be uniform across the surface, and the stresses will neither be normal nor tangential to the surface at a given point. However, the force distribution at a point will have components in the normal and tangential directions giving rise to a normal stress (tensile or compressive) and a tangential stress (shear).

The slice establishes the normal direction to the slice, the x direction. It is next left to establish the y direction, which again is arbitrary. Once this is established, the z direction is fixed.

As before, examine an infinitesimal area $\Delta y \, \Delta z$ surrounding point Q, as shown in Fig. 1-14. The equivalent concentrated force due to the force distribution across this area is ΔF_x, which in general is neither normal nor tangential to the surface (the subscript x is used to designate the normal direction of the first slice). The force ΔF_x has components in the x, y, and z directions, which are ΔF_{xx}, ΔF_{xy}, and ΔF_{xz}, respectively; note that the first

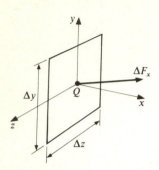

Figure 1-14

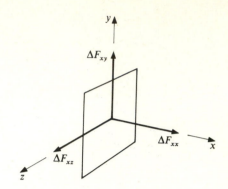

Figure 1-15

subscript gives the direction normal to the surface and the second gives the direction of the force component (see Fig. 1-15). These forces are actually due to the distributed forces in the respective directions. The average distributed force per unit area (average stress) in the x direction is

$$\bar{\sigma}_{xx} = \frac{\Delta F_{xx}}{\Delta A}$$

where $\Delta A = \Delta y \, \Delta z$.

Recalling that stress is actually a point function, we obtain the exact stress in the x direction at point Q by allowing ΔA to approach zero. Thus

$$\sigma_{xx} = \lim_{\Delta A \to 0} \frac{\Delta F_{xx}}{\Delta A}$$

or

$$\sigma_{xx} = \frac{dF_{xx}}{dA} \tag{1-9}$$

Equation (1-9) can be written in integral form:

$$F_{xx} = \int_A \sigma_{xx} \, dA \tag{1-10}$$

where F_{xx} is the total force on the entire exposed surface of Fig. 1-13 in the x direction. F_{xx} can normally be found from the equations of motion. The stress σ_{xx} on ΔA as ΔA approaches zero is illustrated in Fig. 1-16.

Distributed forces or stresses arise from the tangential forces ΔF_{xy} and ΔF_{xz} as well, and since these stresses are tangential, they are shear stresses. The procedure for establishing the shear stresses is the same as before, and the exact values at point Q are

$$\tau_{xy} = \frac{dF_{xy}}{dA} \tag{1-11}$$

$$\tau_{xz} = \frac{dF_{xz}}{dA} \tag{1-12}$$

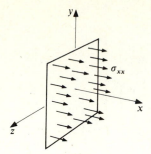

Figure 1-16

The corresponding integral relations are

$$F_{xy} = \int_A \tau_{xy} \, dA \tag{1-13}$$

$$F_{xz} = \int_A \tau_{xz} \, dA \tag{1-14}$$

As before, F_{xy} and F_{xz} are the total forces on the entire exposed surface of Fig. 1-13 (but in the y and z directions, respectively) and can normally be found from the equations of motion. The stresses given by Eqs. (1-11) and (1-12) can be illustrated by Fig. 1-17a and b, respectively.

   The three stresses existing on the exposed surface at point Q can be illustrated together, but to avoid a confusing picture, each stress will be depicted by only one arrow, as shown in Fig. 1-18. It must be kept in mind that each arrow represents a stress (force per unit area) and not a concentrated force. In addition, since by definition σ is a normal stress pointing in the same direction as the corresponding surface normal, double subscripts are not necessary. Thus it is common practice to replace σ_{xx} by the simpler notation σ_x, and this convention will be adopted from now on.

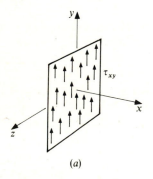

(a)

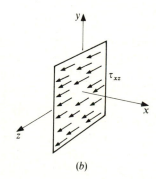

(b)

Figure 1-17

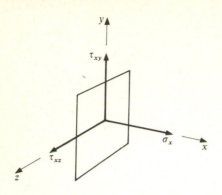

Figure 1-18

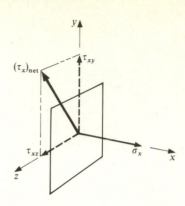

Figure 1-19

On a given surface only one normal stress and *one* shear stress exist. The net tangential force on the surface is $\sqrt{(\Delta F_{xy})^2 + (\Delta F_{xz})^2}$; the net shear stress can therefore be shown to be

$$(\tau_x)_{net} = \sqrt{\tau_{xy}^2 + \tau_{xz}^2} \qquad (1\text{-}15)$$

(see Fig. 1-19).

To describe the state of stress at point Q completely it would be necessary to examine other surfaces by making different planar slices. Since different planar slices would necessitate different coordinates and different free-body diagrams, the stresses on each slice would be, in general, quite different. Then to understand the complete state of stress at point Q, every possible surface intersecting point Q should be examined, but this would require an infinite number of slices surrounding point Q; that is, if point Q were completely isolated from the body, it would be described by an infinitesimal sphere. Naturally, this would be impossible and it is also unnecessary since there is a simple method for accomplishing the same result, called *coordinate transformation*. Although basically simple, it is rather tedious to develop; for two-dimensional analysis it is given in Chap. 2 and for three-dimensional analysis in Chap. 6.

Returning to the general body of Fig. 1-12, a planar slice can be made perpendicular to the y direction and through point Q. The procedure for describing the state of stress on the infinitesimal area $\Delta z \, \Delta x$ is identical to the technique used on the first slice. However, the stresses on the surface whose normal is in the y direction will be σ_y, τ_{yz}, and τ_{yx}. The third orthogonal slice is made perpendicular to the z direction, and the resulting stresses are σ_z, τ_{zx}, and τ_{zy}. Thus the state of stress at point Q described by only three mutually perpendicular surfaces is shown in Fig. 1-20. It will be shown later that this is sufficient to determine the state of stress at any surface intersecting point Q.

To isolate the point Q from the body completely using the orthogonal surfaces, consider the hidden faces of a rectangular parallelepiped of dimensions Δx, Δy, Δz, as shown in Fig. 1-20. If $\Delta x \, \Delta y \, \Delta z$ approaches zero, the stresses on

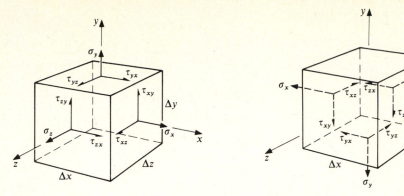

Figure 1-20   **Figure 1-21**

the hidden faces must be equal in magnitude and opposite in direction to the stresses on the visible faces (see Fig. 1-21). Thus, the state of stress at a point using three mutually orthogonal planes can be described by nine distinct values of stress. This state of stress can be written in matrix form, where the stress matrix σ is given by

$$\sigma = \begin{bmatrix} \sigma_x & \tau_{xy} & \tau_{xz} \\ \tau_{yx} & \sigma_y & \tau_{yz} \\ \tau_{zx} & \tau_{zy} & \sigma_z \end{bmatrix} \tag{1-16}$$

If the element has *finite* dimensions, however small, the stresses on opposing faces are not necessarily the same since we are no longer dealing with a point. Consider two points in a body, Q_1 and Q_2, where point Q_2 is located Δx, Δy, and Δz from point Q_1 (see Fig. 1-22). The state of stress for point Q_1 is shown on the hidden faces. When one moves from point Q_1 to point Q_2, the values of stress change in general, where the change is noted by $\Delta\sigma_x$, $\Delta\sigma_y$, etc. In addition, since the element has mass, "body" forces $\bar{F}_x$, $\bar{F}_y$, and $\bar{F}_z$ (force per unit volume) can also exist and are due to force fields such as gravity, magnetic fields, etc. Some materials also exhibit body moments under the influence of magnetic fields, but these cases are extremely rare and will not be discussed here.

Summing moments about the z axis due to the stresses yields

$$\tau_{xy} \, \Delta y \, \Delta z \frac{\Delta x}{2} + (\tau_{xy} + \Delta\tau_{xy}) \, \Delta y \, \Delta z \frac{\Delta x}{2} - \tau_{yx} \, \Delta z \, \Delta x \frac{\Delta y}{2}$$

$$- (\tau_{yx} + \Delta\tau_{yx}) \, \Delta z \, \Delta x \frac{\Delta y}{2} = 0$$

Dividing all terms by $\Delta x \, \Delta y \, \Delta z$ results in

$$\tau_{xy} + \frac{\Delta\tau_{xy}}{2} = \tau_{yx} + \frac{\Delta\tau_{yx}}{2}$$

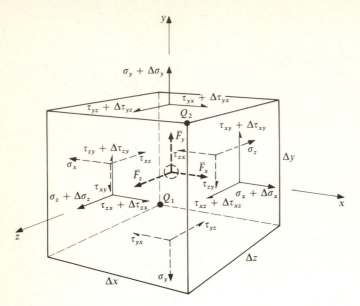

Figure 1-22

If $\Delta x\, \Delta y\, \Delta z$ is allowed to approach zero, $\Delta\tau_{xy}$ and $\Delta\tau_{yx}$ will approach zero; thus

$$\tau_{xy} = \tau_{yx} \tag{1-17a}$$

Summing moments about the other axes will yield similar relationships for the other shear stresses:

$$\tau_{yz} = \tau_{zy} \tag{1-17b}$$

$$\tau_{zx} = \tau_{xz} \tag{1-17c}$$

Thus since the cross shears are equal, it is only necessary to specify six quantities to establish the state of stress for a point, and the stress matrix is

$$\sigma = \begin{bmatrix} \sigma_x & \tau_{xy} & \tau_{zx} \\ \tau_{xy} & \sigma_y & \tau_{yz} \\ \tau_{zx} & \tau_{yz} & \sigma_z \end{bmatrix} \tag{1-18}$$

There are many practical problems where the stresses in the z direction are zero, that is, $\sigma_z = \tau_{yz} = \tau_{zx} = 0$. This is referred to as the state of *biaxial stress* or *plane stress*. The stress matrix can then be written

$$\sigma = \begin{bmatrix} \sigma_x & \tau_{xy} \\ \tau_{xy} & \sigma_y \end{bmatrix} \tag{1-19}$$

and the element surrounding point Q is shown in Fig. 1-23a. The element undergoing biaxial stress is often shown with a two-dimensional diagram, as in Fig. 1-23b, but it is important to realize that the state of stress is two-

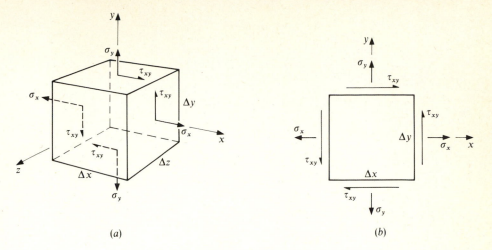

(a)   (b)

Figure 1-23

dimensional whereas the element is still three-dimensional; i.e., the depth of the element shown in Fig. 1-23b is Δz, and the stresses shown are acting over the $\Delta x\,\Delta z$ and $\Delta y\,\Delta z$ surfaces.

1-4 STRAIN

Stresses on the $\Delta x\,\Delta y\,\Delta z$ element cause a change of the element both dimensionally and in shape. The normal stresses cause the element to grow and/or shrink in the x, y, and z directions so that the element remains a rectangular parallellpiped. The shear stresses basically do not cause dimensional changes, but shear causes the element to change shape from a rectangular parallelpiped to a rhombohedron.

Examining the dimensional change on an element with only one normal stress σ_x leads to the observation that the new length in any direction is equal to its original length plus its change in length per unit length times its original length. That is,

$$\Delta x' = \Delta x + \varepsilon_x\,\Delta x \qquad \Delta y' = \Delta y + \varepsilon_y\,\Delta y \qquad \Delta z' = \Delta z + \varepsilon_z\,\Delta z \qquad (1\text{-}20)$$

where ε_x, ε_y, ε_z are the changes in length per unit length in the x, y, or z direction, respectively, called *normal strains* (see Fig. 1-24). There is a direct relationship between strain and stress. Hooke's law for a linear material is simply that the normal strain is directly proportional to the normal stress; i.e.,

$$\varepsilon_x = \frac{\sigma_x}{E}$$

where E is called the *modulus of elasticity* (also referred to as *Young's*

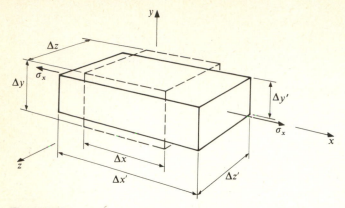

Figure 1-24

modulus). As the element elongates in the direction of the normal stress, contractions in the y and z directions occur. If the material is linear, these contractions are also directly proportional to the normal stress. It is common practice to express the contractions in terms of the primary normal strain, which in this case is ε_x. The proportionality constant relating the contraction to the primary strain is Poisson's ratio ν. If the material is isotropic, i.e., if the material properties are independent of direction, the contractions in the y and z directions are equal and are†

$$\varepsilon_y = -\nu\varepsilon_x = -\frac{\nu}{E}\,\sigma_x \qquad \varepsilon_z = -\nu\varepsilon_x = -\frac{\nu}{E}\,\sigma_x$$

The normal strains caused by σ_y and σ_z are similar to the strains caused by σ_x. The normal strains caused by σ_y are

$$\varepsilon_x = \varepsilon_z = -\frac{\nu\sigma_y}{E} \qquad \varepsilon_y = \frac{\sigma_y}{E}$$

and the normal strains caused by σ_z are

$$\varepsilon_x = \varepsilon_y = -\frac{\nu\sigma_z}{E} \qquad \varepsilon_z = \frac{\sigma_z}{E}$$

For an element undergoing σ_x, σ_y, and σ_z simultaneously, the effect of each stress can be added using the concept of linear superposition. Thus, the general strain-stress relationship for a linear, homogenous, isotropic material is

$$\varepsilon_x = \frac{1}{E}\,[\sigma_x - \nu(\sigma_y + \sigma_z)] \tag{1-21a}$$

$$\varepsilon_y = \frac{1}{E}\,[\sigma_y - \nu(\sigma_z + \sigma_x)] \tag{1-21b}$$

$$\varepsilon_z = \frac{1}{E}\,[\sigma_z - \nu(\sigma_x + \sigma_y)] \tag{1-21c}$$

†Nonisotropic behavior is discussed in Sec. 6-3.

Example 1-2 The stress at a point in a body is

$$\sigma = \begin{bmatrix} 5 & 3 & 2 \\ 3 & -1 & 0 \\ 2 & 0 & 4 \end{bmatrix} (1000) \text{ lb/in}^2$$

Determine the normal strains in the x, y, and z directions if $E = 10 \times 10^6$ lb/in^2, and $v = 0.30$.

SOLUTION

$$\varepsilon_x = \frac{1}{10 \times 10^6} [5 - 0.3(-1+4)](1000) = 410 \times 10^{-6} \text{ in/in} = 410 \ \mu\text{in/in}$$

$$\varepsilon_y = \frac{1}{10 \times 10^6} [-1 - 0.3(5+4)](1000) = -370 \times 10^{-6} \text{ in/in} = -370 \ \mu\text{in/in}$$

$$\varepsilon_z = \frac{1}{10 \times 10^6} [4 - 0.3(5-1)](1000) = 280 \times 10^{-6} \text{ in/in} = 280 \ \mu\text{in/in}$$

If the strains are known, Eqs. (1-21) represent three simultaneous equations in σ_x, σ_y, and σ_z. Solving for the stresses yields

$$\sigma_x = \frac{E}{(1+v)(1-2v)} [(1-v)\varepsilon_x + v(\varepsilon_y + \varepsilon_z)] \qquad (1\text{-}22a)$$

$$\sigma_y = \frac{E}{(1+v)(1-2v)} [(1-v)\varepsilon_y + v(\varepsilon_z + \varepsilon_x)] \qquad (1\text{-}22b)$$

$$\sigma_z = \frac{E}{(1+v)(1-2v)} [(1-v)\varepsilon_z + v(\varepsilon_x + \varepsilon_y)] \qquad (1\text{-}22c)$$

For biaxial stress, $\sigma_z = 0$ and Eqs. (1-21) reduce to

$$\varepsilon_x = \frac{1}{E} (\sigma_x - v\sigma_y) \qquad (1\text{-}23a)$$

$$\varepsilon_y = \frac{1}{E} (\sigma_y - v\sigma_x) \qquad (1\text{-}23b)$$

$$\varepsilon_z = \frac{-v}{E} (\sigma_x + \sigma_y) \qquad (1\text{-}23c)$$

It can be seen from Eqs. (1-23) that for biaxial stress it is only necessary to know two of the strains in order to evaluate the stresses σ_x and σ_y. Normally the two known strains are ε_x and ε_y. Thus, solving for the stresses from the first two equations of Eqs. (1-23) yields

$$\sigma_x = \frac{E}{1-v^2} (\varepsilon_x + v\varepsilon_y) \qquad (1\text{-}24a)$$

$$\sigma_y = \frac{E}{1-v^2} (\varepsilon_y + v\varepsilon_x) \qquad (1\text{-}24b)$$

The change in shape of the element caused by shear stresses can be illustrated first by examining the effect of τ_{xy} alone, as shown in Fig. 1-25. The shear strain γ is a measure of the deviation of the stressed element from a

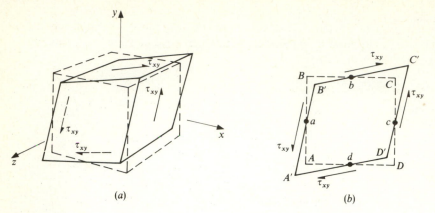

Figure 1-25

rectangular parallelpiped. The shear strain γ_{xy} is defined by

$$\gamma_{xy} = \Delta\angle BAD = \angle BAD - \angle B'A'D'$$

where γ_{xy} is still given in the units of length per unit length (see Sec. 1-5). For a linear, homogeneous, isotropic material, the shear strain is directly related to the shear stress by

$$\gamma_{xy} = \frac{\tau_{xy}}{G} \tag{1-25a}$$

where G is the shear modulus. Similarly, the remaining shear strains are related to the corresponding shear stresses. Thus

$$\gamma_{yz} = \frac{\tau_{yz}}{G} \tag{1-25b}$$

$$\gamma_{zx} = \frac{\tau_{zx}}{G} \tag{1-25c}$$

It can be shown† that the shear modulus is related to the modulus of elasticity and Poisson's ratio by

$$G = \frac{E}{2(1+\nu)} \tag{1-26}$$

Thus Eqs. (1-25) can be rewritten as

$$\gamma_{xy} = \frac{2(1+\nu)}{E}\,\tau_{xy} \tag{1-27a}$$

$$\gamma_{yz} = \frac{2(1+\nu)}{E}\,\tau_{yz} \tag{1-27b}$$

$$\gamma_{zx} = \frac{2(1+\nu)}{E}\,\tau_{zx} \tag{1-27c}$$

†See Appendix B.

Note that in Fig. 1-25*b* the distances *ac* and *bd* remain unchanged when the shear stress τ_{xy} is applied to the element since there are no normal strains in the *x* and *y* directions due to τ_{xy}.

Example 1-3 Determine the shear strains in Example 1-2.

SOLUTION

$$\gamma_{xy} = \frac{2(1+0.3)}{10 \times 10^6} 3000 = 780 \times 10^{-6} \text{ in/in} = 780 \ \mu\text{in/in}$$

$$\gamma_{yz} = 0 \text{ in/in}$$

$$\gamma_{zx} = \frac{2(1+0.3)}{10 \times 10^6} 2000 = 520 \times 10^{-6} \text{ in/in} = 520 \ \mu\text{in/in}$$

Thermal Strains

When an unconstrained solid experiences a temperature change, normal strains develop. Thermal strains for a homogeneous and isotropic material are given by

$$\varepsilon_x = \varepsilon_y = \varepsilon_z = \alpha \ \Delta T \tag{1-28}$$

where α = coefficient of linear expansion, in/(in)(°F) or m/(m)(°C)
ΔT = change in temperature °F or °C

1-5 DEFLECTIONS

The cumulative effect of the strains caused by the varying stresses throughout the member causes gross deflections of the points within the member. Thus, the deflections are directly related to the strains. Since in most practical situations, deflection analyses are restricted to small deflection theory, where two-dimensional analysis in each of the three planes is valid, in this section the discussion is limited to small-deflection theory. (Large-deflection theory is discussed in Chap. 6.) Consider a point Q in a member where the position of point Q before loading of the member is located by the coordinates *x*, *y*, and *z* with respect to an arbitrary origin (see Fig. 1-26). An element of infinitesimal dimensions Δx, Δy, and Δz originating from point Q can be constructed where the corners of the undeformed element are indicated by $QBCD$. Stresses, which in turn cause strains, cause point Q to deflect and the element to change geometrically. The deflections of point Q in the *x* and *y* directions are denoted by *u* and *v*, respectively. The corresponding deflections of points B, C, and D would be identical if the element were rigid and did not rotate, but since we are dealing with an elastic member, the element will change geometrically and will rotate.

The deflection of any point Q within the body can be described by

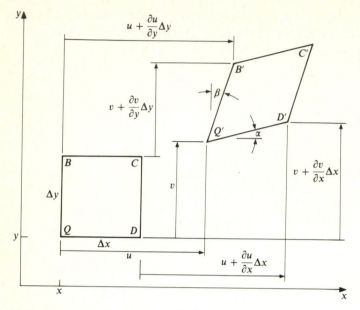

Figure 1-26

continuous functions of x and y. That is,

$$u = u(x, y) \qquad v = v(x, y)$$

The functions can be expanded about any point in terms of a Taylor's series expansion. If u, $\partial u/\partial x$, $\partial^2 u/\partial x^2$, etc., are evaluated for point Q, the deflection for point D in the x direction will be

$$u_D = u + \frac{\partial u}{\partial x}\Delta x + \frac{1}{2!}\frac{\partial^2 u}{\partial x^2}(\Delta x)^2 + \cdots$$

since $\Delta y = 0$ between points Q and D. Likewise, if v, $\partial v/\partial x$, $\partial^2 v/\partial x^2$, etc., are evaluated for point Q, the deflection for point D in the y direction is

$$v_D = v + \frac{\partial v}{\partial x}\Delta x + \frac{1}{2!}\frac{\partial^2 v}{\partial x^2}(\Delta x)^2 + \cdots$$

If Δx is considered very small, it is permissible to neglect the terms $(\Delta x)^2$ or higher. Thus (see Fig. 1-26)

$$u_D = u + \frac{\partial u}{\partial x}\Delta x \qquad v_D = v + \frac{\partial v}{\partial x}\Delta x$$

If the deflections of point B are obtained from a Taylor's series expansion about the point Q, where $\Delta x = 0$ and Δy is considered very small, then (see Fig. 1-26)

$$u_B = u + \frac{\partial u}{\partial y}\Delta y \qquad v_B = v + \frac{\partial v}{\partial y}\Delta y$$

For small-deflection theory, the derivative terms are considered small.† Thus, if $(\partial v/\partial x)\Delta x$ is considered small compared with $\Delta x + (\partial u/\partial x)\Delta x$, then $Q'D' \approx \Delta x + (\partial u/\partial x)\,\Delta x$, the rate of elongation of QD is

$$\varepsilon_x = \frac{Q'D' - QD}{QD} = \frac{[\Delta x + (\partial u/\partial x)\,\Delta x] - \Delta x}{\Delta x} = \frac{\partial u}{\partial x} \qquad (1\text{-}29a)$$

and the strain in the y direction of point Q is the rate of elongation of QB, or

$$\varepsilon_y = \frac{\partial v}{\partial y} \qquad (1\text{-}29b)$$

The change in angle BQD is defined as the shear strain at the point Q and is $\gamma_{xy} = \alpha + \beta$. From Fig. 1-26 it can be seen that

$$\tan \alpha = \frac{(\partial v/\partial x)\,\Delta x}{\Delta x} = \frac{\partial v}{\partial x} \qquad \text{and} \qquad \tan \beta = \frac{(\partial u/\partial y)\,\Delta y}{\Delta y} = \frac{\partial u}{\partial y}$$

However, if the strains are small, $\tan \alpha \approx \alpha$, and, $\tan \beta \approx \beta$. Thus the shear strain can be represented by

$$\gamma_{xy} = \frac{\partial v}{\partial x} + \frac{\partial u}{\partial y} \qquad (1\text{-}29c)$$

The rotation of a line segment located at point Q can be found by determining the amount of rotation of the bisector of the angle BQD. The amount of counterclockwise rotation in the xy plane Θ_{xy} can be shown to be $\Theta_{xy} = \frac{1}{2}(\alpha - \beta)$, or

$$\Theta_{xy} = \frac{1}{2}\left(\frac{\partial v}{\partial x} - \frac{\partial u}{\partial y}\right) \qquad (1\text{-}29d)$$

Considering w to be the deflection of point Q in the z direction and performing a similar analysis in the yz and zx planes results in

$$\varepsilon_z = \frac{\partial w}{\partial z} \qquad (1\text{-}29e)$$

$$\gamma_{yz} = \frac{\partial w}{\partial y} + \frac{\partial v}{\partial z} \qquad (1\text{-}29f)$$

$$\gamma_{zx} = \frac{\partial u}{\partial z} + \frac{\partial w}{\partial x} \qquad (1\text{-}29g)$$

$$\Theta_{yz} = \frac{1}{2}\left(\frac{\partial w}{\partial y} - \frac{\partial v}{\partial z}\right) \qquad (1\text{-}29h)$$

$$\Theta_{zx} = \frac{1}{2}\left(\frac{\partial u}{\partial z} - \frac{\partial w}{\partial x}\right) \qquad (1\text{-}29i)$$

If a structural member is very thin in the z direction, the stress field will be biaxial, where $\sigma_z = \tau_{zx} = \tau_{yz} = 0$. Furthermore, the deflections u and v in the x

---

† This brings up one of the many misnomers in stress analysis. Small-deflection theory actually means small-*strain* theory, since the derivative terms are measures of strain.

and y directions, respectively, can be considered to be functions of x and y only. If the displacement field $u(x, y)$ and $v(x, y)$ is known for such a member, determination of the strain field in the xy plane is quite straightforward using Eqs. (1-29a) to (1-29c). Once the strains are known, determination of the stress field is also quite simple using Eqs. (1-24) and (1-25a) for in this case the problem is one of biaxial stress.

Example 1-4 The displacement field for the thin beam shown in Fig. 1-27 considering bending only is

$$u(x, y) = \frac{P}{EI} \left(Lx - \frac{x^2}{2} \right) y - \frac{vP}{6EI} y^3$$

$$v(x, y) = \frac{-vP}{2EI} (L - x) y^2 - \frac{P}{EI} \left(\frac{Lx^2}{2} - \frac{x^3}{6} \right)$$

where P = applied force
$\quad I$ = area moment of inertia about bending axis
$\quad L$ = length of beam

Determine (a) the vertical deflection and slope at $x = L$, $y = 0$ and (b) the entire stress field.

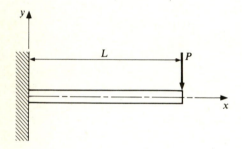

Figure 1-27

SOLUTION (a) The vertical deflection at $x = L$, $y = 0$ is $v(L, 0)$

$$v(L, 0) = -\frac{P}{EI} \left(\frac{L^3}{2} - \frac{L^3}{6} \right) = -\frac{PL^3}{3EI}$$

(check with Appendix D). The slope at $x = L$, $y = 0$ is $\Theta_{xy}(L, 0)$. Thus

$$\Theta_{xy} = \frac{1}{2} \left(\frac{\partial v}{\partial x} - \frac{\partial u}{\partial y} \right)$$

$$= \frac{1}{2} \left\{ \frac{vPy^2}{2EI} - \frac{P}{EI} \left(Lx - \frac{x^2}{2} \right) - \left[\frac{P}{EI} \left(Lx - \frac{x^2}{2} \right) - \frac{vPy^2}{2EI} \right] \right\}$$

$$= \frac{vPy^2}{2EI} - \frac{P}{EI} \left(Lx - \frac{x^2}{2} \right)$$

$$\Theta_{xy}(L, 0) = -\frac{P}{EI} \left(L^2 - \frac{L^2}{2} \right) = -\frac{PL^2}{2EI}$$

(check with Appendix D).

(b) $\qquad \varepsilon_x = \dfrac{\partial u}{\partial x} = \dfrac{P}{EI}(L-x)y \qquad \varepsilon_y = \dfrac{\partial v}{\partial y} = \dfrac{-\nu P}{EI}(L-x)y$

$$\gamma_{xy} = \frac{\partial v}{\partial x} + \frac{\partial u}{\partial y} = \left[\frac{\nu P y^2}{2EI} - \frac{P}{EI}\left(Lx - \frac{x^2}{2}\right)\right] + \left[\frac{P}{EI}\left(Lx - \frac{x^2}{2}\right) - \frac{\nu P y^2}{2EI}\right] = 0$$

Substituting into Eqs. (1-24) and (1-25) yields the stress field

$$\sigma_x = \frac{E}{1-\nu^2}\left[\frac{P}{EI}(L-x)y - \frac{\nu^2 P}{EI}(L-x)y\right] = \frac{P}{I}(L-x)y$$

$$\sigma_y = \frac{E}{1-\nu^2}\left[-\frac{\nu P}{EI}(L-x)y + \frac{\nu P}{EI}(L-x)y\right] = 0$$

$$\tau_{xy} = 0$$

If the stresses or strains are known as functions of x and y, it is a little more difficult to determine the displacement field using Eqs. (1-29a) to (1-29d). This type of problem is discussed in Chap. 6.

1-6 POLAR COORDINATES

In many problems the geometry of the component does not lend itself to the use of a cartesian coordinate system, and it is more practical to use a different coordinate system. Problems like thin- or thick-walled pressure vessels, circular rings, curved beams, and half-plane problems, are more suitable to a polar coordinate system. In Fig. 1-28a, an infinitesimal element is constructed using $r\theta$ coordinates, and the corresponding normal and shear stresses are shown. The depth of the element in the z direction is dz. As before, it is assumed that both Δr and $\Delta\theta$ are approaching zero.

The strain-stress relations are the same as in cartesian coordinates and are given by

$$\varepsilon_r = \frac{1}{E}[\sigma_r - \nu(\sigma_\theta + \sigma_z)] \tag{1-30a}$$

$$\varepsilon_\theta = \frac{1}{E}[\sigma_\theta - \nu(\sigma_z + \sigma_r)] \tag{1-30b}$$

$$\varepsilon_z = \frac{1}{E}[\sigma_z - \nu(\sigma_r + \sigma_\theta)] \tag{1-30c}$$

and

$$\gamma_{r\theta} = \frac{2(1+\nu)}{E}\tau_{r\theta} \tag{1-31a}$$

$$\gamma_{\theta z} = \frac{2(1+\nu)}{E}\tau_{\theta z} \tag{1-31b}$$

$$\gamma_{zr} = \frac{2(1+\nu)}{E}\tau_{zr} \tag{1-31c}$$

The strain-displacement relations in polar coordinates are determined in a manner similar to the development presented for rectangular coordinates, but

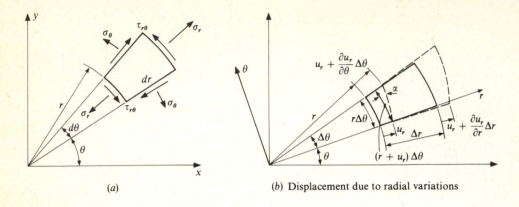

(a)

(b) Displacement due to radial variations

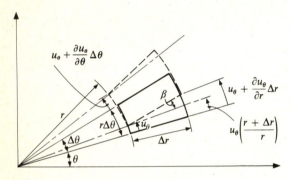

(c) Displacement due to tangential variations

Figure 1-28

due to the complexity of the deformed element, radial and tangential variations are viewed separately. Figure 1-28b and c represents the displacements of the deformed element due to the radial and tangential variations, respectively (shown in dotted lines). The net radial strain is

$$\varepsilon_r = \frac{u_r + (\partial u_r/\partial r)\,\Delta r - u_r}{\Delta r} = \frac{\partial u_r}{\partial r} \tag{1-32a}$$

The net tangential strain is

$$\varepsilon_\theta = \frac{(r + u_r)\,\Delta\theta - r\,\Delta\theta}{r\,\Delta\theta} + \frac{u_\theta + (\partial u_\theta/\partial\theta)\,\Delta\theta - u_\theta}{r\,\Delta\theta}$$

$$= \frac{u_r}{r} + \frac{1}{r}\frac{\partial u_\theta}{\partial\theta} \tag{1-32b}$$

The shear strain $\gamma_{r\theta}$ is equal to $\alpha + \beta$. Thus

$$\gamma_{r\theta} = \frac{u_r + (\partial u_r/\partial\theta)\,\Delta\theta - u_r}{r\,\Delta\theta} + \frac{u_\theta + (\partial u_\theta/\partial r)\,\Delta r - u_\theta(r+\Delta r)/r}{\Delta r}$$

$$= \frac{1}{r}\frac{\partial u_r}{\partial\theta} + \frac{\partial u_\theta}{\partial r} - \frac{u_\theta}{r} \tag{1-32c}$$

The rotation of the element in the counterclockwise direction is

$$\Theta_{r\theta} = \frac{u_\theta}{r} + \tfrac{1}{2}(\beta - \alpha) = \frac{1}{2}\left(\frac{\partial u_\theta}{\partial r} + \frac{u_\theta}{r} - \frac{1}{r}\frac{\partial u_r}{\partial\theta}\right) \tag{1-32d}$$

The strains and rotations in the θz and zr planes are developed in a manner similar to that for rectangular coordinates; they are

$$\varepsilon_z = \frac{\partial u_z}{\partial z} \tag{1-32e}$$

$$\gamma_{\theta z} = \frac{1}{r}\frac{\partial u_z}{\partial\theta} + \frac{\partial u_\theta}{\partial z} \tag{1-32f}$$

$$\gamma_{zr} = \frac{\partial u_r}{\partial z} + \frac{\partial u_z}{\partial r} \tag{1-32g}$$

$$\Theta_{\theta z} = \frac{1}{2}\left(\frac{1}{r}\frac{\partial u_z}{\partial\theta} - \frac{\partial u_\theta}{\partial z}\right) \tag{1-32h}$$

$$\Theta_{zr} = \frac{1}{2}\left(\frac{\partial u_r}{\partial z} - \frac{\partial u_z}{\partial r}\right) \tag{1-32i}$$

A special case of Eqs. (1-32) is encountered in axisymmetric problems (such as circular rings), where variations with respect to θ are zero and thanks to symmetry $u_\theta = 0$ everywhere. Thus, for axisymmetric problems, Eqs. (1-32) reduce to

$$\varepsilon_r = \frac{\partial u_r}{\partial r} \tag{1-33a}$$

$$\varepsilon_\theta = \frac{u_r}{r} \tag{1-33b}$$

$$\gamma_{r\theta} = 0 \tag{1-33c}$$

$$\Theta_{r\theta} = 0 \tag{1-33d}$$

$$\varepsilon_z = \frac{\partial u_z}{\partial z} \tag{1-33e}$$

$$\gamma_{\theta z} = 0 \tag{1-33f}$$

$$\gamma_{zr} = \frac{\partial u_r}{\partial z} + \frac{\partial u_z}{\partial r} \tag{1-33g}$$

$$\Theta_{\theta z} = 0 \tag{1-33h}$$

$$\Theta_{zr} = \frac{1}{2}\left(\frac{\partial u_r}{\partial z} - \frac{\partial u_z}{\partial r}\right) \tag{1-33i}$$

1-7 SUMMARY OF IMPORTANT RELATIONSHIPS

Triaxial Stress

Strain-stress relationships:

$$\varepsilon_x = \frac{1}{E}\left[\sigma_x - \nu(\sigma_y + \sigma_z)\right] \tag{1-21a}$$

$$\varepsilon_y = \frac{1}{E}\left[\sigma_y - \nu(\sigma_z + \sigma_x)\right] \tag{1-21b}$$

$$\varepsilon_z = \frac{1}{E}\left[\sigma_z - \nu(\sigma_x + \sigma_y)\right] \tag{1-21c}$$

$$\gamma_{xy} = \frac{2(1+\nu)}{E}\tau_{xy} \tag{1-27a}$$

$$\gamma_{yz} = \frac{2(1+\nu)}{E}\tau_{yz} \tag{1-27b}$$

$$\gamma_{zx} = \frac{2(1+\nu)}{E}\tau_{zx} \tag{1-27c}$$

Stress-strain relationships:

$$\sigma_x = \frac{E}{(1+\nu)(1-2\nu)}\left[(1-\nu)\varepsilon_x + \nu(\varepsilon_y + \varepsilon_z)\right] \tag{1-22a}$$

$$\sigma_y = \frac{E}{(1+\nu)(1-2\nu)}\left[(1-\nu)\varepsilon_y + \nu(\varepsilon_z + \varepsilon_x)\right] \tag{1-22b}$$

$$\sigma_z = \frac{E}{(1+\nu)(1-2\nu)}\left[(1-\nu)\varepsilon_z + \nu(\varepsilon_x + \varepsilon_y)\right] \tag{1-22c}$$

Shear relations are straightforward from Eqs. (1-27).

Biaxial Stress ($\sigma_z = \tau_{yz} = \tau_{zx} = 0$)

Strain-stress:

$$\varepsilon_x = \frac{1}{E}(\sigma_x - \nu\sigma_y) \tag{1-23a}$$

$$\varepsilon_y = \frac{1}{E}(\sigma_y - \nu\sigma_x) \tag{1-23b}$$

$$\varepsilon_z = -\frac{\nu}{E}(\sigma_x + \sigma_y) \tag{1-23c}$$

$$\gamma_{xy} = \frac{2(1+\nu)}{E}\tau_{xy} \tag{1-27a}$$

Stress-strain:

$$\sigma_x = \frac{E}{1 - \nu^2} (\varepsilon_x + \nu\varepsilon_y) \qquad (1\text{-}24a)$$

$$\sigma_y = \frac{E}{1 - \nu^2} (\varepsilon_y + \nu\varepsilon_x) \qquad (1\text{-}24b)$$

$$\tau_{xy} = \frac{E}{2(1 + \nu)} \gamma_{xy} \qquad (1\text{-}27a)$$

Thermal Strains

$$\varepsilon_x = \varepsilon_y = \varepsilon_z = \alpha \, \Delta T \qquad (1\text{-}28)$$

Strain-Displacements

$$\varepsilon_x = \frac{\partial u}{\partial x} \qquad (1\text{-}29a)$$

$$\varepsilon_y = \frac{\partial v}{\partial y} \qquad (1\text{-}29b)$$

$$\varepsilon_z = \frac{\partial w}{\partial z} \qquad (1\text{-}29e)$$

$$\gamma_{xy} = \frac{\partial v}{\partial x} + \frac{\partial u}{\partial y} \qquad (1\text{-}29c)$$

$$\gamma_{yz} = \frac{\partial w}{\partial y} + \frac{\partial v}{\partial z} \qquad (1\text{-}29f)$$

$$\gamma_{zx} = \frac{\partial u}{\partial z} + \frac{\partial w}{\partial x} \qquad (1\text{-}29g)$$

$$\Theta_{xy} = \frac{1}{2}\left(\frac{\partial v}{\partial x} - \frac{\partial u}{\partial y}\right) \qquad (1\text{-}29d)$$

$$\Theta_{yz} = \frac{1}{2}\left(\frac{\partial w}{\partial y} - \frac{\partial v}{\partial z}\right) \qquad (1\text{-}29h)$$

$$\Theta_{zx} = \frac{1}{2}\left(\frac{\partial u}{\partial z} - \frac{\partial w}{\partial x}\right) \qquad (1\text{-}29i)$$

Polar Coordinates

Strain-stress:

$$\varepsilon_r = \frac{1}{E}\left[\sigma_r - \nu(\sigma_\theta + \sigma_z)\right] \qquad (1\text{-}30a)$$

$$\varepsilon_\theta = \frac{1}{E}\left[\sigma_\theta - \nu(\sigma_z + \sigma_r)\right] \qquad (1\text{-}30b)$$

$$\varepsilon_z = \frac{1}{E}\left[\sigma_z - \nu(\sigma_r + \sigma_\theta)\right] \qquad (1\text{-}30c)$$

$$\gamma_{r\theta} = \frac{2(1+\nu)}{E}\,\tau_{r\theta} \qquad (1\text{-}31a)$$

$$\gamma_{\theta z} = \frac{2(1+\nu)}{E}\,\tau_{\theta z} \qquad (1\text{-}31b)$$

$$\gamma_{zr} = \frac{2(1+\nu)}{E}\,\tau_{zr} \qquad (1\text{-}31c)$$

Strain-displacement:

$$\varepsilon_r = \frac{\partial u_r}{\partial r} \qquad (1\text{-}32a)$$

$$\varepsilon_\theta = \frac{u_r}{r} + \frac{1}{r}\frac{\partial u_\theta}{\partial \theta} \qquad (1\text{-}32b)$$

$$\gamma_{r\theta} = \frac{1}{r}\frac{\partial u_r}{\partial \theta} + \frac{\partial u_\theta}{\partial r} - \frac{u_\theta}{r} \qquad (1\text{-}32c)$$

$$\Theta_{r\theta} = \frac{1}{2}\left(\frac{\partial u_\theta}{\partial r} + \frac{u_\theta}{r} - \frac{1}{r}\frac{\partial u_r}{\partial \theta}\right) \qquad (1\text{-}32d)$$

$$\varepsilon_z = \frac{\partial u_z}{\partial z} \qquad (1\text{-}32e)$$

$$\gamma_{\theta z} = \frac{1}{r}\frac{\partial u_z}{\partial \theta} + \frac{\partial u_\theta}{\partial z} \qquad (1\text{-}32f)$$

$$\gamma_{zr} = \frac{\partial u_r}{\partial z} + \frac{\partial u_z}{\partial r} \qquad (1\text{-}32g)$$

$$\Theta_{\theta z} = \frac{1}{2}\left(\frac{1}{r}\frac{\partial u_z}{\partial \theta} - \frac{\partial u_\theta}{\partial z}\right) \qquad (1\text{-}32h)$$

$$\Theta_{zr} = \frac{1}{2}\left(\frac{\partial u_r}{\partial z} - \frac{\partial u_z}{\partial r}\right) \qquad (1\text{-}32i)$$

Axisymmetric Problems

$$\tau_{r\theta} = \tau_{\theta z} = 0 \qquad u_\theta = 0$$

$$\varepsilon_r = \frac{\partial u_r}{\partial r} \qquad (1\text{-}33a)$$

$$\varepsilon_\theta = \frac{u_r}{r} \qquad (1\text{-}33b)$$

$$\gamma_{r\theta} = 0 \qquad (1\text{-}33c)$$

$$\Theta_{r\theta} = 0 \qquad (1\text{-}33d)$$

$$\varepsilon_z = \frac{\partial u_z}{\partial z} \tag{1-33e}$$

$$\gamma_{\theta z} = 0 \tag{1-33f}$$

$$\gamma_{zr} = \frac{\partial u_r}{\partial z} + \frac{\partial u_z}{\partial r} \tag{1-33g}$$

$$\Theta_{\theta z} = 0 \tag{1-33h}$$

$$\Theta_{zr} = \frac{1}{2}\left(\frac{\partial u_r}{\partial z} - \frac{\partial u_z}{\partial r}\right) \tag{1-33i}$$

EXERCISES

1-1 For the beam shown in Fig. 1-29, determine the support reactions at points B and D.

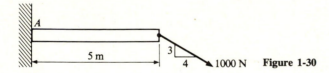

Figure 1-29

1-2 For the beam shown in Fig. 1-30, determine the support reactions at point A.

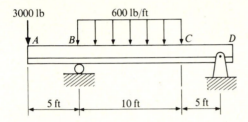

Figure 1-30

1-3 For the bent wire form shown in Fig. 1-31, determine the support reactions at point A.

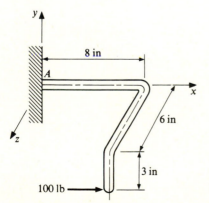

Figure 1-31

1-4 For the frame shown in Fig. 1-32, determine all support and connection forces on each member. The 400-lb·ft couple M is in the plane of the page and applied to member BF. Neglect friction on all mating parts.

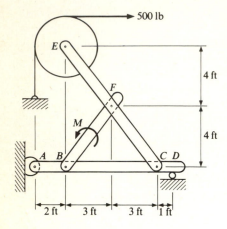

Figure 1-32

1-5 Figure 1-33 depicts a stress distribution across an internal surface of a rectangular rod of height 3 in and depth 1 in. Assuming that the stress distribution does not vary with respect to z, determine the net force in the x direction and the net moment about the z axis. The stress distribution is $\sigma_x = 2000\,y + 500$ lb/in².

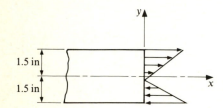

Figure 1-33

1-6 For a particular steel $E = 29 \times 10^6$ lb/in², and $G = 11 \times 10^6$ lb/in². Assuming the material to be homogeneous and isotropic, determine the value of Poisson's ratio.

1-7 The state of stress at a particular point in a body is known to be biaxial, where $\sigma_z = \tau_{yz} = \tau_{zx} = 0$. The material constants for the body are $E = 10 \times 10^6$ lb/in² and $\nu = 0.3$. If $\varepsilon_x = 1000\ \mu$in/in and $\varepsilon_y = -500\ \mu$in/in, determine σ_x, σ_y, and ε_z.

1-8 The state of normal strain at a particular body is $\varepsilon_x = 1000\ \mu$m/m, $\varepsilon_y = -500\ \mu$m/m, and $\varepsilon_z = 200\ \mu$m/m. If the material constants for the body are $E = 7 \times 10^{10}$ N/m² and $\nu = 0.3$, determine the normal stresses σ_x, σ_y, and σ_z in terms of SI and U.S. Customary System units (USCS) (see Appendix A if SI units are unfamiliar).

1-9 When a rectangular beam of cross-sectional area $2bc$ made of a perfectly elastic-plastic material undergoes plastic deformation, the residual stress distribution after the loading is released

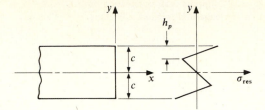

Figure 1-34

appears as shown in Fig. 1-34. Show that the net force and moment of the distribution about the z axis is zero. The stress distribution for *positive* y values is

$$\sigma_{res} = \begin{cases} S_y \dfrac{y}{c}\left[1+\dfrac{h_p}{c}-\dfrac{1}{2}\left(\dfrac{h_p}{c}\right)^2\right] - S_y\dfrac{y}{c-h_p} & 0 < y < c - h_p \\ S_y \dfrac{y}{c}\left[1+\dfrac{h_p}{c}-\dfrac{1}{2}\left(\dfrac{h_p}{c}\right)^2\right] - S_y & c - h_p < y < c \end{cases}$$

where S_y is the yield strength of the material, a constant.

1-10 Figure 1-35 illustrates a thin plate of thickness t. An approximate displacement field which accounts for the displacements due to the weight of the plate is given by

$$u(x, y) = \frac{D_p}{2E}(2bx - x^2 - \nu y^2) \qquad v(x, y) = -\frac{\nu D_p}{E}y(b - x)$$

where D_p is the weight density of the material.

(a) Determine the corresponding biaxial stress field, $\sigma_x(x, y)$, $\sigma_y(x, y)$, and $\tau_{xy}(x, y)$.
(b) Qualitatively draw the deformed shape of the plate.
(c) Determine the rotation of the plate at points A and B.

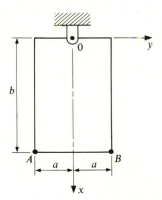

Figure 1-35

1-11 The stress matrix at a particular point in a body is

$$\sigma = \begin{bmatrix} -2 & 1 & -3 \\ 1 & 0 & 4 \\ -3 & 4 & 5 \end{bmatrix} \times 10^7 \text{ N/m}^2$$

Determine the corresponding strains if $E = 20 \times 10^{10}$ N/m^2 and $\nu = 0.3$.

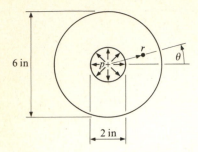

Figure 1-36

1-12 The thin pressurized ring shown in Fig. 1-36 is an *axisymmetric* problem. The normal stresses as functions of r are

$$\sigma_r = 625 - \frac{5625}{r^2} \qquad \sigma_\theta = 625 + \frac{5625}{r^2} \qquad \sigma_z = 0$$

If the material constants are $E = 29 \times 10^6$ lb/in^2 and $\nu = 0.29$, for 1 in $\leq r \leq 3$ in, determine (*a*) the normal strains as functions of r and (*b*) the circumference at $r = 3$ in before and after the application of the pressure p.

TWO

A REVIEW OF ELEMENTARY STRENGTH OF MATERIALS

2-0 INTRODUCTION

As stated earlier, it is assumed that the reader has completed a basic course in strength of materials and is somewhat familiar with the various fundamental modes of loading of simple structural elements. The basic topics normally covered in this first course include direct axial and shear loading, torsional loading of circular rods, and bending and shear loading of long, narrow beams. Other topics normally included are pressurized thin-walled cylinders, buckling of long bars in compression, superposition techniques, and Mohr's circle for stress and strain.

   If any of the subjects discussed in this chapter are unfamiliar, the reader should consult a basic text, as complete derivations will not be given for this elementary material. The basic equations are provided, the main emphasis being on the underlying assumptions used in the derivations, which point out the restrictions to the equations. It is important that the analyst know when an equation is appropriate or, if not, to what extent the restrictions are not met. In this way, the analyst can form a relative degree of confidence in the results.

2-1 AXIAL LOADING

The equations for stress, strain, and deflections of an axially loaded element are based on a uniform stress distribution like that assumed in the examples of compressive and tensile loaded members in Chap. 1. The major restrictions are that:

1. The material of the member must be homogenous.
2. For the section under examination, the cross section of the member must be constant or gradually varying in the axial direction.
3. The points of load applications or support connections must be at reasonable distances from the point of interest.
4. The applied loads and/or support connection(s) must be geometrically positioned perfectly in line with the centroid of the cross section.
5. In compressive loading the axial force is less than the critical buckling force of the rod (see Sec. 2-10).

Axial Stresses

If the weight of the bar in Fig. 2-1 is assumed to be negligible, the stress σ_x at section c-c integrated across the area of c-c must equal the transmitted force P. That is, for Fig. 2-1a, $P = \int \sigma_x \, dA$. If it is assumed that the displacement in the x direction is uniform, ε_x and σ_x will be uniform. This assumption can easily be verified by a simple experiment. Thus, if σ_x is constant across the section,

$$\sigma_x = \frac{P}{A} \tag{2-1}$$

where A is the cross-sectional area at c-c. The state of stress on section c-c can be visualized as shown in Fig. 2-1a by making a break at c-c and isolating either the top or bottom portions. Note that each one of the elements is in equilibrium since $P = \int \sigma_x \, dA$.

When the load is compressive, as shown in Fig. 2-1b,

$$\sigma_x = -\frac{P}{A} \tag{2-2}$$

The major restrictions of Eqs. (2-1) and (2-2) can be viewed independently. First, the material must be homogenous. To see the effect of this, the bar shown in Fig. 2-2a is loaded in tension and consists of two differing materials bonded together. Let one half be made of aluminum, with $E_A = 10 \times 10^6$ lb/in², and the other half of steel, $E_S = 30 \times 10^6$ lb/in². Since the two halves are bonded together, the displacement in the x direction at the joint is the same for both members. This means that the strain in the x direction is also the same for both members. Thus

$$(\varepsilon_x)_{\text{steel}} = (\varepsilon_x)_{\text{aluminum}}$$

Since $\varepsilon_x = \sigma_x/E$, it follows that†

$$\left(\frac{\sigma_x}{E}\right)_{\text{steel}} = \left(\frac{\sigma_x}{E}\right)_{\text{aluminum}}$$

†There is actually a simplification here. Shear stresses will develop across the joint perpendicular to the x direction due to the difference in contraction of the materials. This will also lead to normal stresses in this direction and affect the strains in the x direction.

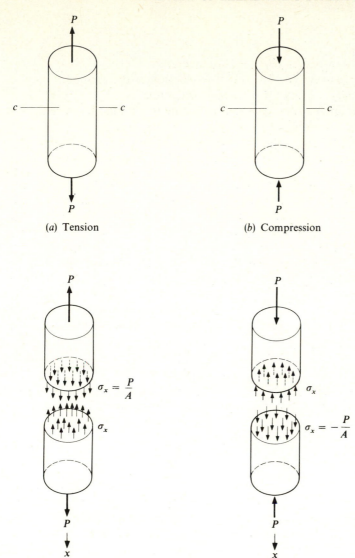

(a) Tension

(b) Compression

$$\sigma_x = \frac{P}{A}$$

$$\sigma_x$$

$$\sigma_x$$

$$\sigma_x = -\frac{P}{A}$$

(c) Tension

(d) Compression

Figure 2-1

which can be rewritten

$$(\sigma_x)_{\text{steel}} = \frac{E_S}{E_A}(\sigma_x)_{\text{aluminum}}$$

Since $E_S = 3E_A$,

$$(\sigma_x)_{\text{steel}} = 3(\sigma_x)_{\text{aluminum}}$$

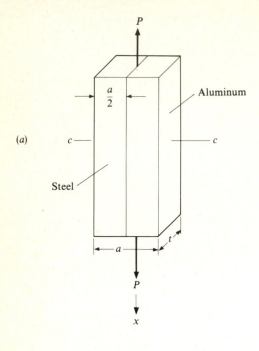

(a)

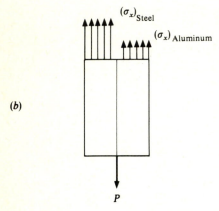

(b)

Figure 2-2

Thus, the stress distribution across section c-c is as shown in Fig. 2-2b and is not uniform across the entire surface.

To determine the values of stress, the stress distribution is integrated across the total area and equated to the net force P. Thus

$$\left(\int \sigma_x \, dA \right)_{\text{steel}} + \left(\int \sigma_x \, dA \right)_{\text{aluminum}} = P$$

$$(\sigma_x)_{\text{steel}} \frac{a}{2} t + (\sigma_x)_{\text{aluminum}} \frac{a}{2} t = P$$

Substitution of $(\sigma_x)_{\text{steel}} = 3(\sigma_x)_{\text{aluminum}}$ yields

$$(\sigma_x)_{\text{aluminum}} = \frac{1}{2}\frac{P}{at} \qquad \text{and} \qquad (\sigma_x)_{\text{steel}} = \frac{3}{2}\frac{P}{at}$$

It should also be noted that if the applied load P is centered between the steel and aluminum, the nonuniform distribution shown will not assure equilibrium in the moment equation. Thus, the distribution found is correct only if the applied load is positioned so that the sum of the moments for the element shown in Fig. 2-2b is zero. The applied load P must be positioned in line with the equivalent force of the stress distribution, as shown in Fig. 2-3. The technique of obtaining the location of the equivalent force of the stress distribution is exactly the same as outlined in Chap. 1, and the reader should verify that $e = a/8$.

   If the cross section of an axially loaded element suddenly changes, the stress distribution will not be uniform where the sudden change in area occurs. To illustrate this, a grid is marked on a rubber plate with a centrally located hole (see Fig. 2-4a). An axial load is then applied to the plate, as shown in Fig. 2-4b. When the grid along line c-c is observed, it can be seen that there is more stretch near the hole than at the outer surface of the plate. The nonuniform displacement behavior indicates the presence of a nonuniform stress field along c-c. A representation of the stress field across section c-c is given in Fig. 2-4c.

   The stress field is generally not uniform at or near points of load application or support connections. Figure 2-5 shows two examples of a rubber plate loaded axially in tension. In Fig. 2-5a, the load is applied through a clamped rigid connector, whereas in Fig. 2-5b the load is applied by a rigid pin through a hole in the plate. In both cases, the nonuniformity of the stress distribution at and near the point of load application can clearly be seen.

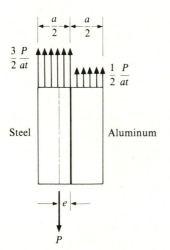

Figure 2-3

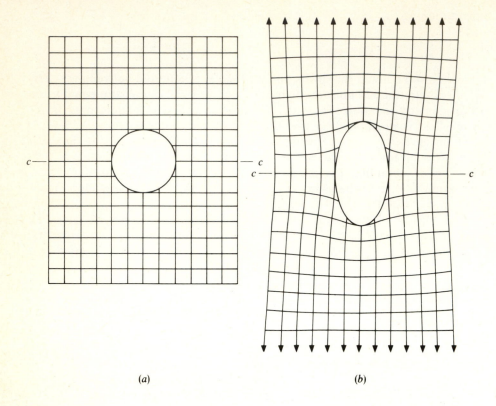

(a)

(b)

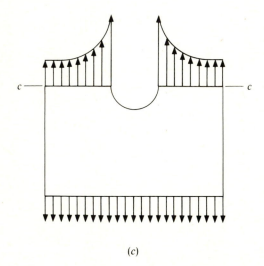

(c)

Figure 2-4

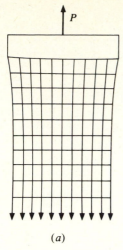

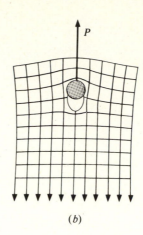

(a)   (b)

Figure 2-5

Strains and Deflections due to Axial Stress

The strains induced by the state of stress from axial loading are obtained using Eqs. (1-21) and (1-25) and are

$$\varepsilon_x = \frac{\sigma_x}{E} \qquad \varepsilon_y = \varepsilon_z = -\frac{\nu\sigma_x}{E} \qquad \gamma_{xy} = \gamma_{yz} = \gamma_{zx} = 0 \tag{2-3}$$

For tensile loading, using Eq. (2-1) the strains can be written as

$$\varepsilon_x = \frac{P}{AE} \qquad \varepsilon_y = \varepsilon_z = -\frac{\nu P}{AE} \qquad \gamma_{xy} = \gamma_{yz} = \gamma_{zx} = 0 \tag{2-4}$$

For compressive loading, from Eq. (2-2) the strains are

$$\varepsilon_x = -\frac{P}{AE} \qquad \varepsilon_y = \varepsilon_z = \frac{\nu P}{AE} \qquad \gamma_{xy} = \gamma_{yz} = \gamma_{zx} = 0 \tag{2-5}$$

The deflection field can be determined by substituting Eqs. (2-4) into Eqs. (1-29a) and (1-29b) and integrating. For the case shown in Fig. 2-6, the geometric conditions that $u = v = 0$ at $x = y = 0$ results in

$$u = \int \varepsilon_x \, dx = \frac{P}{AE} x \tag{2-6a}$$

$$v = \int \varepsilon_y \, dy = -\frac{\nu P}{AE} y \tag{2-6b}$$

where from the boundary conditions, the constants of integration are zero.† If

---

†A simplification has been made here. The partial derivatives of Eqs. (1-29a) and (1-29b) have been replaced by ordinary derivatives. This implies that u is a function only of x and v is a function only of y. Although this is basically true for a member transmitting a constant axial force where the cross section is not changing, it is not true in general. For further discussion, see Sec. 6-5.

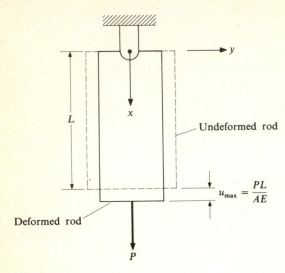

Figure 2-6

the original length of the member in the x direction is L, the maximum value of u is at $x = L$ and is

$$u_{max} = \frac{PL}{AE} \tag{2-7}$$

Example 2-1 For the rod shown in Fig. 2-7, determine the deflection of point B due to the weight of the rod. For the rod, the weight density is ρ, the area is A, and the modulus of elasticity is E.

SOLUTION From the force diagram in Fig. 2-7b, the force transmitted through a section at x is $F = \rho A(L - x)$. Thus the stress is

$$\sigma_x = \frac{F}{A} = \rho(L - x)$$

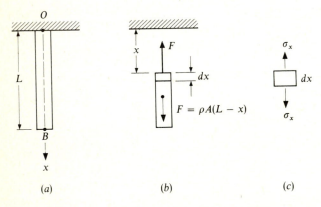

(a)   (b)   (c)

Figure 2-7

The stretch of a dx element is determined from Eq. (2-7), where dx is used instead of L. Thus

$$du = \frac{F}{AE} dx$$

and the total extension of the rod is found by integrating the deflections of each dx element, resulting in

$$u_{max} = \int_0^L \frac{F}{AE} dx = \int_0^L \frac{\rho}{E}(L - x) dx = \frac{\rho L^2}{2E}$$

The deflection can be expressed in terms of the total weight W of the rod. Since $W = \rho AL$, we have

$$u_{max} = \frac{1}{2} \frac{WL}{AE}$$

Example 2-2 Two cables each with a cross-sectional area A support a force P as shown in Fig. 2-8a. Each cable is of length L and modulus of elasticity E. Determine the deflection of point B.

SOLUTION From Fig. 2-8b, equilibrium in the x direction results in $F_{BD} = F_{BC}$. Equilibrium in the y direction yields

$$(F_{BC} + F_{BD}) \cos \beta = P$$

and using the fact that $F_{BD} = F_{BC}$ results in

$$F_{BC} = F_{BD} = \frac{P}{2 \cos \beta} \tag{a}$$

The approach to finding the deflection of point B is first to assume that the cables are not connected and to allow them to deflect elastically dependent on the internal forces (see Fig. 2-8c). Thus, substituting Eq. (a) into Eq. (2-7) yields

$$\delta_{BC} = \delta_{BD} = \frac{[P/(2 \cos \beta)]L}{AE} = \frac{PL}{2AE \cos \beta} \tag{b}$$

The next step is to rotate the elements so that point B of each cable reconnects. If the rotations are very small, the rotation can be approximated by a line drawn perpendicular to the cable, as shown in Fig. 2-8c. Thus the deflection of point B is given by

$$\delta_B = \frac{\delta_{BC}}{\cos \beta} \tag{c}$$

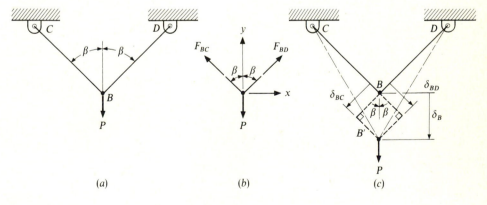

(a)　　　　　(b)　　　　　(c)

Figure 2-8

Substitution of Eq. (*b*) results in

$$\delta = \frac{PL}{2AE \cos^2 \beta} \qquad\qquad (d)$$

2-2 TORSION OF CIRCULAR SECTIONS

Torsion Stresses

A simple illustration of torsional loading of a circular rod is shown in Fig. 2-9*a*, and the idealized model representing the loading of the rod is shown in Fig. 2-9*b*. If a break is made at section *c*-*c*, as shown in Fig. 2-9*c*, the internal force distribution is tangential to the exposed surface so that for equilibrium the total force due to the distribution is zero and the net moment, or torque, about the cylinder axis is equal to the applied torque *T*. The tangential force distribution is therefore in the form of shear stresses. The value of shear stress is given by

$$\tau_{\theta z} = \frac{Tr}{J} \qquad\qquad (2\text{-}8)$$

where T = internal torque at section *c*-*c*
$\quad r$ = position from z axis,
$\quad J$ = area polar moment of inertia about z axis

The derivation of Eq. (2-8) is based on the assumption that deflections are directly proportional to the radial position r, which leads to the result that the shear strains and stresses are directly proportional to r. The area polar moment of inertia for a hollow cylinder of inner radius r_i and outer radius r_o is

$$J = \frac{\pi}{2}(r_o^4 - r_i^4) \qquad\qquad (2\text{-}9)$$

The maximum and minimum values of shear stress are thus

$$\tau_{max} = \frac{Tr_o}{J} \qquad \tau_{min} = \frac{Tr_i}{J} \qquad\qquad (2\text{-}10)$$

For a solid cylinder, $r_i = 0$.

A fixed xyz coordinate system can be used (see Sec. 6-8); or when analyzing combined stresses, a local xyz coordinate system can be used, as shown in Fig. 2-10. Note that for a positive torque T about the axis of the rod (the x axis), the shear stress is given by

$$\tau_{xy} = -\frac{Tr}{J} \qquad\qquad (2\text{-}11)$$

The major restrictions of the formulations on torsional loading of a circular cylinder are the same as those for axial loading; i.e., the material is homogenous, the cross section is not rapidly varying at the section under examination, and applied loads are at reasonable distances from the point of interest and geometrically perfectly positioned.

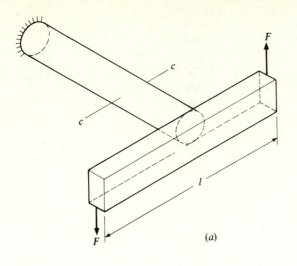

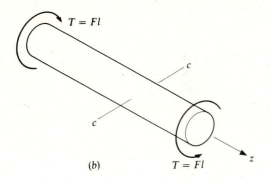

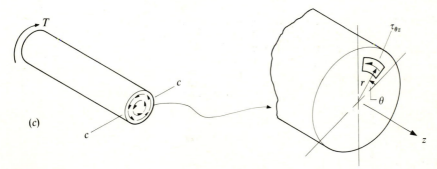

Figure 2-9

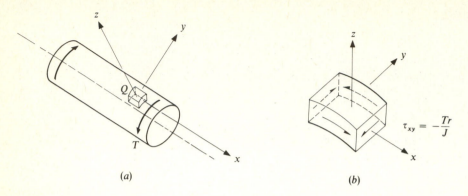

Figure 2-10

Strains and Deflections due to Torque

To obtain the deflection equations, consider (before loading) a line scribed on the outer surface of the cylinder and parallel to the cylinder centerline. If one end of the cylinder is assumed fixed and torque is applied to the other end, the line becomes helical, as shown in Fig. 2-11. Examine an element on the outer surface of the cylinder where one side is located on the scribed line. When the torque is applied, the element then displaces and distorts as shown.

The scribed line forms a helix angle ϕ, which, in radians, is the shear strain γ at $r = r_o$. Therefore,

$$\phi = \gamma \bigg|_{r=r_o} = \frac{2(1+\nu)}{E} \tau \bigg|_{r=r_o}$$

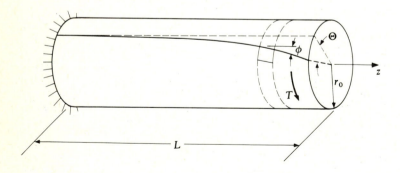

Figure 2-11

and from Eq. (2-10a)

$$\phi = \frac{2(1+\nu)}{E}\frac{Tr_o}{J}$$

Through geometry it can be shown that

$$\frac{\phi}{r_o} = \frac{\Theta}{L}$$

where Θ, in radians, is the angle of twist between the end surfaces. Thus, the angle of twist is given by

$$\Theta = 2(1+\nu)\frac{TL}{EJ} \qquad (2\text{-}12)$$

The above formulations apply only to circular rods. Noncircular rods are covered in Sec. 6-8.

Example 2-3 A step shaft (Fig. 2-12a) transmits a torque of 5000 lb·in. Determine the maximum shear stress and the relative angle of twist between surfaces located at points A and $D(E = 30 \times 10^6 \text{ lb/in}^2, \ \nu = 0.3)$.

SOLUTION For a *solid circular* rod, Eq. (2-10a), is used together with Eq. (2-9), and $r_i = 0$, $r_o = d_o/2$. The resulting equation is

$$\tau_{max} = \frac{16T}{\pi d_o^3} \qquad (2\text{-}13)$$

Thus it can be seen that the maximum shear stress will occur in section AB, where d_o is minimum. Therefore

$$\tau_{max} = \frac{(16)(5000)}{\pi(1)^3} = 25,465 \text{ lb/in}^2 \qquad (a)$$

To obtain the total angle of twist, Eq. (2-12) can be used for sections AB and CD. The angle of twist of B relative to A is

$$\theta_{B/A} = \frac{2(1+0.3)(5000)(8)}{(30 \times 10^6)(\pi/2)(0.5)^4} = 0.0353 \text{ rad} \qquad (b)$$

The angle of twist of D relative to C is

$$\theta_{D/C} = \frac{(2)(1+0.3)(5000)(12)}{(30 \times 10^6)(\pi/2)(1^4)} = 0.0033 \text{ rad} \qquad (c)$$

Since J varies in section BC, Eq. (2-12) must be modified and applied to a differential

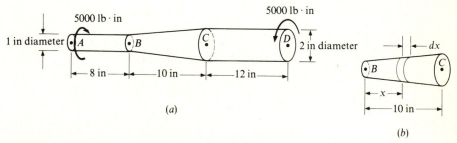

(a)

(b)

Figure 2-12

element as shown in Fig. 2-12b. The radius in section BC varies with respect to x as

$$r_o = (0.05)(10+x) \quad \text{in} \tag{d}$$

Thus

$$J = (9.83 \times 10^{-6})(10+x)^4 \quad \text{in}^4 \tag{e}$$

The angle of twist for the element of length dx is

$$d\theta = \frac{(2)(1+0.3)(5000)\,dx}{(30 \times 10^6)(9.83 \times 10^{-6})(10+x)^4} = \frac{44.1}{(10+x)^4}\,dx \tag{f}$$

The total angle of twist, found by integrating Eq. (f) for section BC, is

$$\theta_{C/B} = \int_0^{10} \frac{44.1}{(10+x)^4}\,dx = -\frac{1}{3}\frac{44.1}{(10+x)^3}\bigg|_0^{10} = -14.7\left(\frac{1}{20^3}-\frac{1}{10^3}\right) = 0.0129\,\text{rad} \tag{g}$$

Adding Eqs. (b), (c), and (g) results in the total angle of twist of the shaft:

$$\theta_{D/A} = \theta_{B/A} + \theta_{C/B} + \theta_{D/C}$$

$$= 0.0353 + 0.0129 + 0.0033 = 0.0515\,\text{rad}$$

$$= (0.0515)(57.3) = 3.0°$$

2-3 BENDING

Transverse loading of a bar, like that shown in Fig. 2-13a, produces bending. If a break is made in section AB and the left portion of the remaining element is isolated, a shear force V_y in the -y direction and bending moment M_z about the

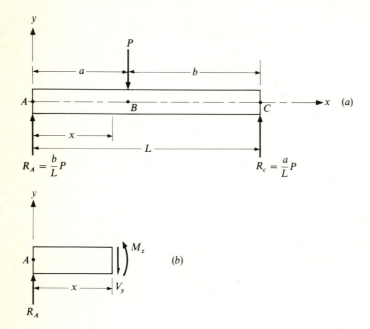

Figure 2-13

z axis are necessary at the exposed surface for equilibrium (see Fig. 2-13b). Since the terms V_y and M_z will arise in the stress, strain, and deflection formulations, a sign convention is necessary. The positive direction for the shear force V_y is established by the positive direction of the shear stress τ_{xy}. The positive direction for the bending moment M_z will be counterclockwise (ccw) on a surface facing the positive x direction or clockwise (cw) on a surface facing the negative x direction (see Fig. 2-14). This establishes a sign convention which may not agree with the reader's basic strength of materials textbook.†

To establish the shear force or bending moment on a surface, first draw V_y and M_z on the free-body diagram in their respective *positive* directions (by definition); then solve for the values using equilibrium equations. For example, return to section AB of Fig. 2-13a, this time showing V_y and M_z in directions positive by *convention* (see Fig. 2-15a):

$$V_y = -R_A = -\frac{b}{L}P \tag{a}$$

$$M_z = R_A x = \frac{b}{L}Px \tag{b}$$

Relationships (a) and (b) are valid in the zone between points A and B of the beam since any value of x $(0 < x < a)$ yields an isolated element of identical appearance, as shown in Fig. 2-15a. However, in zone BC $(a < x < L)$, the isolated element is as shown in Fig. 2-15b, where the applied load P now appears on the isolated element. This requires another set of equations for V_y and M_z. Solving for V_y and M_z in zone BC yields, for $a < x < L$,

$$V_y = -R_A + P = -\frac{b}{L}P + P \tag{c}$$

$$M_z = R_A x - P(x - a) = \frac{b}{L}Px - P(x - a) \tag{d}$$

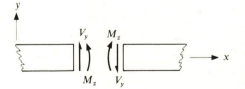

Figure 2-14

---

† The sign convention used in this book agrees with Ref. 1. The convention is consistent with vector mechanics and will be discussed again when three-dimensional problems are dealt with.

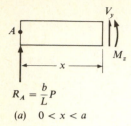

$$R_A = \frac{b}{L} P$$

(a) $0 < x < a$

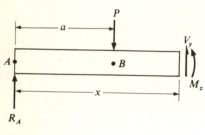

R_A

(b) $a < x < L$

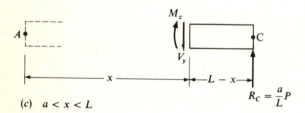

$$R_C = \frac{a}{L} P$$

(c) $a < x < L$

Figure 2-15

Utilizing the fact that $b = L - a$, we can simplify Eqs. (c) and (d) to

$$V_y = \frac{a}{L} P \tag{e}$$

$$M_z = \frac{a}{L} P(L - x) \tag{f}$$

Equations (e) and (f) for zone BC could have been arrived at in a slightly simpler manner if instead of using the left side of the isolated element shown in Fig. 2-15b, the right side had been used, as shown in Fig. 2-15c. V_y and M_z are again drawn in the positive directions as indicated by the convention shown in Fig. 2-14. Solving for V_y and M_z again yields Eqs. (e) and (f).

The shear-force and bending-moment *diagrams* indicating how V_y and M_z vary with respect to position x are presented in Fig. 2-16.

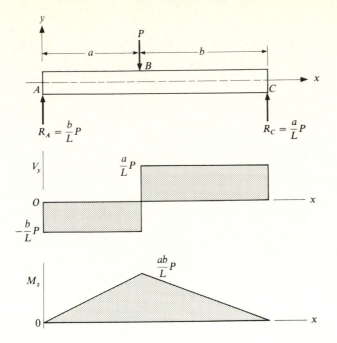

Figure 2-16

Using concentrated forces for the model is an idealization which does not actually occur. The concentrated forces cause discontinuities in values in the V_y curve at $x = 0$, $x = a$, and $x = L$ and discontinuities in slope in the M_z curve at the same values of x. A more realistic physical representation of this problem is shown in Fig. 2-17a, where the load and support forces are applied to the beam using rollers at points A, B, and C. Thus, as force P is increased, the forces on the beam at A, B, and C will be fairly concentrated. The forces will not be entirely concentrated, because when P is gradually increased from zero force, the rollers will deform, tending to flatten out. Thus, instead of a perfectly concentrated force being transmitted at the roller, a distributed force over a very small area exists, as shown in Fig. 2-17b. The net force from the distribution at A is bP/L, at B is P, and at C is aP/L. From the loading of the beam (Fig. 2-17b) the shear-force and bending moment diagrams can be graphed (Fig. 2-17c and d). Comparing the diagrams with Fig. 2-16, one can see that they are quite similar. The maximum values of V_y turn out to be the same for this particular problem, and the idealistic model shown in Fig. 2-16 gives a maximum bending moment slightly larger than that of Fig. 2-17d. Thus, the ideal model will give more conservative results in design applications. One more observation is that there are no discontinuities in the diagrams in Fig. 2-17c and d. The analysis for the model of Fig. 2-17 is much more difficult to perform than that of Fig. 2-16. Since either analysis would yield almost identical results, the simpler analysis assuming concentrated forces is acceptable for the shear and bending analysis.

The mathematical formulations for the shear force and bending moment as functions of position must be expressed in as many equations as there are

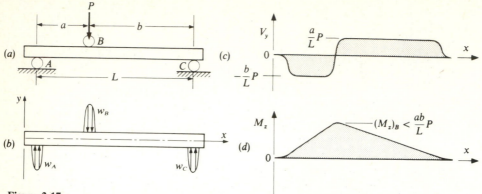

Figure 2-17

zones within the beam with different free-body diagrams.† In the previous example since there were two zones with differing free-body diagrams, two separate sets of equations for V_y and M_z resulted, $(a)(b)$ and $(e)(f)$, respectively. Appendix D gives several examples of equations and diagrams for various beams and loading conditions.

Bending Stresses

The summation of stresses across an internal transverse surface yields the net shear force V_y and bending moment M_z. The stress distribution which produces the bending moment is one of normal stresses in the x direction which yield no net force in the x direction (see Fig. 2-18). The stress distribution at a section located at a particular value of x is given by

$$\sigma_x = -\frac{M_z y}{I_z} \tag{2-14}$$

where y = distance from x axis
I_z = area moment of inertia about z axis

Again, the derivation of the stress formulation is based on a deflection assumption. Here it is assumed that plane surfaces parallel to the yz plane remain plane after bending. The result is normal strains and subsequently normal stresses in the x direction which are linear with respect to y.

The x axis is the bending neutral axis since the normal stress σ_x is zero there. The neutral axis coincides with the centroidal axis of the section, as this will ensure that there is no net force from the stress distribution in the x

---

†For beams with an overabundance of discontinuous loading zones, there is a simpler but more mathematical approach to the many sets of discontinuous equations. Through the use of singularity functions, the load-intensity equation (force per unit length) can be written. Integration yields the shear-force equation, and integration again yields the bending-moment equation. This technique is discussed in Appendix E. However, in this chapter, the method will not be utilized. One must be very careful using this technique, and a great deal of practice should precede any actual application.

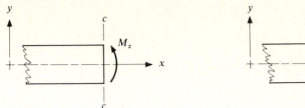

(a) Surface with positive bending moment M_z

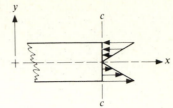

(b) Stress distribution across surface due to M_z

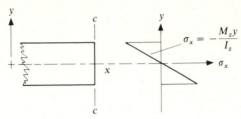

(c) Stress across surface c–c as a function of position y

Figure 2-18

direction. In addition, the applied loading must be such that the net effect of the load causes bending about the z axis only. Thus the equivalent load at any position x must lie in the xy plane. Furthermore, the yz coordinate system is located at the centroid of the cross section, and must be the principal axes of inertia (see Sec. 3-1 and Appendix F).

Example 2-4 Determine the maximum tensile and maximum compressive stresses in the beam shown in Fig. 2-19.

SOLUTION To construct the bending-moment diagram, the reactions at points A and B are determined. The free-body diagram of the beam is shown in Fig. 2-19b. Solving for R_A and R_B yields

$$R_A = 600 \text{ lb} \qquad R_B = 3400 \text{ lb}$$

With this, the bending-moment equation and diagram can be developed. The equation is

$$M_z = \begin{cases} 600x & 0 < x < 50 \\ -200(7x - 500) & 50 < x < 100 \\ 2000(x - 120) & 100 < x < 120 \end{cases}$$

and the diagram is shown in Fig. 2-19b. For positive bending, the tensile stresses are at the beam bottom and compressive stresses are at beam top; for negative bending tensile stresses are at the top and compressive stresses at the bottom.

The next step is to locate the neutral, or centroidal, axis of the section (see Example C-1 in Appendix C). (The reader should consult a text on statics.) This step is shown in Fig. 2-20. Now, since it is known where the neutral axis is, the area moment of inertia about the z axis can be found, as in Fig. 2-21.

For the total section using the parallel-axis theorem (see Appendix C)

$$I_z = 0.25 + (3)(1.0)^2 + 2.25 + (3)(1.0)^2 = 8.5 \text{ in}^4†$$

†The moment of inertia in Fig. 2-21 was found by using the parallel-axis theorem (see Appendix C). A somewhat simpler method for rectangular elements utilizes the formulation for the moment of inertia about the base of a rectangle, which results in

$$I_z = (\tfrac{1}{3})(1)(2.5)^3 + (\tfrac{1}{3})(3)(1.5)^3 - (2)(\tfrac{1}{3})(1)(0.5)^3 = 8.5 \text{ in}^4$$

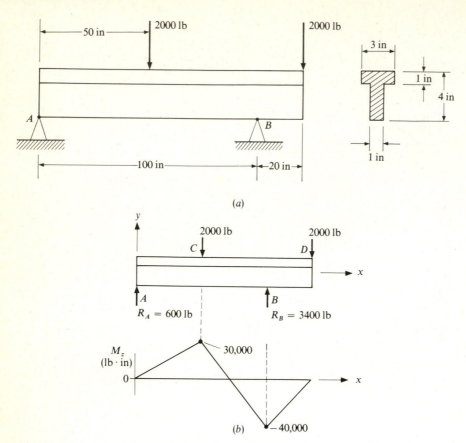

(a)

(b)

Figure 2-19

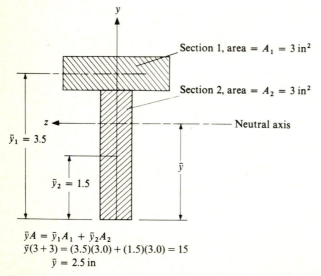

$\bar{y}A = \bar{y}_1 A_1 + \bar{y}_2 A_2$

$\bar{y}(3+3) = (3.5)(3.0) + (1.5)(3.0) = 15$

$\bar{y} = 2.5$ in

Figure 2-20

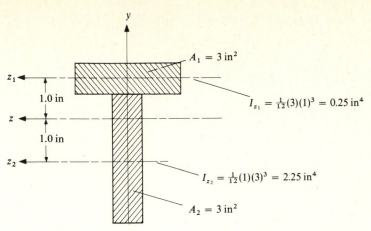

Figure 2-21

The maximum tensile stress can occur at the bottom of the section located at C or the top of the section located at B. That is, from Eq. (2-14) the tensile stresses at C and B respectively, are:

C:
$$(\sigma_x)_{\substack{x=50\,\text{in}\\y=-2.5\,\text{in}}} = -\frac{(30,000)(-2.5)}{8.5} = 8824\ \text{lb/in}^2$$

B:
$$(\sigma_x)_{\substack{x=100\,\text{in}\\y=1.5\,\text{in}}} = -\frac{(-40,000)(1.5)}{8.5} = 7059\ \text{lb/in}^2$$

Obviously, the maximum tensile stress occurs at the bottom of the beam at C. The maximum compression should then be checked at points C and B:

C:
$$(\sigma_x)_{\substack{x=50\,\text{in}\\1.5\,\text{in}}} = -\frac{(30,000)(1.5)}{8.5} = -5294\ \text{lb/in}^2$$

B:
$$(\sigma_x)_{\substack{x=100\,\text{in}\\y=-2.5\,\text{in}}} = -\frac{(40,000)(-2.5)}{8.5} = -11,765\ \text{lb/in}^2$$

Thus, the maximum compressive stress will occur at the bottom of the beam at B.

A common expression for the maximum bending stress is

$$\sigma_{max} = \frac{Mc}{I} \tag{2-15}$$

where $M = M_z$, $I = I_z$, and $c = |y_{max}|$. The term I/c is called the *section modulus* Z, with units of length^3. Thus, Eq. (2-15) can be expressed as

$$\sigma_{max} = \frac{M}{Z} \tag{2-16}$$

Handbooks normally tabulate values of Z for standard cross sections.†

Transverse Shear Stresses

Shearing stresses occur on an internal transverse surface of a beam due to the transverse shear force V_y. The shear force at a particular value of x is V_y, and

†For example, see the AISC (American Institute of Steel Construction) Manual of Steel Construction.

at a particular value of y, say $y = y_1$, the shear stress is given by

$$\tau_{xy} = \frac{V_y Q}{I_z b} \tag{2-17}$$

where $Q = \int_{y_1}^{y_{max}} y \, dA$, called the area moment about the z axis of the section from y_1 to y_{max}

I_z = area moment of inertia about z axis

b = width of beam at $y = y_1$

The derivation of Eq. (2-11) is not directly based on a deflection assumption. A particular shear-stress distribution is necessary to ensure the internal state of equilibrium within the beam. To help understand this, return to Fig. 2-16 and isolate a dx element in zone AB, where the element is bounded above by the top free surface and below by some arbitrary value of y, say y_1. This is illustrated in Fig. 2-22a, where the bending stresses, assumed positive, are also shown.

Since, in general, the bending moment changes with respect to x, so will σ_x. If in Fig. 2-22a the moment is increasing in magnitude with respect to x, the bending stress on the right face will be greater than that on the left face. This will cause a net force in the positive x direction and can only be balanced by the shear stress on the lower surface shown to be a positive τ_{xy}. The net force to the right is due to $(\partial \sigma_x / \partial x)(\Delta x)$. Differentiating Eq. (2-14) with respect to x yields

$$\frac{\partial \sigma_x}{\partial x} = \frac{\partial}{\partial x} \left(-\frac{M_z y}{I_z} \right) = -y \frac{\partial}{\partial x} \left(\frac{M_z}{I_z} \right)$$

Assuming that I_z is constant, this reduces to

$$\frac{\partial \sigma_x}{\partial x} = \frac{-y}{I_z} \frac{\partial M_z}{\partial x}$$

It is well known from elementary strength of materials that $\partial M_z / \partial x = -V_y$;[†] thus

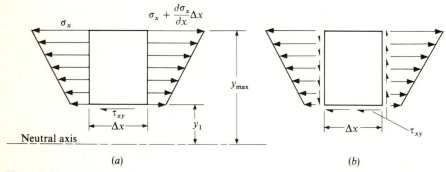

Neutral axis

(a)   (b)

Figure 2-22

---

† Depending on the convention used, the sign associated with V_y may be different. Also, see Sec. E-1 of Appendix E.

$$\frac{\partial \sigma_x}{\partial x} = \frac{V_y y}{I_z}$$

The net force in the x direction is determined by integrating $(\partial \sigma_x / \partial x)\Delta x$ across the area from $y = y_1$ to $y = y_{max}$. Thus

$$\Delta F_x = \int_{y_1}^{y_{max}} \left(\frac{V_y y}{I_z} \Delta x\right) dA$$

Note that since for the element V_y, I_z, and Δx are constant, the force is

$$\Delta F_x = \frac{V_y \Delta x}{I_z} \int_{y_1}^{y_{max}} y \, dA$$

The integral has already been defined as Q, and so the net force due to $d\sigma_x$ is

$$\Delta F_x = \frac{V_y Q}{I_z} \Delta x$$

This is balanced by the shear force from the shear stress τ_{xy} acting over the surface $b \, \Delta x$, where b is the width of the beam at $y = y_1$. Thus

$$\tau_{xy} b \, \Delta x = \frac{V_y Q}{I_z} \Delta x$$

Solving for τ_{xy} yields

$$\tau_{xy} = \frac{V_y Q}{I_z b}$$

Example 2-5 For a beam with a solid rectangular cross section of depth h and width b, derive an expression for the shear-stress distribution if the section is transmitting a transverse shear force V_y.

SOLUTION The cross section of the beam is shown in Fig. 2-23a. To determine the shear stress at a particular value of y, a break is made at $y = y_1$. This isolates the shaded area shown

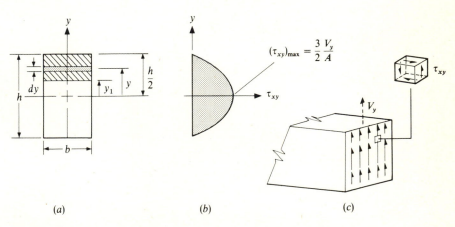

$$(\tau_{xy})_{max} = \frac{3}{2} \frac{V_y}{A}$$

(a)　　　　　　　(b)　　　　　　　(c)

Figure 2-23

in Fig. 2-23a. In order to evaluate the area moment Q, the infinitesimal area $dA = b\, dy$ can be utilized. Thus,

$$Q = \int_{y_1}^{h/2} y\, dA = \int_{y_1}^{h/2} y(b\, dy) = \frac{b}{2}\left[\left(\frac{h}{2}\right)^2 - y_1^2\right]^\dagger$$

The width of the section at $y = y_1$ is b. Substituting Q and b into Eq. (2-17) results in

$$\tau_{xy} = \frac{V_y Q}{I_z b} = \frac{V_y (b/2)[(h/2)^2 - y_1^2]}{I_z b} = \frac{V_y}{2I_z}\left[\left(\frac{h}{2}\right)^2 - y_1^2\right]$$

This equation shows that the transverse shear stress varies in a parabolic fashion with respect to y_1, as shown in Fig. 2-23b. The maximum value of the shear stress occurs when $y_1 = 0$. Substituting $y_1 = 0$ into the above equation yields

$$(\tau_{xy})_{max} = \frac{V_y h^2}{8 I_z}$$

However, since $I_z = bh^3/12$,

$$(\tau_{xy})_{max} = \frac{V_y h^2}{8(bh^3/12)} = \frac{3}{2}\frac{V_y}{bh}$$

Since the total area of the cross section is $A = bh$, the maximum shear stress can be expressed as

$$(\tau_{xy})_{max} = \frac{3}{2}\frac{V_y}{A}$$

Figure 2-23c illustrates how the shear stress varies across the transverse surface of the beam.

For most cross sections, the ratio of Q/b is maximum at the bending neutral axis. For a *rectangular* cross section, the maximum value of shear stress, as determined in the previous example, is

$$(\tau_{xy})_{max} = \frac{3}{2}\frac{V_y}{A} \qquad (2\text{-}18)$$

For a *circular* cross section, the maximum shear stress is

$$(\tau_{xy})_{max} = \frac{4}{3}\frac{V_y}{A} \qquad (2\text{-}19)$$

Verification of Eq. (2-19) is left as an exercise (see Exercise 2-7).

† Note that when the centroid and area of the isolated area are known, determination of the area moment Q can be simplified by using the properties of centroids. It is well known that

$$\bar{y}A = \int_A y\, dA$$

where $\bar{y}$ is the centroid of the area A. For the shaded area shown in Fig. 2-23.

$$\bar{y}_1 A_1 = \int_{A_1} y\, dA = \int_{y_1}^{y_{max}} y\, dA = Q$$

where $\bar{y}_1$ is the centroid of the shaded area A_1. Thus, in this example, $\bar{y}_1 = \frac{1}{2}(h/2 + y_1)$, $A_1 = b(h/2 - y_1)$, and the area moment is

$$Q = \bar{y}_1 A_1 = \frac{1}{2}\left(\frac{h}{2} + y_1\right)\left[b\left(\frac{h}{2} - y_1\right)\right] = \frac{b}{2}\left[\left(\frac{h}{2}\right)^2 - y_1^2\right]$$

Example 2-6 Determine the maximum shearing stress due to the shear force V_y for the beam given in Example 2-4.

SOLUTION The shear-force equations are written and the diagram is constructed in order to determine the maximum shear force on an internal surface of the beam. From Fig. 2-20, the shear-force equations are

$$V_y = \begin{cases} -600 & 0 < x < 50 \\ 1400 & 50 < x < 100 \\ -2000 & 100 < x < 120 \end{cases}$$

and the resulting diagram is illustrated in Fig. 2-24. The maximum shear force occurs throughout the span to the right of B and is $V_{max} = |-2000|$ lb. The maximum shear stress will occur in this span, but at what value of y? The shear stress is greatest with respect to y where Q/b is the greatest value. Since Q is maximum and b is minimum at $y_1 = 0$, Q/b is greatest at $y = 0$. The determination of Q is illustrated in Fig. 2-25.

For the section, at $y_1 = 0$

$$Q_{max} = \int_0^{y_{max}} y\, dA = \sum \bar{y}_i A_i = (1.0)(3.0) + (0.25)(0.5) = 3.13 \text{ in}^3\dagger$$

and $b = 1.0$ in and is the minimum value. Thus

$$(Q/b)_{max} = 3.13/1 = 3.13 \text{ in}^3.$$

Thus, from Eq. (2-14), the maximum shear stress due to V_y is

$$(\tau_{xy})_{max} = \frac{-2000}{8.5}\frac{3.13}{1.0} = -736 \text{ lb/in}^2$$

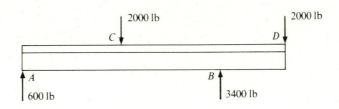

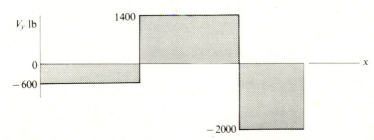

Figure 2-24

†For negative values of y_1, the area moment Q is defined as

$$Q = \int_{y_1}^{y_{max}} y\, dA = -\int_{y_{min}}^{y_1} y\, dA$$

Thus, in Fig. 2-25, the value of Q could have been found more simply by using the area below the z axis. The location to the centroid of the lower section is $\bar{y} = -1.25$ in, and the area is 2.5 in²; thus

$$Q = -(-1.25)(2.5) = 3.13 \text{ in}^3$$

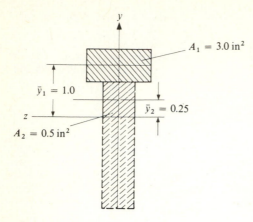

Figure 2-25

Strains and Deflections due to Bending

The strains due to the bending moment and shear force at a point are†

$$\varepsilon_x = \frac{-M_z y}{EI_z} \qquad (2.20a)$$

$$\varepsilon_y = \varepsilon_z = \frac{\nu M_z y}{EI_z} \qquad (2\text{-}20b,c)$$

$$\gamma_{xy} = \frac{2(1+\nu)}{E} \frac{V_y Q}{I_z b} \qquad (2\text{-}20d)$$

In basic strength of materials, the deflection of the centroidal axis due to bending effects only‡ is considered using analytical geometry. Here, the basic formulation for the deflection of the centroidal axis is

$$\frac{1}{\rho} = \frac{d^2 y_c / dx^2}{[1 + (dy_c/dx)^2]^{3/2}} = \frac{M_z}{EI_z} \qquad (2\text{-}21)$$

where y_c is the deflection of the centroidal axis in the y direction and ρ is the radius of curvature of the centroidal axis. For small slopes ($dy_c/dx \le 0.1$), Eq. (2-21) reduces to

$$\frac{d^2 y_c}{dx^2} = \frac{M_z}{EI_z} \qquad (2\text{-}22)$$

The procedure in establishing the deflection shape is first to integrate Eq. (2-22) with respect to x, yielding the equation for the slope of the centroidal axis

---

† Assuming that $\sigma_y = 0$ and that $\sigma_z = \tau_{yz} = \tau_{zx} = 0$. The first assumption neglects the bearing stresses on the beam (see Sec. 3-13). The second assumption, where stresses in the z direction are assumed to be zero, applies only for narrow beams (see Sec. 3-7).

‡ Shear effects are discussed in Chap. 4.

as a function of x:

$$\frac{dy_c}{dx} = \int \frac{M_z}{EI_z} dx + C_1 \tag{2-23}$$

where C_1 is an integration constant. Integrating once more yields

$$y_c = \int \left(\int \frac{M_z}{EI_z} dx \right) dx + C_1 x + C_2 \tag{2-24}$$

The constants C_1 and C_2 then can be evaluated using the geometric end conditions of the beam.

Tables giving the shear and bending-moment diagrams and the deflection and slope equations for beams undergoing various loading conditions appear in Appendix D.

Example 2-7 Verify the slope and deflection of the centroidal axis as functions of x for the cantilever beam D.3 given in Appendix D.

SOLUTION Referring to beam D.3 of Appendix D, we see that the bending-moment equation, which is continuous in x, is given by

$$M_z = -\frac{w}{2}(L - x)^2 \tag{a}$$

Thus, from Eq. (2-23)

$$\frac{dy_c}{dx} = \int \frac{-(w/2)(L-x)^2}{EI_z} dx + C_1 = \frac{-w}{2EI_z}\left(L^2 x - Lx^2 + \frac{x^3}{3}\right) + C_1 \tag{b}$$

and from Eq. (2-24)

$$y_c = \int \frac{-w}{2EI_z}\left(L^2 x - Lx^2 + \frac{x^3}{3}\right) dx + C_1 x + C_2$$

$$= -\frac{w}{2EI_z}\left(\frac{L^2 x^2}{2} - \frac{Lx^3}{3} + \frac{x^4}{12}\right) + C_1 x + C_2 \tag{c}$$

The end conditions for the beam are $y_c = 0$ at $x = 0$ and $dy_c/dx = 0$ at $x = 0$. Substitution of the end conditions into Eqs. (b) and (c) yields $C_1 = C_2 = 0$.

Thus

$$\frac{dy_c}{dx} = -\frac{wx}{6EI_z}(3L^2 - 3Lx + x^2) \tag{d}$$

and

$$y_c = -\frac{wx^2}{24EI_z}(6L^2 - 4Lx + x^2) \tag{e}$$

which can be seen to agree with the results given in Appendix D.

2-4 THIN-WALLED PRESSURE VESSELS

Consider the closed pressurized vessel illustrated in Fig. 2-26. If the wall thickness t of the vessel is small compared with the radius of curvature of the surface, the state of stress at a point can be described by the *membrane equation*.

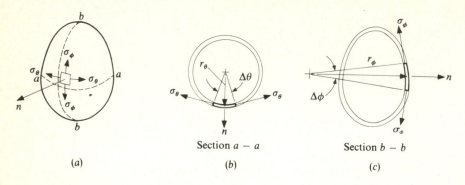

Section $a - a$

Section $b - b$

(a)   (b)   (c)

Figure 2-26

Isolate an element of the wall using coordinates tangent to the principal arcs of curvature and perpendicular to the surface. Thus the dimensions of the element will be $r_\phi\,\Delta\phi$ by $r_\theta\,\Delta\theta$ by t, and the outward normal of the infinitesimal surface is established by n. The state of stress on the element is shown in Fig. 2-26a, where for equilibrium of forces in the n direction

$$-2\sigma_\theta t r_\phi\,\Delta\phi\,\sin\frac{\Delta\theta}{2}-2\sigma_\phi t r_\theta\,\Delta\theta\,\sin\frac{\Delta\phi}{2}+pr_\theta\,\Delta\theta\,r_\phi\,\Delta\phi=0$$

Since $\Delta\theta$ and $\Delta\phi$ are infinitesimal, $\sin(\Delta\theta/2)=\Delta\theta/2$ and $\sin(\Delta\phi/2)=\Delta\phi/2$. Simplifying gives

$$\frac{\sigma_\theta}{r_\theta}+\frac{\sigma_\phi}{r_\phi}=\frac{p}{t}\tag{2-25}$$

The stress in the radial direction σ_r is usually assumed to be zero toward the outer surface of the vessel and $-p$ at the inner surface. To obtain a second relationship for the "membrane" stresses σ_θ and σ_ϕ a symmetric break is usually made in the vessel so that only one of the stresses exists along the break, and the stress is determined using the equilibrium conditions of the isolated element. For example, consider the *circular cylinder* shown in Fig. 2-27a, where $r_\phi=\infty$ and $r_\theta=r$. From Eq. (2-25)

$$\sigma_\theta=\frac{pr_\theta}{t}=\frac{pr}{t}\tag{2-26}$$

where σ_θ is referred to as the *hoop stress*. Cylindrical coordinates (r,θ,z) are used to replace the longitudinal stress of the cylinder σ_ϕ by σ_z. This stress is found by making a break exposing σ_z as shown in Fig. 2-27b. Equilibrium of force in the axial direction yields $\sigma_z(2\pi rt)=p(\pi r^2)$. Thus

$$\sigma_z=\frac{pr}{2t}\tag{2-27}$$

Considering the case of a *spherical cylinder* total symmetry exists, $\sigma_\theta=\sigma_\phi$

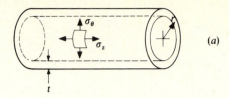

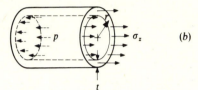

(a)

(b)

Figure 2-27

and $r_\theta = r_\phi = r$, and Eq. (2-25) yields

$$\sigma_\theta = \sigma_\phi = \frac{pr}{2t} \tag{2-28}$$

Example 2-8 A thin closed toroidal shell of thickness t is shown in Fig. 2-28a. Determine (a) the membrane stresses as a function of ϕ and (b) the membrane stresses at point Q under the following conditions: $p = 105$ kN/m^2, $r = 0.3$ m, $R = 1.0$ m, and $t = 10$ mm.

SOLUTION (a) First, make a vertical cylindrical break B-B around the torus and consider the outer element of the break. Secondly, make a break at ϕ around the torus so that the remaining element appears as shown in Fig. 2-28b. The remaining area that the pressure acts on, projected onto a horizontal plane, is $\pi[(R + r \sin \phi)^2 - R^2]$. The total vertical force F on the remaining element due to the pressure is then given by

$$F = p(\pi)[(R + r \sin \phi)^2 - R^2] = \pi pr(2R + r \sin \phi) \sin \phi \tag{a}$$

The force which balances this comes from the vertical component of σ_ϕ acting over the area $2\pi(R + r \sin \phi)t$. Or

$$F = \sigma_\phi [2\pi(R + r \sin \phi)t] \sin \phi \tag{b}$$

Equating Eqs. (a) and (b) and simplifying yields

$$\sigma_\phi = \frac{pr(2R + r \sin \phi)}{2t(R + r \sin \phi)} \tag{c}$$

To determine the other membrane stress, we find the radial dimension

$$r_\theta = r + \frac{R}{\sin \phi} \tag{d}$$

and substitute Eqs. (c) and (d) into Eq. (2-25) with $r = r_\phi$, to give

$$\sigma_\theta = \frac{pr}{2t} \tag{e}$$

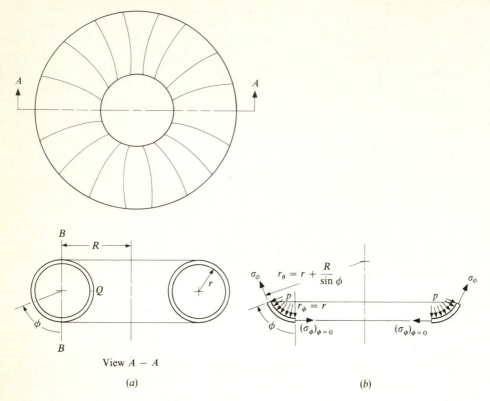

Figure 2-28

(b) It can be shown that σ_ϕ is maximum at point Q, where $\phi = \frac{3}{2}\pi$. For this case, Eq. (c) reduces to

$$\sigma_\phi = \frac{pr}{2t}\frac{2R - r}{R - r} \tag{f}$$

Substituting the numerical values of p, R, r, and t into Eqs. (f) and (e) yields

$$\sigma_\phi = \frac{(105)(0.3)[2(1.0) - 0.3]}{(2)(0.01)(1.0 - 0.3)} = 3825 \text{ kN/m}^2$$

$$\sigma_\theta = \frac{(105)(0.3)}{(2)(0.1)} = 1575 \text{ kN/m}^2$$

2-5 SUPERPOSITION

Linear elasticity is predicated on the assumption that strains are small (this is sometimes mistakenly referred to as small-deflection theory). For linear systems, the action of each force with respect to a particular effect can be analyzed independently, and the results can be added algebraically or as vectors depending on the situation. The particular effect considered can be either internal or support forces, moments, slopes, deflections, stresses, or strains. The advantage of this is that the results of simple loading configurations are known, and the results of complex loading can be found by adding the results of each individual load or load distribution.

Example 2-9 For the steel beam shown in Fig. 2-29 determine the support reactions, shear and bending-moment diagrams, and deflection of the centroidal axis as functions of x and the maximum tensile stress.

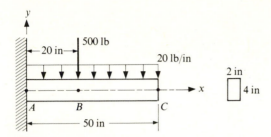

Figure 2-29

SOLUTION The effects of the uniform force distribution and the concentrated force can be viewed independently. The independent results are given in Appendix D and repeated below.

Support reactions

See Fig. 2-30.

$$M_1 = 25{,}000 \text{ lb·in} \qquad M_2 = 10{,}000 \text{ lb·in}$$
$$R_1 = 1000 \text{ lb} \qquad R_2 = 500 \text{ lb}$$

Superposition: $\qquad M = M_1 + M_2 = 35{,}000 \text{ lb·in} \qquad R = R_1 + R_2 = 1500 \text{ lb}$

Shear and bending-moment diagrams

See Fig. 2-31.

Deflections

For beam 1:
$$y_{c_1} = \frac{wx^2}{24EI}(4Lx - x^2 - 6L^2)$$

For beam 2:
$$(y_{c_2})_{AB} = \frac{Px^2}{6EI}(x - 3a) \qquad (y_{c_2})_{BC} = \frac{Pa^2}{6EI}(a - 3x)$$

Superposition:
$$y_{AB} = \frac{wx^2}{24EI}(4Lx - x^2 - 6L^2) + \frac{Px^2}{6EI}(x - 3a) \tag{a}$$

$$y_{BC} = \frac{wx^2}{24EI}(4Lx - x^2 - 6L^2) + \frac{Pa^2}{6EI}(a - 3x) \tag{b}$$

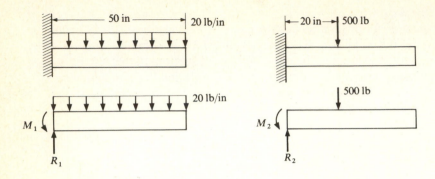

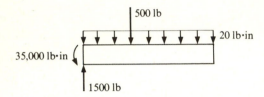

Figure 2-30

where $w = 20$ lb/in, $P = 500$ lb, $E = 30 \times 10^6$ lb/in², $I = 10.7$ in⁴, and $a = 20$ in. The maximum deflection occurs at $x = L = 50$ in, where

$$(y_c)_{max} = \frac{(20)(50^2)}{(24)(30 \times 10^6)(10.7)} [(4)(50)(50) - 50^2 - (6)(50^2)] + \frac{(500)(20^2)}{(6)(30 \times 10^6)(10.7)} [20 - (3)(50)]$$

$$= -0.062 \text{ in}$$

Stress

Maximum tensile stress occurs where the maximum bending moment appears. Since in both beams it occurs at the wall, using Eq. (2-14) yields

$$(\sigma_x)_{max,1} = -\frac{(-25,000)(2)}{10.7} = 4670 \text{ lb/in}^2$$

$$(\sigma_x)_{max,2} = -\frac{(-10,000)(2)}{10.7} = 1870 \text{ lb/in}^2$$

Superposition: $\qquad (\sigma_x)_{max} = (\sigma_x)_{max,1} + (\sigma_x)_{max,2} = 6540 \text{ lb/in}^2$

Using superposition, only like stresses can be added algebraically; i.e., normal stresses can be added to normal stresses or shear stresses to shear stresses algebraically. If the two stresses to be superposed are unlike, e.g., a normal stress and a shear stress, they must be shown separately on a diagram.

Example 2-10 The beam shown in Fig. 2-32 is fixed at one end and loaded at the other end with a transverse load P and a torque T. Neglecting the inaccuracies in the stress formulations at the support end, determine the state of stress at points A, B, C, and D.

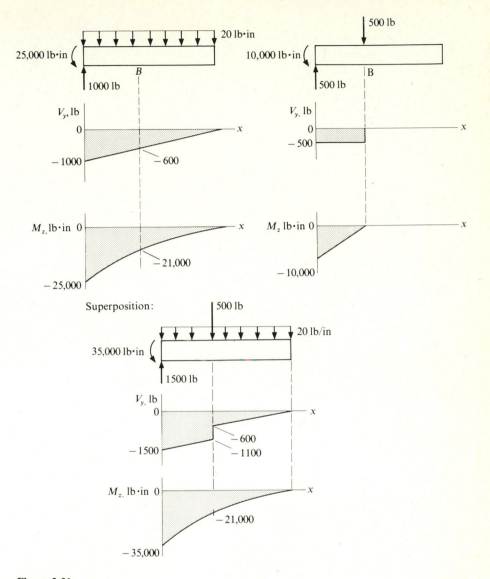

Figure 2-31

SOLUTION The two loading states can be analyzed independently initially. The torque T causes a maximum shear stress τ_1 at $r = d/2$, where

$$\tau_1 = \frac{Td/2}{(\pi/32)d^4} = \frac{16T}{\pi d^3} \qquad (a)$$

The torsional shear stresses at points A, B, C, and D are illustrated in Fig. 2-33a.

The transverse force P causes bending stresses as well as transverse shear stresses. The bending moment at the wall is $M_z = -PL$, and the shear force is $V_y = -P$. The bending

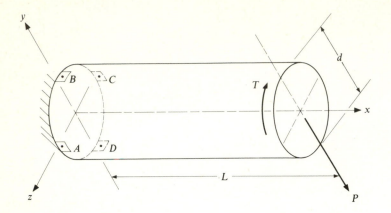

Figure 2-32

stresses are given by

$$\sigma_x = -\frac{M_z y}{I_z}$$

where $I_z = (\pi/64)d^4$. Thus, at points A, B, C, and D, the bending stresses are

$$(\sigma_x)_A = -\frac{(-PL)(0)}{(\pi/64)d^4} = 0 \qquad (\sigma_x)_B = -\frac{(-PL)(d/2)}{(\pi/64)d^4} = \frac{32PL}{\pi d^3}$$

$$(\sigma_x)_C = -\frac{(-PL)(0)}{(\pi/64)d^4} = 0 \qquad (\sigma_x)_D = -\frac{(-PL)(-d/2)}{(\pi/64)d^4} = -\frac{32PL}{\pi d^3}$$

The shear stresses, given by Eq. (2-17), reduce to [see also Eq. (2-19)]

$$(\tau_2)_B = (\tau_2)_D = 0 \qquad (\tau_2)_A = (\tau_2)_C = \frac{4}{3}\frac{V_y}{A} = \frac{-16P}{3\pi d^2}$$

The shear and bending stresses induced by the transverse load are illustrated in Fig. 2-33b. The stresses due to the transverse and torsional loading, determined separately, can now be combined by superposition to reveal the complete state of stress. Since the stresses at points A and C are of the same type, namely shear, the stresses can be added algebraically. Note, however, that at point C the directions of the shear stresses are the same, whereas at point A, the shear stresses are in opposite directions. Since the stresses at points B and D are not of the same type, i.e., both normal and shear stresses exist, the stresses cannot be added algebraically. Thus the combined stresses (illustrated in Fig. 2-33d) are as follows:

Point A: $\qquad\qquad\qquad \sigma_x = 0 \qquad\qquad \tau = \frac{16T}{\pi d^3} - \frac{16P}{3\pi d^2}$

Point B: $\qquad\qquad\qquad \sigma_x = \frac{32PL}{\pi d^3} \qquad \tau = \frac{16T}{\pi d^3}$

Point C: $\qquad\qquad\qquad \sigma_x = 0 \qquad\qquad \tau = \frac{16T}{\pi d^3} + \frac{16P}{3\pi d^2}$

Point D: $\qquad\qquad\qquad \sigma_x = -\frac{32PL}{\pi d^3} \qquad \tau = \frac{16T}{\pi d^3}$

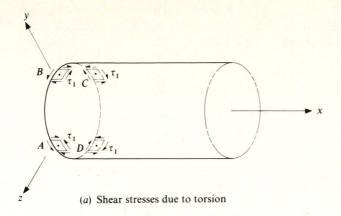

(a) Shear stresses due to torsion

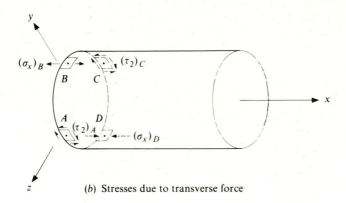

(b) Stresses due to transverse force

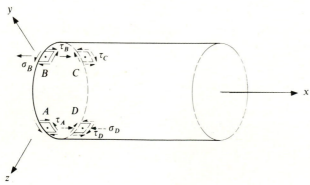

(c) Superposition of combined stresses

Figure 2-33

When the analysis of an element at a point is complete, the state of stress with respect to other internal surfaces at the point can be scrutinized using Mohr's circle, discussed in detail in Sec. 2-7.

2-6 STATICALLY INDETERMINATE PROBLEMS

The method of superposition is very effective in dealing with simple, statically indeterminate problems where support reactions cannot be completely determined from the equations of equilibrium. When the geometry of the system begins to become complex, superposition becomes difficult and energy techniques become more advantageous (see Chap. 4).

When a statically indeterminate problem arises, there will be more unknown support or internal reactions than there are equations of equilibrium. The difference will be the *order* for which the structure is redundant or indeterminate. Thus, additional equations *equal* to this order of indeterminateness are necessary to complete the solution of the support reactions. The additional equations *must* be arrived at through a *deflection* analysis of the structure. The technique for obtaining these additional deflection equations in this book may differ from the reader's past experience, but it works well for all problems, simple as well as complex, and should be studied carefully. The steps are as follows:

Step 1 Solve for all possible unknown reactions of each member of the structure using the equilibrium equations (even in statically indeterminate problems, some but not all, of the reactions may be solvable).

Step 2 Subtract the number of remaining equilibrium equations from the number of remaining unknowns. This is the *order n* for which the structure is indeterminate *and* the number of additional deflection equations necessary.

Step 3 Eliminate *n* of the unknown reactions and solve for the *n* deflections and/or rotations of the structure at the locations of the unknown reactions in terms of the applied loads. At this point the structure should be reduced to a statically determinate problem.

Step 4 Considering the *n* unknown reactions of step 3 as "applied" forces, determine the deflections and/or rotations at the same *n* points of step 3 but this time in terms of the unknown reactions.

Step 5 Using the method of superposition, add the results of steps 3 and 4 at each of the *n* points such that the *n* deflections and/or rotations are compatible with the geometric restrictions of the problem. This results in a set of *n* simultaneous equations in terms of the unknown support reactions.

Step 6 Solve the simultaneous equations to obtain the values of the support reactions.

Step 7 Substitute the results of step 6 into the equilibrium equations of step 1 to solve for the remaining unknown reactions.

It is recommended that these steps be read over again after studying the following examples.

Example 2-11 A step shaft is supported at each end by thrust bearings, as shown in Fig. 2-34a. A force, $F = 1000$, is applied at the step (point B) as shown. Determine the stress in each section of the shaft. The material is steel.

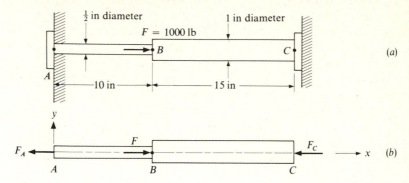

Figure 2-34

SOLUTION The free-body diagram is shown in Fig. 2-34b. Note that there is only one equilibrium equation available (the equilibrium of forces in the x direction) and two unknowns, F_A and F_C. Thus, the member is statically indeterminate to the first order. This means that an additional deflection equation is necessary for a complete solution. A deflection equation which may be obvious is that the elongation of element AB equals the contraction of BC. Then using the fact that the respective axial deflections are related to the internal forces F_A and F_C via Eq. (2-7) results in a second equation in F_A and F_C. The two simultaneous equations in F_A and F_C will then complete the solution.

Although acceptable for this problem, this approach is highly restrictive, and as problems become more complex, the analyst's intuitive reasoning will be overtaxed. With the recommended approach, the steps are as follows.

Step 1 The equilibrium equation is

$$F_A + F_C = F \tag{a}$$

Step 2 There are two unknowns and one equilibrium equation. Thus, the structure is indeterminate of order $n = 2 - 1 = 1$.

Step 3 Since $n = 1$, this means that *one* of the unknown reactions is to be eliminated. Selecting F_A to be eliminated results in the diagram shown in Fig. 2-35a. With F_A eliminated the deflection of point A in the x direction is

$$(\delta_A)_F = \frac{FL_{BC}}{A_{BC}E} = \frac{(1000)(15)}{(\pi/4)(1^2)(30 \times 10^6)} = 6.37 \times 10^{-4} \text{ in} \tag{b}$$

Step 4 The force F_A was eliminated in Step 3. The deflection at point A in the x direction due to F_A, as shown in Fig. 2-35b, is

$$(\delta_A)_{F_A} = -\frac{F_A L_{AB}}{A_{AB}E} - \frac{F_A L_{BC}}{A_{BC}E} = -\frac{F_A}{E}\left(\frac{L_{AB}}{A_{AB}} + \frac{L_{BC}}{A_{BC}}\right)$$

$$= \frac{-F_A}{30 \times 10^6}\left[\frac{10}{(\pi/4)(\frac{1}{2})^2} + \frac{15}{(\pi/4)(1)^2}\right]$$

$$= -(2.33 \times 10^{-6})F_A \tag{c}$$

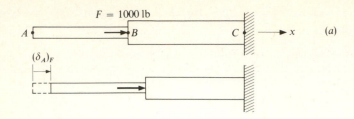

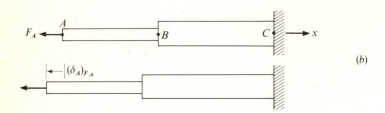

Figure 2-35

Step 5 Using superposition, we see that the total deflection of point A is $(\delta_A)_F + (\delta_A)_{F_A}$. The geometric restriction is that the total deflection of point A is zero. Thus $(\delta_A)_F + (\delta_A)_{F_A} = 0$, and from Eqs. (*b*) and (*c*)

$$\delta_A = 6.37 \times 10^{-4} - (2.33 \times 10^{-6})F_A = 0$$

$$F_A = 273 \text{ lb}$$

Step 6 Substituting $F_A = 273$ lb into Eq. (*a*) yields

$$273 + F_C = 1000$$

$$F_C = 727 \text{ lb}$$

This completes the force analysis of the structure. To obtain the stresses, using Eq. (2-1) for element AB and Eq. (2-2) for element BC gives

$$(\sigma_x)_{AB} = \frac{F_A}{A_{AB}} = \frac{273}{(\pi/4)(\tfrac{1}{2})^2} = 1390 \text{ lb/in}^2$$

$$(\sigma_x)_{BC} = \frac{-F_C}{A_{BC}} = \frac{-727}{(\pi/4)(1)^2} = -926 \text{ lb/in}^2$$

Example 2-12 Using beams D.1 and D.3 of Appendix D, obtain the support reactions of the beam shown in Fig. 2-36.

SOLUTION

Step 1 The equilibrium equations are

$$R_A + R_B = wL \qquad R_B L + M_A = \frac{wL^2}{2} \tag{a}$$

Step 2 There are three unknowns (R_A, R_B, M_A), and two equilibrium equations. Thus, $n = 3 - 2 = 1$.

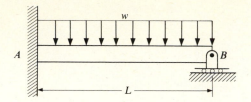

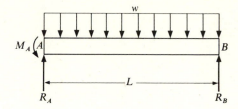

Figure 2-36

Step 3 Eliminate one ($n = 1$) unknown. In order to use beams D.1 and D.3, R_B must be eliminated. With R_B eliminated, the deflection of point B of beam D.3 due to w is determinated from the table in Appendix D:

$$(\delta_B)_w = -\frac{wL^4}{8EI} \qquad (b)$$

Step 4 Considering R_B alone, using beam D.1, we find the deflection of point B to be

$$(\delta_B)_{R_B} = -\frac{(-R_B)L^3}{3EI} = \frac{R_B L^3}{3EI} \qquad (d)$$

Step 5 Superposing beams D.1 and D.3 with the restriction that the total deflection of point B must be zero results in

$$\frac{-wL^4}{8EI} + \frac{R_B L^3}{3EI} = 0$$

$$R_B = \tfrac{3}{8}wL$$

Step 6 Substituting R_B into Eq. (a) yields

$$R_A + \tfrac{3}{8}wL = wL$$

$$R_A = \tfrac{5}{8}wL$$

$$(\tfrac{3}{8}wL)(L) + M_A = \frac{wL^2}{2}$$

$$M_A = \tfrac{1}{8}wL^2$$

The results agree with the equations given in the table for beam D.12.

Example 2-13 A beam fixed at both ends carries a uniformly distributed load w over half of the span, as shown in Fig. 2-37. Determine the support reactions. Assume that the beam is capable of sliding horizontally in one of the supports so that no axial support force exists.

SOLUTION

Step 1 From Fig. 2-37b, the equilibrium equations are

$$(a) \quad R_A + R_B = \frac{wL}{2} \qquad (b) \quad R_B L + M_A - M_B = \frac{wL^2}{8}$$

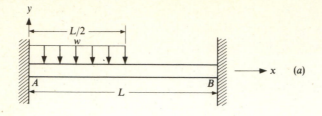

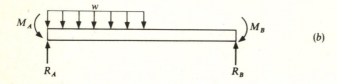

Figure 2-37

Step 2 Since there are four unknowns (R_A, R_B, M_A, M_B) and two equilibrium equations, the beam is statically indeterminate of order $n = 4 - 2 = 2$.

Step 3 Since $n = 2$, two of the reactions are eliminated. If R_B and M_B are eliminated, the structure will be the same as beam D.9 of Appendix D. The resulting two deflections and/or rotations of interest will be the deflection and slope at point B. From beam D.9 these are

$$(\delta_B)_w = \frac{w(L/2)^3}{24EI}\left(\frac{L}{2} - 4L\right) = -\frac{7}{384}\frac{wL^4}{EI} \tag{c}$$

$$(\theta_B)_w = -\frac{w(L/2)^3}{6EI} = -\frac{1}{48}\frac{wL^3}{EI} \tag{d}$$

Step 4 The deflection and slope at point B due to R_B are obtained from beam D.1:

$$(\delta_B)_{R_B} = -\frac{(-R_B)L^3}{3EI} = \frac{R_B L^3}{3EI} \tag{e}$$

$$(\theta_B)_{R_B} = \frac{(-R_B)(L)}{2EI}(L - 2L) = \frac{R_B L^2}{2EI} \tag{f}$$

The deflection and slope at point B due to M_B are obtained from beam D.4:

$$(\delta_B)_{M_B} = \frac{(-M_B)(L)^2}{2EI} = -\frac{M_B L^2}{2EI} \tag{g}$$

$$(\theta_B)_{M_B} = \frac{(-M_B)(L)}{EI} = -\frac{M_B L}{EI} \tag{h}$$

Step 5 The total deflection of point B is zero and is found by superposing Eqs. (c), (e), and (g). Thus

$$\frac{-7}{384}\frac{wL^4}{EI} + \frac{R_B L^3}{3EI} - \frac{M_B L^2}{2EI} = 0$$

which reduces to

$$128R_B L - 192M_B = 7wL^2 \tag{i}$$

The total slope of point B is zero and is found by superposing Eqs. (d), (f), and (h):

$$-\frac{1}{48}\frac{wL^3}{EI} + \frac{R_B L^2}{2EI} - \frac{M_B L}{EI} = 0$$

which reduces to

$$24R_BL - 48M_B = wL^2 \tag{j}$$

Solving Eqs. (i) and (j) simultaneously yields

$$R_B = \tfrac{3}{32}wL \qquad M_B = \tfrac{5}{192}wL^2$$

Step 6 Substituting R_B and M_B into Eqs. (a) and (b) yields

$$R_A + \tfrac{3}{32}wL = \frac{wL}{2}$$

$$R_A = \tfrac{13}{32}wL$$

$$(\tfrac{3}{32}wL)(L) + M_A - \tfrac{5}{192}wL^2 = \frac{wL^2}{8}$$

$$M_A = \tfrac{11}{192}wL^2$$

2-7 MOHR'S CIRCLE FOR STRESS

Mohr's circle is a graphical technique used in coordinate transformation of various state properties. In stress analysis, Mohr's circle is used to describe the state of stress or strain graphically at a particular point with respect to any plane or surface intersecting the point. There is a small ambiguity in the development of the two-dimensional Mohr's circle with respect to the transformation equations and the sign convention for shear stress, which should be examined so that errors in interpretation do not occur.

Most practical engineering problems involve investigating the biaxial stress state. Thus, assume that a general state of biaxial stress is known at a point described by $\Delta x \, \Delta y \, \Delta z$, as shown in Fig. 2-38$a$. In Fig. 2-38$b$, the stress cube is

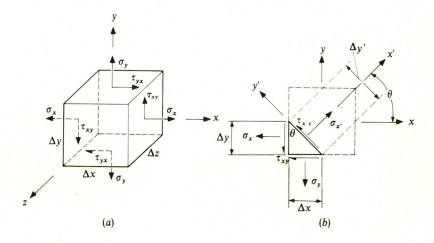

(a)   (b)

Figure 2-38

intersected by a plane exposing a surface whose normal is at an angle θ with respect to the x axis. Since Δx and Δy are arbitrary, they can be redefined as shown in Fig. 2-38b.

Thus, stresses $\sigma_{x'}$ and $\tau_{x'y'}$ are the normal and shear stresses acting at the given point on a surface whose normal is at an angle θ with respect to the x axis. The area of this surface is $\Delta y' \, \Delta z$, and

$$\Delta x = \Delta y' \sin \theta \qquad \Delta y = \Delta y' \cos \theta \qquad (a)$$

Summation of forces in the x' direction yields

$$\sigma_{x'} \, \Delta y' \, \Delta z - \sigma_x \, \Delta y \, \Delta z \cos \theta - \sigma_y \, \Delta x \, \Delta z \sin \theta - \tau_{xy} \, \Delta y \, \Delta z \sin \theta$$
$$- \tau_{xy} \, \Delta x \, \Delta z \cos \theta = 0 \qquad (b)$$

When Eqs. (a) are used and $\Delta y' \, \Delta z$ is factored out, Eq. (b) reduces to

$$\sigma_{x'} = \sigma_x \cos^2 \theta + \sigma_y \sin^2 \theta + 2\tau_{xy} \sin \theta \cos \theta \qquad (2\text{-}29)$$

Similarly, if the forces are summed in the y' direction and Eqs. (a) are used again, the equation for $\tau_{x'y'}$ is found to be

$$\tau_{x'y'} = -(\sigma_x - \sigma_y) \sin \theta \cos \theta + \tau_{xy}(\cos^2 \theta - \sin^2 \theta) \qquad (2\text{-}30)$$

Now knowing σ_x, σ_y, and τ_{xy}, using Eqs. (2-29) and (2-30), we can obtain the state of stress at the point for any surface located by the angle θ. Equations (2-29) and (2-30) can be simplified further by using the trigonometric identities

$$\cos^2 \theta = \frac{1 + \cos 2\theta}{2} \qquad \sin^2 \theta = \frac{1 - \cos 2\theta}{2}$$

$$\sin \theta \cos \theta = \tfrac{1}{2} \sin 2\theta \qquad \cos^2 \theta - \sin^2 \theta = \cos 2\theta$$

Thus, Eqs. (2-29) and (2-30) are modified to

$$\sigma_{x'} = \frac{\sigma_x + \sigma_y}{2} + \frac{\sigma_x - \sigma_y}{2} \cos 2\theta + \tau_{xy} \sin 2\theta \qquad (2\text{-}31)$$

$$\tau_{x'y'} = -\frac{\sigma_x - \sigma_y}{2} \sin 2\theta + \tau_{xy} \cos 2\theta \qquad (2\text{-}32)$$

Example 2-14 Show how the normal and shear stresses vary in the xy plane as a function of θ for a point Q within a rod undergoing an axial tensile force P. The cross-sectional area of the rod is A.

SOLUTION The biaxial state of stress for point Q is shown in Fig. 2-39, where the stresses are

$$\sigma_x = \frac{P}{A} \qquad \sigma_y = \tau_{xy} = 0 \qquad (a)$$

Substituting Eqs. (a) into Eqs. (2-31) and (2-32) yields

$$\sigma_{x'} = \frac{P}{2A}(1 + \cos 2\theta) \qquad (b)$$

$$\tau_{x'y'} = -\frac{P}{2A} \sin 2\theta \qquad (c)$$

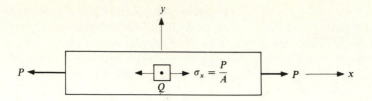

Figure 2-39

For a better physical picture of how the stresses vary around the infinitesimal point Q, Eqs. (b) and (c) can be qualitatively depicted showing all surfaces perpendicular to the xy plane. If these surfaces are at some infinitesimal distance from Q, a cylinder will be formed. The stresses on these surfaces are shown in Fig. 2-40.

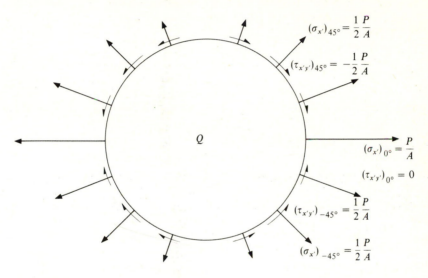

Figure 2-40

Since σ_x, σ_y, and τ_{xy} are assumed known in Eqs. (2-31) and (2-32) the terms

$$A = \frac{\sigma_x + \sigma_y}{2} \qquad B = \frac{\sigma_x - \sigma_y}{2} \qquad \text{and} \qquad C = \tau_{xy}$$

are fixed values for a specific point. Thus, Eqs. (2-31) and (2-32) can be written

$$\sigma_{x'} = A + B \cos 2\theta + C \sin 2\theta \qquad \tau_{x'y'} = -B \sin 2\theta + C \cos 2\theta$$

where A, B, and C are constants. It is a simple task to plot all possible values of $\sigma_{x'}$ and $\tau_{x'y'}$ at the particular point under consideration. This is done graphically with fixed line segments A, B, and C, as shown in Fig. 2-41a. Note that Eqs. (2-31) and (2-32) are always a combination of fixed lines. For $\theta = 0$, x' aligns with x and $\sigma_{x'} = \sigma_x$ and $\tau_{x'y'} = \tau_{xy}$. This agrees graphically, as shown in Fig. 2-41b. Now note that θ is defined positive in the counterclockwise direction with respect to the element (see Fig. 2-38b). As θ increases (counterclockwise), the line segments B and C rotate through an angle of 2θ *clockwise* (see Fig. 2-41c). This action describes a circle on the σ vs. τ graph with its center at point O. Thus it can be seen that all points on the circle depict the normal and shear stresses in the $x'y'$ directions on every surface with a surface normal in the x' direction. When $\theta = 90°$, x' points in the y direction and y' points in the negative x direction. Thus, for $\theta = 90°$, $\sigma_{x'} = \sigma_y$ and $\tau_{x'y'} = -\tau_{xy}$. This point can be seen on the circle in Fig. 2-41c. Since the direction of $\sigma_{y'}$ is $90°$ from the x' direction, on the circle $\sigma_{y'}$ will always be $180°$ from $\sigma_{x'}$. Thus

$$\sigma_{y'} = A - B \cos 2\theta - C \sin 2\theta$$

Substitution of the values of A, B, and C yields

$$\sigma_{y'} = \frac{\sigma_x + \sigma_y}{2} - \frac{\sigma_x - \sigma_y}{2} \cos 2\theta - \tau_{xy} \sin 2\theta \tag{2-33}$$

Adding Eqs. (2-31) and (2-33) results in

$$\sigma_{x'} + \sigma_{y'} = \sigma_x + \sigma_y \tag{2-34}$$

Thus it can be seen that the *sum* of any two mutually perpendicular normal stresses at a point is constant (invariant).

The circle previously described obeys the transformation equations, *but it is not* the commonly referred to Mohr's circle. Since Mohr's circle is generally used in analysis work, it will be adopted from this point on.

The basic difference between Mohr's circle and the circle just described is that when Mohr's circle is used, the standard sign convention for shear stress is replaced by a different convention. Shear stresses pointing clockwise with respect to the center of the element are plotted above the σ axis, whereas shear stresses pointing counterclockwise with respect to the element are plotted below the σ axis. A sign convention is actually not established for shear stresses when Mohr's circle is used; thus the positive convention with respect to the infinitesimal element is still maintained as shown in Fig. 2-42a. The advantage of Mohr's circle is that counterclockwise rotation of the surface normal corresponds to counterclockwise rotation on Mohr's circle. For the state of stress illustrated in Fig. 2-42a, Mohr's circle is as shown in Fig. 2-42b.

Now, as θ (corresponding to the x' direction) increases in the counterclockwise direction with respect to the x axis of the element, the point on Mohr's circle moves counterclockwise the amount 2θ. This point corresponds to the state of stress on the surface perpendicular to the x' direction. The sign convention for σ is the same as before. However, the shear stress $\tau_{x'y'}$ will be positive (with respect to the $x'y'$ coordinate system) if counterclockwise and negative if clockwise. Thus, together with the angular convention, the sign of the shear stress with respect to the $x'y'$ coordinate system can be represented as shown in Fig. 2-42b. Positive shear stresses are plotted below the σ axis and negative shear stresses above.

(a)

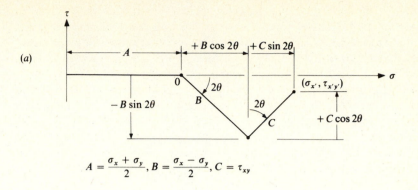

$$A = \frac{\sigma_x + \sigma_y}{2}, B = \frac{\sigma_x - \sigma_y}{2}, C = \tau_{xy}$$

(b)

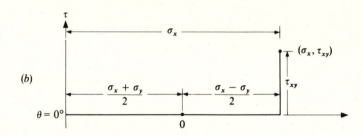

(c)

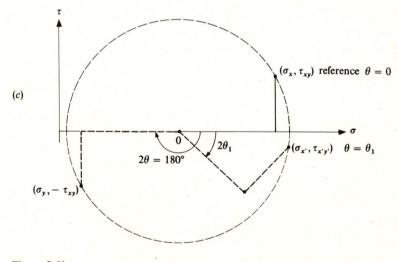

Figure 2-41

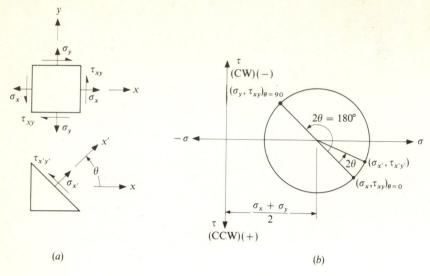

(a)

(b)

Figure 2-42

Example 2-15 (a) Draw Mohr's circle for the state of stress shown in Fig. 2-43a.

(b) Determine the state of stress on a surface whose normal x' is located at an angle θ = 15° counterclockwise from the x axis.

(c) Complete a rectangular element associated with the x' and y' axes of part (b).

(d) Establish the location of surfaces where the shear stresses are zero and determine the values of the normal stresses (principal stresses) at these surfaces.

(e) Establish the location of surfaces where the shear stress is maximum and determine the state of stress on these surfaces.

SOLUTION (a) Mohr's circle can be constructed without using any of the transformation equations.

Step 1 The state of stress for θ = 0° is σ = +6000 kN/m² and τ = 3000 kN/m² clockwise and is reference point A on Mohr's circle (Fig. 2-43b).

Step 2 Since the y axis is 90° from the reference x axis, the state of stress for θ = 90° is σ = −2000 kN/m² and τ = 3000 kN/m² counterclockwise and is point B in Fig. 2-43b.

Step 3 Since points A and B represent states of stress on surfaces whose normals are 90° apart (θ), A and B must be 180° apart (2θ) on Mohr's circle. This means that the line AB must be the diameter of the circle, where point C, shown in Fig. 2-43b, is the center. Due to the antisymmetric positions of points A and B above and below the σ axis respectively, the center is located on the σ axis at a distance whose value is the average of σ_x and σ_y. Thus

$$\sigma_{av} = \frac{\sigma_x + \sigma_y}{2} = \frac{6000 + (-2000)}{2} = 2000 \text{ kN/m}^2$$

Step 4 With a compass, or freehand, draw Mohr's circle centered at point C and intersecting points A and B.

(b) To determine the state of stress on a surface whose normal is located at an angle θ counterclockwise (or clockwise) with respect to the reference x axis, simply rotate through an angle of 2θ counterclockwise (or clockwise) from the reference line CA on Mohr's circle. In this part of the example, the state of stress is desired for a surface located at θ = 15°

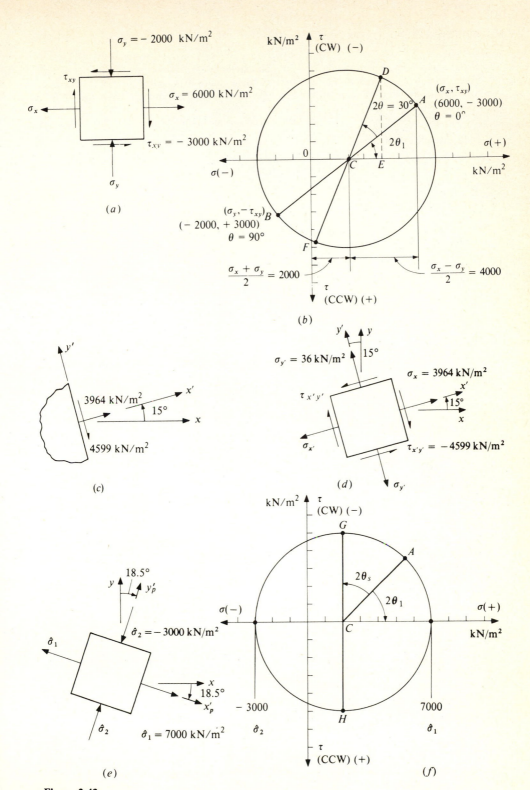

$\sigma_y = -2000 \text{ kN/m}^2$

τ_{xy}

σ_x

$\sigma_x = 6000 \text{ kN/m}^2$

$\tau_{xy} = -3000 \text{ kN/m}^2$

σ_y

(a)

kN/m^2 τ (CW) (−)

$2\theta = 30°$ A

(σ_x, τ_{xy}) (6000, −3000) $\theta = 0°$

$2\theta_1$

$\sigma(+)$ kN/m^2

$\sigma(-)$

$(\sigma_y, -\tau_{xy})B$ (−2000, +3000) $\theta = 90°$

$\dfrac{\sigma_x + \sigma_y}{2} = 2000$

$\dfrac{\sigma_x - \sigma_y}{2} = 4000$

τ (CCW) (+)

(b)

3964 kN/m^2 $15°$

4599 kN/m^2

(c)

$\sigma_{y'} = 36 \text{ kN/m}^2$ $15°$

$\sigma_x = 3964 \text{ kN/m}^2$ $15°$

$\tau_{x'y'}$

$\sigma_{x'}$

$\tau_{x'y'} = -4599 \text{ kN/m}^2$

$\sigma_{y'}$

(d)

$18.5°$

$\hat{\sigma}_2 = -3000 \text{ kN/m}^2$

$\hat{\sigma}_1$

$18.5°$

$\hat{\sigma}_2$

$\hat{\sigma}_1 = 7000 \text{ kN/m}^2$

(e)

kN/m^2 τ (CW) (−)

G

A

$2\theta_s$

$2\theta_1$

$\sigma(-)$

$\sigma(+)$ kN/m^2

C

-3000

H

$\hat{\sigma}_2$

7000

$\hat{\sigma}_1$

τ (CCW) (+)

(f)

Figure 2-43

83

counterclockwise. Thus, rotating $2\theta = 30°$ counterclockwise from reference line CA establishes point D on Mohr's circle. The corresponding values of σ and τ for $\theta = 15°$ are then found by determining the values of OE and ED, respectively. The radius, of the circle CA is

$$CA = \sqrt{4000^2 + 3000^2} = 5000 \text{ kN/m}^2$$

and the angle $2\theta_1$ is

$$2\theta_1 = \tan^{-1} \frac{3000}{4000} = 36.9°$$

Therefore, from Fig. 2-43b,

$$\sigma_{x'} = 2000 + 5000 \cos (30° + 36.9°) = 3964 \text{ kN/m}^2 \ (+)$$
$$\tau_{x'y'} = 5000 \sin (30° + 36.9°) = 4599 \text{ kN/m}^2 \ (\text{cw})(-)$$

This state of stress, corresponding to a normal rotated 15° counterclockwise from the x axis, is shown in Fig. 2-43c.

(c) To establish the state of stress on a surface whose normal is in the y' direction of part (b), simply rotate 180° from point D to point F since the y' axis is 90° from the x' axis. The state of stress described by point F on Mohr's circle is

$$\sigma_{y'} = 2000 - 5000 \cos (30° + 36.9°) = 36 \text{ kN/m}^2 \ (+)$$
$$\tau_{x'y'} = 5000 \sin (30° + 36.9°) = 4599 \text{ kN/m}^2 \ (\text{ccw})$$

The state of stress on a rectangular element described by the $x'y'$ coordinate system is shown in Fig. 2-43d.

(d) To locate an axis of principal stress, i.e., where the shear stress is zero, simply rotate from the reference line CA to a point which intersects the σ axis. In this example, if rotation is $2\theta_1 = 36.9°$ clockwise from line CA, an axis of principal stress (called the principal axis) is found where the value of the normal stress $\hat{\sigma}_1$ is[†]

$$\hat{\sigma}_1 = 2000 + 5000 = 7000 \text{ kN/m}^2$$

Rotating 180° from this point on Mohr's circle establishes the stress on the surface which is 90° from the first principal axis. It can be seen that the shear stress for this surface is also zero. Thus, a second principal axis is found where the value of the normal stress $\hat{\sigma}_2$ is[‡]

$$\hat{\sigma}_2 = 2000 - 5000 = -3000 \text{ kN/m}^2$$

(e) Points G and H (see Fig. 2-43f) on Mohr's circle establish maximum shear values on surfaces with normals in the xy plane. Point G is at an angle of $2\theta_s$ counterclockwise from reference line CA, and it can be seen that

$$2\theta_1 + 2\theta_s = 90° \qquad \text{or} \qquad \theta_1 + \theta_s = 45°$$

[†] Most strength of materials books use σ_1 and σ_2 rather than $\hat{\sigma}_1$ and $\hat{\sigma}_2$. The reason for this deviation from standard practice is that the analysis of stress must *always* be considered to be three-dimensional. As will be demonstrated later, there will always be three principal stresses, called σ_1, σ_2, and σ_3. Although quite arbitrary, it is common practice to order the principal stresses such that $\sigma_1 > \sigma_2 > \sigma_3$. In the case of biaxial stress, one of the principal stresses is always equal to zero. For example, in Fig. 2-43a, the shear stress on the surface in the plane of the page is zero; thus, by definition, the normal stress on this surface is a principal stress with a value of zero. Until all *three* principal stresses are known, it is uncertain which ones to label σ_1, σ_2, and σ_3.

[‡] The three principal stresses in this example are 7000, -3000, and 0 kN/m^2. Thus, using the standard practice that $\sigma_1 > \sigma_2 > \sigma_3$, we have

$$\sigma_1 = 7000 \text{ kN/m}^2 \qquad \sigma_2 = 0 \text{ kN/m}^2 \qquad \sigma_3 = -3000 \text{ kN/m}^2$$

Thus, $\theta_s = 45° - 18.5° = 26.5°$, and

$$\sigma_{av} = 2000 \text{ kN/m}^2 \ (+) \qquad (\tau_{xy})_{max} = 5000 \text{ kN/m}^2 \ (cw)$$

Point H is 180° from point G; hence it establishes a surface 90° from the surface established by point G. The state of stress established by point H is

$$\sigma_{av} = 2000 \text{ kN/m}^2 \ (+) \qquad (\tau_{xy})_{max} = 5000 \text{ kN/m}^2 \ (ccw)$$

The equations for the principal stresses can be found from the transformation equations or Mohr's circle:

$$\hat{\sigma}_1 = \frac{\sigma_x + \sigma_y}{2} + \sqrt{\left(\frac{\sigma_x - \sigma_y}{2}\right)^2 + \tau_{xy}^2} \qquad (2\text{-}35a)$$

$$\hat{\sigma}_2 = \frac{\sigma_x + \sigma_y}{2} - \sqrt{\left(\frac{\sigma_x - \sigma_y}{2}\right)^2 + \tau_{xy}^2} \qquad (2\text{-}35b)$$

The maximum shear stress† is given by

$$\tau_{max} = \sqrt{\left(\frac{\sigma_x - \sigma_y}{2}\right)^2 + \tau_{xy}^2} \qquad (2\text{-}36a)$$

and the normal stresses on the surfaces containing the maximum shear stresses are given by

$$\sigma_{av} = \frac{\sigma_x + \sigma_y}{2} \qquad (2\text{-}36b)$$

2-8 MOHR'S CIRCLE FOR STRAIN

Strains transform in a manner similar to the transformation of stresses. Assume that ε_x, ε_y, and γ_{xy} are known for a particular point Q shown in Fig. 2-44. The

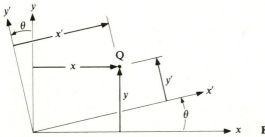

Figure 2-44

---

† Note that this is the maximum shear stress for surfaces perpendicular to the xy plane and *may not* be the maximum shear stress for the element. This is discussed in detail in Sec. 5-1.

strains ε_x, ε_y, and γ_{xy} are with respect to the xy coordinate system. From Eqs. (1-29)

$$\varepsilon_x = \frac{\partial u}{\partial x} \tag{1-29a}$$

$$\varepsilon_y = \frac{\partial v}{\partial y} \tag{1-29b}$$

$$\gamma_{xy} = \frac{\partial v}{\partial x} + \frac{\partial u}{\partial y} \tag{1-29c}$$

The strains corresponding to the $x'y'$ coordinate system are thus

$$\varepsilon_{x'} = \frac{\partial u'}{\partial x'} \tag{2-37a}$$

$$\varepsilon_{y'} = \frac{\partial v'}{\partial y'} \tag{2-37b}$$

$$\gamma_{x'y'} = \frac{\partial v'}{\partial x'} + \frac{\partial u'}{\partial y'} \tag{2-37c}$$

where u' and v' are displacements in the x' and y' directions, respectively. The chain rule can be applied to Eqs. (2-37) resulting in

$$\varepsilon_{x'} = \frac{\partial u'}{\partial x'} = \frac{\partial u'}{\partial x}\frac{\partial x}{\partial x'} + \frac{\partial u'}{\partial y}\frac{\partial y}{\partial x'} \tag{2-38a}$$

$$\varepsilon_{y'} = \frac{\partial v'}{\partial y'} = \frac{\partial v'}{\partial x}\frac{\partial x}{\partial y'} + \frac{\partial v'}{\partial y}\frac{\partial y}{\partial y'} \tag{2-38b}$$

$$\gamma_{x'y'} = \frac{\partial v'}{\partial x'} + \frac{\partial u'}{\partial y'} = \left(\frac{\partial v'}{\partial x}\frac{\partial x}{\partial x'} + \frac{\partial v'}{\partial y}\frac{\partial y}{\partial x'}\right)$$
$$+ \left(\frac{\partial u'}{\partial x}\frac{\partial x}{\partial y'} + \frac{\partial u'}{\partial y}\frac{\partial y}{\partial y'}\right) \tag{2-38c}$$

From Fig. 2-44 it can be seen that

$$x = x' \cos\theta - y' \sin\theta \qquad y = x' \sin\theta + y' \cos\theta \tag{2-39}$$

Figure 2-45 shows how displacements are measured with respect to the two coordinate systems. It can be seen that

$$u' = u \cos\theta + v \sin\theta \qquad v' = -u \sin\theta + v \cos\theta \tag{2-40}$$

Substituting Eqs. (2-39) and (2-40) into Eq. (2-38a) yields

$$\varepsilon_{x'} = \left(\frac{\partial u}{\partial x}\cos\theta + \frac{\partial v}{\partial x}\sin\theta\right)\cos\theta + \left(\frac{\partial u}{\partial y}\cos\theta + \frac{\partial v}{\partial y}\sin\theta\right)\sin\theta$$

$$= \frac{\partial u}{\partial x}\cos^2\theta + \frac{\partial v}{\partial y}\sin^2\theta + \left(\frac{\partial v}{\partial x} + \frac{\partial u}{\partial y}\right)\sin\theta\cos\theta$$

Finally, substitution of Eqs. (1-29) results in

$$\varepsilon_{x'} = \varepsilon_x \cos^2\theta + \varepsilon_y \sin^2\theta + \gamma_{xy} \sin\theta\cos\theta$$

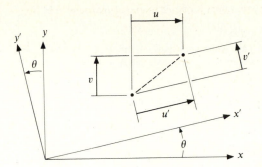

Figure 2-45

Using trigonometric identities, as in the stress transformations, yields

$$\varepsilon_{x'} = \frac{\varepsilon_x + \varepsilon_y}{2} + \frac{\varepsilon_x - \varepsilon_y}{2} \cos 2\theta + \frac{\gamma_{xy}}{2} \sin 2\theta \qquad (2\text{-}41a)$$

Performing the same steps on Eqs. (2-38b) and (2-38c) yields

$$\varepsilon_{y'} = \frac{\varepsilon_x + \varepsilon_y}{2} - \frac{\varepsilon_x - \varepsilon_y}{2} \cos 2\theta - \frac{\gamma_{xy}}{2} \sin 2\theta \qquad (2\text{-}41b)$$

$$\frac{\gamma_{x'y'}}{2} = - \frac{\varepsilon_x - \varepsilon_y}{2} \sin 2\theta + \frac{\gamma_{xy}}{2} \cos 2\theta \qquad (2\text{-}41c)$$

Equations (2-41) are quite similar to Eqs. (2-31) to (2-33) for stress transformations. If the interchanges

$$\sigma_x \leftrightarrow \varepsilon_x \qquad \sigma_{x'} \leftrightarrow \varepsilon_{x'}$$

$$\sigma_y \leftrightarrow \varepsilon_y \qquad \sigma_{y'} \leftrightarrow \varepsilon_{y'}$$

$$\tau_{xy} \leftrightarrow \frac{\gamma_{xy}}{2} \qquad \tau_{x'y'} \leftrightarrow \frac{\gamma_{x'y'}}{2}$$

are made, the two sets of transformation equations are identical. Thus, a Mohr's circle for strain can be developed, where ε vs. $\gamma/2$ is plotted.[†] The corresponding equations for the principal strains are

$$\hat{\varepsilon}_1 = \frac{\varepsilon_x + \varepsilon_y}{2} + \sqrt{\left(\frac{\varepsilon_x - \varepsilon_y}{2}\right)^2 + \left(\frac{\gamma_{xy}}{2}\right)^2} \qquad (2\text{-}42a)$$

$$\hat{\varepsilon}_2 = \frac{\varepsilon_x + \varepsilon_y}{2} - \sqrt{\left(\frac{\varepsilon_x - \varepsilon_y}{2}\right)^2 + \left(\frac{\gamma_{xy}}{2}\right)^2} \qquad (2\text{-}42b)$$

[†] As can be seen, γ does not transform between coordinate systems whereas $\gamma/2$ does. For this reason, in the theory of elasticity, shear strain is usually defined by $\gamma/2$. That is, in the theory of elasticity, the shear strain using xy coordinates is given as

$$\varepsilon_{xy} = \frac{1}{2}\left(\frac{\partial v}{\partial x} + \frac{\partial u}{\partial y}\right)$$

which is $\gamma_{xy}/2$.

Example 2-16 For a biaxial state of stress, the values for strain at a point are found to be

$$\varepsilon_x = +800 \ \mu\text{in/in} \qquad \varepsilon_y = +200 \ \mu\text{in/in} \qquad \gamma_{xy} = +800 \ \mu\text{in/in}$$

Determine the location of the principal axes and the values of the principal strains and stresses using Mohr's circle for strain. Assume that the material is aluminum with $E_a = 10 \times 10^6 \ \text{lb/in}^2$, $\nu_a = 0.33$.

SOLUTION Since γ_{xy} is positive, it must be counterclockwise for the x face and clockwise for the y face. Therefore, Mohr's circle for strain is as shown in Fig. 2-46. The angle θ_1 from the x axis to the axis containing the principal strain ε_1 is found from the circle to be

$$2\theta_1 = \tan^{-1} \tfrac{400}{300} = 53.1°$$

$$\theta_1 = 26.6° \ (\text{cw})$$

Also, the principal strains are found to be

$$\hat{\varepsilon}_1 = 500 + \sqrt{300^2 + 400^2} = 1000 \ \mu\text{in/in}$$

$$\hat{\varepsilon}_2 = 500 - \sqrt{300^2 + 400^2} = 0 \ \mu\text{in/in}$$

From Eqs. (1-24)

$$\hat{\sigma}_1 = \frac{E_a}{1 - \nu_a^2} (\hat{\varepsilon}_1 + \nu_a \hat{\varepsilon}_2) \qquad \hat{\sigma}_2 = \frac{E_a}{1 - \nu_a^2} (\hat{\varepsilon}_2 + \nu_a \hat{\varepsilon}_1)$$

Thus

$$\hat{\sigma}_1 = \frac{(10 \times 10^6)(1000 \times 10^{-6})}{1 - (0.33)^2} = +11,200 \ \text{lb/in}^2$$

$$\hat{\sigma}_2 = \frac{10 \times 10^6}{1 - (0.33)^2} (0.33)(1000 \times 10^{-6}) = +3700 \ \text{lb/in}^2$$

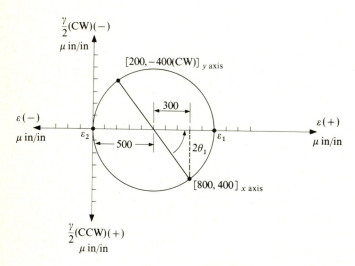

Figure 2-46

2-9 BENDING OF CIRCULAR RODS IN TWO PLANES

In Sec. 2-3 beams bending in only one plane were discussed. Design problems often occur where bending is not isolated in one plane, and the bending may appear rather complex. These cases can be handled effectively using vector algebra, but since vector notation is not used in this book, two-plane analysis will be developed. If a section is undergoing bending about both the y and z axes, as shown in Fig. 2-47a, with moments M_y and M_z respectively, the total moment exerted on the section $M_{z'}$ will be $\sqrt{M_y^2 + M_z^2}$ about an axis located at an angle β of $\tan^{-1}(M_y/M_z)$.

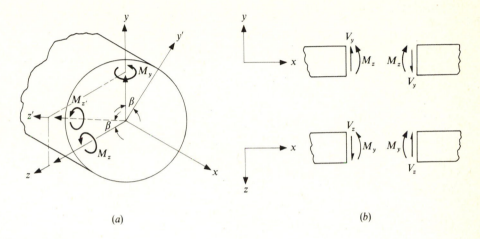

(a)   (b)

Figure 2-47

The positive sign convention for bending in two planes is shown in Fig. 2-47b. Note that the convention for the analysis in the xy plane is identical to that used in Sec. 2-3, and the convention used for the yz plane is slightly different; however, this is consistent with vector notation.

The analysis which follows is restricted to circular shafts, since circular shafts have total symmetry with respect to the x axis and the stress equation for bending about the z' axis is basically the same as Eq. (2-14) and is given by

$$\sigma_x = \frac{-M_{z'}y'}{I_{z'}} \tag{2-43}$$

where $\qquad M_{z'} = \sqrt{M_y^2 + M_z^2} \qquad I_{z'} = \frac{\pi}{64}d^4$

and y' is the perpendicular distance from the z' axis. For two-plane analysis, the stress equation can also be written

$$\sigma_x = \frac{M_y z}{I_y} - \frac{M_z y}{I_z}$$ (2-44)†

where z is the position from the y axis.

Equations (2-43) and (2-44) are valid only if the yz axes are principal axes of inertia (see Sec. 3-1). Equation (2-43) is further restricted to cross sections where $I_y = I_z$, which is the case for a circular cross section. The analysis of two-plane bending of beams of noncircular cross sections is discussed in Sec. 3-1. Deflections in each plane are found in a manner identical to that of Sec. 2-3. The deflection of the centroidal axis in the z direction z_c is given by

$$\frac{d^2 z_c}{dx^2} = -\frac{M_y}{EI_y}$$ (2-45)

Example 2-17 Figure 2-48a shows a rotating shaft mounted in bearings and a torque is being transmitted between pulleys located at B and C. If the shaft has a diameter of 1.5 in, determine at the instant shown the location and state of stress of the element undergoing the greatest stress conditions. Consider the end conditions on the shaft as simply supported.

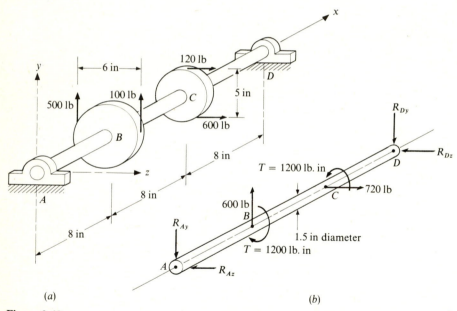

(a)   (b)

Figure 2-48

†Note, from Fig. 2-47a, that $y' = y \cos \beta - z \sin \beta$. Thus Eq. (2-43) can be written

$$\sigma_x = -\frac{1}{I_{z'}} (M_z \cdot y \cos \beta - M_z \cdot z \sin \beta)$$ (a)

Furthermore, $M_{z'} = M_z/(\cos \beta)$, $M_{z'} = M_y/(\sin \beta)$, and $I_y = I_z = I_{z'}$. Thus Eq. (a) reduces to $\sigma_x = M_y z/I_y - M_z y/I_z$, which is Eq. (2-44).

SOLUTION The net effect of the shaft loading is shown in Fig. 2-48b. For equilibrium,

$$R_{Ay} = 400 \text{ lb} \qquad R_{Az} = 240 \text{ lb} \qquad R_{Dy} = 200 \text{ lb} \qquad R_{Dz} = 480 \text{ lb}$$

Considering bending only, the loading in the xy and xz planes is shown in Fig. 2-49. The maximum moment will occur either at point B or C. Since the maximum moment is given by $M_{z'} = \sqrt{M_y^2 + M_z^2}$, at points B and C

$$M_B = \sqrt{(-3200)^2 + 1920^2} = 3730 \text{ lb·in}$$

$$M_C = \sqrt{(-1600)^2 + 3840^2} = 4160 \text{ lb·in}$$

Thus it can be seen that the maximum bending moment occurs at point C. The cross section at point C whose surface normal is pointing in the positive x direction is shown in Fig. 2-50a. This surface is at a value of x slightly less than 16 in so as to include the transmitted torque T.

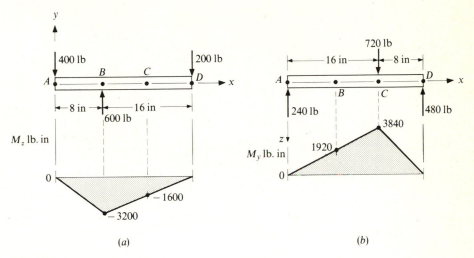

(a)   (b)

Figure 2-49

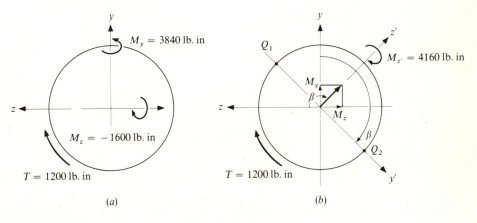

(a)   (b)

Figure 2-50

The z' axis is located by the angle β, where $\beta = \tan^{-1}(M_y/M_z)$. Thus

$$\beta = \tan^{-1}\frac{3840}{-1600} \qquad \beta = 112.6°$$

The $y'z'$ axes are shown in Fig. 2-50b. Because of bending, the maximum tensile stress occurs at point Q_1, where $y' = -d/2$, and is

$$(\sigma_x)_{Q_1} = -\frac{(4160)(-1.5/2)}{(\pi/64)(1.5)^4} = 12,600 \text{ lb/in}^2$$

The maximum compressive stress due to bending is at point Q_2, where $y' = d/2$, and is

$$(\sigma_x)_{Q_2} = -\frac{(4160)(1.5/2)}{(\pi/64)(1.5)^4} = -12,600 \text{ lb/in}^2$$

The shear stress at points Q_1 and Q_2 due to the torque T is determined from Eq. (2-10a):

$$\tau = \frac{(1200)(1.5/2)}{(\pi/32)(1.5)^4} = 1810 \text{ lb/in}^2$$

Thus the state of stress at points Q_1 and Q_2 is as shown in Fig. 2-51.

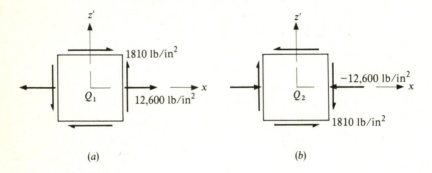

(a)   (b)

Figure 2-51

2-10 BUCKLING OF RODS IN COMPRESSION

Consider a rod fixed at one end and free on the other end with a compressive axial force, as shown in Fig. 2-52a. Linear elasticity theory will predict no bending deflection in this case. However, since buckling is an instability phenomenon, assume that the rod is perturbed slightly, as might occur from a slight vibration of the wall. Thus, the beam at an instant would appear as in Fig. 2-52b. If the load is small, the disturbance will decay and the rod will return to the original state as shown in Fig. 2-52a, but if the load is large enough, the disturbance will grow as the force causes increased bending and the rod begins to buckle. Linear elasticity theory will predict the value of the load which initiates buckling. However, subsequent deflections as a function of load must be arrived at through nonlinear theory. In the analysis which follows, it is assumed that the yz axes are principal axes of inertia (see Appendix F) and

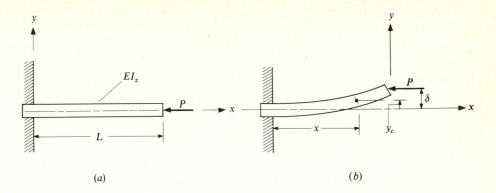

Figure 2-52

that the moment of inertia I_y about the y axis is greater than I_z. Thus, if buckling occurs, the rod will bend about the z axis.

Returning to Fig. 2-52b and neglecting the weight of the rod, we see that the bending moment at any point in the rod is

$$M_z = P(\delta - y_c)^\dagger$$

and using the deflection equation $d^2 y_c / dx^2 = M_z / EI_z$ results in

$$\frac{d^2 y_c}{dx^2} + \frac{P}{EI_z} y_c = \frac{P}{EI_z} \delta \tag{2-46}$$

Equation (2-46) represents a second-order linear nonhomogeneous differential equation for which the solution is

$$y_c = C_1 \sin kx + C_2 \cos kx + \delta$$

where C_1 and C_2 are constants of integration and $k = \sqrt{P/EI_z}$.

To determine C_1 and C_2, geometric conditions that $y_c = 0$ and $dy_c / dx = 0$ at $x = 0$ yield

$$C_1 = 0 \qquad \text{and} \qquad C_2 = -\delta$$

Thus
$$y_c = \delta(1 - \cos kx) \tag{2-47}$$

When $x = L$, $y_c = \delta$, and there are two cases in which Eq. (2-47) is valid, namely $\delta = 0$ and $\cos kL = 0$. If $\delta = 0$, the rod will return to its original shape and buckling does not start. If $\cos kL = 0$, a deflection value is possible for equilibrium. There is a distinct force which will cause this. For $\cos kL = 0$

$$kL = \frac{n\pi}{2} \qquad n = 1, 3, 5, \ldots$$

---

† Note that, the bending moment is no longer a function of initial geometry but is strongly coupled to deflections.

or, since $k = \sqrt{P/EI_z}$,

$$P = \left(\frac{n\pi}{2L}\right)^2 EI_z \qquad n = 1, 3, 5, \ldots$$

The first opportunity for the rod to buckle is when $n = 1$; thus the critical load is

$$P_{cr} = \frac{\pi^2 EI_z}{4L^2} \tag{2-48}†$$

and since $k = \pi/2L$, the deflection curve of the rod is

$$y_c = \delta\left(1 - \cos\frac{\pi x}{2L}\right) \tag{2-49}$$

However, no additional information can be found from linear theory, and the value of δ cannot be determined.

The critical load depends on the geometric end conditions of the rod. Equations (2-48) and (2-49) can be written for various end conditions using Eq. (2-50) and Table 2-1, where

$$P_{cr} = \frac{\pi^2 EI_z}{(KL)^2} \tag{2-50}$$

From Eq. (2-50), for a given material the critical load is dependent on the length and moment of inertia of the cross section about the z axis, as illustrated in Fig. 2-53a. There is an upper bound on P_{cr}, which is based on either the yield

Table 2-1

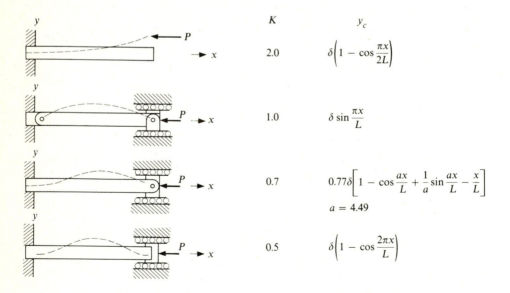

	K	y_c
	2.0	$\delta\left(1 - \cos\dfrac{\pi x}{2L}\right)$
	1.0	$\delta \sin\dfrac{\pi x}{L}$
	0.7	$0.77\delta\left[1 - \cos\dfrac{ax}{L} + \dfrac{1}{a}\sin\dfrac{ax}{L} - \dfrac{x}{L}\right]$ $a = 4.49$
	0.5	$\delta\left(1 - \cos\dfrac{2\pi x}{L}\right)$

† The critical force for buckling is often referred to as the *Euler force*.

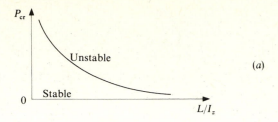

(a)

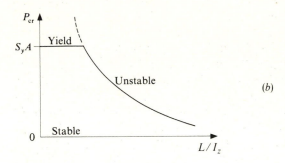

(b)

Figure 2-53

strength or ultimate strength under pure compression. If the failure criterion is based on the yield strength S_y, the upper bound for P_{cr} will be S_yA, where A is the area of the cross section of the rod (see Fig. 2-53b).

There are a number of design equations available which reduce these limits for safety purposes, e.g., the secant formula, Johnson's equations, Rankine's equation, as well as various code equations. Most elementary strength of materials textbooks list these equations, which the reader is urged to review.

Example 2-18 A column is fixed at one end and is only allowed to bend in the xy plane (see Fig. 2-54a). The bending rigidity of the beam is EI_z. A spring of spring constant k_s lb/in is attached to the other end. Determine the critical load P_{cr} for buckling if $EI_z = 10^5$ lb·in², $L = 30$ in, and $k_s = 1000$ lb/in. Neglect the weight of the beam.

SOLUTION The bending moment for any position x (see Fig. 2-54b) is given by

$$M_z = P_{cr}(\delta - y_c) - k_s\delta(L - x) \tag{a}$$

Since $M_z = EI_z\, d^2y_c/dx^2$, Eq. (a) reduces to

$$\frac{d^2y_c}{dx^2} + k_1^2y_c = (k_1^2 - k_2L)\delta + k_2x\delta \tag{b}$$

where

$$k_1 = \sqrt{\frac{P_{cr}}{EI_z}} \quad \text{and} \quad k_2 = \frac{k_s}{EI_z}$$

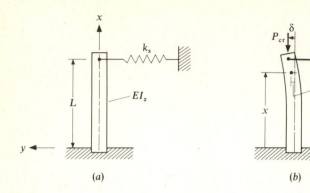

Figure 2-54

The solution of the differential equation (b) is

$$y_c = C_1 \sin k_1 x + C_2 \cos k_1 x + \frac{(k_1^2 - k_2 L)\delta}{k_1^2} + \frac{k_2}{k_1^2} x\delta \qquad (c)$$

Equation (c) is subjected to the boundary condition that $y_c = dy_c/dx = 0$ at $x = 0$. After rearranging, this results in

$$C_2 = -(1 - \frac{k_2}{k_1^2} L)\delta \qquad (d)$$

$$C_1 = -\frac{k_2}{k_1^3}\delta \qquad (e)$$

Substitution back into Eq. (c) yields

$$y_c = \frac{k_2}{k_1^3}\delta(k_1 x - \sin k_1 x) + \delta\left(1 - \frac{k_2}{k_1^2} L\right)(1 - \cos k_1 x) \qquad (f)$$

For buckling, $y_c = \delta$ at $x = L$. Thus from Eq. (f),

$$\delta = \frac{k_2}{k_1^3}\delta(k_1 L - \sin k_1 L) + \delta\left(1 - \frac{k_2}{k_1^2} L\right)(1 - \cos k_1 L)$$

which can be shown to reduce to

$$\tan k_1 L = k_1 L\left[1 - \frac{(k_1 L)^2}{k_2 L^3}\right] \qquad (g)$$

provided $k_1 L \neq 0$.

Since $L = 30$ and

$$k_2 = \frac{k_s}{EI_z} = \frac{100}{10^5} = 10^{-3}$$

Eq. (g) becomes

$$\tan k_1 L = k_1 L\left[1 - \frac{(k_1 L)^2}{27}\right] \qquad (h)$$

The lowest nonzero value of $k_1 L$ which satisfies Eq. (h) is $k_1 L = 4.13$. Since $k_1 L = L\sqrt{P_{cr}/EI_z}$,

$$P_{cr} = \left(\frac{k_1 L}{L}\right)^2 EI_z = \left(\frac{4.13}{30}\right)^2 10^5 = 1900 \text{ lb}$$

EXERCISES

2-1 The axially loaded composite bar shown in Fig. 2-55 has a rectangular cross section. The force F is to be positioned so that the stress in each section is uniform. If the thickness of the bar is t and the moduli of bonded materials 1 and 2 are E_1 and E_2, respectively, determine (a) the normal stress in each member and (b) the value of e such that the force F causes uniform stress in each material. Consider $E_1 > E_2$.

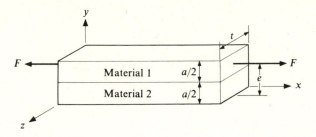

Figure 2-55

2-2 A circular steel rod of length L rotates with a constant angular velocity ω in a horizontal plane about an axis through its midpoint, as shown in Fig. 2-56. The weight density ρ of steel is 0.284 lb/in³. If $d = 1.0$ in, $L = 20$ in, $E = 29 \times 10^6$ lb/in², and $\omega = 52$ rad/s, determine (a) the normal stress σ_x as a function of x; (b) the maximum normal stress, $(\sigma_x)_{max}$; (c) the increase in the length of the rod. *Hint:* First, *prove* that the internal axial force at any position x is given by $(\rho/2g)\omega^2 A (L^2 - x^2)$, where A is the cross-sectional area in square inches and g is the gravity constant in inches per second per second.

Figure 2-56

2-3 In Fig. 2-55 consider that the the top material is aluminum ($E_a = 10 \times 10^6$ lb/in²) and the lower material is steel ($E_s = 30 \times 10^6$ lb/in²). If $a = 8$ in, $t = \frac{1}{2}$ in, and the normal strain is *not* uniform and given by $\varepsilon_x = 100y\,\mu$in/in, determine the magnitude and location of the load F.

2-4 A *rigid* bar AC is hinged to a vertical wall and supported by steel cable BD, as shown in Fig. 2-57. Determine the maximum allowable value of the applied load P such that the tensile stress in the cable does not exceed 70 MN/m². The cross-sectional area of the cable is 625 mm². Also, determine the resulting vertical deflection of point C due to P. The modulus of elasticity of the cable material is 20×10^{10} N/m².

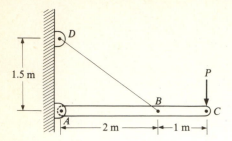

Figure 2-57

2-5 Figure 2-58 illustrates a steel step shaft loaded in torsion. In the 1.0-in-diameter section, a $\frac{3}{8}$-in-diameter hole is drilled to a depth of 2.0 in. If $E = 30 \times 10^6$ lb/in² and $\nu = 0.3$, determine (a) the maximum shear stresses in the three differing cross sections, AB, BC, and CD and (b) the total angle of twist of the shaft at D.

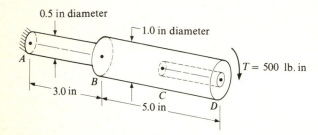

Figure 2-58

2-6 The beam of Exercise 1-1 is repeated in Fig. 2-59, where an enlarged view of the cross section is included. Determine (a) the maximum tensile and compressive bending stresses and (b) the maximum shear stress due to the maximum transverse shear force.

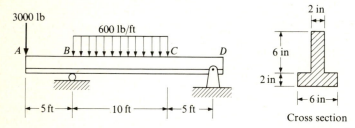

Cross section

Figure 2-59

2-7 Consider a beam with a solid circular cross section of diameter d transmitting a transverse shear force V_y.

(*a*) Derive an expression for the shear-stress distribution across the transverse surface.

(*b*) Verify that the maximum shear stress occurs at $y_1 = 0$ and that Eq. (2-19) is valid.

(*c*) Physically show the shear stresses on the transverse surface and discuss any possible discrepancies in the distribution.

2-8 For the diamond-shaped cross section in Fig. 2-60, bending occurs about the z axis, and the surface contains a net shear force V_y in the positive y direction. Since the maximum τ_{xy} shear stress due to V_y occurs where Q/b is a maximum, prove that τ_{xy} is maximum at $y = \pm h/8$.

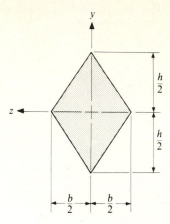

Figure 2-60

2-9 Using the double-integration method, verify the deflection and slope equations as functions of x for beam D.6 given in Appendix D.

2-10 Using the double-integration method, verify the deflection and slope equations as functions of x for beam D.5 given in Appendix D.

2-11 A pressure vessel consists of a circular cylinder capped at both ends by welding on two hemispherical caps. The vessel is then subjected to an internal pressure of 7.0 MN/m². If the inner diameter and thickness of each member are 0.5 m and 12 mm, respectively, determine the membrane stresses of each member.

2-12 A cylindrical water tank 40 ft high and 5 ft in diameter is completely full of water. Determine the tangential stress at the bottom of the tank if it is fabricated of 0.25-in-thick steel plates.

2-13 Using the method of linear superposition and the appropriate beams in Appendix D, determine the deflection and slope at point A of the beam in Exercise 2-6. The material of the beam is steel, where $E_s = 30 \times 10^6$ lb/in². *Hint:* The slope in section AB due to the uniform load of 600 lb/ft only is constant.

2-14 An aluminum beam AC ($E_a = 10 \times 10^6$ lb/in²) is attached to a vertical wall and supported by a steel cable CD ($E_s = 30 \times 10^6$ lb/in²), as shown in Fig. 2-61. The beam is loaded by a vertical force $P = 1000$ lb at point B. If the beam has a rectangular cross section of 2 by 1 in and the cable area is 0.10 in², determine the support reactions at points A and D and the maximum normal stresses in the beam and cable if point A is considered to be supported (a) by a pin connection and, (b) by a totally fixed connection.

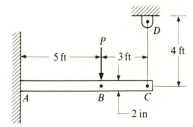

Figure 2-61

2-15 If in Exercise 2-5 instead of a torque being applied at end D consider ends A and D to be fixed and a torque of 8000 lb-in to be applied at point B. Determine (a) the torsional reactions at A and D, (b) the maximum shear stress, and (c) the maximum angle of twist at B.

2-16 A common problem in design is that of a circular shaft of diameter d transmitting a bending moment M and a torsional load T simultaneously. Using Mohr's circle prove that

$$\tau_{max} = \frac{16}{\pi d^3}\sqrt{M^2 + T^2}$$

and

$$\sigma_{max} = \frac{16}{\pi d^3}(M + \sqrt{M^2 + T^2}) \qquad \sigma_{min} = \frac{16}{\pi d^3}(M - \sqrt{M^2 + T^2})$$

2-17 For pure torsion on a circular rod, the maximum shear stress occurs at the maximum radius, where the value of stress is given by Eq. (2-10a). Construct an element located at this position using a localized xyz coordinate system, where x is parallel to the shaft axis and z is directed radially outward from the shaft axis. Thus, the state of stress at the point is biaxial in the xy plane, where $\sigma_x = \sigma_y = 0$ and $\tau_{xy} = -Tr_0/J$. In a fashion similar to that used in Example 2-14, construct a polar diagram of the normal and shear stresses on surfaces corresponding to $15°$ increments relative to the x axis from 0 to $360°$. On the diagram show the correct value and direction of each stress using different colored lines to differentiate the shear and normal stresses.

2-18 For each of the following biaxial states of stress draw the corresponding stress element properly oriented with respect to the x axis and construct the Mohr's circle diagram. Considering the analysis to be restricted to surfaces perpendicular to the xy plane for each case, (1) draw the stress element containing the principal stresses properly oriented to the x axis showing the values and directions of the principal stresses and (2) draw a stress element containing the maximum shear stress properly oriented to the x axis showing the values and directions of the normal and shear stresses for the surfaces containing the maximum shear stresses.

(a) $\sigma_x = 10{,}000 \text{ lb/in}^2$, $\sigma_y = \tau_{xy} = 0 \text{ lb/in}^2$
(b) $\tau_{xy} = 5000 \text{ lb/in}^2$, $\sigma_x = \sigma_y = 0 \text{ lb/in}^2$
(c) $\sigma_x = 10{,}000 \text{ lb/in}^2$, $\sigma_y = 5000 \text{ lb/in}^2$, $\tau_{xy} = 0 \text{ lb/in}^2$
(d) $\sigma_x = \sigma_y = 5000 \text{ lb/in}^2$, $\tau_{xy} = 0 \text{ lb/in}^2$
(e) $\sigma_x = 8000 \text{ lb/in}^2$, $\sigma_y = 0 \text{ lb/in}^2$, $\tau_{xy} = -3000 \text{ lb/in}^2$
(f) $\sigma_x = 10{,}000 \text{ lb/in}^2$, $\sigma_y = -2000 \text{ lb/in}^2$, $\tau_{xy} = -8000 \text{ lb/in}^2$

2-19 A point undergoing a biaxial state of stress is isolated by three surfaces, as shown in Fig. 2-62. Determine the values of σ, τ, and τ'.

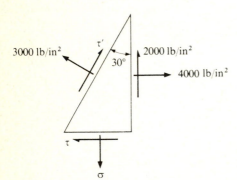

Figure 2-62

2-20 A plate of thickness t has a tapered section, as shown in Fig. 2-63. Assume that the normal stress in the x direction can still be expressed by Eq. (2-1). That is, $\sigma_x = P/ht$, where h is the width of the plate at the given section. At point A prove that (a) the maximum principal stress is

$$\sigma_1 = \sigma_x/\cos^2 \theta$$

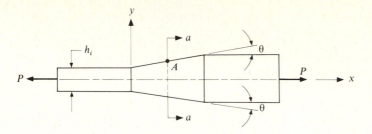

Figure 2-63

and (b) the stresses σ_y and τ_{xy} are given by

$$\sigma_y = \sigma_x \tan^2 \theta \qquad \text{and} \qquad \tau_{xy} = \sigma_x \tan \theta$$

Hint: In addition to an element with surfaces perpendicular to the x and y directions, consider an element at point A with surfaces parallel and perpendicular to the free surface.

2-21 For a given point in a mechanical element the state of stress is biaxial, and the corresponding strains are found to be $\varepsilon_x = 1000 \, \mu\text{m/m}$, $\varepsilon_y = -200 \, \mu\text{m/m}$, and $\gamma_{xy} = -800 \, \mu\text{m/m}$. The material is steel, where $E = 20 \times 10^{10} \, \text{N/m}^2$ and $\nu = 0.29$.

    (a) Draw Mohr's circle of strain and determine the principal strains. Using Hooke's law, calculate the corresponding principal stresses.

    (b) Determine the stresses σ_x, σ_y, and τ_{xy} using Hooke's law, draw Mohr's circle of stress, and determine the principal stresses. Compare the results with part (a).

2-22 The steel shaft shown in Fig. 2-64 is supported in flexible bearings at its ends, and the two pulleys shown are keyed to the shaft. Draw complete stress elements for the four points A, B, C, and D and draw each of the corresponding Mohr's circles.

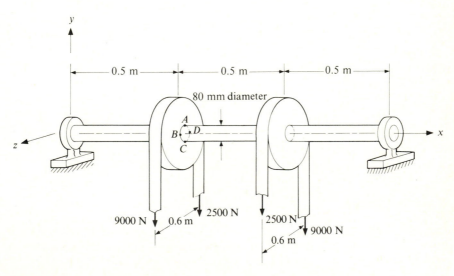

Figure 2-64

2-23 As shown in Fig. 2-65, a torque of $T = 750$ lb·in applied through the shaft of gear E is sufficient to drive the roller chain at point B. The chain sprocket has a pitch diameter of 6 in and transmits a force of 500 lb, as shown. The pitch diameters of gears C and E are 10 and 5 in, respectively, and the contact force between the gears is transmitted through the pitch angle $\phi = 20°$. Locate the point on the simply supported main shaft AD which contains the maximum tensile bending stress and the maximum torsional shear stress. Draw the corresponding stress element indicating the stress values and directions.

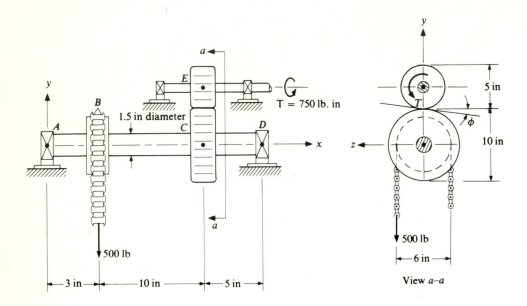

View a–a

Figure 2-65

REFERENCES

1. Crandall, S. H., N. C. Dahl, and T. J. Lardner: "An Introduction to the Mechanics of Solids," 2d ed., McGraw-Hill, New York, 1972. A reasonably consistent and well-defined book containing elementary and advanced formulations. The level is intermediate to advanced.
2. Nash, W. A.: "Strength of Materials," 2d ed., Schaum's Outline Series, McGraw-Hill, New York, 1972. A good basic review book consistent with a typical first course in strength of materials. The book contains many example problems where some basic formulations are developed. The level is elementary to intermediate.
3. Timoshenko, S., and D. H. Young: "Elements of Strength of Materials," 5th ed., Van Nostrand, Princeton, N.J., 1968. A classic first book in strength of materials, first published in 1935. Relies on physical insight more than other books on the subject. In addition, the notation and sign conventions used are not completely consistent with current practice. However, this book is still a reasonably good review book or textbook in an introductory course. The level is elementary to intermediate.
4. Shigley, J. E.: "Applied Mechanics of Materials," McGraw-Hill, New York, 1976. A short elementary book containing only the fundamental subjects of a first course of strength of

materials, but the material is covered in better than average detail, a tendency toward design applications being evident. Included also, to a small extent, is the use of SI units. The level is elementary to intermediate.

5. Popov, E. P.: "Mechanics of Materials," 2d ed., Prentice-Hall, Englewood Cliffs, N.J., 1976. A good introductory book in strength of materials containing basic to intermediate formulations. The second edition contains many examples and problems using an apparently equal mix of SI and USCS units. The level is intermediate.

THREE

TOPICS FROM ADVANCED STRENGTH OF MATERIALS

3-0 INTRODUCTION

This chapter is devoted to extending the analyst's knowledge and capability farther into the area of classical problems and their solutions. The techniques used in elementary strength of materials are expanded to cover additional practical problems of a more advanced nature. Further discussion of beams in bending is presented, including the topics of unsymmetric bending, more about transverse shear stresses, and shear flow in thin sections, shear center-loading axis, moving loads, composite beams, wide beams, and curved beams. Various specialized topics such as plates in bending, thick-walled cylinders and rotating disks, contact stresses, stress concentrations, and plastic behavior are also discussed.

3-1 UNSYMMETRICAL LOADING OF BEAMS

Considering the xy plane to be the loading plane for a beam in bending, if the y and z axes are principal axes of inertia of the beam cross section,† bending occurs about the z axis and the normal stress σ_x is given by the flexure formula

$$\sigma_x = -\frac{M_z y}{I_z} \tag{3-1}$$

---

† The method of locating the principal axes of inertia of a cross section is discussed in Appendix F.

However, if the y and z axes are not principal axes of inertia, Eq. (3-1) is not valid and bending will not occur about the z axis. Since the bending equation is valid for bending about a principal axis, for unsymmetrical loading it is only necessary to determine the components of bending about each principal axis. The bending equation can then be applied using the components, and the results can be added algebraically using superposition. Thus, if the principal axes are denoted by m and n, and if M_m and M_n are the components of the bending moment and I_m and I_n the moments of inertia about the m and n axes, respectively, the bending stress is given by

$$\sigma_x = \frac{M_m n}{I_m} - \frac{M_n m}{I_n} \tag{3-2}$$

Note that this is basically the same as Eq. (2-44); the sign convention for bending in the xm and xn planes is shown in Fig. 3-1.

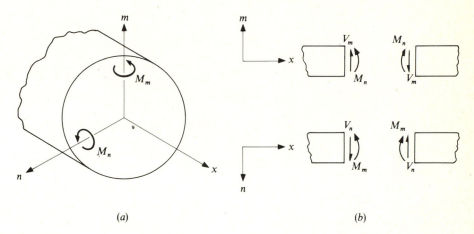

(a) $\qquad\qquad\qquad\qquad\qquad\qquad\qquad\qquad\qquad (b)$

Figure 3-1

Example 3-1 Determine the maximum tensile stress developed in the unsymmetrically loaded beam in Fig. 3-2.

SOLUTION Here, since the y and z axes are not principal axes, the loading must be resolved into components along the principal axes, as shown in Fig. 3-2b. Bending about the n axis is caused by the loading in the mx plane, whereas bending about the m axis is caused by loading in the nx plane. Following the established convention, the respective bending-moment diagrams are shown in Fig. 3-2c.

The next step is to determine the moments of inertia

$$I_m = (\tfrac{1}{12})(2)(1^3) = 0.167 \text{ in}^4 \qquad I_n = (\tfrac{1}{12})(1)(2^3) = 0.667 \text{ in}^4$$

The bending stress is found using Eq. (3-2). The maximum stress occurs at the wall, where the moments are maximum, and when m and n are maximum positive values. Thus, for the

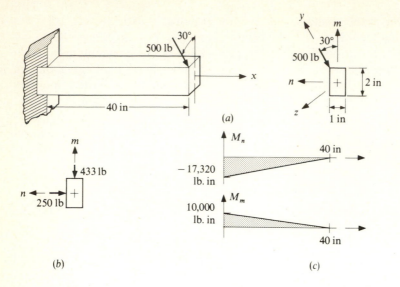

Figure 3-2

location $x = 0$, $m = 1.0$, and $n = 0.5$, the stress is

$$(\sigma)_{max} = \frac{(+10,000)(0.5)}{0.167} - \frac{(-17,320)(1.0)}{0.667} = 56,000 \text{ lb/in}^2$$

Thanks to the symmetry of the cross section, the maximum compressive stress occurs at the wall and at a point diagonally opposite the point of maximum tensile stress. The maximum compressive stress at this point is $-56,000$ lb/in².

In Example 3-1 it was easy to see where the maximum stresses would occur, since the maximum values of both terms in Eq. (3-2) occurred at a common point, but for some cross sections it is not always possible to see where the combination of the two stresses yields the greatest value.

To locate the points of maximum stress it would be useful to determine the location of the neutral plane. The maximum perpendicular positions from this plane will be the points of maximum tensile and compressive stresses. For unsymmetric bending, the neutral plane does not necessarily coincide with the axis the net moment acts on. Since the neutral plane is where $\sigma_x = 0$, from Eq. (3-2)

$$\frac{M_m n}{I_m} - \frac{M_n m}{I_n} = 0 \qquad \text{or} \qquad \frac{m}{n} = \frac{M_m}{M_n} \frac{I_n}{I_m}$$

This locates the neutral plane with respect to the n axis. Defining an angle α clockwise from the n axis, we have $\tan \alpha = m/n$ or

$$\tan \alpha = \frac{M_m}{M_n} \frac{I_n}{I_m} \tag{3-3}$$

Example 3-2 Determine the location of the neutral plane in Example 3-1.

SOLUTION

$$\tan \alpha = \frac{(10,000)(0.667)}{(-17,320)(0.167)}$$

$$\alpha = -66.6°$$

It can be seen in Fig. 3-3 that although the xz plane is not the neutral plane, points A and B are the maximum distances from the neutral plane, where (as determined in Example 3-1) the maximum tensile and compressive stresses occur.

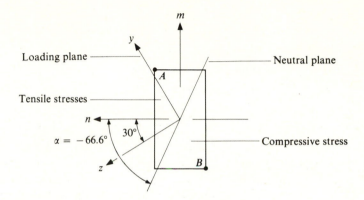

Figure 3-3

When dealing with standard unsymmetrical sections, such as angle sections, the location of the principal axes can be found in handbooks.†

3-2 FURTHER DISCUSSION OF TRANSVERSE SHEAR STRESSES

For transverse loading in the xy plane, the shear stress resulting from the shear force V_y is given by

$$\tau_{xy} = \frac{V_y Q}{I_z b} \tag{2-17}$$

where V_y = total shear force in y direction on surface located at position x
Q = area moment defined by $\int_{y_1}^{y_{max}} y \, dA$ and is a function of y_1
b = width (in z direction) of section at $y = y_1$

† For example, for steel sections, consult the AISC Handbook (American Institute of Steel Construction).

Recalling the derivation of Eq. (2-17) it is assumed that the cross section of the beam does not vary with respect to x and that the shear stress is uniform across the width of the beam.

Under certain conditions, Eq. (2-17) is incomplete and may provide misleading results. The free-surface conditions of the beam must also be considered in analyzing the stress distribution (see Sec. 7-2). For example, consider a circular beam in bending. The circular cross section shown in Fig. 3-4a is transmitting a bending moment M_z and a transverse shear force V_y. The shear stress at position y_1 as predicted by Eq. (2-17) is determined by first evaluating Q for the area shown in Fig. 3-4b. This results in

$$Q = \int_{y_1}^{y_{max}} y \, dA = \int_{y_1}^{d/2} y \left[2\sqrt{\left(\frac{d}{2}\right)^2 - y^2} \right] dy = \frac{2}{3}\left[\left(\frac{d}{2}\right)^2 - y_1^2 \right]^{3/2}$$

Thus, the shear stress as indicated by Eq. (2-17) is

$$\tau_{xy} = \frac{V_y Q}{I_z b} = \frac{V_y \, \frac{2}{3}[(d/2)^2 - y_1^2]^{3/2}}{I_z \, 2[(d/2)^2 - y_1^2]^{1/2}} = \frac{V_y}{3I_z}\left[\left(\frac{d}{2}\right)^2 - y_1^2 \right] \tag{a}$$

and is shown in Fig. 3-4c. This stress distribution, although relatively accurate, does not conform to the surface conditions at points A and B. For example, consider point A. Since the outer surface is stress-free, the complete state of stress at point A must be as shown in Fig. 3-4d. The net shear stress $(\tau_x)_{net}$ must be tangent to the free surface and thus is not in the y direction, as Eq. (2-17) seems to imply. For the shear stress at point A to be tangent to the free surface, a shear stress $-\tau_{zx}$, in the negative z direction must also be present, as shown in Fig. 3-4e.

Exact analysis is beyond the scope of this discussion. The exact solution for τ_{xy} and τ_{zx} is (Ref. 1, p. 213)

$$\tau_{xy} = \frac{1}{8}\frac{3 + 2\nu}{1 + \nu}\frac{V_y}{I_z}\left(r^2 - z^2 - \frac{1 - 2\nu}{3 + 2\nu} y^2 \right) \tag{b}$$

$$\tau_{zx} = -\frac{1}{8}\frac{1 + 2\nu}{1 + \nu}\frac{V_y}{I_z} yz \tag{c}$$

Fortunately, the approximate results of Eq. (2-17) are most accurate along the surface at $y_1 = 0$, where τ_{xy} is a maximum. Note also that at this location the geometric boundary conditions are satisfied. At $y_1 = 0$, the average stress predicted by Eq. (a) reduces to $(\tau_{xy})_{max} = 1.333 \, V_y/A$, where $A = \pi d^2/4$. For $\nu = 0.3$ the exact solution for τ_{xy} is $1.385 \, V_y/A$ at the center of the section, and $1.231 \, V_y/A$ at $y = 0$, $z = d/2$. Thus, for the value of the maximum shear stress, there is less than a 4 percent error in Eq. (2-17).

Since Eq. (2-17) is relatively accurate, the total shear stress at any point can be approximated by a simple technique. Returning to Fig. 3-4, straight lines are constructed at points A and B tangent to the free surfaces as shown in Fig. 3-4f. The constructed lines intersect at point C, and the net shear stresses at points A and B will be in the direction of the lines as shown. When the approximate

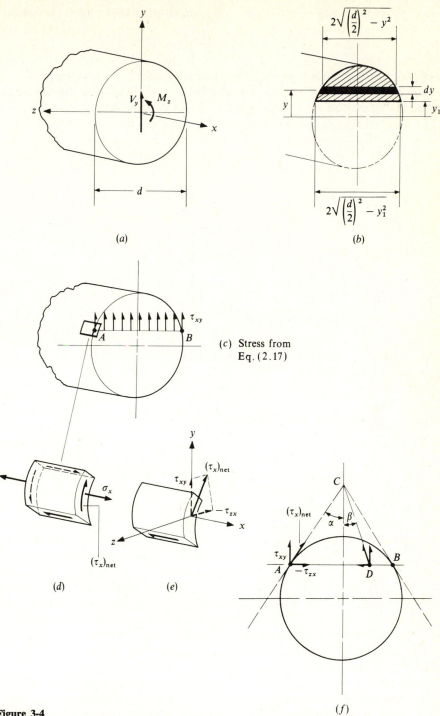

(c) Stress from Eq. (2.17)

Figure 3-4

value of τ_{xy} from Eq. (2-17) is used, τ_{zx} and $(\tau_x)_{net}$ at point A, for example, are

$$\tau_{zx} \approx - \tau_{xy} \tan \alpha \qquad (\tau_x)_{net} \approx \frac{\tau_{xy}}{\cos \alpha}$$

To approximate the shear stress at an interior point along $y = y_1$, a line is drawn from the point to point C. This approximates the direction of the net shear stress at the point (see point D of Fig. 3-4f). As before, the value of τ_{xy} from Eq. (2-17) is used as an approximation. Thus, for point D in Fig. 3-4f,

$$\tau_{zx} \approx \tau_{xy} \tan \beta \qquad (\tau_x)_{net} \approx \frac{\tau_{xy}}{\cos \beta}$$

3-3 SHEAR FLOW IN BEAMS WITH THIN SECTIONS

Another case where the results of Eq. (2-17) must be modified occurs in the analysis of the transverse shear stresses in beams with thin walls, e.g., I beams, channels, angles, or boxes. For example, consider the I beam shown in Fig. 3-5a. The shear-stress distribution predicted by Eq. (2-17) is shown in Fig. 3-5a and plotted as a function of y in Fig. 3-5b. The maximum shear stress occurs at $y = 0$ since at this point the area moment Q is a maximum and the beam width b is a minimum. At $y = \pm (h/2 - t_f)$ a discontinuity in the shear stress seems to occur as the beam width changes instantly from the web thickness to the flange width, but in actuality there is no such discontinuity. Recall that Eq. (2-17) is based on the assumption that the shear stress is uniform in the z direction. Thus at $y = \pm (h/2 - t_f)$ on the flange, Eq. (2-17) is predicting a uniform shear stress along the entire flange. However, on the free surfaces of the flange, τ_{xy} must be zero, as shown in Fig. 3-5c. In addition, since the outer surfaces of the flanges, where $y = h/2$, are free from tangential forces, τ_{xy} must be zero on these surfaces as well. If the flange thickness is small and τ_{xy} is zero along the free surfaces of the flange, τ_{xy} will remain relatively small throughout the flange [much smaller than Eq. (2-17) predicts]. Thus τ_{xy} is mostly contained within the web, and since the geometric boundary conditions are satisfied in the web, Eq. (2-17) is relatively accurate in the web [except near $y = \pm (h/2 - t_f)$]. Thus, Eq. (2-17) is adequate in determining the maximum transverse shear stress for the section.

Since the free surfaces of the flange are horizontal, shear stresses τ_{zx} can be present. For an I-beam, it will be shown that these stresses are less than in the web. There are cases, however, when the shear stresses in a nonvertical wall are appreciable and must be considered. In addition, as demonstrated in Sec. 3-4, shear stresses in thin-walled beams may give rise to twisting if the section is not symmetric about the y axis.

Consider the section of the flange of the I beam shown in Fig. 3-5d. Recalling the derivation of Eq. (2-17), we see that if the bending moment is changing with respect to the x direction, the normal stress due to bending σ_x is

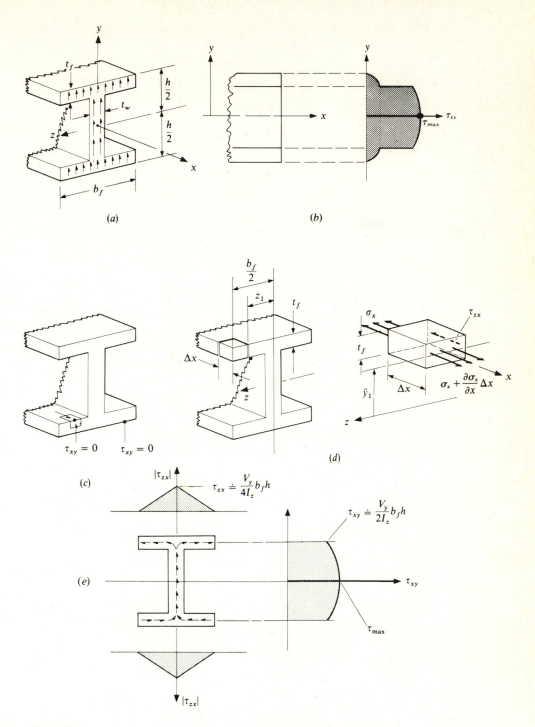

Figure 3-5

also changing and the net force in the x direction caused by this change integrated across the shaded area of the section is

$$\Delta F_x = \int_A \left(\frac{\partial \sigma_x}{\partial x} \Delta x \right) dA$$

Again recalling the derivation of Eq. (2-17), we see that this reduces to

$$\Delta F_x = \frac{V_y Q}{I_z} \Delta x$$

For the flange, this must be balanced by $\tau_{zx} t_f \, \Delta x$. Thus

$$\tau_{zx} t_f \, \Delta x = \frac{V_y Q}{I_z} \Delta x$$

Solving for τ_{zx} results in

$$\tau_{zx} = \frac{V_y Q}{I_z t_f}$$

Note that this equation is quite similar to Eq. (2-17). The area moment Q is the same as before, where $Q = \int_{y_1}^{y_{max}} y \, dA = \bar{y}_1 A_1$ for the isolated area containing the bending stresses. For example, the centroid of the shaded area shown in Fig. 3-5d is

$$\bar{y}_1 = \tfrac{1}{2}(h - t_f)$$

and the area is

$$A_1 = t_f \left(\frac{b_f}{2} - z_1 \right)$$

Therefore the area moment is

$$Q = \bar{y}_1 A_1 = \tfrac{1}{2}(h - t_f) \left[t_f \left(\frac{b_f}{2} - z_1 \right) \right]$$

Substituting this into the equation for τ_{zx} yields

$$\tau_{zx} = \frac{V_y Q}{I_z t_f} = \frac{V_y (\tfrac{1}{2})(h - t_f)[t_f(b_f/2 - z_1)]}{I_z t_f} = \frac{V_y}{2 I_z}(h - t_f)\left(\frac{b_f}{2} - z_1 \right)$$

Thus, τ_{zx} in the flange is a linear function of z and reaches a maximum value when $z_1 = 0$. Thanks to symmetry, the stress distributions in the other flange segments are the same except for a sign change for two of the segments. The resulting shear-stress distribution in the web and flanges is shown in Fig. 3-5e. It is common practice to calculate each stress distribution to the centerlines of the web and flanges so that the values in the shear stress are continuous. Also, note in Fig. 3-5e that the shear stress appears as if it were flowing through the section. For this reason the terminology *shear flow* is commonly used. To substantiate the idea of flow it is interesting to note that the shear stress in the web at $y = \pm (h/2 - t_f)$ is

$$\tau_{xy} = \frac{V_y Q}{I_z t_w} = \frac{V_y}{I_z t_w} \left\{ \frac{b_f}{2} \left[\left(\frac{h}{2} \right)^2 - \left(\frac{h}{2} - t_f \right)^2 \right] \right\} = \frac{V_y}{2 I_z} \frac{t_f}{t_w} b_f (h - t_f)$$

When the web and flange walls are considered to be of the same thickness and very small compared with the section height h, the above reduces to

$$\tau_{xy} = \frac{V_y}{2I_z} b_f h$$

The shear stress in the flange at $z_1 = 0$ with $t_f \ll h$ is

$$\tau_{zx} = \frac{V_y}{4I_z} b_f h$$

Thus the shear stress in the flange at approximately the same point in the web is one-half that of the web stress. This is what would be expected in channel flow, where the flow would divide equally if the width of the channels were equal.

From the example of the I beam, Eq. (2-17) can be modified for thin-walled beams. In general, a wall may be parallel to neither the y nor the z axis, but Eq. (2-17) can be used to determine the net shear stress $(\tau_x)_{net}$ on a wall:

$$(\tau_x)_{net} = \frac{V_y Q}{I_z t} \tag{3-4}$$

where $(\tau_x)_{net}$, the shear stress on a thin wall of thickness t, is in the direction of the wall. Other examples using Eq. (3-4) are given in Sec. 3-4.

3-4 SHEAR CENTER OF THIN-WALLED BEAMS

If the cross section of a thin-walled beam is not symmetrical with respect to the vertical y axis, Eq. (3-4) may not completely ensure equilibrium. For example, consider a cantilever beam with a channel cross section loaded as shown in Fig. 3-6a. The y and z axes are the centroidal axes of the cross section. Here it is assumed that the applied force P is placed along the y axis. This is normally done to avoid twisting the beam. However, in this case, consider the section of the channel isolated at $x = x_1$. In Fig. 3-6a for clarity the bending stresses are not shown and only the shear stresses predicted by Eq. (3-4) are illustrated. To demonstrate that this distribution does not ensure equilibrium, the shear stresses are integrated across the areas of the web and flanges; the resulting net force on each surface is as shown in Fig. 3-6b. Summing moments about the x axis, we see that there is a net torque remaining on the element. Thus, the element is not in equilibrium. This means that the predicted stress distribution is incomplete. In order to ensure equilibrium, a torsional shear stress (Fig. 3-6c) is added by superposition to the shear stress induced by the shear flow. Torsion of noncircular cross sections is discussed in Sec. 6-8.

The stress distribution shown in Fig. 3-6b will be correct, and no additional torsional shear stresses will be induced provided the applied load is positioned so that the net torque is zero. This is done by moving the load P so that it intersects a line defined as the shear center axis, as shown in Fig. 3-6d. The value of e is determined by summing moments about the x axis to obtain equilibrium.

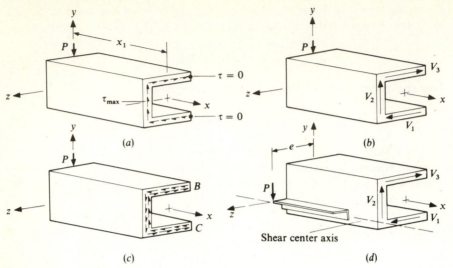

Figure 3-6

Example 3-3 The channel section shown in Fig. 3-7a is bending about the z axis and undergoes a net shear force V_y in the y direction. Determine (a) the shear flow throughout the section, (b) the total shear force in each section, and (c) the shear center through which the plane of loading should intersect.

SOLUTION The moment of inertia about the z axis is approximated by considering the section made up of three rectangles, as shown in Fig. 3-7b†; then

$$I_z = \tfrac{2}{3}a^3t + 2(\tfrac{1}{12}bt^3 + bta^2)$$

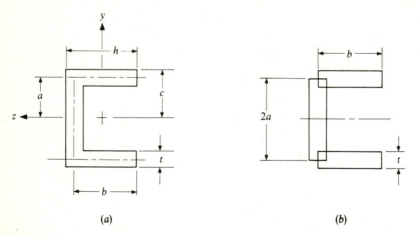

(a)

(b)

Figure 3-7

† This will also ensure continuity in the value of the shear stress at the intersection of the centerlines of the web and flange.

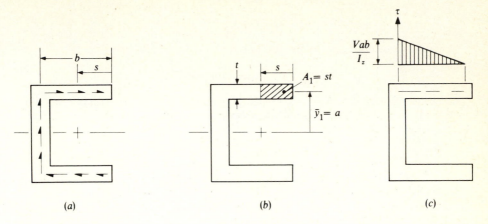

Figure 3-8

However, if t is much less than a and b, then I_z can be approximated by

$$I_z \approx \tfrac{2}{3}a^3t + 2a^2bt$$

For the shear-stress distribution in the top flange, consider the variable position s, as shown in Fig. 3-8a, where $0 \le s \le b$. This isolates the shaded area shown in Fig. 3-8b, of which the area moment is

$$Q = \bar{y}_1 A_1 = a(st)$$

Substituting this into Eq. (3-4), we find the shear-stress distribution in the top flange to be

$$\tau = \frac{V_y Q}{I_z t} = \frac{V_y\, a(st)}{I_z\ \ t} = \frac{V_y a}{I_z} s \tag{a}$$

Thus, the stress distribution in the top flange is linear, as shown in Fig. 3-8c.

The stress distribution in the side web can be found by determining Q for the section shown in Fig. 3-9a. Here, for convenience, the variable is changed to y instead of s. Q is approximated by using the two areas shown in Fig. 3-9b, where

$$Q = \sum \bar{y}A = a(bt) + \frac{a+y}{2}(a-y)t$$

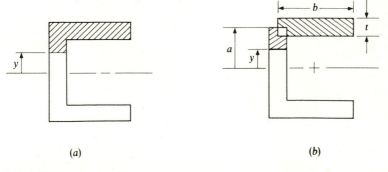

(a)   (b)

Figure 3-9

or

$$Q = abt + \frac{t}{2}(a^2 - y^2)$$

Thus the stress distribution in the side web is

$$\tau = \frac{V_y}{I_z}[ab + \tfrac{1}{2}(a^2 - y^2)] \tag{b}$$

which can be seen to be a parabolic function, as shown in Fig. 3-10a. Thanks to symmetry about the z axis, the distribution of shear stress is symmetric.

To determine the net shear force across each section, as shown in Fig. 3-10b, it is necessary to integrate $\int \tau \, dA$ across the surface. By symmetry, $V_1 = V_3$. The area dA for the top flange is $t \, ds$, and the shear stress $\tau = V_y as/I_z$. Thus, the shear force on the top flange is

$$V_3 = \int_0^b \left(\frac{V_y}{I_z} as\right) t \, ds = \frac{V_y}{2I_z} ab^2 t \tag{c}$$

To create the shear center, it is not necessary to find V_2. However, as a check, the shear stress in the web will be integrated, where $dA = t \, dy$ and τ is given by Eq. (b). Thus

$$V_2 = 2\int_0^a \frac{V_y}{I_z}[ab + \tfrac{1}{2}(a^2 - y^2)]t \, dy = \frac{V_y}{I_z}(2a^2 bt + \tfrac{2}{3}a^3 t) \tag{d}$$

Since $I_z \approx \tfrac{2}{3}a^3 t + 2a^2 bt$, $V_2 = V_y$.

The shear center can be located by summing moments about point D of Fig. 3-11. For equilibrium,

$$V_y e = 2\left(\frac{V_y}{2I_z} ab^2 t\right)a$$

Thus

$$e = \frac{a^2 b^2 t}{I_z}$$

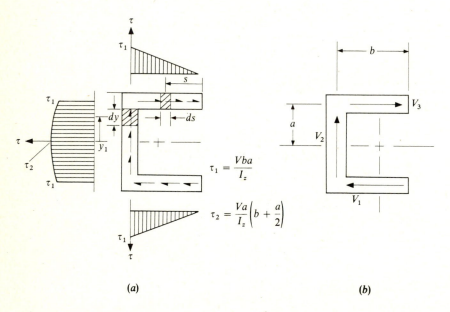

(a)   (b)

Figure 3-10

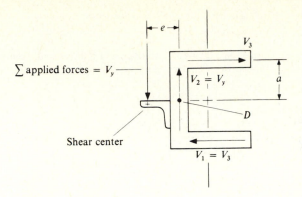

$\sum$ applied forces $= V_y$

Shear center

Figure 3-11

However, since $I_z \approx \frac{2}{3}a^3t + 2a^2bt$,

$$e = \frac{a^2b^2t}{\frac{2}{3}a^3t + 2a^2bt} = \frac{3}{2}\frac{b^2}{a + 3b}$$

Example 3-4 For the shear-stress distribution shown in Fig. 3-12a, assuming that the applied loads act through the shear center, determine the shear-stress distribution and total shear force

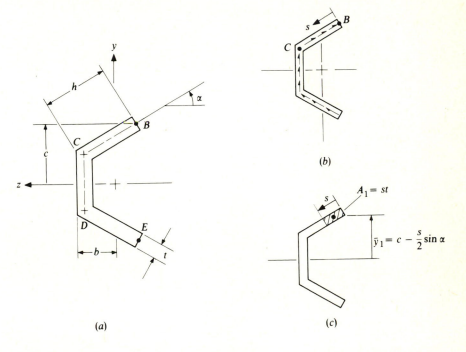

(a)

(b)

(c)

Figure 3-12

transmitted through section BC. Bending is about the z axis, and the net shear force is V_y in the y direction.

SOLUTION The shear flow is illustrated in Fig. 3-12b. Starting from the free end, point B, establish an arbitrary position s on web BC. The area moment Q (see Fig. 3-12c) is

$$Q = \bar{y}_1 A_1 = \left(c - \frac{s}{2}\sin\alpha\right)st$$

Thus, from Eq. (3-4) the shear stress is

$$\tau = \frac{Vs}{I_z}\left(c - \frac{s}{2}\sin\alpha\right)$$

To determine the total force on section BC, integration of $\tau\, dA$ is performed, where $dA = t\, ds$. Thus

$$V_{BC} = \int_0^h \frac{V_y}{I_z}s\left(c - \frac{s}{2}\sin\alpha\right)t\, ds = \frac{V_y}{6I_z}th^2(3c - h\sin\alpha)$$

3-5 MOVING LOADS

As a vehicle (crane, truck, etc.) moves slowly over a beam contained within a structure, the wheel loads travel over the beam as a group. Neglecting dynamic effects, it is possible quickly to determine what position of the vehicle will produce the maximum bending moment in the beam. Figure 3-13a illustrates a simply supported beam in which a group of loads, $W_1, W_2, \ldots, W_n$, are moving

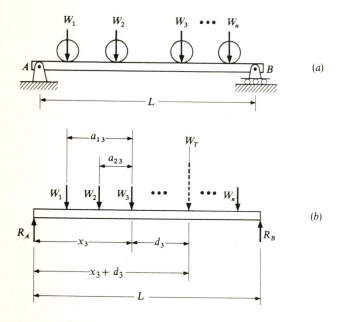

Figure 3-13

across. Since the maximum bending moment occurs under one of the concentrated forces, it is necessary to find the position for which the moment is maximum under each wheel and determine the largest value. There is a simple technique for determining each position. To develop the technique, the position of the moving loads for which the moment under the load W_3 is maximum is found.

Let x_3 denote the position of load W_3 from point A when the maximum moment occurs beneath this load. The resultant, or equivalent force of the distribution of forces on the beam, is W_T, and at the position of interest it is positioned a distance $x_3 + d_3$ from point A (see Fig. 3-13b). The reaction at point A is

$$R_A = \frac{L - x_3 - d_3}{L} W_T$$

and the bending moment under load W_3 is

$$(M_z)_3 = R_A x_3 - W_1 a_{13} - W_2 a_{23}$$

or

$$(M_z)_3 = \frac{(L - x_3 - d_3) W_T x_3}{L} - W_1 a_{13} - W_2 a_{23}$$

To determine the value of x_3 for which $(M_z)_3$ is maximum, let $d(M_z)_3/dx_3 = 0$. Thus

$$\frac{d}{dx_3} (M_z)_3 = \frac{W_T}{L} (L - d_3 - 2x_3) = 0$$

or

$$x_3 = \frac{L - d_3}{2}$$

This means that the midpoint between W_3 and W_T intersects the centerline of the beam when the moment is maximum under load W_3. Substituting x_3 into the equation for $(M_z)_3$ results in the maximum value of

$$(M_z)_3 = \frac{(L - d_3)^2}{4L} W_T - W_1 a_{13} - W_2 a_{23}$$

The next step is to check the remaining wheels to determine under which wheel the bending moment will be the greatest. The equations can be generalized to speed the procedure up. The position for which the moment under wheel i is maximum is given by

$$x_i = \frac{L - d_i}{2} \tag{3-5}$$

and the value of this maximum moment is

$$(M_z)_i = \frac{(L - d_i)^2}{4L} W_T - \sum_{j=1}^{i-1} W_j a_{ji} \tag{3-6}$$

where $(M_z)_i$ = maximum bending moment under wheel i
d_i = distance from wheel i to resultant W_T

If wheel i is to the right of W_T, d_i is negative. If $i = 1$, the summation term vanishes since $a_{11} = 0$.

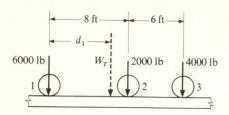

Figure 3-14

Example 3-5 Three moving wheels with their loads are shown in Fig. 3-14. Find the maximum bending moment developed in a 24-ft beam as the wheels move across the beam.

SOLUTION The resultant W_T is

$$W_T = 6000 + 2000 + 4000 = 12,000 \text{ lb}$$

and the location d_1 from wheel 1 is

$$12,000 d_1 = (2000)(8) + (4000)(14)$$

$$d_1 = 6 \text{ ft}$$

From this, $d_2 = -2$ ft and $d_3 = -8$ ft. When Eq. (3-6) is applied, the maximum moments under each wheel are

$$M_{z1} = \frac{(24-6)^2}{(4)(24)}(12,000) = 40,500 \text{ lb·ft}$$

$$M_{z2} = \frac{(24+2)^2}{(4)(24)}(12,000) - (6000)(8) = 36,500 \text{ lb·ft}$$

$$M_{z3} = \frac{(24+8)^2}{(4)(24)}(12,000) - (6000)(14) - (2000)(6) = 32,000 \text{ lb·ft}$$

Thus the maximum bending moment occurs under wheel 1 with a value of 40,500 lb·ft. The position for which this happens is when wheel 1 is

$$x_1 = \frac{L - d_1}{2} = \frac{24 - 6}{2} = 9 \text{ ft}$$

from the left support.

3-6 COMPOSITE BEAMS IN BENDING

In Chap. 2, axial loading of a composite bar was investigated. Recall that in order to avoid bending it is necessary to apply the loading through an axis which is not necessarily the geometrical centroid of the total cross section because the discontinuity in the stress distribution that arises from the differing moduli of elasticity of the two materials, where the material with the higher modulus develops greater stresses. This is advantageous in many cases as certain structural materials are inexpensive but have certain strength shortcomings. With composites, large quantities of the low-modulus material can be used in the low-stressed areas and small quantities of the high-modulus material can be used in the high-stressed areas, e.g., reinforced-concrete beams using steel rods and steel-clad wood beams.

A simple procedure when dealing with composite beams in bending is to transform the cross section of the composite beam into a cross section of one material (called the *equivalent cross section*) so that the strains are identical to the composite. Once the stresses are determined for the equivalent cross section, the stresses are then transformed to correspond to the composite. (This could have been done for the axially loaded rod, but the simplicity of axial loading made it unnecessary.)

Consider the case of a beam composed of two materials, of moduli E_1 and E_2, where $E_2 > E_1$ (see Fig. 3.15a). Let the equivalent single-material cross

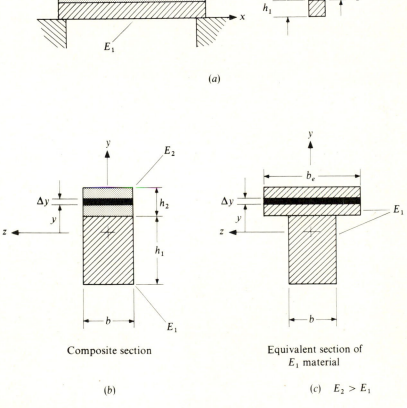

(a)

Composite section

(b)

Equivalent section of E_1 material

(c) $E_2 > E_1$

Figure 3-15

section be entirely made of the material of modulus E_1. In order for the equivalent cross section to have the same bending rigidity as the true composite it is necessary that more E_1 material replace the E_2 material. However, since the strains of the equivalent cross section are to be identical to the composite, the depth of the section relative to the bending axis cannot be altered, as strains are a function of position relative to this axis. Parallel to the bending axis the strains are uniform. Thus, the width of the beam can be increased without affecting the strains (see Fig. 3-15c). At position y, the strains in the x direction must be the same for the equivalent section as for the composite. Thus

$$(\varepsilon_x)_e = \varepsilon_x$$

where the subscript e stands for the equivalent section. From Hooke's law

$$\frac{(\sigma_x)_e}{E_1} = \frac{\sigma_x}{E_2}$$

Once the stresses are determined in the equivalent section $(\sigma_x)_e$, the actual stresses in the E_2 material are given by

$$\sigma_x = \frac{E_2}{E_1}(\sigma_x)_e \tag{3-7}$$

For equilibrium, the forces on the areas $b\,\Delta y$ and $b_e\,\Delta y$ of the composite and equivalent sections must also be identical. Thus

$$(\sigma_x)_e b_e\,\Delta y = \sigma_x b\,\Delta y$$

or

$$b_e = \frac{\sigma_x}{(\sigma_x)_e} b$$

but from Eq. (3-7), $\sigma_x/(\sigma_x)_e = E_2/E_1$. Thus

$$b_e = \frac{E_2}{E_1} b \tag{3-8}$$

which confirms the necessity of increasing the width of the top section since $E_2 > E_1$.

Example 3-6 A beam section composed of aluminum ($E_a = 10 \times 10^6$ lb/in²) and steel ($E_s = 30 \times 10^6$ lb/in²) are bonded together as shown in Fig. 3-16a. If the section is undergoing a positive bending moment of $M_z = 10{,}000$ lb·in about a horizontal axis, determine the resulting stress distribution.

SOLUTION Let the total cross section be completely made of aluminum. Thus, for the bottom section, using Eq. (3-8), the equivalent width is

$$b_e = \frac{30 \times 10^6}{10 \times 10^6}(1) = 3 \text{ in}$$

and thus the equivalent aluminum section is as shown in Fig. 3-16b. Next, the centroidal axis

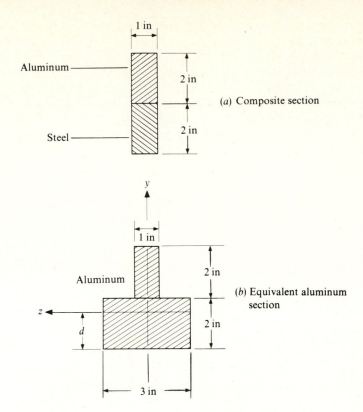

(a) Composite section

(b) Equivalent aluminum
section

Figure 3-16

of the equivalent cross section is determined. From Fig. 3-16b

$$d(6+2) = (1)(6) + (3)(2)$$

$$d = 1.50 \text{ in}$$

Now the moment of inertia of the totally aluminum cross section about the centroidal axis $(I_z)_e$ is determined. Using the parallel-axis theorem, we have

$$(I_z)_e = (\tfrac{1}{12})(3)(2^3) + (6)(0.50)^2 + (\tfrac{1}{12})(1)(2^3) + (2)(1.50)^2 = 8.667 \text{ in}^4$$

The stress distribution in the equivalent section is determined by using the flexure formula $\sigma_x = -M_z y/(I_z)_e$, where $M_z = 10,000$ lb·in. The stress at the top surface is

$$(\sigma_x)_e|_{y=2.5} = -\frac{(10,000)(2.5)}{8.667} = -2885 \text{ lb/in}^2$$

at the interface

$$(\sigma_x)_e|_{y=0.5} = -\frac{(10,000)(0.5)}{8.667} = -577 \text{ lb/in}^2$$

and at the bottom

$$(\sigma_x)_e|_{y=-1.5} = -\frac{(10,000)(-1.5)}{8.667} = 1731 \text{ lb/in}^2$$

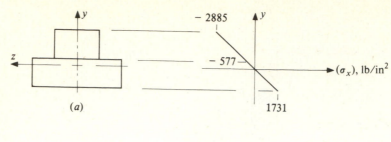

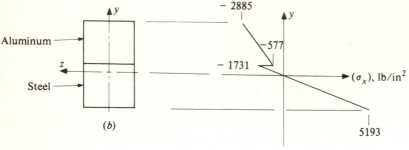

Figure 3-17

The stress distribution for the equivalent section is shown in Fig. 3-17a. However, the stress in the lower half is for the equivalent material and must be transformed using Eq. (3-7) to correspond to the actual material. Thus for the steel at the interface

$$\sigma_x|_{y=-0.5} = (3)(-577) = -1731 \text{ lb/in}^2$$

and at the bottom

$$\sigma_x|_{y=-1.5} = (3)(1731) = 5193 \text{ lb/in}^2$$

The actual stress distribution is as shown in Fig. 3-17b.

Example 3-7 Rotating the section investigated in Example 3-6 90°, as shown in Fig. 3-18a, determine the stress distribution for bending about a horizontal axis of the section if $M_z = 10,000 \text{ lb·in}$.

SOLUTION The moment of inertia of the equivalent section is

$$(I_z)_e = (\tfrac{1}{12})(8)(1^3) = 0.667 \text{ in}^4$$

Thus

$$(\sigma_x)_e|_{y=\pm0.5} = -\frac{(10,000)(\pm0.5)}{0.667} = \mp7500 \text{ lb/in}^2$$

This corresponds to the half made of aluminum. For the steel

$$(\sigma_x)_s|_{y=\pm0.5} = (3)(\mp7500) = \mp22,500 \text{ lb/in}^2$$

The stress distributions in each half are presented in Fig. 3-18c.

Note: In this problem the loading plane xy must be located so that the moment about the y axis is zero. The location of the centroidal y axis of the composite is found by considering bending about the y axis. This was actually done in Example 3-6. The y axis should be located 1.5 in from the right-hand side of Fig. 3-18a.

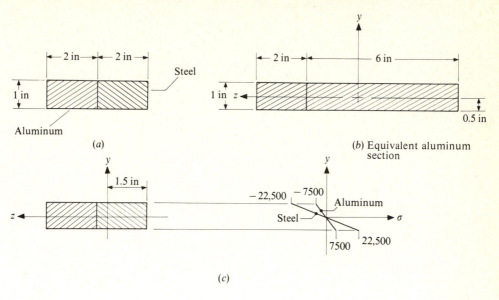

(a)

(b) Equivalent aluminum section

(c)

Figure 3-18

3-7 WIDE BEAMS

In elementary strength of materials, the equations for deflections are based on the assumption that the beam is narrow. For a beam that is very wide compared with the depth, deflections are actually less than predicted with narrow-beam equations. For a narrow beam, lateral (z-direction) expansions or contractions are relatively free to take place when bending in the xy plane occurs. For a wide beam, lateral deflections are prevented, which effectively makes the beam stiffer than that predicted by narrow-beam theory.

Consider the beam shown in Fig. 3-19, where $b \gg h$. If it is assumed that

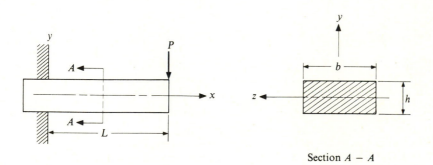

Section $A - A$

Figure 3-19

lateral deformation is prevented at $z = 0$, then, $\varepsilon_z \approx 0$ at $z = 0$. Substituting this into the strain-stress relationships given by Eqs. (1-21) results in

$$\varepsilon_z = \frac{1}{E} [\sigma_z - v(\sigma_x + \sigma_y)] \approx 0$$

Also, since h is small, $\sigma_y \approx 0$, and therefore

$$\sigma_z \approx v\sigma_x \qquad \text{at } z = 0 \qquad (3\text{-}9)$$

Substituting $\sigma_y = 0$ and $\sigma_z = v\sigma_x$ into Eq. (1-21a) results in

$$\varepsilon_x = \frac{1 - v^2}{E} \sigma_x \qquad (3\text{-}10)$$

The bending stress σ_x is given by Eq. (3-1). Thus

$$\varepsilon_x = -\frac{1 - v^2}{EI_z} M_z y \qquad (3\text{-}11)$$

Comparing Eqs. (3-11) and (2-20a) shows that there is a reduction in strain by the factor $1 - v^2$.[†] This correction can be substituted into the deflection equation, yielding

$$\frac{d^2 y_c}{dx^2} \approx (1 - v^2) \frac{M_z}{EI_z} \qquad (3\text{-}12)$$

Thus the results obtained for narrow-beam theory (given in Appendix D) apply to wide beams provided the results are multiplied by $1 - v^2$.

3-8 CURVED BEAMS: TANGENTIAL STRESSES

Consider the segment of curved beam undergoing pure bending as shown in Fig. 3-20. For simple bending in the $r\theta$ plane, the cross section must have an axis of symmetry in this plane. Assume, as in straight-beam theory, that a plane cross section remains plane after bending.

The tangential strain at any position r is then

$$\varepsilon_\theta = \frac{b'b}{ab} = -\frac{(r_n - r)\,d\theta}{r\theta}$$

and the corresponding tangential stress is

$$\sigma_\theta = E\varepsilon_\theta = -\frac{E(r_n - r)}{r}\frac{d\theta}{\theta} \qquad (3\text{-}13)$$

Since the net force in the tangential direction is zero,

$$\int \sigma_\theta \, dA = \frac{E\,d\theta}{\theta} \int \frac{r_n - r}{r} \, dA = 0$$

[†] This correction factor will be discussed further in Chap. 6, where plane-stress and plane-strain problems are covered.

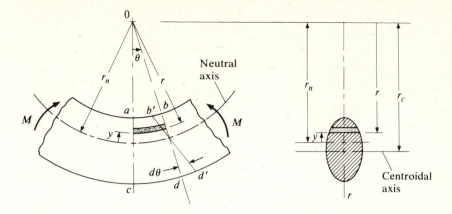

Figure 3-20

In general, $E\, d\theta/\theta \neq 0$; therefore, the integral must be zero. Thus

$$\int \frac{r_n - r}{r}\, dA = r_n \int \frac{dA}{r} - A = 0$$

or

$$r_n = \frac{A}{\displaystyle\int \frac{dA}{r}} \qquad\qquad (3\text{-}14)$$

For a curved beam, the neutral and centroidal axes are not coincident, and it can be shown that it is always true that $r_n < r_c$.

> **Example 3-8** Determine the location of the neutral axis and the distance e for the curved bar of rectangular cross section shown in Fig. 3-21.
>
> SOLUTION The location of the centroidal axis is $r_c = 175$ mm. Establishing $dA = 10\, dr$ and $A = (10)(50) = 500$ mm² leads to
>
> $$r_n = \frac{500}{10 \displaystyle\int_{150}^{200} \frac{dr}{r}} = \frac{50}{\ln 200 - \ln 150} = \frac{50}{\ln 1.333} = 173.80 \text{ mm}$$

Thus

$$e = 175.00 - 173.80 = 1.20 \text{ mm}$$

In order to develop the stress distribution, the net moment from σ_θ must be equated to M. Thus

$$M = -\int y\sigma_\theta\, dA = E\frac{d\theta}{\theta} \int \frac{(r_n - r)^2}{r}\, dA$$

$$= E\frac{d\theta}{\theta}\left(r_n^2 \int \frac{dA}{r} - 2r_n \int dA + \int r\, dA \right)$$

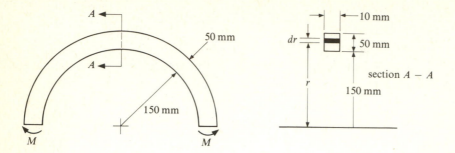

Figure 3-21

But $\int dA = A$, $\int r\, dA = r_c A$, and from Eq. (3-14), $\int dA/r = A/r_n$. Therefore

$$M = E\frac{d\theta}{\theta}(-r_n A + r_c A)$$

Solving for $E\, d\theta/\theta$, letting $r_c - r_n = e$, yields

$$E\frac{d\theta}{\theta} = \frac{M}{Ae}$$

substituting this into Eq. (3-13) results in

$$\sigma_\theta = -\frac{M(r_n - r)}{Aer} \qquad (3\text{-}15)$$

Equation (3-15) can be written in terms of y. Substitution of $y = r_n - r$ gives

$$\sigma_\theta = -\frac{My}{Ae(r_n - y)} \qquad (3\text{-}16)\dagger$$

Table 3-1 presents the formula for calculating $\int dA/r$ for various shapes commonly used in curved beams.

Note that the term e in Eqs. (3-15) and (3-16) is usually a low value obtained from the difference of two much higher-valued terms $r_c - r_n$. Thus, calculation of r_c and r_n must be very accurate so that e is at least known to three significant figures. Note also that Eq. (3-15) or (3-16) gives the normal stresses due only to bending effects. Normally in a curved-beam problem there is an additional normal force. Care must be taken not to neglect the additional normal stress that arises from this force.

---

† Reference 2 (table 5, p. 149) tabulates correction factors for standard cross sections. With them the straight-beam equation can be used, and the result is multiplied by the appropriate correction factor (see Exercise 3-14).

Table 3-1

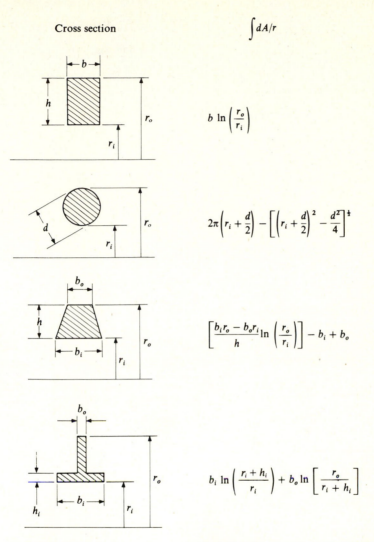

Cross section	$\int dA/r$
	$b \ln\left(\dfrac{r_o}{r_i}\right)$
	$2\pi\left(r_i + \dfrac{d}{2}\right) - \left[\left(r_i + \dfrac{d}{2}\right)^2 - \dfrac{d^2}{4}\right]^{\frac{1}{2}}$
	$\left[\dfrac{b_i r_o - b_o r_i}{h}\ln\left(\dfrac{r_o}{r_i}\right)\right] - b_i + b_o$
	$b_i \ln\left(\dfrac{r_i + h_i}{r_i}\right) + b_o \ln\left[\dfrac{r_o}{r_i + h_i}\right]$

Example 3-9 The clamp body shown in Fig. 3-22 undergoes a force F of 1000 lb. Determine the normal-stress distribution across section a-a and the maximum tensile and compressive stresses.

SOLUTION First, locate the centroidal axis. Considering the web and flange as separate rectangles, we have

$$0.5625 r_c = (0.25)(1.25)(4.875) + (0.25)(1.00)(4.125)$$

$$r_c = 4.5417\dagger$$

† It will be shown later that five significant figures will be necessary for e to be accurate to within three significant places.

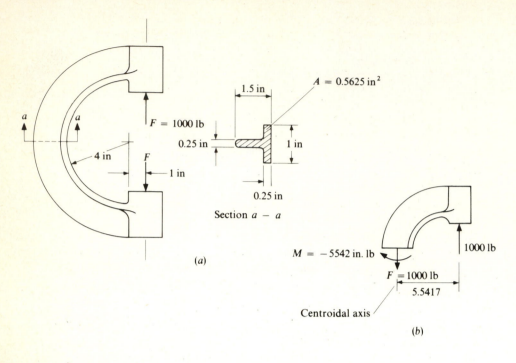

$A = 0.5625$ in^2

$F = 1000$ lb

Section $a - a$

(a)

$M = -5542$ in. lb

$F = 1000$ lb

5.5417

Centroidal axis

1000 lb

(b)

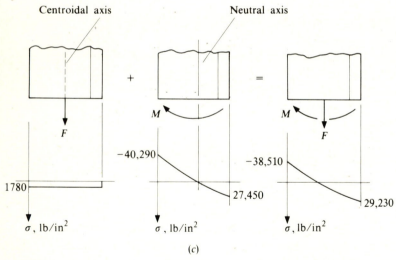

Centroidal axis

Neutral axis

$+$

$=$

M

M

F

F

$-40,290$

$-38,510$

1780

27,450

29,230

σ, lb/in^2

σ, lb/in^2

σ, lb/in^2

(c)

Figure 3-22

After making a break at section a-a, the direct transmitted force F is placed *on the centroidal axis as shown.* This is done so that a uniform stress distribution can be substituted for the force F. The neutral axis in pure bending is found from Table 3-1:

$$r_n = \frac{(1.00)(0.25) + (0.25)(1.25)}{1.00 \ln [(4.00 + 0.25)/4.00] + 0.25 \ln [5.5/(4.00 + 0.25)]} = 4.4971 \text{ in}$$

Thus

$$e = 4.5417 - 4.4971 = 0.0446 \text{ in}$$

Because of bending, the maximum tensile stress occurs at the point closest to the center, where $y = 4.497 - 4.000 = 0.497$; from Eq. (3-16) this maximum is

$$(\sigma_\theta)_{r_i} = -\frac{(-5542)(0.497)}{(0.5625)(0.0446)(4.497 - 0.497)} = 27,450 \text{ lb/in}^2$$

The maximum compressive stress due to bending occurs at $y = 4.497 - 5.500 = -1.003$ and is

$$(\sigma_\theta)_{r_o} = -\frac{(-5542)(-1.003)}{(0.5625)(0.0446)(4.497 + 1.003)} = -40,290 \text{ lb/in}^2$$

The direct stress due to F is $\sigma = F/A$, or

$$\sigma = \frac{1000}{0.5625} = 1780 \text{ lb/in}^2$$

The *net* stresses at these points are found by superposition of the axial and bending stresses and are

$$(\sigma_\theta)_{r_i} = 27,450 + 1780 = 29,230 \text{ lb/in}^2$$

$$(\sigma_\theta)_{r_o} = -40,290 + 1780 = -38,510 \text{ lb/in}^2$$

3-9 CURVED BEAMS; RADIAL STRESSES

Radial stresses† σ_r also develop in curved beams and may be significant if the width of the cross section at the neutral axis is small. Considering pure bending again, isolate a section of Fig. 3-20, as shown in Fig. 3-23. If σ_θ is taken as positive, the net force σ_θ exerts in the negative radial direction is

$$\sin \Delta\theta \int \sigma_\theta \, dA$$

where the integral is evaluated from $r = r_i$ to r_1. Since $\Delta\theta$ is infinitesimal, then $\sin \Delta\theta = \Delta\theta$ and the force reduces to

$$\Delta\theta \int \sigma_\theta \, dA$$

The force in the positive radial direction due to σ_r is $\sigma_r b r_1 \Delta\theta$, where b is the thickness of the section at r_1. Summing forces in the radial direction yields

$$\sigma_r = \frac{1}{b r_1} \int_{r_i}^{r_1} \sigma_\theta \, dA$$

---

† In the derivation of σ_θ, it was assumed that $\varepsilon_\theta = \sigma_\theta / E$. Since σ_r exists, this causes a slight error in Eq. (3-15).

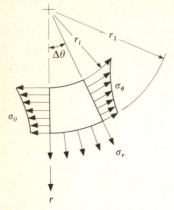

Figure 3-23

Substitution of Eq. (3.15) and simplifying results in

$$\sigma_r = -\frac{M}{Abr_1e}\left(r_n \int_{r_i}^{r_1} \frac{dA}{r} - A_1\right) \tag{3-17}$$

where A_1 is the area of the cross section from r_i to r_1.

Example 3-10 Determine the radial stress distribution in zone a-a of the clamp of Example 3-9.

SOLUTION For the inner flange, evaluate the integral $\int dA/r$ with $r_i = 4$ in, $r_o = r_1$, and $b = 1$ in using the rectangular cross section in Table 3-1:

$$\int \frac{dA}{r} = 1 \ln \frac{r_1}{4}$$

The area from r_i to r_1 is

$$A_1 = (1)(r_1 - 4)$$

Substituting these terms into Eq. (3-17) with $b = 1$ results in the radial stress distribution in the flange:

$$\sigma_r = -\frac{M}{Ar_1e}\left(r_n \ln \frac{r_1}{4} - r_1 + 4\right) \tag{a}$$

In the web area, the section isolated at r_1 is a T section (see Table 3-1) with $b_i = 1$ in, $h_i = 0.25$ in, $r_i = 4$, $r_o = r_1$, and $b = b_o = 0.25$ in. Thus

$$\int \frac{dA}{r} = 1 \ln\left(1 + \frac{0.25}{4}\right) + (0.25)\ln\frac{r_1}{4 + 0.25} = 0.06062 + \tfrac{1}{4}\ln\frac{r_1}{4.25}$$

and

$$A_1 = (1)(0.25) + (r_1 - 4.25)(0.25) = \frac{r_1}{4} - 0.8125$$

Substitution of $\int dA/r$ and A_1 into Eq. (3-17) with $b = 0.25$ yields the radial stress distribution in the web,

$$\sigma_r = -\frac{4M}{Ar_1e}\left(\frac{r_n}{4}\ln\frac{r_1}{4.25} + 0.06062r_n - \frac{r_1}{4} + 0.8125\right) \tag{b}$$

From Example 3-9 $M = -5542$ lb·in, $A = 0.5625$ in^2, $e = 0.0446$ in, and $r_n = 4.4971$ in. Thus Eqs. (a) and (b) reduce to:

Flange:
$$\sigma_r = (2.21 \times 10^5)\left(\frac{4.4971}{r_1} \ln \frac{r_1}{4} + \frac{4}{r_1} - 1\right) \qquad (c)$$

Web:
$$\sigma_r = (8.84 \times 10^5)\left(\frac{1.1244}{r_1} \ln \frac{r_1}{4.25} + \frac{1.0850}{r_1} - \frac{1}{4}\right) \qquad (d)$$

The maximum radial stress in the flange occurs at $r_1 = 4.25$ and is†

$$(\sigma_r)_{\text{flange max}} = 1180 \text{ lb/in}^2$$

The maximum radial stress in the web occurs at $r_1 = 4.4016$ where $d\sigma_r/dr = 0$. This value is

$$(\sigma_r)_{\text{web max}} = 4820 \text{ lb/in}^2$$

The stress distribution is shown in Fig. 3-24.

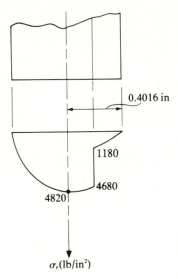

0.4016 in

1180

4680

4820

σ_r(lb/in^2)

Figure 3-24

3-10 BENDING OF THIN PLATES

In Sec. 3-7 lateral narrow-beam theory was extended to cover the deflections of wide beams. A wide beam in the context of this chapter is considered to be a rectangular plate supported on one edge or on two opposing edges. The

---

† This value is actually low, because in the derivation it is assumed that the radial stress acts over the entire thickness. As in shear flow in an I beam, the boundary conditions at the free surface of the flange dictate that $\sigma_r = 0$ on the free surfaces of the flange at $r_1 = 4.25$ in. The radial stress in the web at $r_1 = 4.25$ in is the nominal stress value at this point and is found to be $\sigma_r = 4680$ lb/in^2. The stress is higher due to the rapid change in cross section; however, if the material is ductile and the load is static, the stress concentration can be neglected.

analysis of laterally loaded plates with different geometric or support configurations is highly complex mathematically. In Ref. 3, Ref. 4 (pp. 429–505), and Ref. 5 (pp. 324–413) detailed analysis is provided for such conditions as:

1. Rectangular plates supported on four edges
2. Circular plates supported along the outer edge
3. Circular plates with central holes supported along the outer and/or inner edges

For an engineering approximation, a somewhat simplified approach is common, but it should be understood from the start that the accuracy of the approximation may be poor. The method of approximation is based on the assumption that at a given surface isolation, the total moment acting on the surface acts uniformly across the surface. That is, the moment per unit width along the surface is constant. This assumption is true for narrow beams and relatively valid for a long rectangular plate when two opposing edges along the plate width are not supported. However, for cases like those listed in the previous paragraph, this assumption may or may not be very good. Although the equations obtained by this approximation normally predict bending stresses which are lower than the actual, the equations can still be useful in design. As will be seen in Sec. 3-15, if the plate is made of a ductile material and begins to yield at a localized area where the stresses are higher than predicted, the stress begins to shift toward a more uniform distribution. Thus, the approximate equations are useful provided some localized yielding is acceptable. If not, or if the plate is brittle or undergoing cyclic loading, an exact analysis is necessary.

Circular Plate Approximation

Consider a flat circular plate of diameter d and uniform thickness h subjected to a uniform lateral load of p_o force per unit area (see Fig. 3-25a). If the edges are considered to be simply supported, dividing the net support force $p_o \pi d^2/4$ by the circumference πd yields the support force per unit length $p_o d/4$. The bending moment will be greatest at the center of the plate. Thus, half of the plate is isolated, as shown in Fig. 3-25b. The equivalent force of the applied load is $p_o \pi d^2/8$ and is located at the centroid of the semicircle $2d/3\pi$. The equivalent force of the support reaction is also $p_o \pi d^2/8$, but it is located at the centroid of the semicircular line segment d/π. The net moment M on the exposed internal surface is

$$M = \frac{p_o \pi d^2}{8} \left(\frac{d}{\pi} - \frac{2d}{3\pi} \right) = \frac{p_o d^3}{24}$$

Assuming that the bending stress does not vary along line ABC, Eq. (2.15) can be used. Thus the maximum bending stress is given by,

$$\sigma_{max} = \frac{Mc}{I} = \frac{(p_o d^3/24)h/2}{\frac{1}{12}dh^3} = 0.25 p_o \left(\frac{d}{h} \right)^2 \tag{3-18}$$

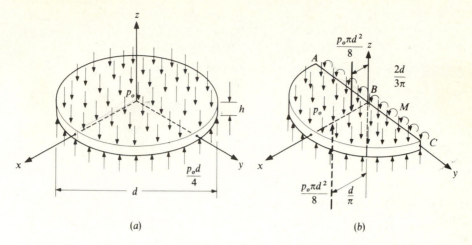

Figure 3-25

The assumption that the bending stress does not vary along ABC is quite arbitrary. Exact analysis of the plate shows that the stress along ABC at $z = -h/2$ varies as (Ref. 4, p. 439)

$$\sigma = \tfrac{3}{32}p_o\left(\frac{d}{h}\right)^2(3+\nu)\left[1-4\left(\frac{y}{d}\right)^2\right] \tag{3-19}$$

which is maximum at point B where $y = 0$. Considering ν to be 0.3, we see that the maximum bending stress is

$$\sigma_{max} = 0.309 p_o\left(\frac{d}{h}\right)^2 \tag{3-20}$$

Thus in this case, the approximation was in error only by 19 percent.

If the lateral load was a concentrated force P located at the center of the plate, the support loading would be $P/\pi d$ force per unit length and the moment along line ABC would be

$$M = \frac{P}{2}\frac{d}{\pi} = \frac{Pd}{2\pi}$$

The maximum bending stress in this case is approximated by

$$\sigma_{max} = \frac{Mc}{I} = \frac{(Pd/2\pi)d/2}{\tfrac{1}{12}dh^3} = \frac{3P}{\pi h^2} \tag{3-21}$$

The exact analysis of a simply supported circular plate with a centrally located concentrated force predicts infinite stress at the center. A concentrated force is not physically acceptable in the exact analysis. The net force P must be distributed over a finite area no matter how small. If the force is considered to be uniformly distributed over a circular area of diameter d_o, the approximate analysis yields

$$\sigma_{max} = \frac{P}{\pi} \frac{3d - 2d_o}{h^2 d} \tag{3-22}$$

whereas the exact analysis predicts a maximum bending stress at the center of

$$\sigma_{max} = \frac{3P}{2\pi h^2} \left[1 + (1 + \nu) \ln \frac{d}{d_o} - \frac{1 - \nu}{4} \left(\frac{d_o}{d} \right)^2 \right] \tag{3-23}$$

Thus, the inaccuracy of the approximation depends on the ratio of d_o/d. For example, if $d_o/d = 0.01$ and $\nu = 0.3$, the actual stress is 3.5 times greater than the approximation.

Rectangular Plate Approximation

For a rectangular plate, the support-force distribution at the edges is not uniform. As a matter of fact, the corners tend to curl up, and if they are restrained, the reaction forces and internal bending moments change directions. The critical section to investigate is therefore the diagonal. Consider the simply supported rectangular plate with a uniform lateral load of p_o force per unit area, as shown in Fig. 3-26a. Isolate half the plate by exposing the diagonal surface. The net force due to p_o is $p_o ab/2$ and is located at a perpendicular distance from the diagonal of $e/3$. The net support reactions R_1 and R_2 are both located at a perpendicular distance from the diagonal of $e/2$. Although the

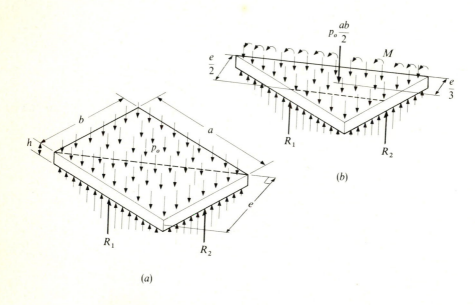

(b)

(a)

Figure 3-26

individual values of R_1 and R_2 are unknown, their sum must be $p_oab/2$. Thus, summing moments yields the moment acting across the diagonal

$$M = (R_1 + R_2)\frac{e}{2} - \frac{p_oab}{2}\frac{e}{3} = \frac{p_oab}{2}\left(\frac{e}{2} - \frac{e}{3}\right) = \tfrac{1}{12}p_oabe$$

From geometry the distance e is found to be

$$e = \frac{ab}{\sqrt{a^2 + b^2}}$$

Thus, the bending moment is

$$M = \frac{p_oa^2b^2}{12\sqrt{a^2 + b^2}}$$

The length of the diagonal is $\sqrt{a^2 + b^2}$. Therefore, the area moment of inertia about the bending axis is

$$I = \tfrac{1}{12}h^3\sqrt{a^2 + b^2}$$

Considering the stress to be uniform in the direction of the diagonal results in a maximum bending stress of

$$\sigma_{max} = \frac{Mc}{I} = \frac{(p_oa^2b^2/12\sqrt{a^2 + b^2})h/2}{\tfrac{1}{12}h^3\sqrt{a^2 + b^2}} = \frac{p_oa^2b^2}{2h^2(a^2 + b^2)} \qquad (3\text{-}24)$$

Equation (3-24) is reasonably accurate when the dimensions a and b are of the same order of magnitude. For example, consider a square plate, where $b = a$. In this case, Eq. (3-24) reduces to

$$\sigma_{max} = 0.25p_o\left(\frac{a}{h}\right)^2 \qquad (3\text{-}25)$$

A numerical solution using the complete thin-plate equations converges to a maximum bending moment per unit width of $0.0479p_oa^2$ at the center of the plate (with $\nu = 0.3$).[†] The moment of inertia per unit length along the diagonal is $h^3/12$. Thus this solution indicates a maximum bending stress at the center of

$$\sigma_{max} = \frac{0.0479p_oa^2(h/2)}{\tfrac{1}{12}h^3} = 0.287p_o\left(\frac{a}{h}\right)^2$$

which is about 15 percent greater than that given by Eq. (3-25).

When one of the dimensions is very large compared with the other, Eq. (3-24) is not very good. If, for example, $a \gg b$, the supports on the short sides are practically ineffective; hence the plate behaves much like a beam of length b. In this case, the maximum bending moment is $wL^2/8$, where $w = p_oa$ is the force per unit length and the length L equals b. Thus

$$M = \tfrac{1}{8}p_oab^2$$

† For the exact numerical solution, see Ref. 3, p. 120.

and the maximum bending stress is

$$\sigma_{max} = \frac{(\frac{1}{8}p_o ab^2)h/2}{\frac{1}{12}ah^3} = 0.75p_o\left(\frac{b}{h}\right)^2 \tag{3-26}†$$

For a concentrated force P located at the center of the rectangular plate, the approximate method yields a maximum bending stress of

$$\sigma_{max} = \frac{3}{2}\frac{Pab}{h^2(a^2+b^2)} \tag{3-27}‡$$

Example 3-11 Approximate the maximum bending stress for a circular plate with a central hole if the plate is loaded laterally with a uniform load p_o force per unit area and simply supported at the inner edge (see Fig. 3-27a).

SOLUTION When a surface is isolated as shown in Fig. 3-27b, the net force on the plate half is $\pi p_o(r_o^2 - r_i^2)/2$. The centroid location of the applied force is

$$r_p = \frac{(\pi/2)r_o^2(\frac{4}{3})r_o/\pi - (\pi/2)r_i^2(\frac{4}{3})r_i/\pi}{(\pi/2)(r_o^2 - r_i^2)} = \frac{4}{3\pi}\frac{r_o^3 - r_i^3}{r_o^2 - r_i^2}$$

For the support force, the centroid of the half-circular line segment r_s is $2r_i/\pi$. Thus the

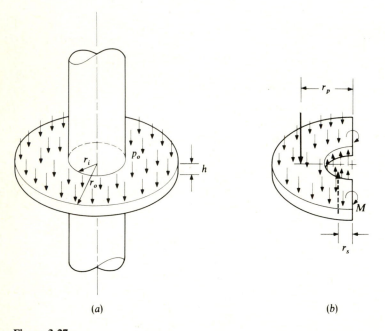

(a)   (b)

Figure 3-27

† This agrees exactly with the numerical solution of Ref. 3, p. 120, with $b/a = \infty$.
‡ As for a circular plate with a concentrated force, the exact analysis requires that the force be distributed over a finite area. Reference 3, p. 135, gives results for the case when the force is distributed over a rectangular area smaller than the dimensions of the plate.

bending moment across the exposed surface is

$$M = \frac{\pi}{2} p_o (r_o^2 - r_i^2)\left(\frac{4}{3\pi} \frac{r_o^3 - r_i^3}{r_o^2 - r_i^2} - \frac{2r_i}{\pi}\right)$$

which reduces to

$$M = \frac{p_o}{3} (r_i + 2r_o)(r_o - r_i)^2$$

The moment of inertia about the bending axis is $2(r_o - r_i)h^3/12$. Thus the approximate maximum bending stress is

$$\sigma_{max} = \frac{p_o}{3} \frac{(r_i + 2r_o)(r_o - r_i)^2 h/2}{2(r_o - r_i)h^3/12} = p_o \frac{(r_i + 2r_o)(r_o - r_i)}{h^2}$$

For ratios of r_o/r_i near unity, this result agrees closely with the exact results.[†] For example, when $r_o/r_i = 1.25$, the exact solution gives a maximum bending stress of $0.66p_o r_o^2/h^2$. The approximate solution shown here gives a slightly lower value of $0.56p_o r_o^2/h^2$. The error for this ratio of r_o/r_i is approximately 15 percent, but as the ratio of r_o/r_i increases, the error increases drastically. For example, when $r_o/r_i = 5$, the exact result is nearly 3 times that of the approximation.

3-11 THICK-WALLED CYLINDERS AND ROTATING DISKS

Consider a cylinder axially symmetric about the z axis, as shown in Fig. 3-28a. The cylinder could be pressurized at $r = r_i$ and/or $r = r_o$, or it could be rotating about the z axis. These conditions cause internal stresses. Thanks to symmetry, however, the shear stresses $\tau_{r\theta}$ will not develop[‡] and σ_θ will be constant for any value of θ at a given position r. Therefore, an element isolated at position (r, θ) will appear as in Fig. 3-28b.

Consider the element having mass with mass density ρ and rotating at a constant angular velocity ω rad/s; the equation of motion in the radial direction is

$$\sum F_r = ma_r = -\rho r\, d\theta\, dr\, dz\, r\omega^2 = -\rho\omega^2 r^2\, d\theta\, dr\, dz$$

The forces in the r direction are obtained from Fig. 3-28b by multiplying the stresses by their respective areas. Thus

$$\sum F_r = \left(\sigma_r + d\sigma_r\right)(r + dr)\, d\theta\, dz - \sigma_r r\, d\theta\, dz - 2\sigma_\theta\, dr\, dz \sin\frac{d\theta}{2}$$

Equate the two expressions for $\sum F_r$. Since $d\theta$ is infinitesimal, $\sin(d\theta/2) = d\theta/2$, and when the higher-order term in $d\sigma_r\, dr\, d\theta\, dz$ is neglected, the result is

$$\sigma_r + r\frac{d\sigma_r}{dr} - \sigma_\theta = -\rho\omega^2 r^2 \tag{3-28}$$

In order to solve Eq. (3-28) the tangential stress σ_θ must be eliminated. If

[†] For the exact solution, see Ref. 3, p. 62.
[‡] See Sec. 1-6, $\gamma_{r\theta} = 0$.

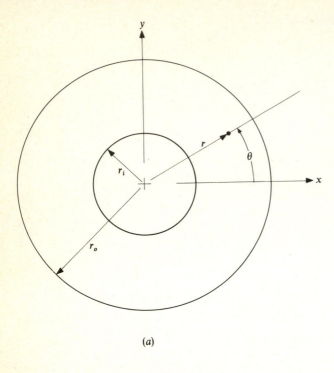

(a)

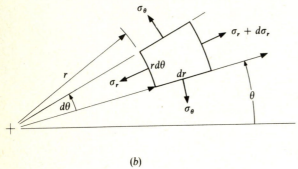

(b)

Figure 3-28

the disk is thin in the z direction,[†] $\sigma_z \approx 0$; thus

$$\varepsilon_r = \frac{\partial u_r}{\partial r} = \frac{1}{E}(\sigma_r - \nu\sigma_\theta) \tag{3-29}$$

$$\varepsilon_\theta = \frac{u_r}{r} = \frac{1}{E}(\sigma_\theta - \nu\sigma_r) \tag{3-30}$$

[†] A disk thick in the z direction will be discussed in Chap. 6.

Solving for u_r in Eq. (3-30) and differentiating yields

$$\frac{du_r}{dr} = \frac{1}{E}\left[r\frac{d\sigma_\theta}{dr} + \sigma_\theta - \nu\left(r\frac{d\sigma_r}{dr} + \sigma_r\right)\right] \tag{3-31}\dagger$$

Equating Eqs. (3-29) and (3-31) results in

$$r\frac{d\sigma_\theta}{dr} - \nu r\frac{d\sigma_r}{dr} + (1+\nu)(\sigma_\theta - \sigma_r) = 0 \tag{3-32}$$

From Eq. (3-28)

$$\sigma_\theta = r\frac{d\sigma_r}{dr} + \sigma_r + \rho\omega^2 r^2 \tag{3-33}$$

Differentiation with respect to r results in

$$\frac{d\sigma_\theta}{dr} = r\frac{d^2\sigma_r}{dr^2} + 2\frac{d\sigma_r}{dr} + 2\rho\omega^2 r \tag{3-34}$$

Substituting Eqs. (3-33) and (3-34) into Eq. (3-32) yields

$$r\frac{d^2\sigma_r}{dr^2} + 3\frac{d\sigma_r}{dr} = -(3+\nu)\rho\omega^2 r \tag{3-35}$$

Case 1 ($\omega = 0$)

If the cylinder is pressurized, as shown in Fig. 3-29 with $\omega = 0$, Eq. (3-35)

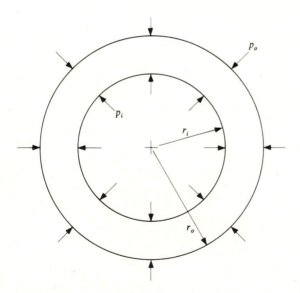

Figure 3-29

---

† Since r is the only independent variable, the partial differential $\partial/\partial r$ can be replaced by an ordinary differential d/dr.

reduces to

$$r\frac{d^2\sigma_r}{dr^2} + 3\frac{d\sigma_r}{dr} = 0 \qquad (3\text{-}36)$$

The form of the solution of Eq. (3-36) is $\sigma_r = Cr^n$. Substituting this into Eq. (3-36) yields $n = 0$ and $n = -2$. Thus the solution is

$$\sigma_r = C_1 + \frac{C_2}{r^2} \qquad (3\text{-}37)$$

Substituting this into Eq. (3-33) with $\omega = 0$ yields

$$\sigma_\theta = C_1 - \frac{C_2}{r^2} \qquad (3\text{-}38)$$

Applying the conditions that

$$\sigma_r = \begin{cases} -p_i & \text{at } r = r_i \\ -p_o & \text{at } r = r_o \end{cases}$$

to Eq. (3-37) results in

$$C_1 = \frac{p_i r_i^2 - p_o r_o^2}{r_o^2 - r_i^2} \qquad C_2 = \frac{(r_i r_o)^2(p_o - p_i)}{r_o^2 - r_i^2}$$

Thus

$$\sigma_r = \frac{p_i r_i^2 - p_o r_o^2 + (r_i r_o/r)^2(p_o - p_i)}{r_o^2 - r_i^2} \qquad (3\text{-}39)$$

and

$$\sigma_\theta = \frac{p_i r_i^2 - p_o r_o^2 - (r_i r_o/r)^2(p_o - p_i)}{r_o^2 - r_i^2} \qquad (3\text{-}40)$$

Example 3-12 Determine the stress distribution in a cylinder with an inner diameter of 2.0 in and outer diameter of 6.0 in with $p_i = 5000$ lb/in^2 and $p_o = 0$.

SOLUTION From Eq. (3-39)

$$\sigma_r = \frac{(5000)(1^2) + [(1)(3)/r]^2(-5000)}{3^2 - 1^2} = 625 - \frac{5625}{r^2} \qquad (a)$$

and from Eq. (3-40)

$$\sigma_\theta = 625 + \frac{5625}{r^2} \qquad (b)$$

The stress distributions are shown in Fig. 3-30.

For the case of a *solid disk*, where $r_i = 0$, it can be shown that Eqs. (3-39) and (3-40) reduce to

$$\sigma_r = -p_o \qquad (3\text{-}41)$$

and

$$\sigma_\theta = -p_o \qquad (3\text{-}42)$$

This means that the state of stress is constant throughout the disk.
If the cylinder has closed or capped ends and is under the influence of

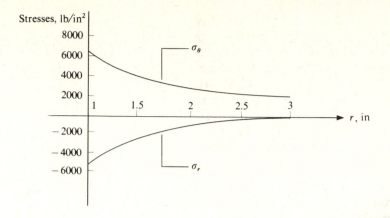

Figure 3-30

internal and external pressures p_i and p_o, respectively, longitudinal stress σ_z in the z direction will develop. Assuming that this stress is distributed uniformly and that p_o acts on the closed ends, we find that the application of the conditions of equilibrium yields†

$$\sigma_z = \frac{p_i r_i^2 - p_o r_o^2}{r_o^2 - r_i^2} \tag{3-43}$$

Press fits The press fit is a classical example of pressurized thick-walled cylinders. The interface pressure between two elements press-fitted together cannot be determined from the equilibrium equations since the problem is statically indeterminate. Thus the solution must be obtained from deflection analysis.

Consider two elements to be pressed together, as shown in Fig. 3-31, where $b_i > b_o$.

The radial displacement is u_r, which (assuming that $\sigma_z = 0$) is given by

$$u_r = \frac{r}{E}(\sigma_\theta - \nu\sigma_r) \tag{3-44}$$

Let the radial deflection u_r for the inner member at $r = b_i$ be $(u_r)_{b_i}$ and for the outer member at $r = b_o$, be $(u_r)_{b_o}$. The reader should verify that

$$(u_r)_{b_o} - (u_r)_{b_i} = b_i - b_o = \delta \tag{3-45}$$

where δ is the *radial* interference. When the two parts are mated, a pressure p develops at the interface. The pressure p corresponds to p_o for the inner element and p_i for the outer element. Thus, for the inner element at $r = b_i$, Eq.

---

† Up to this point of the analysis, determination of σ_r and σ_θ has been on the basis that $\sigma_z = 0$. However, the reader should verify that the effects of σ_z do not influence the analysis if σ_z is constant.

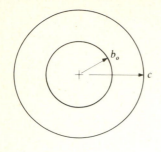

Outer element, material E_o, ν_o Inner element, material E_i, ν_i

Figure 3-31

(3-40) yields

$$(\sigma_\theta)_{bi} = -p\,\frac{b_i^2 + a^2}{b_i^2 - a^2} \tag{3-46}$$

whereas for the outer element at $r = b_o$

$$(\sigma_\theta)_{bo} = p\,\frac{c^2 + b_o^2}{c^2 - b_o^2} \tag{3-47}$$

Since b_i and b_o are almost equal, let $b_o = b_i = b$ [except in Eq. (3.45)]. Then Eqs. (3-46) and (3-47) can be rewritten as

$$(\sigma_\theta)_{bi} = -p\,\frac{b^2 + a^2}{b^2 - a^2} \tag{3-48}$$

and

$$(\sigma_\theta)_{bo} = p\,\frac{c^2 + b^2}{c^2 - b^2} \tag{3-49}$$

The radial stress at the interface of each cylinder is simply the negative of the interface pressure, namely

$$(\sigma_r)_{bi} = (\sigma_r)_{bo} = -p \tag{3-50}$$

Substituting Eqs. (3-48) and (3-50) into Eq. (3-44) yields

$$(u_r)_{bi} = \frac{-b}{E_i}\,p\left(\frac{b^2 + a^2}{b^2 - a^2} - \nu_i\right) \tag{3-51}$$

Likewise, substituting Eqs. (3-49) and (3-50) into Eq. (3-44) yields

$$(u_r)_{bo} = \frac{b}{E_o}\,p\left(\frac{c^2 + b^2}{c^2 - b^2} + \nu_o\right) \tag{3-52}$$

Since $(u_r)_{bo} - (u_r)_{bi} = \delta$, subtracting Eq. (3-51) from Eq. (3-52), setting the result equal to δ, and solving for p yields

$$p = \frac{\delta}{b\left[\dfrac{1}{E_o}\left(\dfrac{c^2 + b^2}{c^2 - b^2} + \nu_o\right) + \dfrac{1}{E_i}\left(\dfrac{b^2 + a^2}{b^2 - a^2} - \nu_i\right)\right]} \tag{3-53}$$

where δ = radial interference of the mating parts, $b_i - b_o$

$\quad\quad b$ = approximate common radius

$\quad\quad c$ = outer radius of outer part

$\quad\quad a$ = inner radius of inner part

E_o, E_i = moduli of elasticity of outer and inner members respectively

ν_o, ν_i = Poisson's ratio of outer and inner members respectively

Example 3-13 An aluminum cylinder ($E_a = 7 \times 10^{10}$ N/m², $\nu_a = 0.33$) with an outer radius of 150 mm and inner radius of 100 mm, is to be press-fitted over a steel cylinder ($E_s = 20 \times 10^{10}$ N/m², $\nu_s = 0.29$) with an outer radius of 100.25 mm and inner radius of 50 mm. Determine the interface pressure p.

SOLUTION

$$\delta = 0.25 \text{ mm} \quad b \approx 100 \text{ mm} \quad c = 150 \text{ mm} \quad a = 50 \text{ mm}$$

$$E_o = 7 \times 10^{10} \text{ N/m}^2 \quad E_i = 20 \times 10^{10} \text{ N/m}^2 \quad \nu_o = 0.33 \quad \nu_i = 0.29$$

Thus, from Eq. (3-53) the interface pressure is

$$p = \frac{0.25}{100\left[\frac{1}{7 \times 10^{10}}\left(\frac{150^2 + 100^2}{150^2 - 100^2} + 0.33\right) + \frac{1}{20 \times 10^{10}}\left(\frac{100^2 + 50^2}{100^2 - 50^2} - 0.29\right)\right]} = 5.13 \times 10^7 \text{ N/m}^2$$

Case 2 ($\omega \neq 0$, $p_i = p_o = 0$)

The solution to Eq. (3-35) is

$$\sigma_r = C_1 + \frac{C_2}{r^2} - \frac{3 + \nu}{8}\rho\omega^2 r^2 \tag{3-54}$$

substituting this into Eq. (3-33) yields

$$\sigma_\theta = C_1 - \frac{C_2}{r^2} - \frac{1 + 3\nu}{8}\rho\omega^2 r^2 \tag{3-55}$$

Applying the conditions that $\sigma_r = 0$ at $r = r_i$ and $r = r_o$ yields

$$\sigma_r = \frac{3 + \nu}{8}\rho\omega^2\left[r_i^2 + r_o^2 - \left(\frac{r_i r_o}{r}\right)^2 - r^2\right] \tag{3-56}$$

$$\sigma_\theta = \frac{3 + \nu}{8}\rho\omega^2\left[r_i^2 + r_o^2 + \left(\frac{r_i r_o}{r}\right)^2 - \frac{1 + 3\nu}{3 + \nu}r^2\right] \tag{3-57}$$

The maximum radial stress $(\sigma_r)_{\text{max}}$ occurs at $r = \sqrt{r_i r_o}$, where the value is

$$(\sigma_r)_{\text{max}} = \frac{3 + \nu}{8}\rho\omega^2(r_o^2 - r_i^2) \tag{3-58}$$

The maximum tangential stress $(\sigma_\theta)_{\text{max}}$ occurs at the inner surface, where $r = r_i$, and is

$$(\sigma_\theta)_{\text{max}} = \frac{\rho\omega^2}{4}[(3 + \nu)r_o^2 + (1 - \nu)r_i^2] \tag{3-59}$$

If the disk is solid, $r_i = 0$ and Eqs. (3-56) and (3-57) reduce to

$$\sigma_r = \frac{3 + \nu}{8} \rho\omega^2(r_o^2 - r^2) \tag{3-60}$$

$$\sigma_\theta = \frac{\rho\omega^2}{8} [(3 + \nu)r_o^2 - (1 + 3\nu)r^2] \tag{3-61}$$

Case 3 ($\omega \neq 0$, $p_i \neq 0$, $p_o \neq 0$)

There are examples, e.g., rotating disks press-fitted onto shafts, where the results of cases 1 and 2 can be combined by superposition of Eqs. (3-39) and (3-40) with Eqs. (3-56) and (3-57), respectively.

Example 3-14 A disk with $r_o = 6.000$ in and $r_i = 1.000$ in is press-fitted onto a shaft of radius 1.003 in. Both members are steel with $E = 30 \times 10^6$ lb/in², $\nu = 0.29$, and with weight densities of 0.28 lb/in³.
    Determine (a) the stress distribution of the disk at $n = 5000$ r/min and (b) the speed for which the interface pressure goes to zero.

SOLUTION (a) First find the interference at 5000 r/min:

$$\omega = \frac{2\pi}{60}(5000) = 523.6 \text{ rad/s}$$

and the mass density is

$$\rho = \frac{0.28}{386} = 7.25 \times 10^{-4} \text{ lb·s}^2/\text{in}^4$$

Because of rotation, the stresses at the outer radius of the shaft are [from Eqs. (3-60) and (3-61)] $\sigma_r = 0$ and

$$\sigma_\theta = \frac{\rho\omega^2}{4}(1 - \nu)r_o^2 = \frac{(7.25 \times 10^{-4})(523.6)^2}{4}(1 - 0.29)(1.003)^2 = 35.5 \text{ lb/in}^2$$

The radial displacement, given by Eq. (3-30), is

$$u_r = \frac{r}{E}(\sigma_\theta - \nu\sigma_r) = \frac{1.003}{30 \times 10^6}(35.5) = 1.19 \times 10^{-6} \text{ in}$$

The radius of the shaft rotating alone is then approximately†

$$(r_o)_{\text{shaft}} = 1.003 \text{ in}$$

Next, the same procedure is applied to the disk at its inner radius. Thus, at $r = r_i$, $\sigma_r = 0$ because of rotation and

$$\sigma_\theta = \frac{3 + \nu}{8}\rho\omega^2 \left[r_i^2 + r_o^2 + \left(\frac{r_i r_o}{r_i}\right)^2 - \frac{1 + 3\nu}{3 + \nu} r_i^2 \right]$$

$$= \frac{3 + 0.29}{8}(7.25 \times 10^{-4})(523.6)^2 \left[1^2 + 6^2 + 6^2 - \frac{1 + (3)(0.29)}{3 + 0.29} 1^2 \right] = 5920 \text{ lb/in}^2$$

† The assumption of a thin disk is actually being violated, but the error here is negligible since the stresses in the shaft are small and will be localized in a zone comparable to the width of the disk.

The radial displacement at the inner radius is

$$u_r = \frac{r\sigma_\theta}{E} = \frac{(1)(5920)}{30 \times 10^6} = 1.97 \times 10^{-4} \text{ in}$$

Thus, the inner radius of the disk due to rotation is

$$(r_i)_{\text{disk}} = 1.000 + 1.97 \times 10^{-4} = 1.000197 \text{ in}$$

Hence, the radial interference at $n = 500$ r/min is

$$\delta = 1.003 - 1.00020 = 0.00280$$

The interface pressure at 5000 r/min is found using Eq. (3-53). Thus at 5000 r/min the pressure is

$$p = \frac{0.00280}{1.0\left[\dfrac{1}{30 \times 10^6}\left(\dfrac{6^2 + 1^2}{6^2 - 1^2} + 0.29\right) + \dfrac{1}{30 \times 10^6}\left(\dfrac{1^2 + 0^2}{1^2 - 0^2} - 0.29\right)\right]} = 40{,}830 \text{ lb/in}^2$$

With the interface pressure, Eqs. (3-39) and (3-40) can be used to find the stresses in the disk due to the interference at 500 r/min. Using $p_i = p$ and $p_o = 0$, we find

$$\sigma_r = \frac{40{,}830}{6^2 - 1^2}\left[1^2 - \left(\frac{6}{r}\right)^2\right] = 1166 - 42{,}000\left(\frac{1}{r}\right)^2 \tag{a}$$

and

$$\sigma_\theta = 1166 + 42{,}000\left(\frac{1}{r}\right)^2 \tag{b}$$

The stresses in the disk due to rotation are given by Eqs. (3-56) and (3-57) and are

$$\sigma_r = \frac{3 + 0.29}{8}(7.25 \times 10^{-4})(523.6)^2\left[1^2 + 6^2 - \left(\frac{6}{r}\right)^2 - r^2\right]$$

$$= 3025 - 2944\left(\frac{1}{r}\right)^2 - 82r^2 \tag{c}$$

and

$$\sigma_\theta = 3025 + 2944\left(\frac{1}{r}\right)^2 - 46r^2 \tag{d}$$

Finally, the stresses of Eqs. (a) and (c) and (b) and (d) are combined, resulting in

$$\sigma_r = 4191 - 44{,}940\left(\frac{1}{r}\right)^2 - 82r^2$$

$$\sigma_\theta = 4191 + 44{,}940\left(\frac{1}{r}\right)^2 - 46r^2$$

(b) The speed for which the interface pressure goes to zero is when $(r_o)_{\text{shaft}} = (r_i)_{\text{disk}}$. This occurs when $(u_r)_{\text{disk}} - (u_r)_{\text{shaft}} = 0.003$ at the interface.

Using Eq. (3-59), we find the tangential stress at the inner radius of the disk:

$$\sigma_\theta = \frac{7.25 \times 10^{-4}}{4}\omega^2[(3 + 0.29)(6^2) + (1 - 0.29)(1^2)] = 2.16 \times 10^{-2}\,\omega^2$$

Thus the radial displacement of the disk at the inner radius (with $\sigma_r = 0$) is

$$(u_r)_{\text{disk}} = \frac{r}{E}\sigma_\theta = \frac{2.16 \times 10^{-2}}{30 \times 10^6}\omega^2$$

The tangential stress at the outer radius of the shaft is found using Eq. (3-61):

$$\sigma_\theta = \frac{7.25 \times 10^{-4}}{8}\omega^2\{(3 + 0.29)(1.003)^2 - [1 + (3)(0.29)](1.003)^2\} = 1.30 \times 10^{-4}\,\omega^2$$

and the radial displacement of the shaft at the outer radius is

$$(u_r)_{\text{shaft}} = \frac{r}{E}\sigma_\theta = \frac{1.30 \times 10^{-4}}{30 \times 10^6}\omega^2$$

Thus

$$\frac{2.16 \times 10^{-2}}{30 \times 10^6}\omega^2 - \frac{1.30 \times 10^{-4}}{30 \times 10^6}\omega^2 = 0.003$$

Solving for ω yields

$$\omega = 2046 \text{ rad/s}$$

and the speed is

$$n = \frac{60}{2\pi}2046 = 19,540 \text{ r/min}$$

The maximum stresses at this speed are determined from Eqs. (3-58) and (3-59) to be $(\sigma_r)_{\text{max}} = 43,710$ lb/in² and $(\sigma_\theta)_{\text{max}} = 90,890$ lb/in².

3-12 TORSION OF THIN-WALLED TUBES

Thin-walled tubes are very efficient when a torsion problem is being considered and weight is a premium. Consider the tube shown in Fig. 3-32a. An element is

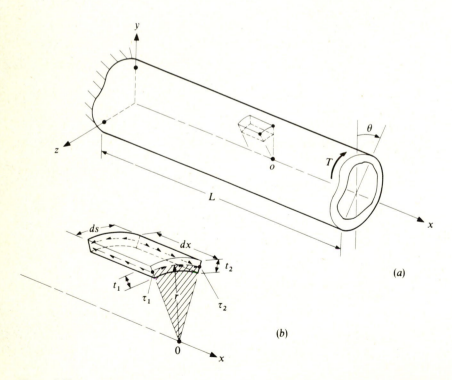

(a)

(b)

Figure 3-32

isolated as shown in Fig. 3-32b, where the possibility of a varying thickness has been assumed. If the forces are summed in the x direction, then for equilibrium, $\tau_2 t_2\, dx - \tau_1 t_1\, dx = 0$, or

$$\tau_2 t_2 = \tau_1 t_1 = \tau t$$

which indicates that at any point on the tube the product τt is constant. The force on the surface whose normal is in the x direction is $\tau t\, ds$. Thus the net torque about point 0 is $r\tau t\, ds$. Integrating this about the total perimeter of the tube yields

$$T = \int r\tau t\, ds$$

Since τt is constant,

$$T = \tau t \int r\, ds$$

The term $r\, ds$ can be seen to be twice the area of the shaded triangle of Fig. 3-32b. Thus the integral $\int r\, ds$ is twice the area enclosed by the *median* line of the tube, called A_m. Thus $T = 2\tau t A_m$ or

$$\tau = \frac{T}{2A_m t} \tag{3-62}$$

To compute the angle of twist, recall from Chap. 2 that for a cylindrical rod

$$\theta = \gamma \frac{L}{r}$$

where γ is the shear strain at a radial position r. When this is applied to the ds element of Fig. 3-32b, the angle the dA_m area rotates is

$$(\theta)_{dA_m} = \frac{2(1+\nu)}{E}\,\tau\frac{L}{r} = \frac{(1+\nu)TL}{EA_m tr}$$

If it is assumed that the average angle of twist of each dA_m section is the angle of twist of the entire section,

$$\theta = \frac{1}{A_m}\int \frac{(1+\nu)TL}{EA_m tr}\, dA_m$$

However, $dA_m = r\, ds/2$, and therefore the above reduces to

$$\theta = \frac{(1+\nu)TL}{2EA_m^2}\int_0^{S_m}\frac{ds}{t} \tag{3-63}$$

where S_m is the mean circumference of the tube.

Example 3-15 Estimate the shear stress and total angle of twist of the thin-walled circular cylinder shown in Fig. 3-33. The steel tube is 0.5 m long and is transmitting a torque of 1000 N·m. The material constants are $E = 20 \times 10^{10}$ N/m² and $\nu = 0.29$.

SOLUTION The mean radius r is constant at $r = (125 - 5)/2 = 60$ mm. Thus

$$A_m = \pi r^2 = \pi(0.060)^2 = 1.131 \times 10^{-2} \text{ m}^2$$

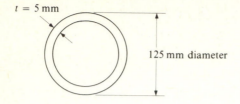

$t = 5\,mm$

125 mm diameter

Figure 3-33

Substituting this into Eq. (3-62) yields

$$\tau = \frac{T}{2A_m t} = \frac{1000}{(2)(0.01131)(0.005)} = 8.84 \times 10^6\,N/m^2$$

Since the thickness t is constant, the angle of twist is

$$\theta = \frac{(1+\nu)TL}{2EA_m^2 t} \int_0^{S_m} ds$$

The integral is simply $\int_0^{S_m} ds = 2\pi r$. Thus

$$\theta = \frac{\pi(1+\nu)TLr}{EA_m^2 t} = \frac{\pi(1+0.29)(1000)(0.5)(0.06)}{(20\times10^{10})[(1.131)^2\times10^{-4}](0.005)} = 9.50\times10^{-4}\,rad$$

The reader is urged to check the solution against that given in elementary strength of materials using the conventional torsion equations.† The exact equation for the shear stress in a circular cylinder undergoing torsion gives a maximum value on the outer radius of $9.19\times10^6\,N/m^2$ and an average value at the mean radius of $8.83\times10^6\,N/m^2$.

Example 3-16 The cross section shown in Fig. 3-34 is subjected to a torque of 50,000 lb·in. Estimate the shear stress in each wall and determine the integral $\int ds/t$ for the cross section.

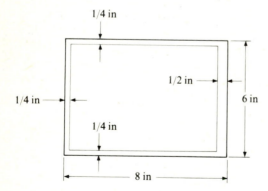

1/4 in

1/2 in

1/4 in

1/4 in

6 in

1/4 in

8 in

Figure 3-34

† Note that in dealing with thin-walled cylinders, if the outer diameter d_o is large and close in size to the inner diameter d_i, the error in calculating the polar moment of inertia may be large if a slide rule is used. This problem is alleviated if $d_o^4 - d_i^4$ is expanded so that

$$J = \frac{\pi}{16} t(d_o + d_i)(d_o^2 + d_i^2)$$

SOLUTION The area enclosed by the wall median line is

$$A_m = (8 - \tfrac{1}{8} - \tfrac{1}{4})(6 - \tfrac{1}{8} - \tfrac{1}{8}) = 43.8 \text{ in}^2$$

The shear stress on the $\tfrac{1}{2}$-in wall is

$$(\tau)_{1/2 \text{ in}} = \frac{T}{2A_m t} = \frac{50,000}{(2)(43.8)(1/2)} = 1140 \text{ lb/in}^2$$

and on the $\tfrac{1}{4}$-in walls

$$(\tau)_{1/4 \text{ in}} = \frac{50,000}{(2)(43.8)(1/4)} = 2280 \text{ lb/in}^2$$

In evaluating the integral $\int ds/t$ note that t is constant for each wall; thus,

$$\int \frac{ds}{t} = \frac{2(8 - \tfrac{1}{8} - \tfrac{1}{4})}{\tfrac{1}{4}} + \frac{6 - \tfrac{1}{8} - \tfrac{1}{8}}{\tfrac{1}{4}} + \frac{6 - \tfrac{1}{8} - \tfrac{1}{8}}{\tfrac{1}{2}} = 95.5$$

3-13 CONTACT STRESSES

When a roller or sphere is in line or point contact with another body, the stresses cannot be determined without considering deflection. In most analyses, deflection does not appreciably alter the results of the analysis of stress. For example, when a beam in bending is considered, the force analysis and subsequent stress analysis are made on the model neglecting the fact that the beam deflects. Consider a load P to be transmitted from the roller in Fig. 3-35 to the flat surface. If deflections are ignored, the transmitted force area is zero and the stresses would be infinite. Hence, the conclusion would be that no force can be transmitted. This is obviously ridiculous and indicates that the deflections of the members must be taken into account.

Contact-stress problems occur often in design where forces are transmitted through contact from one machine element to another, e.g., gearing, roller bearings, cams, and pin joints in linkages (see Fig. 2-5b). There are two major types of contact problems, line contact and point contact. Since the analysis of these problems is rather complex, this section gives only a simple and restrictive introduction to the topic.†

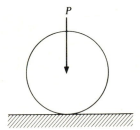

Figure 3-35

---

† See Ref. 2, pp. 342–378; Ref. 4, pp. 163–186; Ref. 5, pp. 513–530; Ref. 6, pp. 163–171.

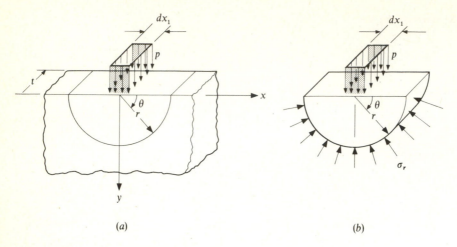

Figure 3-36

Consider a contact pressure p acting over an infinitesimal surface area $t \, dx_1$, as shown in Fig. 3-36a. Isolate a surface of the solid at a distance r, as shown in Fig. 3-36b. Thanks to symmetry, no shear stress exists on this surface. The radial stress distribution is given by†

$$\sigma_r = -\frac{2p \, dx_1}{\pi r} \sin \theta \qquad (3\text{-}64)$$

It can be shown that the vertical component of the force due to σ_r integrated from θ of 0 to π balances the force from $pt \, dx_1$. In addition, further element isolation will uncover the fact that the tangential stress σ_θ is zero at every point except directly under the contact pressure itself. Since in general this is restricted to a very small zone, the tangential stress will be neglected.

The stresses with respect to the xy coordinate system are found using the transformation equations developed in Chap. 2. Thus the radial stress transforms to

$$\sigma_x = \sigma_r \cos^2 \theta = -\frac{2p \, dx_1}{\pi r} \cos^2 \theta \sin \theta \qquad (3\text{-}65a)$$

$$\sigma_y = \sigma_r \sin^2 \theta = -\frac{2p \, dx_1}{\pi r} \sin^3 \theta \qquad (3\text{-}65b)$$

$$\tau_{xy} = \sigma_r \sin \theta \cos \theta = -\frac{2p \, dx_1}{\pi r} \sin^2 \theta \cos \theta \qquad (3\text{-}65c)$$

Using xy coordinates in place of $r\theta$, we can transform Eqs. (3-63) using the

---

† The derivation of Eq. (3-64) is most effectively arrived at by considering stress functions in polar coordinates (see Sec. 6-7).

relations

$$r = \sqrt{x^2 + y^2} \qquad \cos\theta = \frac{x}{\sqrt{x^2 + y^2}} \qquad \sin\theta = \frac{y}{\sqrt{x^2 + y^2}}$$

Thus

$$\sigma_x = -\frac{2}{\pi} p \, dx_1 \frac{x^2 y}{x^2 + y^2} \tag{3-66a}$$

$$\sigma_y = -\frac{2}{\pi} p \, dx_1 \frac{y^3}{x^2 + y^2} \tag{3-66b}$$

$$\tau_{xy} = -\frac{2}{\pi} p \, dx_1 \frac{xy^2}{x^2 + y^2} \tag{3-66c}$$

In order to be able to deal with an arbitrary load distribution across the contact surface, consider the load p to vary as a function of x_1, $p(x_1)$. The force $p(x_1)t \, dx_1$ is shown in Fig. 3-37.

The stresses in this case are found simply by substituting $x - x_1$ for x in Eqs. (3-66). Thus

$$\sigma_x = -\frac{2}{\pi} p \, dx_1 \frac{(x - x_1)^2 y}{[(x - x_1)^2 + y^2]^2} \tag{3-67a}$$

$$\sigma_y = -\frac{2}{\pi} p \, dx_1 \frac{y^3}{[(x - x_1)^2 + y^2]^2} \tag{3-67b}$$

$$\tau_{xy} = -\frac{2}{\pi} p \, dx_1 \frac{(x - x_1)y^2}{[(x - x_1)^2 + y^2]^2} \tag{3-67c}$$

If the load distribution is known, integration of Eqs. (3-67) with respect to x_1 will establish the state of stress at any point (x,y). The integration, however, is not always so simple, and only a uniform load distribution will be presented here.

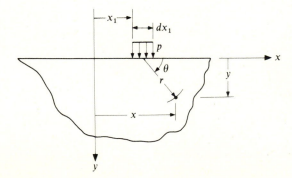

Figure 3-37

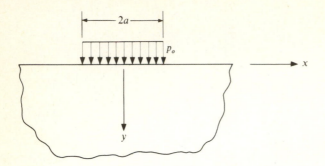

Figure 3-38

Example 3-17 Determine the stress distribution for a constant pressure distribution of $p(x_1) = p_o$ extending from $-a < x < a$, as shown in Fig. 3-38.†

SOLUTION

$$\sigma_x = -\frac{2}{\pi} p_o y \int_{-a}^{a} \frac{(x - x_1)^2 \, dx_1}{[(x - x_1)^2 + y^2]^2} \quad ‡$$

Let $u = x - x_1$; thus $du = -dx_1$ and

$$\sigma_x = +\frac{2}{\pi} p_o y \int_{x+a}^{x-a} \frac{u^2 \, du}{(u^2 + y^2)^2}$$

Integration yields

$$\sigma_x = \frac{2}{\pi} p_o y \left[-\frac{u}{2(u^2 + y^2)} + \frac{1}{2y} \tan^{-1} \frac{u}{y} \right] \Bigg|_{x+a}^{x-a}$$

Substituting the limits of integration and simplifying algebraically yields

$$\sigma_x = -\frac{p_o}{\pi} \left\{ \frac{2ay(x^2 - a^2 - y^2)}{[(x - a)^2 + y^2][(x + a)^2 + y^2]} + \tan^{-1} \frac{x + a}{y} - \tan^{-1} \frac{x - a}{y} \right\}$$

In a similar manner,

$$\sigma_y = \frac{p_o}{\pi} \left\{ \frac{2ay(x^2 - a^2 - y^2)}{[(x - a)^2 + y^2][(x + a)^2 + y^2]} + \tan^{-1} \frac{x - a}{y} - \tan^{-1} \frac{x + a}{y} \right\}$$

and

$$\tau_{xy} = \frac{-(4/\pi)p_o axy^2}{[(x - a)^2 + y^2][(x + a)^2 + y^2]}$$

Whenever two mating parts come in contact with each other, both parts will deform due to the equal and opposite contact force distributions. In the remainder of this section, only the analysis of two cylinders in contact is presented. The results of other cases are given at the end of this section.

Consider the general case of two cylinders of differing radii R_1 and R_2 each made of different materials with moduli of elasticity E_1 and E_2 and Poisson's

† Reference 4 presents this problem using polar coordinates.
‡ Note that x and y are fixed points, and integration is with respect to x_1.

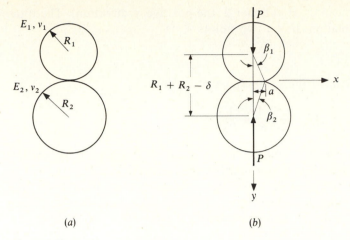

(a)   (b)

Figure 3-39

ratios ν_1 and ν_2, respectively (see Fig. 3-39a). Loads P are placed on the cylinders causing them to press against each other. This results in the deformed configuration shown in Fig. 3-39b.

From geometry, $\sin \beta_1 = a/R_1$, and $\sin \beta_2 = a/R_2$. If β_1 and β_2 are considered to be small, $\sin \beta_1 \approx \beta_1$ and $\sin \beta_2 \approx \beta_2$ and thus

$$\beta_1 \approx \frac{a}{R_1} \qquad \beta_2 \approx \frac{a}{R_2}$$

After the load P is applied, the distance between centers is $R_1 + R_2 - \delta$, which can be written

$$R_1 + R_2 - \delta = R_1 \cos \beta_1 + R_2 \cos \beta_2$$

Again, if β_1 and β_2 are small,

$$\cos \beta_1 \approx 1 - \frac{\beta_1^2}{2} = 1 - \frac{1}{2}\left(\frac{a}{R_1}\right)^2 \qquad \cos \beta_2 \approx 1 - \frac{\beta_2^2}{2} = 1 - \frac{1}{2}\left(\frac{a}{R_2}\right)^2$$

Thus
$$R_1 + R_2 - \delta \approx R_1\left[1 - \frac{1}{2}\left(\frac{a}{R_1}\right)^2\right] + R_2\left[1 - \frac{1}{2}\left(\frac{a}{R_2}\right)^2\right]$$

$$\delta \approx \frac{a^2}{2}\left(\frac{1}{R_1} + \frac{1}{R_2}\right) \tag{3-68}$$

The term δ is the relative displacement of the center distance of the cylinders, and for other points within the cylinders δ is not the magnitude of relative elastic deformation. Since deformation will be primarily confined near the y axis, consider two points (one on each cylinder) located between the centers of the cylinders and in close proximity to the y axis. Let v_1 be the elastic displacement of the point in cylinder 1 toward the center of cylinder 1 (the negative y direction). Likewise, let v_2 be the elastic displacement of the point in cylinder 2 toward the center of cylinder 2 (the positive y direction). The total relative elastic deformation of the two points δ_e is

$$\delta_e = \delta - v_1 - v_2$$

where points 1 and 2 have the same value of x. Thus

$$\delta_e = \frac{a^2}{2}\left(\frac{1}{R_1}+\frac{1}{R_2}\right) - v_1 - v_2 \tag{3-69}$$

The elastic displacement v_1 and v_2 are functions of x and y. However, for Eq. (3-67) these displacements are primarily constrained to $-a < x < a$, where a is relatively small. If we consider a point in either body at position $x = a$ and y at some arbitrary point h_i (which is normally much greater than a), then $v_1 = v_1(a,h_1)$ and $v_2 = v_2(a,h_2)$. The rate of approach of the two points should be stationary with respect to a. Therefore

$$\frac{\partial}{\partial a}\,\delta_e = a\left(\frac{1}{R_1}+\frac{1}{R_2}\right) - \frac{\partial}{\partial a}\,(v_1 + v_2) = 0 \tag{3-70}$$

Thus it is necessary to determine the rate of change of the vertical displacements with respect to a, considering h_1 and h_2 arbitrary but large compared with a. If the contact-pressure distribution were known, the displacements could be arrived at by first determining the stress distribution as in Example 3-15, using the strain-stress relations to obtain the strain in the y direction, and finally, integrating the strain to determine the displacement v.

The pressure distribution is not known and is rather complicated to develop. On a physical level, it appears reasonable that the pressure distribution will be elliptical like that shown in Fig. 3-40. Thus assume that

$$p = \frac{p_o}{a}\sqrt{a^2 - x_1^2} \tag{3-71}$$

where p_o is the maximum pressure at $x_1 = y = 0$.

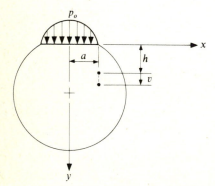

Figure 3-40

Since obtaining the stress distribution resulting from this load distribution by the method used in Example 3-17 is too complicated, an alternative scheme is presented.

The displacement v can be determined considering the infinitesimal load in Fig. 3-36. Recall the normal stresses from Eqs. (3-66); substitution into the strain-stress relationship for ε_y yields

$$\varepsilon_y = \frac{1}{E}(\sigma_y - \nu\sigma_x) = -\frac{2p}{\pi E}\,dx_1\left[\frac{y^3}{(x^2+y^2)^2} - \nu\frac{x^2 y}{(x^2+y^2)^2}\right]$$

Now, if x is kept constant ($x = a$), the vertical displacement of a point at $y = h$ ($h \gg a$) can be found by integrating ε_y from $y = 0$ to $y = h$. Thus

$$v = -\frac{2p}{\pi E}\,dx_1\left[\int_0^h \frac{y^3\,dy}{(a^2+y^2)^2} - \nu a^2\int_0^h \frac{y\,dy}{(a^2+y^2)^2}\right]$$

Integration yields

$$v = -\frac{2p}{\pi E}\,dx_1\left[\ln\sqrt{\frac{a^2+h^2}{a^2}} - \frac{h^2}{2(a^2+h^2)} - \frac{\nu}{2}\frac{h^2}{a^2+h^2}\right]$$

For $h \gg a$, the above is approximated by

$$v \approx -\frac{2p}{\pi E}\,dx_1\left[\ln\frac{h}{a} - \frac{1}{2}(1+\nu)\right]$$

The rate of change of v with respect to a is

$$\frac{\partial v}{\partial a} \approx \frac{2p}{\pi E a}\,dx_1 \tag{3-72}$$

Now for a distributed pressure $p(x_1)$ recall that $x - x_1$ is substituted for x (see Fig. 3-37). Thus substitute $a - x_1$ for a in Eq. (3-72) and integrate. This results in

$$\frac{\partial v}{\partial a} = \frac{2}{\pi E}\int_{-a}^a \frac{p(x_1)\,dx_1}{a - x_1} \tag{3-73}$$

Substituting Eq. (3-71) into (3-73) yields

$$\frac{\partial v}{\partial a} = \frac{2p_o}{\pi E a}\int_{-a}^a \frac{\sqrt{a^2-x_1^2}}{a - x_1}\,dx_1$$

Multiplying and dividing the term in the integral by $\sqrt{a^2-x_1^2}$ and simplifying gives

$$\frac{\partial v}{\partial a} = \frac{2p_o}{\pi E a}\left(a\int_{-a}^a \frac{dx_1}{\sqrt{a^2-x_1^2}} + \int_{-a}^a \frac{x_1}{\sqrt{a^2-x_1^2}}\,dx_1\right)$$

The first integral can be shown to reduce to

$$\int_{-a}^a \frac{dx_1}{\sqrt{a^2-x_1^2}} = \pi$$

and the second integral reduces to zero. Thus

$$\frac{\partial v}{\partial a} = 2\frac{p_o}{E}$$

Therefore, for cylinders 1 and 2

$$\frac{\partial v_1}{\partial a} = 2\frac{p_o}{E_1} \qquad \frac{\partial v_2}{\partial a} = 2\frac{p_o}{E_2}$$

Substituting this into Eq. (3-70) yields

$$a\left(\frac{1}{R_1} + \frac{1}{R_2}\right) - 2p_o\left(\frac{1}{E_1} + \frac{1}{E_2}\right) = 0$$

Solving for a results in

$$a = \frac{2p_o R_1 R_2 (E_1 + E_2)}{E_1 E_2 (R_1 + R_2)} \tag{3-74}$$

The maximum pressure p_o is found by integrating the pressure distribution (given by Eq. 3-71) over the contact area and equating this to the net force P:

$$P = \int_{-a}^{a} \frac{p_o}{a} \sqrt{a^2 - x_1^2}\, t\, dx_1 = \frac{\pi}{2} p_o a t$$

Therefore, the maximum pressure is given by

$$p_o = \frac{2}{\pi}\frac{P}{at}$$

When this is substituted into Eq. (3-74), the half width of the contact zone becomes

$$a = 2\sqrt{\frac{P}{\pi t}\frac{R_1 R_2 (E_1 + E_2)}{E_1 E_2 (R_1 + R_2)}} \tag{3-75}$$

and the maximum pressure is

$$p_o = \sqrt{\frac{P}{\pi t}\frac{E_1 E_2 (R_1 + R_2)}{R_1 R_2 (E_1 + E_2)}} \tag{3-76}$$

The analysis is based on the assumption that the thickness t is small compared with the radii such that $\sigma_z \approx 0$. If t is large, replace E by $E/(1 - \nu^2)$. Recall that this was done in Sec. 3-1 for wide beams. Thus, if t is large,

$$a = 2\sqrt{\frac{P}{\pi t}\frac{R_1 R_2[(1 - \nu_2^2)E_1 + (1 - \nu_1^2)E_2]}{E_1 E_2 (R_1 + R_2)}} \tag{3-77}$$

$$p_o = \sqrt{\frac{P}{\pi t}\frac{E_1 E_2 (R_1 + R_2)}{R_1 R_2[(1 - \nu_2^2)E_1 + (1 - \nu_2^2)E_2]}} \tag{3-78}$$

For most metals, Poisson's ratio is about 0.3. Substituting $\nu_1 = \nu_2 = 0.3$ into

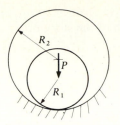

Figure 3-41

Eqs. (3-77) and (3-78) yields

$$a = 1.076\sqrt{\frac{P}{t}\frac{R_1R_2(E_1+E_2)}{E_1E_2(R_1+R_2)}} \tag{3-79}$$

and

$$p_o = 0.591\sqrt{\frac{P}{t}\frac{E_1E_2(R_1+R_2)}{R_1R_2(E_1+E_2)}} \tag{3-80}$$

If cylinder 1 is a pin and cylinder 2 is *concave* of radius R_2, as shown in Fig. 3-41, substitute $-R_2$ for R_2 in the preceding equations. If cylinder 2 is a flat surface, $R_2 = \infty$ and the equations must be rewritten in a different form.†

The analysis for spheres in contact is similar to the analysis for cylinders except that the radial stress distribution is spherical instead of cylindrical and the stresses are three-dimensional. The results are

$$a = 0.909\sqrt[3]{P\frac{R_1R_2[(1-\nu_2^2)E_1+(1-\nu_1^2)E_2]}{E_1E_2(R_1+R_2)}} \tag{3-81}$$

and

$$p_{max} = 0.580\sqrt[3]{P\frac{E_1E_2(R_1+R_2)}{R_1R_2[(1-\nu_2^2)E_1+(1-\nu_1^2)E_2]}} \tag{3-82}$$

where a is the radius of the contact zone. If $\nu_1 = \nu_2 = 0.3$,

$$a = 0.881\sqrt[3]{P\frac{R_1R_2(E_1+E_2)}{E_1E_2(R_1+R_2)}} \tag{3-83}$$

$$p_{max} = 0.618\sqrt[3]{P\left[\frac{E_1E_2(R_1+R_2)}{R_1R_2(E_1+E_2)}\right]^2} \tag{3-84}$$

The state of stress within the elements is triaxial, and material failures normally start below the contact surface because most materials fail by distortion or a form of shear stress under the influence of a primarily

† Note that

$$\frac{R_1R_2}{R_1+R_2} = \frac{R_1}{R_1/R_2+1}$$

Thus, if $R_2 = \infty$, $R_1R_2/(R_1+R_2) = R_1$.

compressive state of stress.† For example, for cylinders or spheres in contact the maximum shear stress is approximately $0.3p_{max}$ at a depth of the same order of magnitude as the contact dimension a.

3-14 STRESS CONCENTRATIONS

It was demonstrated in Sec. 2-1 that where sudden changes in cross section occur, the stress distribution will have large gradients at localized points. A classical example of this is a plate loaded in tension which contains a centrally located hole (see Fig. 3-42a). In Chap. 2, a rubber model was used to demonstrate the nonuniform behavior of the tensile stress at a transverse section taken through the plate intersecting the center of the hole. A photoelastic‡ part of thickness $t = 0.125$ in, $w = 1.50$ in, $d = 0.50$ in, and $P = 60$ lb was tested, and the photoelastic fringes are as shown in Fig. 3-42b. In zone A-A, the behavior is uniform, as might be expected, whereas in zone B-B, the behavior is in no way uniform. From the photoelastic analysis, the stresses at points a, b, and c were found to be

zone A-A: $\qquad\qquad\qquad \sigma_a = 320$ lb/in²

zone B-B: $\qquad\quad \sigma_b = 320$ lb/in² $\qquad \sigma_c = 1130$ lb/in²

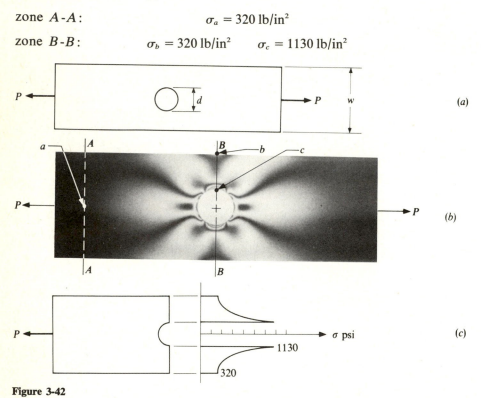

Figure 3-42

† See Chap. 5.

‡ Photoelasticity is discussed at some length in Chap. 7.

Analytical approaches to stress-concentration problems are beyond the scope of this text, but the literature provides the results of analytical and experimental investigations of standard types of problems (see Refs. 7 and 8). The results are often given in graphical form and use the *stress concentration factor K_t*, defined as

$$K_t = \frac{\sigma_{max}}{\sigma_{nom}} \tag{3-85}$$

where σ_{nom} is the nominal stress determined from the simple strength of materials equations. Generally, the nominal stress is determined by using the net cross section, e.g., section *B-B* in the previous example. For the example under discussion the nominal stress in section *B-B* is 480 lb/in². Thus, the stress-concentration factor is

$$K_t = 1130/480 = 2.35$$

The stress-concentration factor for a plate containing a centrally located hole in which the plate is loaded in tension depends on the ratio of d/w. Figure 3-43 illustrates the dependence of K_t on the geometry of the plate (see Ref. 8, pt. 5).

Other charts are available for various geometric and loading configurations. Appendix G gives a few charts that apply to the fundamental forms of geometry and loading conditions.

If a part is made of a ductile material and it is loaded by gradually applied static loading, the stress-concentration factor is seldom used in design equations because the zone where the stress is above the nominal value is generally very small and permanent deformation in this small zone normally can be tolerated. If yielding does occur, the stress distribution changes and tends toward a uniform stress distribution (see Sec. 3-15). In the area where yielding occurs there is no danger of fracture for a ductile material, but if the possibility of a brittle fracture exists, the stress-concentration factor should be considered. In addition, materials often thought of as being ductile can fail in a brittle manner under certain conditions, e.g., cyclic loading, rapid application of static

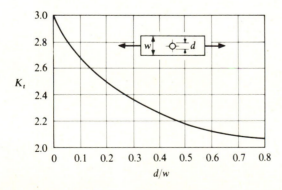

Figure 3-43

loads, and loading at low temperatures. The effects on a ductile material due to hardening, hydrogen embrittlement, and welding may also accelerate this type of failure. Thus care should be exercised when using the nominal stress in design.

In practice, for general configurations, the stress-concentration factor is determined experimentally. A method commonly used is photoelasticity. However, intuitive methods such as the *flow analogy* are often helpful to the designer faced with the task of reducing stress concentrations.

When dealing with the situation where it is absolutely necessary to reduce the cross section abruptly, the resulting stress concentration can often be minimized by a further reduction of material. This is completely contrary to the common advice "if it is not strong enough, make it bigger." To explain this, a brief explanation of the flow analogy is given.

There is a similarity between the velocity of the flow of an ideal fluid in a channel and the flow of stresses in an element with the same geometry as the channel. For example, considering a uniform flow in the channel of Fig. 3-44, the flow velocity v everywhere is constant and uniform. Likewise, the stresses in the uniformly loaded plate of Fig. 3-44 are constant and uniform. Note that σ is analogous to v and the boundary of the channel is analogous to the boundary of the loaded part. Consider the problem of a plate uniformly loaded at the ends but containing a centrally located hole. The flow analogy would be as shown in Fig. 3-45a, where the hole in the plate is represented by a solid circular boundary in the center of the channel.

Along section A-A the flow is uniform and constant, but as the particles in each streamline approach section B-B, the path adjusts to move around the

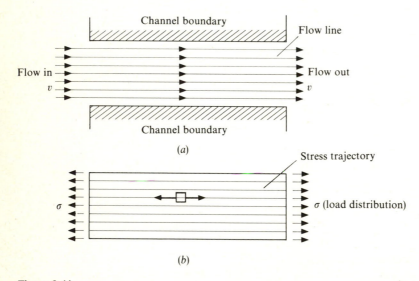

(a)

(b)

Figure 3-44

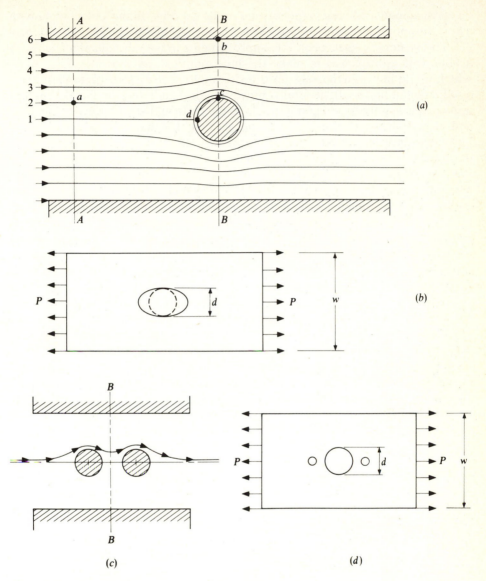

Figure 3-45

obstacle. In order to do this, particles on streamlines close to streamline 1 make the greatest adjustment and begin to accelerate until they reach section B-B, where they then begin deceleration. Thus, at section B-B maximum velocity is reached at point c. The velocity at point c is maximum since particles along streamline 1 have undergone the maximum acceleration. Note that when particles along this streamline reach point d, the particles theoretically take on an instantaneous velocity perpendicular to the net flow. In the loaded plate with a hole, this point has a compressive stress perpendicular to the load direction.

This analogy suggests possible improvements in the example of a plate with a hole. One approach is to make the hole elliptical, as shown in Fig. 3-45*b*, which would improve the velocity transition into section *B*-*B* (note that this is a reduction of material); generally, however, this is not very practical. Another approach is simply to add another hole to the plate in line with the original hole and close to it, as shown in Fig. 3-45*c*. At first this makes no sense to some designers since it is a reduction of more material and if one hole weakens the part, obviously two holes will make things worse. One thing to keep in mind is that the first hole increased stress in two ways: (1) by reducing the cross section and (2) by changing the stress distribution, which was the worse offender. Addition of a second hole does not change the reduction of cross section unless the hole is larger than the first one. The second hole improves the flow transition, which will reduce the stress concentration. There is an optimum spacing between holes for the greatest reduction of the stress concentration. Another method is to add two smaller holes on both sides of the original hole, as shown in Fig. 3-45*d*. This approximates the behavior of an eliptical hole. A third method is to replace the hole with a slot of width *d*.†

Some other examples of situations where high stress concentrations occur and possible methods of improvements are given in Fig. 3-46. Note that in each case improvement is made by reducing the amount of material.

This is not a hard-and-fast rule, however; most reductions in high stress concentrations are made by removing material in adjacent areas, where the stresses are minimum. This causes the maximum stresses to shift toward this area, thereby increasing the stress in previously low-stressed areas and decreasing the stress in previously high-stressed areas.

3-15 PLASTIC BEHAVIOR

There are many situations in design where plastic behavior is tolerated. In fact, for structures that are highly statically indeterminate, some standard analyses actually allow the maximum stresses to exceed the elastic limit, and the designer may not be aware of this (see Sec. 5-4). As stated in Sec. 3-14, it is sometimes permissible to tolerate plastic behavior in small localized areas undergoing high stresses. If plastic behavior is initiated, the stress distribution will change since the plastic area no longer acts according to Hooke's law.

For mathematical description, curve fitting can be employed to model the stress-strain behavior of a given material. Some examples are shown in Fig. 3-47. The limiting stress value for elastic behavior is σ_e. This section will deal only with an elastic perfectly plastic (EPP) material, which corresponds to most structural steels.

---

† For a comparison of the four suggested improvements, see Ref. 7.

When the stress in an element within a part made of an EPP material reaches the elastic limit, the stress in the element will remain constant as the loading of the structure increases. The stress then is in the plastic region, and the element will fracture when $\varepsilon = \varepsilon_f$. The element retains its usefulness in the plastic region; i.e., it still is carrying part of the load on the structure. However, in the plastic range the element cannot carry any stress greater than σ_e. This means that additional loads on the structure must be carried by other elements of the structure. As an example, consider a plate containing a hole where the

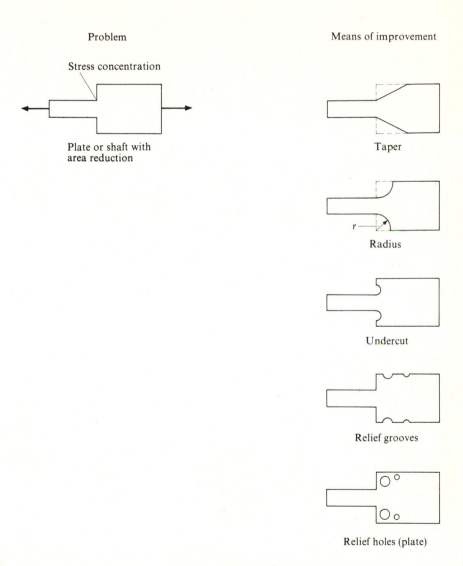

Figure 3-46 (*continued on next page*)

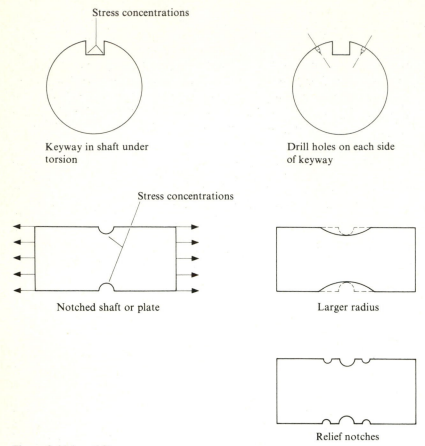

Stress concentrations

Keyway in shaft under torsion

Drill holes on each side of keyway

Stress concentrations

Notched shaft or plate

Larger radius

Relief notches

Figure 3-46 (*cont'd.*)

plate is loaded in pure tension. As shown in Sec. 3-14, the stress at the edge of the hole is maximum. If the load is increased until the stress at this point reaches σ_e, the resulting stress distribution is the same shape as that shown in Fig. 3-42c (repeated in Fig. 3-48a). Let P_e represent the tensile force on the plate when this occurs. The load can be increased beyond P_e; however, the stress at the edge of the hole will no longer increase if the material is EPP. The stress increases at other points, however, and elements near the edge of the hole increase to σ_e, as shown in Fig. 3-48b. This process continues until the strain at the edge of the hole reaches ε_f, in which case a fracture will start. A theoretical limit is the point where the entire structure reaches a fully plastic condition, as shown in Fig. 3-48c, where the entire cross section has reached σ_e. The load in which the entire section becomes completely plastic is called P_p.†

† This limit is not really possible, as at this point the strain goes to infinity. Thus the true limit would be slightly less than P_p.

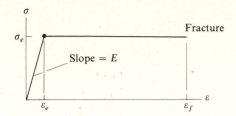

(a) Elastic—perfectly plastic (E—PP)

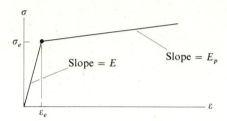

(b) Elastic—linear plastic (E—LP)

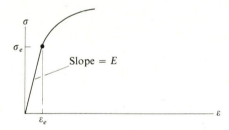

(c) Elastic—plastic (general)

Figure 3-47

This load can be calculated easily since at this limit the stress is uniform and the equation $\sigma = P/A$ can be used. Thus the limit load is

$$P_p = \sigma_e A = \sigma_e(w - d)t \tag{3-86}$$

where t is the plate thickness.†

If the stress concentration at the hole is 2.5 for example, $P_p = 2.5P_e$. That is, the maximum load carrying capacity of the plate is 2.5 times greater than the load for which plastic behavior starts. This is quite an increase in usefulness of the plate if inelastic behavior can be tolerated in the design. As mentioned earlier, there are cases where plastic behavior is not desired, such as cyclic

† Reference 2, pp. 517–522 presents an approximate method for the stress distribution when $P_e < P < P_p$.

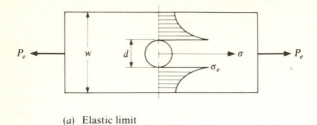

(a) Elastic limit

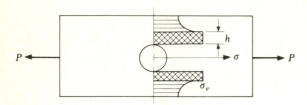

(b) $P > P_e$ h = plastic zone

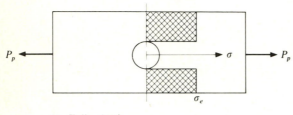

(c) Fully plastic

Figure 3-48

loading, brittle materials, etc. Also, once the material is loaded beyond the yield point, there will be permanent deformation and possibly residual stresses after the part is unloaded.

As another example of plastic behavior, consider a rectangular beam in pure bending, as shown in Fig. 3-49a. The bending equation (3-1) is valid until the maximum stress reaches σ_e (see Fig. 3-49b), where

$$\sigma_e = \frac{M_e c}{\frac{1}{12}b(2c)^3}$$

or
$$M_e = \tfrac{2}{3}bc^2\sigma_e \tag{3-87}$$

M_e is the maximum moment in which the entire section of the beam behaves in the elastic range. If the moment is increased beyond M_e, plastic behavior starts, as shown in Fig. 3-49c. The depth of the plastic zone h_p will increase as M increases until theoretically $h_p = c$ (see Fig. 3-49d). At this point, the entire

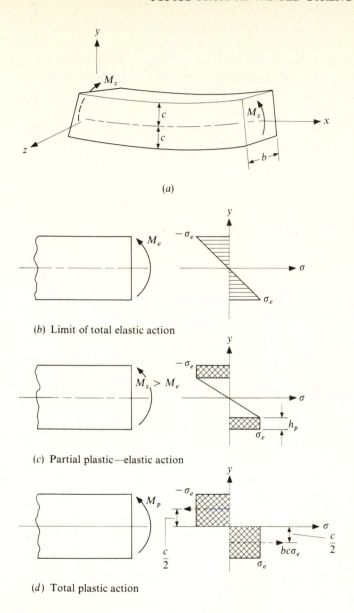

(a)

(b) Limit of total elastic action

(c) Partial plastic—elastic action

(d) Total plastic action

Figure 3-49

section is plastic, and the section cannot accommodate any further increase in load. The moment at this point is thus the limit moment M_p.†

The limit moment is quite simple to determine. Since each distribution is constant, the equivalent forces are as shown in Fig. 3-49d. The moment due to these forces is

$$M_p = \sigma_e bc^2 \tag{3-88}$$

Shape factor The ratio of M_p/M_e is called the *shape factor* (SF) of the cross section. Thus, for a rectangular beam the shape factor is

$$SF = \frac{\sigma_e bc^2}{\frac{2}{3}\sigma_e bc^2} = 1.5 \tag{3-89}$$

The shape factor in bending indicates how effective the cross section is in terms of the elastic limit of the beam; i.e., with large shape factors much of the material is considerably below the elastic limit when the outer fibers reach σ_e. Thus much of the section is not being effectively utilized. If the shape factor is close to unity, a good deal of the material is close to σ_e when the outer fibers reach σ_e, and hence the section is very effective in bending. To illustrate this, the shape factors of three sections are given in Table 3-2.

Table 3-2

Cross section	SF
Rectangular	1.5
Circular	1.7
16 WF 36 I beam	1.13

Material handbooks give information which enables the designer to determine the limit moment of a particular section. A common way of expressing the maximum bending stress is $\sigma_{max} = Mc/I$, where c is y_{max}. The term I/c is called the *section modulus* Z. Thus, $\sigma_{max} = M/Z$. Material handbooks tabulate values of Z for various standard cross sections. In addition, the "plastic" section modulus, Z_p is also normally tabulated. If $\sigma_{max} = \sigma_e$, the elastic moment is simply $M_e = \sigma_e Z$ and the limit moment is $M_p = \sigma_e Z_p$. For example, for a rectangular cross section, $Z = \frac{2}{3}bc^2$ and $Z_p = bc^2$.

Shifting of the neutral axis In an elastic analysis if the maximum tensile stress does not equal the maximum compressive stress, the neutral axis will shift when plastic behavior occurs. Examples of this are axially loaded symmetric cross sections in bending or beams in bending where the section is asymmetric about the z axis. For example, consider the section in Fig. 3-50a. The centroidal axis is shown for the section, and the moment of inertia can be shown to be

---

† Again, this limit corresponds to infinite strains and will not be reached. Thus, the true limit is slightly less than M_p.

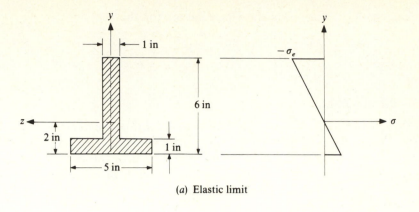

(a) Elastic limit

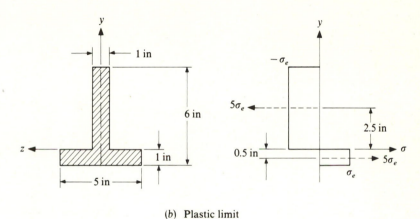

(b) Plastic limit

Figure 3-50

$I_z = 33.33$ in⁴. The maximum value for y is 4 in. Thus when the maximum stress reaches the elastic limit,

$$\sigma_e = \frac{M_e(4)}{33.33} \quad \text{or} \quad M_e = 8.33\sigma_e$$

If the section becomes totally plastic, where $M_z = M_p$, then if the neutral axis remains the centroidal axis, the net force above the centroidal axis is $4\sigma_e$ whereas the net force below the axis is $6\sigma_e$. Thus there is an unbalance of forces along the beam axis. To balance the net force, the neutral axis of the section moves toward an axis where the *areas* balance. For the example under discussion, when $M_z = M_p$, the neutral axis moves downward to the top of the flange, as shown in Fig. 3-50b. This is where the areas are equal above and below the neutral axis.

The net moment on the section due to the stress distribution is $15\sigma_e$, which is the limit moment M_p. This yields the shape factor for the section:

$$\text{SF} = \frac{M_p}{M_e} = \frac{15\sigma_e}{8.33\sigma_e} = 1.8$$

Depth of the Plastic Zone

For simple stress distributions, it is relatively easy to determine the depth of the plastic zone as a function of the bending moment. With this, one can also determine either the residual stresses or the permanent set of the beam once the load is released. Returning to the rectangular beam shown in Fig. 3-49c, we see that the stress distribution for positive values of y is

$$\sigma_x = \begin{cases} -\dfrac{\sigma_e}{c-h_p}\, y & 0 < y < c - h_p \\[2ex] -\sigma_e & c - h_p < y < c \end{cases} \tag{3-90}$$

The moment due to this stress distribution will contribute to one-half of the total moment acting on the section. Thus,

$$M_z = -2 \int_0^c y\sigma_x \, dA$$

where the negative sign is necessary due to the sign convention. Substituting the stress distribution with $dA = b\,dy$ yields

$$M_z = -2\left[\int_0^{c-h_p}\left(-\frac{\sigma_e}{c-h_p}\,y\right) yb\,dy + \int_{c-h_p}^c -\sigma_e yb\,dy\right]$$

Integration and simplification yield

$$M_z = \tfrac{2}{3}bc^2\sigma_e\left[1 + \frac{h_p}{c} - \frac{1}{2}\left(\frac{h_p}{c}\right)^2\right]$$

Since $I_z = \tfrac{2}{3}bc^3$

$$M_z = \frac{\sigma_e I_z}{c}\left[1 + \frac{h_p}{c} - \frac{1}{2}\left(\frac{h_p}{c}\right)^2\right] \tag{3-91}$$

Note that when $h_p = 0$, $M_z = \sigma_e I_z/c$, or in other words $M_z = M_e$. Also, when $h_p = c$, $M_z = \tfrac{3}{2}\sigma_e I_z/c$, or in other words $M_z = M_p$.

If M_z is known and $M_e < M_z < M_p$, the depth of the plastic zone can be found by rewriting Eq. (3-91) as

$$\left(\frac{h_p}{c}\right)^2 - 2\frac{h_p}{c} + 2\left(\frac{M_z c}{\sigma_e I_z} - 1\right) = 0 \tag{3-92}$$

which is a quadratic equation in h_p/c. The one root of this equation which is acceptable is

$$\frac{h_p}{c} = 1 - \sqrt{3 - \frac{2M_z c}{\sigma_e I_z}} \tag{3-93}$$

Since $M_e = \pm \sigma_e I_z / c$, Eq. (3-93) can be written

$$\frac{h_p}{c} = 1 - \sqrt{3 - 2 \left| \frac{M_z}{M_e} \right|}$$

(3-94)

Once h_p is established, the maximum strain for the particular value of M_z can be determined. The maximum strain occurs at the outer fibers of the beam, but Hooke's law cannot be applied at this point since the outer fibers are in the plastic region. For $|y| < c - h_p$, the beam is still elastic, where

$$\sigma_x = - \frac{\sigma_e}{c - h_p} y$$

Since Hooke's law is valid in this zone,

$$\varepsilon_x = \frac{\sigma_x}{E} = - \frac{\sigma_e}{E(c - h_p)} y \qquad 0 \le |y| < c - h_p$$

If it is still assumed that plane surfaces perpendicular to the centroidal axis remain plane up to this point, the linear strain distribution is valid for $|y| > c - h_p$. Thus, the strain across the entire surface is

$$\varepsilon_x = - \frac{\sigma_e}{E(c - h_p)} y \qquad 0 \le |y| \le c$$

(3-95)

The maximum strain occurs at $y = \pm c$, and at a magnitude of

$$\varepsilon_{max} = \frac{\sigma_e c}{E(c - h_p)}$$

(3-96)

Residual Stresses

If the load is released after plastic deformation takes place, residual stresses will develop as well as residual strains and deformations. Assume that a rectangular beam has been loaded so that $M_e < M_z < M_p$. The depth of the plastic zone h_p can be determined from Eq. (3-93). The strain in the plastic zone can be calculated from Eq. (3-95). Strains greater than ε_e, however, are in the plastic zone, as shown in Fig. 3-51. As the beam is unloaded, points on the beam

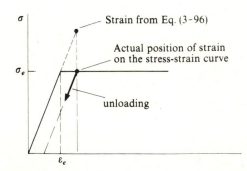

Strain from Eq. (3-96)

Actual position of strain on the stress-strain curve

unloading

Figure 3-51

that are in the elastic zone *tend* to return on the linear part of the $\sigma\varepsilon$ curve. However, points on the beam that are in the plastic zone *tend* to return along a line parallel to the linear curve from the initial position (see Fig. 3-51).

Thus, the unloading for either point tends to follow a linear path. This is equivalent to superimposing a linear unloading stress distribution over the initial stress distribution, as shown in Fig. 3-52.

The moment from the stress distribution, $\sigma = ky$, must cancel the initial moment since the net moment on the surface after unloading is zero. Since the loading moment is M_z, a linear stress distribution for $-M_z$ is

$$\sigma = \frac{M_z y}{I_z}$$

Therefore $k = M_z/I_z$.

Adding the distribution of $\sigma = ky$ to that in Eq. (3-90) yields the residual stress (see Fig. 3-52). For positive values of y

$$\sigma_{res} = \begin{cases} \left(\dfrac{M_z}{I_z} - \dfrac{\sigma_e}{c - h_p}\right)y & 0 \le y \le c - h_p & (3\text{-}97a) \\[2em] \dfrac{M_z}{I_z}y - \sigma_e & c - h_p \le y \le c & (3\text{-}97b) \end{cases}$$

The maximum residual stress can occur at either $y = c$ or $y = c - h_p$. A plot of the residual stresses at these points as a function of h_p/c is given in Fig. 3-53. For values of h_p/c less than 0.585, the maximum residual stress occurs at $|y| = c$, and for h_p/c greater than 0.585 the maximum stress occurs at $|y| = c - h_p$.

The *residual strains* are determined in a similar fashion. The initial strain is given by Eq. (3-95). Since unloading is assumed to be linear, the unloading strain is

$$\varepsilon_x = \frac{ky}{E} = \frac{M_z y}{EI_z}$$

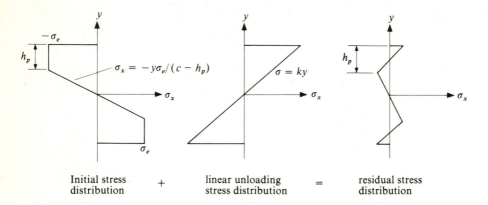

| Initial stress distribution | + | linear unloading stress distribution | = | residual stress distribution |

Figure 3-52

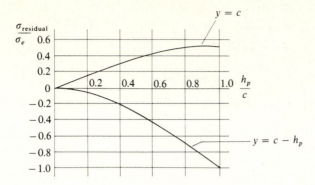

Figure 3-53

Adding this to the initial strain results in

$$\varepsilon_{\text{res}} = \frac{1}{E}\left(\frac{M_z}{I_z} - \frac{\sigma_e}{c - h_p}\right)y \qquad (3\text{-}98)$$

provided that $M_z > M_e$.

Recall that h_p is given by Eq. (3-94); then Eq. (3-98) can be written explicitly as a function of M_z, giving

$$\varepsilon_{\text{res}} = \frac{1}{E}\left(\frac{M_z}{I_z} - \frac{\sigma_e}{c\sqrt{3 - 2|M_z/M_e|}}\right)y \qquad (3\text{-}99)$$

The permanent deflection of the beam is obtained from the deflection equation, which can be written as

$$\frac{d^2 y_c}{dx^2} = -\frac{\varepsilon_{\text{res}}}{y} \qquad (3\text{-}100)$$

where y_c would be the permanent deflection of the centroidal axis. However, for most cases solutions are difficult to obtain; since ε_{res} is normally discontinuous, and since M_z is normally a function of x, integration of ε_{res} is rather tedious. Energy techniques are much more suited to this type of problem (see Sec. 4-11). A simple example will be given here where ε_{res} is not discontinuous and M_z is not a function of x.

Example 3-18 The steel cantilever beam shown in Fig. 3-54 is loaded with a pure moment. If $M = 1.2M_e$ and $\sigma_e = 1.4 \times 10^8$ N/m^2, determine the permanent displacement at the end after the load is released. $E_s = 20.5 \times 10^{10}$ N/m^2.

SOLUTION Since the bending moment is a constant function of x, the residual strains are constant. From Eq. (3-94)

$$\frac{h_p}{c} = 1 - \sqrt{3 - \frac{(2)(1.2M_e)}{M_e}} = 0.225$$

From Eq. (3-98), substitution of $M_z = 1.2M_e$, using the fact that $M_e/I_z = \sigma_e/c$, results in

$$\varepsilon_{\text{res}} = \frac{\sigma_e}{E}\left(\frac{1.2}{c} - \frac{1}{c - h_p}\right)y$$

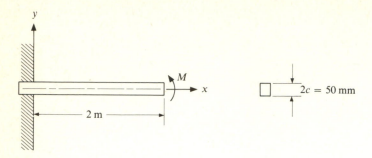

Figure 3-54

Simplifying and substituting the values of h_p, c, σ_e, and E yield

$$\varepsilon_{res} = -2.467 \times 10^{-3} \, y \text{ m/m}$$

Substituting this into Eq. (3-100) results in

$$\frac{d^2 y_c}{dx^2} = 2.467 \times 10^{-3}$$

Integrating this twice gives

$$y_c = 1.233 \times 10^{-3} \, x^2 + C_1 x + C_2$$

Imposing the boundary conditions that $y_c = dy_c/dx = 0$ at $x = 0$ results in $C_1 = C_2 = 0$. Thus, $y_c = 1.233 \times 10^{-3} \, x^2$. The maximum permanent set occurs at $x = 2$ m, and is

$$(y_c)_{\text{max set}} = (1.233 \times 10^{-3})(2^2) = 4.93 \times 10^{-3} \text{ m} = 4.93 \text{ mm}$$

Residual Stresses and Fatigue

Residual stresses can be beneficial when a mechanical element is under the influence of oscillating stresses. Fatigue failures begin at locations of high shear and tensile stresses. High stresses normally occur on the outer surfaces of a mechanical element, and consequently this is where fatigue failures generally begin. The high tensile stresses can be reduced if a compressive residual stress is present before actual loading. One mechanical method of doing this is *shot peening*, in which a rain of metallic shot (such as cast-iron pellets) impinges at high speed on the surface of the mechanical element, inducing a compressive stress on a small outer layer of the element.

If the cyclic loading is primarily in one direction, residual stresses obtained by presetting the element may be more effective.

Example 3-19 Consider a beam of rectangular cross section loaded in bending so that the bending moment cycles from zero to $M_z = 4000$ lb·in (see Fig. 3-55). Determine the maximum tensile stress and compare this with the maximum tensile stress at the same point if a preset moment of $M_z = 1.2M_e$ is applied before the service load. The elastic limit of the material is 20,000 lb/in².

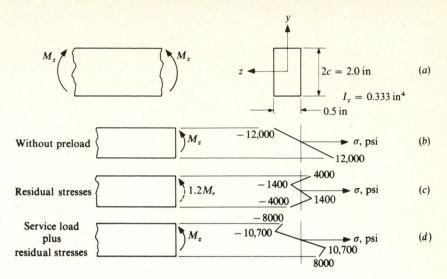

Figure 3-55

SOLUTION Without any preset, the maximum tensile stress occurs at the bottom fibers and is

$$(\sigma_x)_{max} = \frac{(4000)(1)}{0.333} = 12{,}000 \text{ lb/in}^2$$

The complete stress distribution is shown in Fig. 3-55b.

To reduce the maximum tensile stress by presetting, the preset moment should be applied in the *same* direction as the service moment. From Eq. (3-94) the depth of the plastic zone is

$$\frac{h_p}{c} = 1 - \sqrt{3 - \frac{(2)(1.2 M_e)}{M_e}} = 0.225 \qquad h_p = (0.225)(1.0) = 0.225 \text{ in}$$

The moment necessary to develop h_p is

$$M_z = 1.2 M_e = (1.2)(\tfrac{2}{3}) bc^2 \sigma_e = (1.2)(\tfrac{2}{3})(0.5)(1^2)(20{,}000) = 8000 \text{ lb·in}$$

Thus the residual stress distribution, given by Eqs. (3-97), for positive values of y is

$$\sigma_{res} = \begin{cases} -1810y & 0 \le y \le 0.775 \text{ in} \\ 24{,}000y - 20{,}000 & 0.775 \text{ in} \le y \le 1.0 \text{ in} \end{cases}$$

and is shown in Fig. 3-55c.

Adding the service stress to the residual stress results in the maximum stress distribution when $M_z = 4000$ lb·in. This is illustrated in Fig. 3-55d.

Thus it can be seen that the maximum tensile stress on the outer fiber is reduced from 12,000 to 8000 psi. This will give a drastic improvement in performance with respect to fatigue. Note, however, that at $y = -0.775$ in, the tensile stress has increased to 10,700 lb/in². This is not as bad as it seems. First, the stress is still lower than the original maximum of 12,000 lb/in². Second, the maximum stress now occurs internally and not at the surface, where surface effects contribute to fatigue failure (see Sec. 5-5). Note that the preset technique is *not* to be used if the service moment is purely reversing. In this example, if the service moment were purely reversing between 4000 and −4000 lb·in, the preset would cause a tensile stress of 16,000 lb/in² at the top surface when $M_z = -4000$ lb·in.

EXERCISES

3-1 For the beam shown in Fig. 3-56, the applied 1-kN force is in the $-y$ direction through the centroid of the cross section. Determine the locations and values of the maximum tensile and compressive bending stresses.

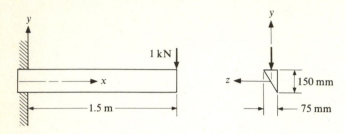

Figure 3-56

3-2 A beam is fabricated as shown in Fig. 3-57 by welding 1-in-thick steel sections. If the section undergoes a maximum bending moment of $-100{,}000$ lb·in about the z axis, determine the locations and values of the maximum tensile and compressive bending stresses.

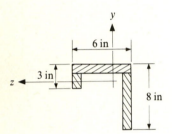

Figure 3-57

3-3 A beam with a square cross section is in bending, where the net bending moment M_z is about the z axis, as shown in Fig. 3-58. Prove that Eq. (3-1) can be applied for this case. That is, the stress at any point located a distance y from the z axis is given by $-M_z y/I_z$. Also prove that the maximum bending stress is given by

$$\sigma_{max} = 6\sqrt{2}\,\frac{M_z}{a^3}\cos\left(\frac{\pi}{4}-\beta\right)$$

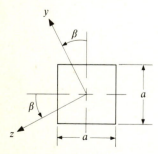

Figure 3-58

3-4 For the beam shown in Fig. 3-59, (*a*) determine the locations and values of the maximum tensile and compressive bending stresses and (*b*) estimate the net torque about the *x* axis at the wall considering the displacements due to bending ($E = 30 \times 10^6$ lb/in²).

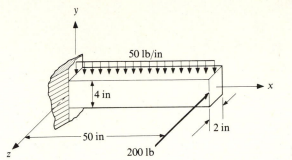

50 lb/in

4 in

50 in

2 in

200 lb

Figure 3-59

3-5 If bending is about the *z* axis of the cross section shown in Fig. 3-60, determine the location of the shear center.

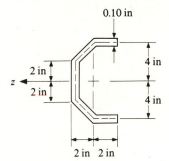

0.10 in

4 in

2 in

2 in

4 in

2 in 2 in

Figure 3-60

3-6 The channel section shown in Fig. 3-61 undergoes bending about both the *y* and *z* axes. The corresponding net transverse shear forces on the section are $V_y = 1800$ N and $V_z = -2400$ N. Determine how the shear stress varies throughout the entire section due to the transverse shear forces. Assume that the applied forces act through the shear center of the cross section.

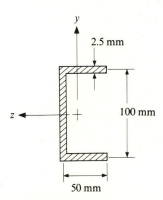

2.5 mm

100 mm

50 mm

Figure 3-61

3-7 The four-wheel set shown in Fig. 3-62 moves slowly across a 40-ft-long simply supported beam. Find the maximum bending moment and the position of the wheel set under worst conditions.

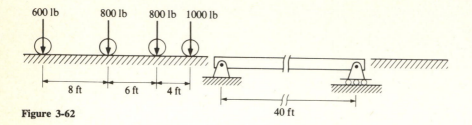

Figure 3-62

3-8 The four-wheel set given in Exercise 3-7 is moving slowly over a 20-ft-long simply supported beam. Find the maximum bending moment and the position of the wheel set under worst conditions. Note that under worst conditions, it is not necessarily true that all wheels will be on the beam.

3-9 The composite beam shown in Fig. 3-63 is made up of 2 by 4 in spruce with a modulus of elasticity of 1.5×10^6 lb/in² and two $\frac{1}{4}$-in steel plates attached to the top and bottom of the wood section. If the bending moment about the z axis is 10,000 lb·in, determine the stress distribution in the wood and in the steel due to bending. Compare the results of the maximum stresses in the wood of this exercise with the case without steel plates. Assume nominal dimensions for the wood.

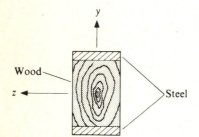

Figure 3-63

3-10 A composite cantilever beam 2 m long is made of an aluminum core 50 mm wide and 100 mm deep completely enclosed by steel 10 mm thick all around. The steel and aluminum are totally bonded together. If the maximum allowable stress in the aluminum is 6×10^7 N/m², determine the maximum allowable concentrated force which the beam can support at the free end. For the aluminum use $E_a = 7 \times 10^{10}$ N/m² and for steel $E_s = 20 \times 10^{10}$ N/m².

3-11 A composite beam is made of an aluminum core of diameter d bonded to an outer steel shell of thickness t. Replace the aluminum by an equivalent amount of steel and show that the net moment of inertia of the equivalent steel cross section about the bending axis z is given by

$$I_z = \frac{\pi}{64} [(d + 2t)^4 - \tfrac{2}{3}d^4]$$

For the aluminum use $E_a = 10 \times 10^6$ lb/in² and for steel $E_s = 30 \times 10^6$ lb/in².

3-12 For the composite beam shown in Fig. 3-64, determine the maximum bending stresses in the aluminum and steel. Where should the 1000-lb load be placed so that bending is confined to the xy plane? $E_s = 30 \times 10^6$ lb/in², $E_a = 10 \times 10^6$ lb/in².

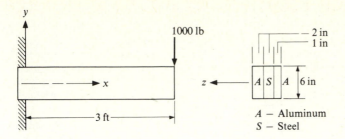

Figure 3-64

3-13 For the curved beam shown in Fig. 3-65, determine the tangential and radial stress distributions along section A-A.

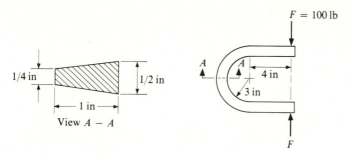

Figure 3-65

3-14 For Exercise 3-13, determine the normal stress distribution across section A-A using the straight-beam equation for bending. At the inner and outer radii of section A-A determine the correction factors K_i and K_o, respectively, which when multiplied by the solutions using straight-beam theory agree with that obtained in Exercise 3-13 using curved-beam theory.

3-15 A steel cylinder is subjected to an internal pressure of $35 \times 10^6 \, \text{N/m}^2$ and an external pressure of $70 \times 10^6 \, \text{N/m}^2$. The inner diameter of the cylinder is 50 mm, and the outer diameter is 150 mm. $E_s = 20 \times 10^{10} \, \text{N/m}^2$. (*a*) Determine and graph the radial and tangential stress distribution and (*b*) determine the radial displacements of the outer and inner surfaces.

3-16 A steel cylinder having a nominal internal diameter of 4 in and an external diameter of 12 in is press-fitted over a steel shaft having a diameter of 4 in. If the radial interference is 0.005 in, calculate the normal stress of the greatest magnitude in (*a*) the cylinder and (*b*) the shaft.

3-17 Two cylinders are to be press-fitted together so that the principal stress with the greatest absolute value is not to exceed 60 percent of the yield strength of the material. Both members are made of a steel with a yield strength of 80,000 lb/in², $E = 30 \times 10^6$ lb/in², and $\nu = 0.29$. The nominal diameters of the cylinders are 1.0, 1.5, and 2.0 in.

    (*a*) Determine the maximum acceptable radial interference and the corresponding interface pressure.

    (*b*) Under the conditions of part (*a*) plot the stress distributions for each member.

    (*c*) If under the conditions of part (*a*) the assembly is further subjected to an internal pressure of 5000 lb/in², determine the resulting stress distributions in both members using the method of superposition.

3-18 Due to friction and the interface pressure p between two press-fitted circular members, relative movement in the axial or tangential directions is impeded. Considering dry friction and the coefficient of friction between the mating parts f, show (a) that the force F necessary to cause axial motion between the mating parts is $F = 2\pi ftbp$ and (b) that the maximum allowable torque which can be transmitted from one member to the other is given by $T_{max} = 2\pi ftb^2p$, where t is the width of the narrowest member and b is the interference radius.

3-19 In Exercise 3-16, a steel cylinder is press-fitted over a steel shaft. The assembly can be used to transmit torque as shown in Fig. 3-66. If the torque T at the outer radius of the cylinder can be applied uniformly, prove that the corresponding uniformly distributed tangential load (force per unit area) is $\tau_c = T/2\pi tc^2$. For equilibrium, show that a uniformly distributed tangential load at the inner radius of the cylinder is necessary, where $\tau_b = T/2\pi tb^2$. Finally, prove that the shear-stress distribution in the cylinder is given by $\tau_{r\theta} = T/2\pi tr^2$.

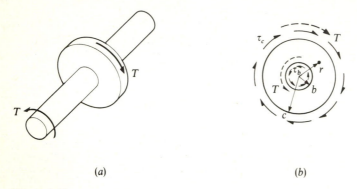

(a) (b)

Figure 3-66

3-20 Using the results of Exercises 3-16 and 3-19, show the stress element and completely describe the state of stress at the inner radius of the cylinder if a torque of 10,000 lb·in is being transmitted between the members. The width of the cylinder is 1 in, and the coefficient of friction between the members is $f = 0.05$.

3-21 A thin-walled tube of wall thickness t is to be used to transmit a torque T. The following cross-sectional shapes are under consideration: (a) circular; (b) square; (c) equilateral triangle. Considering each section to have the same weight per unit length, determine and compare the average shear stress and the angle of twist per unit length for each section. As an approximation, consider each section to have the same mean peripheral length l.

3-22 Figure 3-48a shows a plate with a centrally located hole. If $w = 100$ mm, $d = 20$ mm, $t = 10$ mm, and the elastic limit of the material is 3×10^8 N/m², determine the maximum axial force P_e which can be applied if the material is to remain in the elastic range.

3-23 Prove that the shape factor in bending of a beam with a circular cross section is 1.7.

3-24 For the diamond-shaped cross section shown in Exercise 2-8 prove that the shape factor in bending about axis z is 2.0. Explain qualitatively why the shape factor for a diamond-shaped cross section is greater than that for a circular cross section.

3-25 The elastic limit of the EPP material of the section shown in Fig. 3-67 is 350 MN/m². Considering bending about the horizontal axis, determine the limit bending moment M_p and the shape factor.

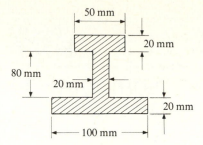

Figure 3-67

3-26 A cantilever beam of rectangular cross section b by $2c$ is shown in Fig. 3-68 and loaded so that $PL = 1.2M_e$. If the material is EPP and the elastic limit is 50,000 lb/in², determine (a) the maximum theoretical depth of the plastic zone and (b) the minimum value of x for which the depth of the plastic zone is zero.

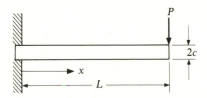

Figure 3-68

3-27 A steel beam with rectangular cross section 2 in thick and 4 in high is loaded so that $M_z = 1.3M_e$. The load is then released. Determine (a) the stress distribution before the load is released; (b) the maximum strain before the load is released; and (c) the residual stress distribution after the load is released. Assume that the material is EPP and that $\sigma_e = 60,000$ lb/in².

REFERENCES

1. Sokolnikoff, I. S.: "Mathematical Theory of Elasticity," 2d ed., McGraw-Hill, New York, 1956.
2. Seely, F.B., and A. M. Smith: "Advanced Mechanics of Materials," 2d ed., Wiley, New York, 1952.
3. Timoshenko, S., and S. Woinowsky-Krieger: "Theory of Plates and Shells," 2d ed., McGraw-Hill, New York, 1959.
4. Volterra, E., and J. H. Gaines: "Advanced Strength of Materials," Prentice-Hall, Englewood Cliffs, N.J., 1971.
5. Roark, R. J., and W. C. Young: "Formulas for Stress and Strain," 5th ed. McGraw-Hill, New York, 1975.
6. Spotts, M. F.: "Mechanical Design Analysis," Prentice-Hall, Englewood Cliffs, N.J., 1964.
7. Peterson, R. E.: "Stress Concentration Factors," Wiley, New York, 1974.
8. Peterson, R. E.: Design Factors for Stress Concentration, *Mach. Des.*, vol. 23, pt. 1, no. 2, p. 169, February 1951; pt. 2, no. 3, p. 161, March 1951; pt. 3, no. 5, p. 159, May 1951; pt. 4, no. 6, p. 173, June 1951; pt. 5, no. 7, p. 155, July 1951.

ENERGY TECHNIQUES IN
STRESS ANALYSIS

4-0 INTRODUCTION

Many practical engineering problems involve the combination of a large system of simple elements into a complex and often highly statically indeterminate structure. If a structure is statically indeterminate, it is necessary to use deflection theory first in order to determine support reactions. Once the reactions are determined, a complete analysis of the structure can be performed. Even if the structure is statically determinate, it may be necessary to perform a deformation analysis, as deflections at various points of the structure may pertain to the overall design requirements.

Deflection analysis using geometric approaches and superposition techniques are advantageous for simple systems, but if the system is complex, these approaches become difficult and cumbersome. To illustrate this point a simple example is analyzed.

Example 4-1 Cables BC and BD, shown in Fig. 4-1 form angles of 30° and 45°, respectively, to the vertical. Determine the vertical and horizontal displacement of point B if a vertical force P of 2000 lb is applied at B. The cables can be considered to be of solid steel ($E = 30 \times 10^6$ lb/in²) each with a cross-sectional area of 0.2 in².

SOLUTION The first step is to determine the forces in each cable. From Fig. 4-1b, summation of forces in the x and y directions results in

$$F_{BC} = 1464 \text{ lb} \qquad F_{BD} = 1035 \text{ lb} \qquad (a)$$

The next step is to imagine the pin at B removed and cables BC and BD allowed to stretch under the internal forces F_{BC} and F_{BD}, respectively. Then, by rotating the cables, they can be

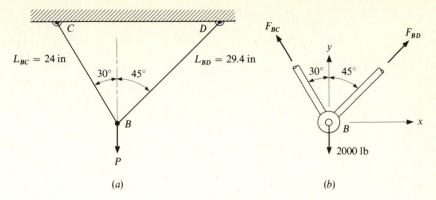

$L_{BC} = 24$ in

$L_{BD} = 29.4$ in

2000 lb

(a)   (b)

Figure 4-1

"rejoined." This must be where point B actually displaces to. If the rotations are very small, rotation of BC and BD can be approximated by a perpendicular movement, as shown in Fig. 4-2. The deflections of each cable are

$$\delta_{BC} = \frac{F_{BC}L_{BC}}{AE} = \frac{(1464)(24)}{(0.2)(30 \times 10^6)} = 5.86 \times 10^{-3} \text{ in}$$

and
$$\delta_{BD} = \frac{F_{BD}L_{BD}}{AE} = \frac{(1035)(29.4)}{(0.2)(30 \times 10^6)} = 5.07 \times 10^{-3} \text{ in}$$

to obtain the deflection of point B, relationships between δ_{BC}, δ_{BD} and $(\delta_B)_V$, $(\delta_B)_H$ must be developed. Obtaining the necessary relationships is not always simple, but after some head

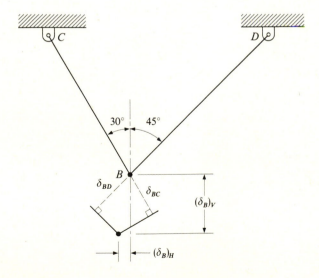

Figure 4-2

scratching one can see that

$$(\delta_B)_V = \frac{\delta_{BC}}{\cos 30} + (\delta_B)_H \tan 30 \qquad (b)$$

and

$$(\delta_B)_V = \frac{\delta_{BD}}{\cos 45} - (\delta_B)_H \tan 45 \qquad (c)$$

Setting the two equations equal with

$$\delta_{BC} = 5.86 \times 10^{-3} \text{ in} \qquad \delta_{BD} = 5.07 \times 10^{-3} \text{ in}$$

yields

$$(\delta_B)_H = 0.26 \times 10^{-3} \text{ in}$$

Substituting this into either of Eqs. (b) or (c) yields

$$(\delta_B)_V = 6.91 \times 10^{-3} \text{ in}$$

Thus it can be seen that even problems which appear rather simple can be aggravating when analyzed using geometry of deformation. Most of the energy techniques do not utilize overall geometry and, in practice, can simplify the analysis tremendously. To illustrate a simple energy technique, let us return to Example 4-1.

Example 4-2 Determine the vertical drop of point B in Example 4-1 by equating the total work done by the force P to the sum of the work P performs on each cable.

SOLUTION The work done by gradually applying the force P of 2000 lb is simply

$$W = \tfrac{1}{2}P(\delta_B)_V = 1000(\delta_B)_V$$

The work applied to each cable is

$$W_{BC} = \tfrac{1}{2}F_{BC}\delta_{BC} = \frac{1}{2}\frac{F_{BC}^2 L_{BC}}{AE} = 4.29 \text{ in·lb}$$

$$W_{BD} = \tfrac{1}{2}F_{BD}\delta_{BD} = \frac{1}{2}\frac{F_{BD}^2 L_{BD}}{AE} = 2.62 \text{ in·lb}$$

The total work is the sum of the work applied to each cable, and so

$$W = W_{BC} + W_{BD}$$

$$1000(\delta_B)_V = 4.29 + 2.62$$

$$(\delta_B)_V = 6.91 \times 10^{-3} \text{ in}$$

Thus it can be seen that a work or energy approach can simplify things considerably.†

In order to determine the horizontal reaction of point B, an artifice must be employed to avoid the fact that no work is done by P moving horizontally.

Example 4-3 Determine the horizontal deflection of point B in Example 4-1 by the approach used in Example 4-2.

---

† The answer obtained in Example 4-2 is identical to that of Example 4-1. Recall that in Example 4-1 an approximation for small rotations was made. This same assumption applies to the approach used in Example 4-2. For example, the work in cable BC is not exactly $\tfrac{1}{2}F_{BC}\delta_{BC}$ but is slightly less due to the rotation. However, as long as rotations are small, this error is negligible.

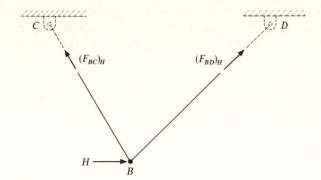

Figure 4-3

SOLUTION Since there is no horizontal force at point B, there is no work due to the horizontal deflection $(\delta_B)_H$. However, it is possible first to place an exceedingly small horizontal force H at point B where the force is so small that it causes negligible deflections (see Fig. 4-3). Then, when the vertical force P is placed on the structure, the constant horizontal force H will do work. The work due to H is†

$$W_H = H(\delta_B)_H$$

The forces in cables BC and BD due to H alone are found by static analysis and are

$$(F_{BC})_H = 0.732H \qquad (F_{BD})_H = -0.897H‡$$

The additional work obtained from these forces on the cables arises from the deflections caused by the actual applied force. The additional work terms are§

$$(W_{BC})_H = (F_{BC})_H \delta_{BC} = (0.732H)(5.86 \times 10^{-3}) = 4.29H \times 10^{-3}$$

and $\qquad (W_{BD})_H = (F_{BD})_H \delta_{BD} = (-0.897H)(5.07 \times 10^{-3}) = -4.55H \times 10^{-3}$

The work due to H equals the work H performs on the elements of the system¶

$$W_H = (W_{BC})_H + (W_{BD})_H$$

$$H(\delta_B)_H = 4.29H \times 10^{-3} - 4.55H \times 10^{-3}$$

$$(\delta_B)_H = -0.26 \times 10^{-3} \text{ in}^{\|}$$

Note that the procedure followed in Examples 4-2 and 4-3 avoided analyzing the overall geometry of deformation of the structure, which was necessary in Example 4-1, where Eqs. (b) and (c) were developed.

Since energy techniques are quite powerful, this chapter will attempt to provide a detailed exploration of the major techniques applied to the field of

---

† The $\frac{1}{2}$ is omitted since the force H remains constant throughout the deflection $(\delta_B)_H$.

‡ Do not be concerned here with a compressive force in a cable. Recall that H is actually an artificial force and is basically infinitesimal in value.

§ Again, since the forces $(F_{BC})_H$ and $(F_{BD})_H$ are present before deflections, the work expression will not contain the $\frac{1}{2}$ term.

¶ This technique is restricted to an infinitesimal value for H and is not true for a finite value of H.

$\|$ The negative sign indicates that the deflection is in the direction opposite H.

stress analysis. In order to develop and illustrate the techniques, the analyses will be demonstrated on some simple problems in the beginning to avoid overburdening the reader with complex systems at the start. Thus, in some cases, the particular energy technique may not appear that impressive, since, as stated earlier, geometric and superposition approaches are generally more practical on simple, single-element systems.

The major energy techniques discussed in this chapter are:

1. The strain-energy theorem
2. The complementary-energy theorem
3. Special applications of the complementary-energy theorem
    a. Castigliano's theorem applied to linear systems
    b. The dummy-load method
4. Rayleigh's technique
5. The Rayleigh-Ritz technique

4-1 WORK

When a force P is gradually applied at a particular point Q on a structure (see Fig. 4-4), the deflection of the point δ in the direction of the applied force increases at a rate directly related to P.

The total deflection of point Q is δ_T, the vector sum of δ and δ'. However, the work performed by P is

$$W = \int_0^{\delta_1} P \, d\delta \tag{4-1}$$

where δ is in the direction of P and the integral of Eq. (4-1) is merely the area under the P-vs.-δ curve evaluated up to δ_1. If the relationship between P and δ is linear, that is, $P = k\delta$, where k is a constant, then from Eq. (4-1) the work is

$$W = \tfrac{1}{2}k\delta_1^2 = \tfrac{1}{2}P_1\delta_1 \tag{4-2}$$

If P is constant, $P = P_1$; while if a deflection process takes place,

$$W = P_1\delta_1 \tag{4-3}$$

4-2 STRAIN ENERGY

Uniaxial Case

Consider an infinitesimal element within a member which will develop a final uniaxial stress of σ_x (see Fig. 4-5). If the behavior of the material is linear, the force due to σ_x ($F_x = \sigma_x \, \Delta y \, \Delta z$) displaces in a linear fashion the amount

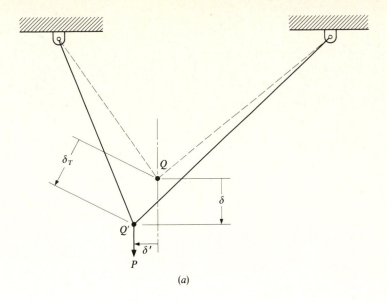

(a)

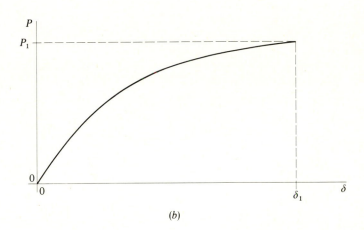

(b)

Figure 4-4

$\delta_x = \varepsilon_x \, \Delta x$. From Eq. (4-2) the total work on the element is

$$W = \tfrac{1}{2}F_x\delta_x = \tfrac{1}{2}\sigma_x\varepsilon_x(\Delta x \; \Delta y \; \Delta z)$$

The work per unit volume w is determined by dividing the work by the volume $\Delta x \; \Delta y \; \Delta z$. Thus

$$w = \tfrac{1}{2}\sigma_x\varepsilon_x \qquad\qquad (4\text{-}4)$$

where the units of w are inch-pounds per cubic inch (USCS) and newton-meters per cubic meter or joules per cubic meter (SI). If the material is

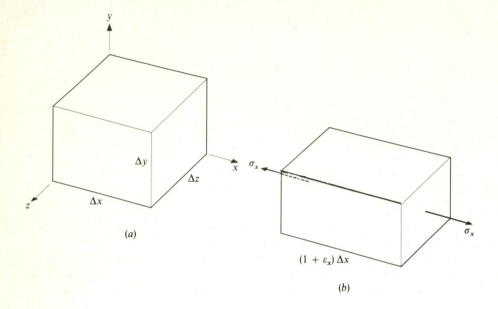

Figure 4-5

perfectly elastic where it exhibits no energy losses, the total work (or work per unit volume) increases the potential energy of the element U, called the strain energy (or strain energy per unit volume u). Thus

$$u = w = \tfrac{1}{2}\sigma_x\varepsilon_x$$

Since the stress is uniaxial, in the elastic range $\varepsilon_x = \sigma_x/E$ and the resulting strain energy per unit volume is

$$u = \frac{1}{2E}\,\sigma_x^2 \tag{4-5}$$

Additional Normal Stresses

If the normal stresses σ_y and σ_z are also present,

$$u = \tfrac{1}{2}\sigma_x\varepsilon_x + \tfrac{1}{2}\sigma_y\varepsilon_y + \tfrac{1}{2}\sigma_z\varepsilon_z$$

and using the general strain-stress relationships [Eqs. (1-21)], we see that the strain energy per unit volume is

$$u = \frac{1}{2E}\,[\sigma_x^2 + \sigma_y^2 + \sigma_z^2 - 2\nu(\sigma_x\sigma_y + \sigma_y\sigma_z + \sigma_z\sigma_x)] \tag{4-6}$$

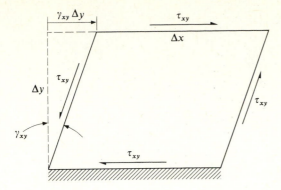

Figure 4-6

Shear Stress

The work due to τ_{xy} can be seen in Fig. 4-6. The force $\tau_{xy} \, \Delta x \, \Delta z$ moves a distance $\gamma_{xy} \, \Delta y$, which results in a work per unit volume of $\frac{1}{2}\gamma_{xy}\tau_{xy}$. When Eq. (1-27a) is used, the strain energy per unit volume is

$$u = \frac{1+\nu}{E}\tau_{xy}^2 \tag{4-7}$$

General State of Stress

Combining Eq. (4-6) and the strain energy per unit volume due to all the shear stresses yields, for the general case,

$$u = \frac{1}{2E}[\sigma_x^2 + \sigma_y^2 + \sigma_z^2 - 2\nu(\sigma_x\sigma_y + \sigma_y\sigma_z + \sigma_z\sigma_x)$$
$$+ 2(1+\nu)(\tau_{xy}^2 + \tau_{yz}^2 + \tau_{zx}^2)] \tag{4-8}$$

Equation (4-8) can be written in terms of the principal stresses σ_1, σ_2, and σ_3, which is basically the same element transformed to eliminate the shear stresses. Thus, the strain energy per unit volume for the general case can also be expressed as

$$u = \frac{1}{2E}[\sigma_1^2 + \sigma_2^2 + \sigma_3^2 - 2\nu(\sigma_1\sigma_2 + \sigma_2\sigma_3 + \sigma_3\sigma_1)] \tag{4-9}$$

Biaxial State of Stress

For this case (see Fig. 4-7) $\sigma_z = \tau_{yz} = \tau_{zx} = 0$, and Eq. (4-8) reduces to

$$u = \frac{1}{2E}[\sigma_x^2 + \sigma_y^2 - 2\nu\sigma_x\sigma_y + 2(1+\nu)\tau_{xy}^2] \tag{4-10}$$

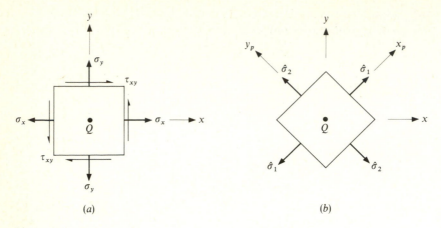

Figure 4-7

or, in terms of principal stresses (with $\hat{\sigma}_3 = 0$), from Eq. (4-9)

$$u = \frac{1}{2E}(\hat{\sigma}_1^2 + \hat{\sigma}_2^2 - 2\nu\hat{\sigma}_1\hat{\sigma}_2) \tag{4-11}$$

In many cases of biaxial stress, $\sigma_y = 0$, and Eq. (4-10) reduces to

$$u = \frac{1}{2E}[\sigma_x^2 + 2(1 + \nu)\tau_{xy}^2] \tag{4-12}$$

4-3 TOTAL STRAIN ENERGY IN BARS WITH SIMPLE LOADING CONDITIONS

Axial Loading

The stress in a bar undergoing axial loading (see Fig. 4-8) is $\sigma_x = \pm P/A$ throughout the bar, where the plus sign indicates tension and minus compres-

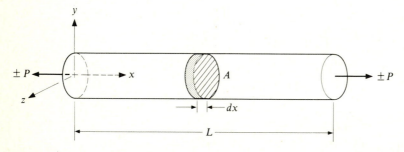

Figure 4-8

sion. Thus, integrating the strain energy per unit volume [Eq. (4-5)] throughout the bar yields the total strain energy. The total strain energy is therefore

$$U_a = \frac{1}{2} \int \frac{1}{E} \left(\frac{P}{A}\right)^2 dV$$

However, $dV = A\,dx$; thus

$$U_a = \frac{1}{2} \int_0^L \frac{P^2\,dx}{AE} \tag{4-13}$$

If P, A, and E are constant, Eq. (4-13) reduces to

$$U_a = \frac{1}{2} \frac{P^2 L}{AE} \tag{4-14}$$

which, as can be seen, is the total work performed by P since $W = \frac{1}{2}P\delta$ and $\delta = PL/AE$.

Torsional Loading

The shear stress induced by torsion is a function of radial position, that is, $\tau = Tr/J$. Establish a dV element in which the shear stress is constant (see Fig. 4-9). Thus from Eq. (4-7)

$$U_t = \int \frac{1+\nu}{E} \left(\frac{Tr}{J}\right)^2 dx\,dA$$

In general, except for r^2 all the finite terms within the integral will be functions of x. Thus integrating with respect to dA first results in

$$U_t = \int \left[\frac{1+\nu}{E} \left(\frac{T}{J}\right)^2 \int r^2\,dA\right] dx$$

However, since $\int r^2\,dA = J$,

$$U_t = \int_0^L \frac{(1+\nu)T^2}{EJ}\,dx \tag{4-15}$$

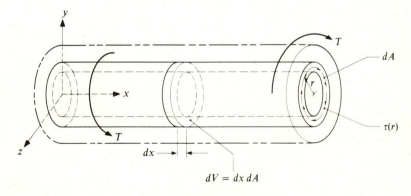

$$dV = dx\,dA$$

Figure 4-9

If v, T, E, and J are not functions of x, Eq. (4-15) reduces to

$$U_t = (1 + v)\frac{T^2 L}{EJ} \tag{4-16}$$

Transverse Loading

See Fig. 4-10a. Since the normal bending stresses and transverse shear stresses vary with respect to x and y in general, a dV element undergoing a constant state of stress is as shown in Fig. 4-10b. Since this volume is undergoing one normal stress

$$\sigma_x = -\frac{M_z y}{I_z}$$

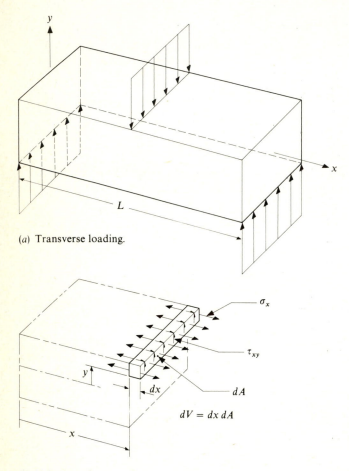

(a) Transverse loading.

(b) Volume under constant stress.

Figure 4-10

and one shear stress

$$\tau_{xy} = \frac{V_y Q}{I_z b}$$

from Eq. (4-12), the strain energy will be

$$U = \frac{1}{2} \int \frac{1}{E} \left[\left(\frac{M_z y}{I_z} \right)^2 + 2(1 + \nu) \left(\frac{V_y Q}{I_z b} \right)^2 \right] dx \, dA$$

or

$$U = \frac{1}{2} \int \frac{1}{E} \left(\frac{M_z}{I_z} \right)^2 y^2 \, dx \, dA + \int \frac{1 + \nu}{E} \left(\frac{V_y}{I_z} \right)^2 \left(\frac{Q}{b} \right)^2 dx \, dA$$

The first integral is the part of the total strain energy due to the bending moment; it will be designated U_b. The second integral is the part due to the shear force, designated U_s. Thus

$$U = U_b + U_s \tag{4-17}$$

Since M_z/I_z is a function of x alone, integration of U_b with respect to dA is performed first. Since $\int y^2 \, dA = I_z$, the total strain energy due to bending is

$$U_b = \frac{1}{2} \int_0^L \frac{M_z^2}{EI_z} \, dx \tag{4-18}$$

If E and I_z are constant,

$$U_b = \frac{1}{2EI_z} \int_0^L M_z^2 \, dx \tag{4-19}$$

An alternate form for Eqs. (4-18) and (4-19) comes from Eq. (2-22), where $M_z = EI_z(d^2 y_c/dx^2)$. Thus, the alternate form is

$$U_b = \frac{1}{2} \int_0^L EI_z \left(\frac{d^2 y_c}{dx^2} \right)^2 dx \tag{4-20}$$

or if E and I_z are constant,

$$U_b = \frac{EI_z}{2} \int_0^L \left(\frac{d^2 y_c}{dx^2} \right)^2 dx \tag{4-21}$$

The second integral in Eq. (4-17), U_s, is evaluated in a similar manner. Since V_y/I_z is a function of x alone, integration with respect to dA can be performed first as was done for U_b, but the integral $\int (Q/b)^2 \, dA$ must be evaluated for the particular shape of cross section under investigation. It is common practice to assume a constant shear stress across the cross section, $\tau_{xy} = V_y/A$. This introduces a small error, and in light of the fact that for most beams $U_b \gg U_s$, this error becomes quite negligible. Therefore, if $\tau_{xy} = V_y/A$ is used instead of $\tau_{xy} = V_y Q/I_z b$, the strain energy due to the shear force V_y is

$$U_s = \int_0^L \frac{(1 + \nu) V_y^2}{EA} \, dx \tag{4-22}$$

If ν, E, and A are constant,

$$U_s = \frac{1 + \nu}{EA} \int_0^L V_y^2 \, dx \tag{4-23}$$

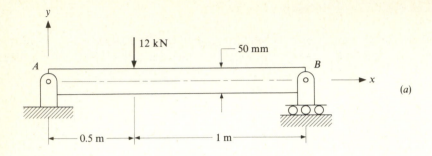

Figure 4-11

Example 4-4 Determine the total strain energy of the steel beam shown in Fig. 4-11. The cross section is rectangular, 25 mm wide, and 50 mm deep. $E_s = 20.5 \times 10^{10}$ N/m²; $\nu = 0.3$.

SOLUTION First, solving for the reactions at A and B gives

$$R_A = 8 \text{ kN} \qquad R_B = 4 \text{ kN}$$

The next step is to determine the shear-force and bending-moment equations:

$$V_y = \begin{cases} -8 \text{ kN} & 0 < x < 0.5 \\ 4 \text{ kN} & 0.5 < x < 1.5 \end{cases}$$

$$M_z = \begin{cases} 8x \text{ kN·m} & 0 < x < 0.5 \\ 8x - 12(x - 0.5) \text{ kN·m} & 0.5 < x < 1.5 \end{cases}$$

When Eq. (4-19) is used, the bending energy is given by

$$U_b = \frac{1}{2EI_z}\left\{ \int_0^{0.5} (8000x)^2 \, dx + \int_{0.5}^{1.5} [8000x - 12,000(x - 0.5)]^2 \, dx \right\}$$

Integration of this and substitution of $E = 20.5 \times 10^{10}$ N/m² and $I_z = \frac{1}{12}(0.025)(0.05)^3 = 2.604 \times 10^{-7}$ m⁴ results in

$$U_b = 74.9 \text{ N·m}$$

The strain energy due to the shear force V_y is found using Eq. (4-23), which yields

$$U_s = \frac{1 + 0.3}{(20.5 \times 10^{10})(0.05)(0.025)}\left[\int_0^{0.5} (-8000)^2 \, dx + \int_{0.5}^{1.5} 4000^2 \, dx \right] = 0.244 \text{ N·m}$$

Thus

$$U = U_b + U_s \approx 75 \text{ N·m}$$

Note that the strain energy due to bending is considerably greater than the shear energy. This is usually the case for beams unless they are very short. Thus, in most cases, the shear energy in bending problems is normally ignored.

4-4 THE STRAIN-ENERGY THEOREM

Consider a series of forces P_i gradually applied to a structure. At each point of load application i let δ_i be the component of the total deflection of point i along the line of action of P_i. Therefore, the total work applied is

$$W = \sum_{i=1}^{n} \int P_i \, d\delta_i$$

If energy is conserved, the total work performed on the structure increases the total strain energy of the structure. Thus $U = W$, and

$$U = \sum_{i=1}^{n} \int P_i \, d\delta_i \tag{4-24}$$

Now, fixing the points of application of all the forces except one specific point i, allow point i to deflect an infinitesimal amount $\Delta\delta_i$ in the direction of P_i by applying an infinitesimal force ΔP_i. The additional work ΔW will cause an infinitesimal change in strain energy of the system ΔU. Thus

$$\Delta W = P_i \, \Delta\delta_i + \int_{0}^{\Delta\delta_i} \Delta P_i \, d\delta_i = \Delta U†$$

The work term in the integral is of the order of the product of two infinitesimal terms and thus can be neglected. Therefore, $\Delta U = P_i \, \Delta\delta_i$, or

$$P_i = \frac{\Delta U}{\Delta\delta_i} \tag{4-25}$$

If $\Delta\delta_i$ is now allowed to approach zero, Eq. (4-25) reduces to

$$P_i = \frac{\partial U}{\partial\delta_i} \tag{4-26}$$

Equation (4-26) actually represents n equations since $i = 1, 2, \ldots, n$.

To use Eq. (4-26), it is necessary to describe the strain energy as a function of the deflections δ_i, which may be difficult to do as the geometry of deformation is necessary. This is the disadvantage of Eq. (4-26).

Equation (4-26) is also applicable to problems where the force-deflection relation is nonlinear provided energy is conserved and rotations are small.

Example 4-5 Two cables shown in Fig. 4-12a are made of a nonlinear material whose stress-strain behavior is $\sigma = E\varepsilon - K\varepsilon^2$, where $E = 30 \times 10^6$ lb/in² and $K = 10 \times 10^9$ lb/in². The cross-sectional area A of each cable is 0.2 in², and both have a length L of 5 ft. Determine the vertical deflection δ_B of point B after a load P_B of 2000 lb is applied.

SOLUTION Thanks to symmetry, the strain energies in each cable are equal. Thus, it is only necessary to evaluate one cable, for example, BC. The total energy is then

$$U = 2 \int F_{BC} \, d\delta_{BC}$$

† Here P_i was considered constant through the deflection $\Delta\delta_i$. The force ΔP_i is gradually applied throughout the loading period, resulting in the integral term.

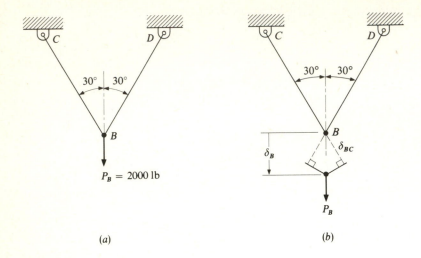

(a)   (b)

Figure 4-12

Since $F_{BC} = \sigma A = (E\varepsilon - K\varepsilon^2)_{BC} A$ and $\varepsilon_{BC} = \delta_{BC}/L$, we have

$$F_{BC} = \left[E\frac{\delta_{BC}}{L} - K\left(\frac{\delta_{BC}}{L}\right)^2 \right] A \tag{a}$$

$$U = 2A \int_0^{\delta_{BC}} \left[E\frac{\delta_{BC}}{L} - K\left(\frac{\delta_{BC}}{L}\right)^2 \right] d\delta_{BC} = 2AL \left[\frac{E}{2}\left(\frac{\delta_{BC}}{L}\right)^2 - \frac{K}{3}\left(\frac{\delta_{BC}}{L}\right)^3 \right]$$

Now from Eq. (4-24)

$$P_B = \frac{\partial U}{\partial \delta_B} = 2A \left[E\left(\frac{\delta_{BC}}{L}\right) - K\left(\frac{\delta_{BC}}{L}\right)^2 \right] \frac{\partial \delta_{BC}}{\partial \delta_B} \tag{b}$$

It is now necessary to obtain a relationship between δ_{BC} and δ_B using geometry of deformation. From Fig. 4-12b it can be seen that

$$\delta_{BC} = \delta_B \cos 30 \tag{c}$$

Therefore

$$\frac{\partial \delta_{BC}}{\partial \delta_B} = \cos 30 \tag{d}$$

Substituting Eqs. (c) and (d) into Eq. (b) yields

$$P_B = 2A \left[E\frac{\delta_B}{L} - K\cos 30 \left(\frac{\delta_B}{L}\right)^2 \right] \cos^2 30$$

Rearranging results in

$$\left(\frac{\delta_B}{L}\right)^2 - \frac{E}{K\cos 30}\frac{\delta_B}{L} + \frac{P_B}{2AK\cos^3 30} = 0 \tag{e}$$

Solving Eq. (e) yields

$$\delta_B = \frac{EL}{2K\cos 30} \pm \frac{L}{2} \left[\left(\frac{E}{K\cos 30}\right)^2 - \frac{4P_B}{2AK\cos^3 30} \right]^{1/2} \tag{f}$$

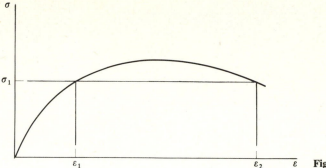

Figure 4-13

Substituting in numerical values and solving for δ_B results in

$$\delta_B = 0.0142 \text{ in} \qquad \text{or} \qquad 0.194 \text{ in}$$

Two solutions are obtained because the stress-strain relationship given is double-valued in strain, as can be seen in Fig. 4-13. Thus for a given stress in cables BC and BD, it is possible to have either strain ε_1 or ε_2. However, since the applied force is gradually increased to $P_B = 200$ lb and does not exceed this value, only the lower strain (and displacement) value is acceptable. Thus

$$\delta_B = 0.0142 \text{ in}$$

Note that in Example 4-5 the equilibrium force equations were not used in obtaining the solution. Although in this example, it would be quite easy to solve for F_{BC} and F_{BD} using the static equilibrium equations, it would not be a simple matter if the problem were statically indeterminate. When the strain-energy theorem is used, the solution of the forces comes out of the analysis without any difficulty. If δ_B is substituted into Eq. (c),

$$\delta_{BC} = 0.0142 \cos 30 = 0.0123 \text{ in}$$

Substituting δ_{BC} into Eq. (a) yields

$$F_{BC} = \left[(30 \times 10^6) \frac{0.0123}{60} - (10 \times 10^9) \left(\frac{0.0123}{60} \right)^2 \right] (0.2) = 1150 \text{ lb}$$

The method in evaluating forces in statically indeterminate problems is basically the same as that illustrated in Example 4-5.

Example 4-6 Using the strain-energy method, determine the cable forces of the symmetric structure shown in Fig. 4-14. The length, area, and modulus of cable 1 are L_1, A_1, and E_1 and of cable 2 are L_2, A_2, and E_2; etc. The materials are linear.

SOLUTION The total strain energy is

$$U = \int F_1 \, d\delta_1 + 2 \int F_2 \, d\delta_2 + 2 \int F_3 \, d\delta_3 \qquad (a)$$

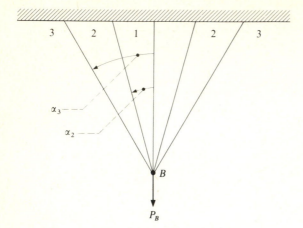

Figure 4-14

Since the force-deflection relationship for each cable is

$$F_i = \left(\frac{AE}{L}\right)_i \delta_i \tag{b}$$

Eq. (a) becomes

$$U = \left(\frac{AE}{L}\right)_1 \frac{\delta_1^2}{2} + \left(\frac{AE}{L}\right)_2 \delta_2^2 + \left(\frac{AE}{L}\right)_3 \delta_3^2 \tag{c}$$

The deflection of point B is the deflection δ_1 of cable 1, and so

$$P_B = \frac{\partial U}{\partial \delta_1} = \left(\frac{AE}{L}\right)_1 \delta_1 + 2\left(\frac{AE}{L}\right)_2 \delta_2 \frac{\partial \delta_2}{\partial \delta_1} + 2\left(\frac{AE}{L}\right)_3 \delta_3 \frac{\partial \delta_3}{\partial \delta_1} \tag{d}$$

As in Example 4-5, a geometric relationship is established:

$$\delta_2 = \delta_1 \cos \alpha_2 \qquad \delta_3 = \delta_1 \cos \alpha_3 \tag{e}$$

and

$$\frac{\partial \delta_2}{\partial \delta_1} = \cos \alpha_2 \qquad \frac{\partial \delta_3}{\partial \delta_1} = \cos \alpha_3 \tag{f}$$

Substituting Eqs. (e) and (f) into Eq. (d) results in

$$P_B = \left[\left(\frac{AE}{L}\right)_1 + 2\left(\frac{AE}{L}\right)_2 \cos^2 \alpha_2 + 2\left(\frac{AE}{L}\right)_3 \cos^2 \alpha_3\right] \delta_1$$

Solving for δ_1 yields

$$\delta_1 = \frac{P_B}{K} \tag{g}$$

where

$$K = \left(\frac{AE}{L}\right)_1 + 2\left(\frac{AE}{L}\right)_2 \cos^2 \alpha_2 + 2\left(\frac{AE}{L}\right)_3 \cos^2 \alpha_3 \tag{h}$$

Substituting δ_1 into Eq. (b) yields the force in cable 1. Thus

$$F_1 = \left(\frac{AE}{L}\right)_1 \frac{P}{K} \tag{i}$$

Substituting δ_1 into Eqs. (e) and then substituting into Eq. (b) yields the forces in cables 2 and 3:

$$F_2 = \left(\frac{AE}{L}\right)_2 \frac{P}{K} \cos \alpha_2 \qquad F_3 = \left(\frac{AE}{L}\right)_3 \frac{P}{K} \cos \alpha_3 \tag{j}$$

The approach outlined in Example 4-6 can be applied just as easily to any number of cables. The one disadvantage of the strain-energy theorem comes from the fact that geometry of deformation is necessary, and in many cases some difficulties arise in obtaining the deflection relationships. The complementary-energy theorem circumvents this problem.

4-5 THE COMPLEMENTARY-ENERGY THEOREM

The definition of the complementary work performed by a gradually applied force P undergoing a deflection δ is

$$W_c = \int_0^{P_1} \delta \, dP \tag{4-27}$$

where δ is the component of the total deflection in the direction of P. The complementary work is illustrated by the shaded area of Fig. 4-15. Defining the complementary energy of the system Φ as being equal to the complementary work, we have

$$\Phi = \int_0^{P_1} \delta \, dP \tag{4-28}$$

If a number of forces P_i are applied to a structure and each P_i deflects δ_i, where δ_i is the component of the total deflection of P_i in the direction of the line of action of the force, the total complementary energy of the system is

$$\Phi = \sum_{i=1}^{n} \int \delta_i \, dP_i \tag{4-29}$$

where n is the total number of forces. If an infinitesimal force of ΔP_i is placed at point i, the increase in complementary energy is

$$\Delta \Phi = \delta_i \, \Delta P_i \qquad \text{or} \qquad \delta_i = \frac{\Delta \Phi}{\Delta P_i}$$

Allowing ΔP_i to approach zero results in

$$\delta_i = \frac{\partial \Phi}{\partial P_i} \tag{4-30}$$

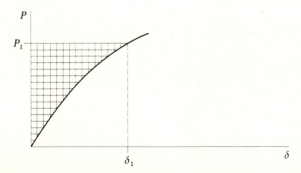

Figure 4-15

As before with Eq. (4-26), Eq. (4-30) actually represents n equations, since $i = 1, 2, \ldots, n$.

It can be seen in Eq. (4-30) that the complementary energy must be written in terms of the applied forces, which are arrived at through the equilibrium equations. Thus, to obtain the deflections, Eq. (4-30) is used without resorting to the overall geometry of deformation.

If rotations within a structure are to be considered, it can also be shown that

$$\theta_i = \frac{\partial \Phi}{\partial M_i} \tag{4-31}$$

is a simple extension of Eq. (4-30), where θ_i is the rotation at point i in the direction of a concentrated moment M_i applied at point i.

Example 4-7 Repeat Example 4-1 using the complementary-energy theorem.

SOLUTION For a cable made of a linear elastic material, the complementary energy is equal to the strain energy. Thus

$$\Phi = U = \frac{1}{2} \frac{P^2 L}{AE}$$

For the system of cables in Example 4-1, the total complementary energy is

$$\Phi = \frac{1}{2} \frac{F_{BC}^2 L_{BC}}{AE} + \frac{1}{2} \frac{F_{BD}^2 L}{AE}_{BD}$$

But
$$F_{BC} = 0.732P \qquad F_{BD} = 0.518P$$

Therefore
$$\Phi = \frac{0.535 P^2 L_{BC}}{2AE} + \frac{0.268 P^2 L_{BD}}{2AE}$$

The deflection of point B in the vertical direction is given by Eq. (4-30) and is

$$(\delta_B)_V = \frac{\partial \Phi}{\partial P} = \frac{0.536 P L_{BC}}{AE} + \frac{0.268 P L_{BD}}{AE} = \frac{P}{AE} (0.536 L_{BC} + 0.268 L_{BD})$$

Substituting numerical values results in

$$(\delta_B)_V = \frac{2000}{(0.2)(30 \times 10^6)} [(0.536)(24) + (0.268)(29.4)] = 6.91 \times 10^{-3} \text{ in}$$

To determine the horizontal deflection of point B, a horizontal dummy load H must be applied at point B so that Eq. (4-30) can be used (see Fig. 4-16).† The total forces in cables BC and BD are
$$F_{BC} = 0.732P + 0.732H \qquad F_{BD} = 0.518P - 0.897H$$

The total complementary energy is therefore

$$\Phi = \frac{1}{2AE} (0.732P + 0.732H)^2 L_{BC} + \frac{1}{2AE} (0.518P - 0.897H)^2 L_{BD}$$

Using Eq. (4-30) yields‡

$$(\delta_B)_H = \frac{\partial U}{\partial H} = \frac{(0.732)(0.732P + 0.732H) L_{BC}}{AE} - \frac{(0.897)(0.518P - 0.897H) L_{BD}}{AE}$$

---

† It is not necessary that H be infinitesimal as in Example 4-3. As a matter of fact, if a finite H did exist on the structure, this analysis would yield the correct solution. However, for this example, at the end of the analysis, H will be set equal to zero.

‡ It is unnecessary to square the terms in Φ before differentiation.

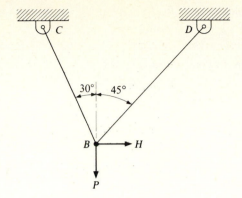

Figure 4-16

However, $H = 0$; therefore

$$(\delta_B)_H = \frac{P}{AE}[(0.732)^2 L_{BC} - (0.897)(0.518)L_{BD}]$$

$$= \frac{2000}{(0.2)(30 \times 10^6)}[(0.732)^2(24) - (0.897)(0.518)(29.4)]$$

$$= -0.26 \times 10^{-3} \text{ in}\dagger$$

The complementary-energy theorem is also suitable for nonlinear load-vs.-deflection relations provided rotations are small.

Example 4-8 Repeat Example 4-5 but for cables made of a material whose stress-strain behavior is $\varepsilon = (1/E)\sigma + C\sigma^2$, where $E = 10 \times 10^6$ lb/in² and $C = 4 \times 10^{-12}$ (lb/in²)⁻².

SOLUTION As in Example 4-5, from symmetry the complementary energies of the cables are identical. Thus, the total complementary energy can be written in terms of one cable, say cable BC. Thus

$$\Phi = 2\Phi_{BC} = 2\int \delta_{BC}\, dF_{BC}$$

Since $\delta_{BC} = (\varepsilon L)_{BC}$, where $\varepsilon_{BC} = (\sigma/E + C\sigma^2)_{BC}$ and $\sigma_{BC} = (F/A)_{BC}$, we have

$$\Phi = 2L \int \left[\frac{F_{BC}}{EA} + C\left(\frac{F_{BC}}{A}\right)^2\right] dF_{BC}$$

Integration yields

$$\Phi = 2LA\left[\frac{1}{2E}\left(\frac{F_{BC}}{A}\right)^2 + \frac{C}{3}\left(\frac{F_{BC}}{A}\right)^3\right]$$

From equilibrium

$$F_{BC} = \frac{P_B}{2 \cos 30}$$

therefore

$$\Phi = 2L\left(\frac{P_B^2}{8EA \cos^2 30} + \frac{CP_B^3}{24A^2 \cos^3 30}\right)$$

---

† Again, the negative sign indicates that the deflection is in the direction opposite H.

Using Eq. (4-30), we see that the deflection of point B is

$$\delta_B = \frac{\partial \Phi}{\partial P_B} = \frac{P_B L}{2EA \cos^2 30} + \frac{CP_B^2 L}{4A^2 \cos^3 30}$$

Substitution of the numerical values from Example 4-5 results in

$$\delta_B = \frac{(2000)(60)}{(2)(10 \times 10^6)(0.2)(0.866)^2} + \frac{(4 \times 10^{-12})(2000^2)(60)}{(4)(0.2)^2(0.866)^3} = 0.049 \text{ in}$$

4-6 CASTIGLIANO'S THEOREM

As demonstrated in Example 4-7, if the load-displacement relation is *linear*, the complementary energy equals the strain energy. Thus, Eqs. (4-30) and (4-31) can be written

$$\delta_i = \frac{\partial U}{\partial P_i} \tag{4-32}$$

and

$$\theta_i = \frac{\partial U}{\partial M_i} \tag{4-33}$$

Equations (4-32) and (4-33) are statements of Castigliano's theorem as applied to problems where the force-deflection relation is linear. Since many practical engineering problems involve linear load-displacement relations and the forms of the strain energy are known for simple cases (see Sec. 4-3), the method of Castigliano provides a simple and straightforward approach to a complex collection of simple elements.

Example 4-9 For the structure shown in Fig. 4-17, determine the vertical deflection of point B. All members have a cross-sectional area A; members 1, 2, 4, and 5 are of length L, and members 3, 6, and 7 are of length $1.41L$.

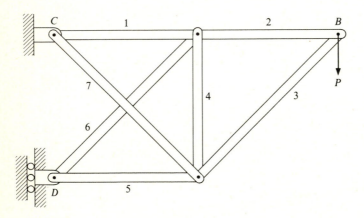

Figure 4-17

SOLUTION The total strain energy is the sum of the strain energies of all the members. Thus

$$U = U_1 + U_2 + U_3 + U_5 + U_6 + U_7$$

Since each member is loaded axially, the energy in the ith member is

$$U_i = \frac{1}{2} \frac{F_i^2 L_i}{AE} \qquad i = 1, 2, \ldots, 7$$

Thus it is necessary to determine the axial load in each member F_i using static analysis. Assuming tension positive for each member, we have

$$F_1 = P \qquad F_2 = P \qquad F_3 = -1.41P \qquad F_4 = 0$$
$$F_5 = -2P \qquad F_6 = 0 \qquad F_7 = 1.41P$$

The total energy is then given by

$$U = \frac{1}{2} \frac{P^2 L}{AE} + \frac{1}{2} \frac{P^2 L}{AE} + \frac{1}{2} \frac{(-1.41P)^2 (1.41L)}{AE} + 0 + \frac{1}{2} \frac{(-2P)^2 L}{AE} + 0 + \frac{1}{2} \frac{(1.41P)^2 (1.41L)}{AE}$$

Simplifying results in

$$U = \frac{11.6}{2} \frac{P^2 L}{AE}$$

From Eq. (4-32), the vertical deflection $(\delta_B)_V$ of point B is $\partial U / \partial P$. Thus

$$(\delta_B)_V = \frac{\partial U}{\partial P} = 11.6 \frac{PL}{AE}$$

In Example 4-9 in order to determine the horizontal deflection of point B using Eq. (4-32), it would be necessary to place a dummy force at B in the horizontal direction.

Example 4-10 Using Castigliano's method, determine the horizontal deflection of point B in Example 4-9.

SOLUTION Place a horizontal dummy force H at point B as shown in Fig. 4-18. A static analysis of the forces in the members caused by H alone can be added by superposition to the

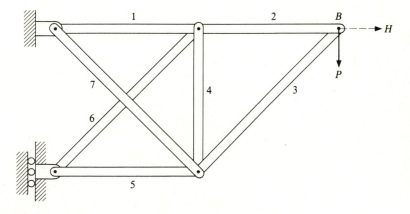

Figure 4-18

forces caused by P, as determined in Example 4-9. Thus

$$F_1 = P + H \qquad F_2 = P + H \qquad F_3 = -1.41P \qquad F_4 = 0$$

$$F_5 = -2P \qquad F_6 = 0 \qquad F_7 = 1.41P$$

The forces can then be substituted into the strain-energy relations to obtain U_1, U_2, etc. The displacement of B in the horizontal direction is then calculated from Eq. (4-32), where,

$$(\delta_B)_H = \frac{\partial U}{\partial H} = 2\frac{(P + H)L}{AE}$$

However, $H = 0$, and the final result is

$$(\delta_B)_H = 2\frac{PL}{AE}$$

The displacement of any point in the structure in any direction can be found by (1) placing a dummy force at that point in the direction desired, (2) determining the strain energy of the system due to all applied forces and the dummy force, (3) substituting the total strain energy into Eq. (4-32) and performing the partial differentiation, and (4) equating the dummy force to zero.

The following examples illustrate the method for various applications of statically determinate problems.

Example 4-11 Determine the vertical deflection and slope at point A of the end-loaded cantilever beam shown in Fig. 4-19. The stiffness of the beam is EI_z. (Neglect transverse shear stresses.)

SOLUTION Since P is applied at point A and in the vertical direction,

$$\delta_A = \frac{\partial U}{\partial P}$$

The strain-energy formulation for a beam in bending is

$$U_b = \frac{1}{2EI} \int_0^L M_z^2 \, dx$$

The bending moment as a function of x is

$$M_z = -Px \qquad 0 < x < L$$

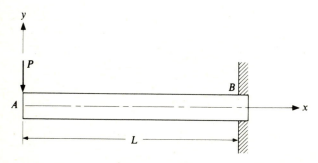

Figure 4-19

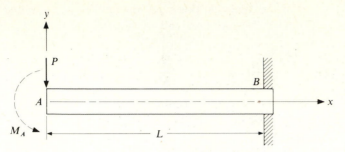

Figure 4-20

Therefore
$$U_b = \frac{1}{2EI_z} \int_0^L (-Px)^2 \, dx = \frac{P^2 L^3}{6EI_z}$$

Hence
$$\delta_A = \frac{\partial U_b}{\partial P} = \frac{PL^3}{3EI_z}$$

To obtain the slope at point A, it is necessary to apply a dummy moment M_A at A so that Eq. (4-33) can be used. This is shown in Fig. 4-20. The bending moment as a function of x is then

$$M_z = -Px - M_A \qquad 0 < x < L$$

$$U_b = \frac{1}{2EI_z} \int_0^L (-Px - M_A)^2 \, dx = \frac{1}{2EI_z} \left(\frac{P^2 L^3}{3} + PM_A L^2 + M_A^2 L \right)$$

To find the slope, Eq. (4-33) is used:

$$\theta_A = \frac{\partial U_b}{\partial M_A} = \frac{1}{2EI_z} (PL^2 + 2M_A L)$$

However, since M_A is a dummy moment, $M_A = 0$ and

$$\theta_A = \frac{PL^2}{2EI_z}$$

Since θ_A is positive, the beam at point A has rotated in the same direction as M_A.

Note that the applied *concentrated* forces or moments necessary for Eqs. (4-32) and (4-33) are not functions of x. This means that the partial differentiation with respect to the forces or moments can be performed before integration. That is, since,

$$U_b = \frac{1}{2} \int \frac{1}{EI_z} M_z^2 \, dx$$

differentiation with respect to a force, say P_i, when dealing with Eq. (4-32) can be written as

$$\delta_i = \frac{\partial U_b}{\partial P_i} = \int \frac{1}{EI_z} \left(M_z \frac{\partial M_z}{\partial P_i} \right) dx \qquad (4\text{-}34)$$

or if Eq. (4-33) is being used, a similar equation can be written for rotations:

$$\theta_i = \frac{\partial U_b}{\partial M_i} = \int \frac{1}{EI_z} \left(M_z \frac{\partial M_z}{\partial M_i} \right) dx \qquad (4\text{-}35)$$

This greatly simplifies part of the analysis. For example, apply this to finding θ_A of Example 4-11. Since final differentiation will be with respect to M_A, Eq. (4-35) becomes

$$\theta_A = \frac{\partial U_b}{\partial M_A} = \frac{1}{EI_z} \int M_z \frac{\partial M_z}{\partial M_A} \, dx \qquad (a)$$

where EI_z is considered to be constant. Since $M_z = -Px - M_A$, we have $\partial M_z / \partial M_A = -1$. Substituting these terms into Eq. (a) results in

$$\theta_A = \frac{\partial U_b}{\partial M_A} = \frac{1}{EI_z} \int_0^L (-Px - M_A)(-1) \, dx$$

Once differentiation is completed, the dummy load can be set equal to zero. Thus $M_A = 0$, and the above reduces to

$$\theta_A = \frac{1}{EI_z} \int_0^L (-Px)(-1) \, dx = \frac{PL^2}{2EI_z}$$

The advantages of Eqs. (4-34) and (4-35) are twofold: (1) when a dummy load is necessary, it can be set equal to zero earlier in the analysis, making the integration easier; (2) the integration of Eqs. (4-34) and (4-35) is easier than Eqs. (4-32) and (4-33) when the moment M_z is a function of more than one load.

Example 4-12 Using Castigliano's theorem, determine the vertical deflection at point A of the cantilever beam shown in Fig. 4-21a.† The stiffness of the beam is $EI_z = 10 \times 10^4 \, \text{N·m}^2$, $P = 500 \, \text{N}$, $L = 1 \, \text{m}$, and $b = 0.2 \, \text{m}$. (Neglect shear effects.)

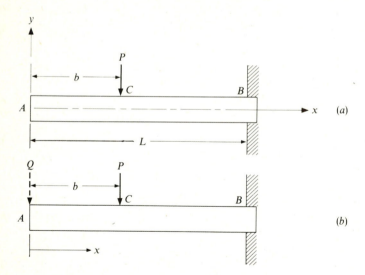

Figure 4-21

† Note that for cantilever beams the expression for the bending moment is usually slightly simpler if the variable x originates from the free end rather than the wall.

SOLUTION Since there is no force at A, a dummy force Q is introduced. The bending-moment equation is then

$$M_z = \begin{cases} -Qx & 0 < x < b \qquad (a) \\ -Qx - P(x - b) & b < x < L \qquad (b) \end{cases}$$

Since

$$\delta_A = \frac{\partial U}{\partial Q}\bigg|_{Q=0}$$

it is necessary to find $\partial M_z / \partial Q$. From Eqs. (a) and (b)

$$\frac{\partial M_z}{\partial Q} = -x \qquad 0 < x < L$$

Thus

$$\delta_A = \frac{\partial U}{\partial Q} = \frac{1}{EI_z} \left\{ \int_0^b (-Qx)(-x)\, dx + \int_b^L [-Qx - P(x - b)](-x)\, dx \right\}$$

However, since $Q = 0$,

$$\delta_A = \frac{\partial U}{\partial Q}\bigg|_{Q=0} = \frac{1}{EI_z} \int_b^L P(x - b)x\, dx$$

and integration yields

$$\delta_A = \frac{1}{EI_z} \left[\frac{P}{3}(L^3 - b^3) - \frac{Pb}{2}(L^2 - b^2) \right]$$

which reduces to

$$\delta_A = \frac{P}{6EI_z}(2L^3 - 3bL^2 + b^3) = \frac{500}{(6)(10 \times 10^4)}[(2)(1^3) - (3)(0.2)(1^2) + (0.2)^3]$$

$$= 1.17 \times 10^{-3}\ \text{m} = 1.17\ \text{mm}$$

Example 4-13 Using Castigliano's theorem, determine the deflection at the midspan of the uniformly loaded simply supported beam shown in Fig. 4-22a. The stiffness of the beam is EI_z. (Neglect shear effects.)

SOLUTION Again there is no applied (concentrated) force at the point where deflection is desired. Thus, a dummy load Q is placed at midspan, as shown in Fig. 4-22b. Solving for the reactions yields

$$R_A = R_B = \tfrac{1}{2}(wL + Q)$$

The bending-moment equation is

$$M_z = \begin{cases} \tfrac{1}{2}(wL + Q)x - \tfrac{1}{2}wx^2 & 0 < x < \dfrac{L}{2} \\[2mm] \tfrac{1}{2}(wL + Q)x - \tfrac{1}{2}wx^2 - Q\left(x - \dfrac{L}{2}\right) & \dfrac{L}{2} < x < L \end{cases}$$

However, thanks to symmetry, the energy in the first half of the beam is the same as the other half. Thus, it is only necessary to integrate over half the beam and double the results. Therefore

$$M_z = \tfrac{1}{2}(wL + Q)x - \tfrac{1}{2}wx^2 \qquad 0 < x < \dfrac{L}{2}$$

and

$$\frac{\partial M_z}{\partial Q} = \frac{x}{2}$$

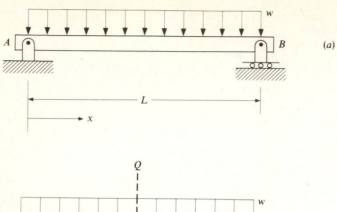

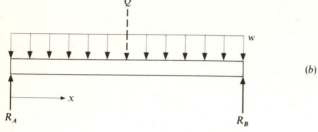

Figure 4-22

The deflection at midspan is $\delta_Q = \partial U/\partial Q|_{Q=0}$. Thus

$$\delta_Q = \frac{\partial U}{\partial Q}\bigg|_{Q=0} = \frac{2}{EI_z}\int_0^{L/2} [\tfrac{1}{2}(wL+0)x - \tfrac{1}{2}wx^2]\frac{x}{2}\,dx$$

Integration yields

$$\delta_Q = \frac{5}{384}\frac{wL^4}{EI_z}$$

Example 4-14 For the beam shown in Fig. 4-23, determine the deflection of point A using Castigliano's theorem. The stiffness of the beam is $EI_z = 20 \times 10^6$ lb·in².

SOLUTION Substitute P_A and P_B for the loads at points A and B as shown.† The deflection at point A is

$$\delta_A = \frac{\partial U}{\partial P_A}$$

The bending-moment equations are

$$M_z = \begin{cases} -P_A x & 0 < x < 30 \\ -P_A x - P_B(x-30) & 30 < x < 50 \end{cases}$$

and

$$\frac{\partial M_z}{\partial P_A} = -x \qquad 0 < x < 50$$

† Note that although the two applied forces are equal, it is important that they be considered distinctly different by virtue of the fact that they are acting at two different locations.

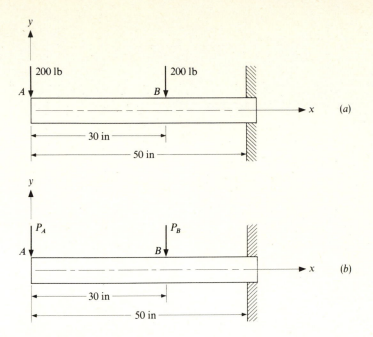

Figure 4-23

Thus
$$\delta_A = \frac{\partial U}{\partial P_A} = \frac{1}{EI_z} \left\{ \int_0^{30} (-P_A x)(-x)\, dx + \int_{30}^{50} [-P_A x - P_B(x - 30)](-x)\, dx \right\}$$

At this point, once differentiation is performed, it is acceptable to write $P_A = P_B = P$. Integration of the above yields

$$\delta_A = 5.033 \times 10^4 \frac{P}{EI_z}$$

Substituting $P = 200$ lb and $EI_z = 20 \times 10^6$ lb·in² results in

$$\delta_A = 0.503 \text{ in}$$

Example 4-15 Using Castigliano's theorem, determine the deflection of point A of the step shaft shown in Fig. 4-24. The moment of inertia of the beam between points A and B is I_1, and from point B to C the moment of inertia is $I_2 = 2I_1$. The entire beam is made of a material with a modulus of elasticity of E.

SOLUTION The bending moment is

$$M_z = -Px \qquad 0 < x < L$$

Therefore
$$\frac{\partial M_z}{\partial P} = -x$$

Since the *moment of inertia* is a discontinuous function of x, the integration must be divided into two parts. Thus

$$\delta_A = \frac{\partial U}{\partial P} = \frac{1}{EI_1} \int_0^{L/2} (-Px)(-x)\, dx + \frac{1}{EI_2} \int_{L/2}^{L} (-Px)(-x)\, dx$$

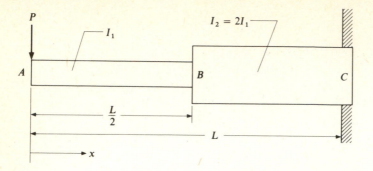

Figure 4-24

Evaluating the integrals and substituting $I_2 = 2I_1$ yields

$$\delta_A = \frac{3}{16} \frac{PL^3}{EI_1}$$

Example 4-16 For the wire form shown in Fig. 4-25a determine the deflection of point B in the direction of the applied force P. (Neglect direct shear effects.) The diameter of the wire is d, the modulus of elasticity is E, and Poisson's ratio is ν.

SOLUTION Using *free-body diagrams* the reader should verify that the reactions in elements BC, CD, and GD are as shown.

The strain energy in BC is due to bending alone:

$$\frac{\partial U_{BC}}{\partial P} = \frac{1}{EI_z} \int_0^a (-Py)(-y)\, dy = \frac{Pa^3}{3EI_z} \tag{a}$$

The strain energy in CD is due to torsion and bending,†

$$\frac{\partial U_{CD}}{\partial P} = \frac{2(1+\nu)}{JE}(Pa)(a)(b) + \frac{1}{EI_z}\int_0^b (-Px)(-x)\, dx$$

$$= \frac{2(1+\nu)Pa^2b}{JE} + \frac{Pb^3}{3EI_z} \tag{b}$$

The strain energy in DG is due to an axial load and bending in two planes:‡

$$\frac{\partial U_{DG}}{\partial P} = \frac{Pc}{AE} + \frac{1}{EI_z}\int_0^c (Pb)(b)\, dz + \frac{1}{EI_z}\int_0^c (Pa)(a)\, dz$$

$$= \frac{Pc}{AE} + \frac{Pb^2c}{EI_z} + \frac{Pa^2c}{EI_z} \tag{c}$$

† Note, for torsion, $U_t = [(1+\nu)/JE]T^2L$. Therefore

$$\frac{\partial U_t}{\partial P} = \frac{2(1+\nu)}{JE} T \frac{\partial T}{\partial P} L$$

‡ Likewise, for axial loading

$$U_a = \frac{1}{2}\frac{F^2L}{AE}$$

therefore

$$\frac{\partial U_a}{\partial P} = \frac{F}{AE}\frac{\partial F}{\partial P} L$$

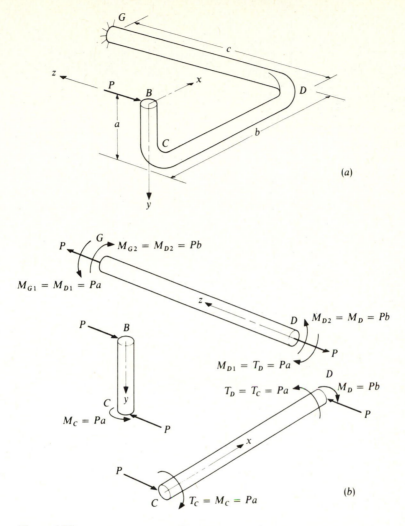

Figure 4-25

The deflection of point B in the direction of P is

$$(\delta_B)_P = \frac{\partial U}{\partial P} = \frac{\partial U_{BC}}{\partial P} + \frac{\partial U_{CD}}{\partial P} + \frac{\partial U_{DG}}{\partial P}$$

Therefore after some rearranging and using the fact that for a circular section $J = 2I_z$, the sum of Eqs. (a) (b), and (c) reduces to

$$(\delta_B)_P = \frac{P}{EI_z}\left[\tfrac{1}{3}(a^3 + b^3) + c(a^2 + b^2) + (1 + \nu)a^2b + c\frac{I_z}{A}\right]$$

Example 4-17 Rods BC and CD, shown in Fig. 4-26, are pin-connected to each other at point C and pin-connected to the ground. If for each member the length is L, the cross-sectional area is A, and the modulus of elasticity is E, determine the vertical deflection of point C when load P is applied.

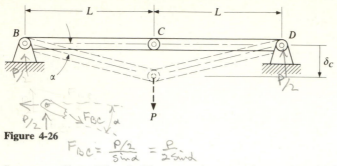

Figure 4-26

$$F_{BC} = \frac{P/2}{\sin\alpha} = \frac{P}{2\sin\alpha}$$

SOLUTION Since BC and CE are two force members, only axial forces are transmitted. Also, since the system is symmetric, the strain energies in each member are identical and the total strain energy is twice the energy in, say, member BC. Thus

$$U = 2U_{BC} = \frac{F_{BC}^2 L}{AE} \tag{a}$$

The force in BC is obtained from applying the equilibrium equations in Fig. 4-26. Thus

$$F_{BC} = \frac{P}{2\sin\alpha}$$

If α is small, $\sin\alpha \approx \alpha$, where α is in radians. Thus F_{BC} can be approximated by

$$F_{BC} \approx \frac{P}{2\alpha} \tag{b}$$

The angle α, however, is dependent on the value of P. Thus, F_{BC} will not be a linear function of P. Hence, since $\delta_C = \partial U/\partial P$, δ_C will *not* be a linear function of P. This means that Castigliano's method *cannot* be used, as a linear load-deflection relationship is necessary.

It will be shown later that $\alpha = (P/AE)^{1/3}$. Thus, from Eq. (b),

$$F_{BC} = \tfrac{1}{2}(P^2 AE)^{1/3}$$

and

$$U = \frac{L}{4}\frac{(P^2 AE)^{2/3}}{AE} = \frac{L}{4}\left(\frac{P^4}{AE}\right)^{1/3}$$

If Castigliano's method is used,

$$\delta_C = \frac{\partial U}{\partial P} = \frac{L}{3}\left(\frac{P}{AE}\right)^{1/3}$$

which is *incorrect*, as δ_C is not a linear function of P.

The solution to this problem is best obtained by a geometric approach. From Fig. 4-26 it can be seen that

$$L + \delta_{BC} = \sqrt{L^2 + \delta_C^2} = L\sqrt{1 + \left(\frac{\delta_C}{L}\right)^2} \tag{c}$$

The square-root term can be expanded into a series using the binomial expansion theorem. If $\delta_C/L \ll 1$, terms of $(\delta_C/L)^4$ and higher can be ignored as they will be negligible. Thus

$$\sqrt{1 + \left(\frac{\delta_C}{L}\right)^2} = \left[1 + \left(\frac{\delta_C}{L}\right)^2\right]^{1/2} \approx 1 + \frac{1}{2}\left(\frac{\delta_C}{L}\right)^2$$

and substitution into Eq. (c) yields

$$L + \delta_{BC} \approx L + \frac{L}{2}\left(\frac{\delta_C}{L}\right)^2$$

Canceling L out results in

$$\left(\frac{\delta_C}{L}\right)^2 \approx \frac{2\delta_{BC}}{L} \tag{d}$$

Since $\delta_C/L \approx \alpha$ and $\delta_{BC} = F_{BC}L/AE$, Eq. (d) becomes

$$\alpha^2 \approx \frac{2F_{BC}}{AE}$$

Thus, since $F_{BC} \approx P/2\alpha$,

$$\alpha^3 \approx \frac{P}{AE} \qquad \text{or} \qquad \left(\frac{\delta_C}{L}\right)^3 \approx \frac{P}{AE}$$

Solving for δ_C yields

$$\delta_C \approx L\left(\frac{P}{AE}\right)^{1/3} \qquad \text{correct solution}$$

which is 3 times that found by Castigliano's method.

The deflections of curved beams can be handled very effectively using Castigliano's theorem, especially when the radius of curvature of the beam is much larger than the lateral dimensions of the beam cross section. When this occurs, it is acceptable to use straight-beam equations for energy, to ignore direct axial and transverse shear forces, and to consider only bending. Several examples follow to illustrate the approach.

Example 4-18 The curved beam shown in Fig. 4-27a has a radius of curvature R, modulus E, and a sectional moment of inertia I_z about an axis out of the page directed through the centroid of an area section. Determine the horizontal and vertical deflections of point A considering only bending due to the application of the horizontal force P.

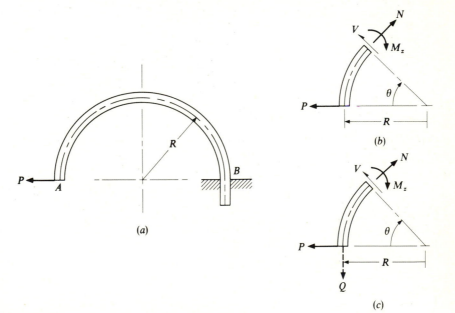

(a)

(b)

(c)

Figure 4-27

SOLUTION It is necessary to establish how the bending moment varies throughout the beam. When the variable θ is used, as shown in Fig. 4-27b, the bending moment can be determined as a function of θ. If this is done, an infinitesimal length of beam will be $R \, d\theta$. Substituting $R \, d\theta$ for dx in Eq. (4-34) gives

$$\delta_i = \frac{\partial U}{\partial P_i} = \int \frac{M_z}{EI_z} \frac{\partial M_z}{\partial P_i} R \, d\theta \qquad (a)$$

Summing moments at the center of the break at θ yields

$$M_z = -PR \sin \theta \qquad 0 < \theta < \pi \qquad (b)$$

The horizontal deflection $(\delta_A)_H$ of point A is given by $\partial U / \partial P$; thus $\partial M_z / \partial P$ is desired. From Eq. (b)

$$\frac{\partial M_z}{\partial P} = -R \sin \theta \qquad (c)$$

Substitution of Eqs. (b) and (c) into Eq. (a) results in

$$(\delta_A)_H = \frac{\partial U}{\partial P} = \int_0^\pi \frac{1}{EI_z} (-PR \sin \theta)(-R \sin \theta) R \, d\theta$$

Since EI_z, P, and R are constants, the integral simplifies to

$$(\delta_A)_H = \frac{PR^3}{EI_z} \int_0^\pi \sin^2 \theta \, d\theta \qquad (d)$$

After using the trigonometric identity $\sin^2 \theta = (1 - \cos 2\theta)/2$, integration of Eq. ($d$) reduces it to

$$(\delta_A)_H = \frac{\pi}{2} \frac{PR^3}{EI_z}$$

For the vertical deflection $(\delta_A)_V$ of point A a dummy force Q must be added, as shown in Fig. 4-27c, where

$$(\delta_A)_V = \frac{\partial U}{\partial Q} \bigg|_{Q=0} \qquad (e)$$

The bending moment as a function of θ is then given by

$$M_z = -PR \sin \theta + QR(1 - \cos \theta) \qquad 0 < \theta < \pi \qquad (f)$$

and since $\partial M_z / \partial Q$ will be needed,

$$\frac{\partial M_z}{\partial Q} = R(1 - \cos \theta) \qquad (g)$$

Now that differentiation has been completed, Eqs. (f) and (g) can be substituted into Eq. (e) with $Q = 0$. This results in

$$(\delta_A)_V = \int_0^\pi \frac{1}{EI_z} (-PR \sin \theta)[R(1 - \cos \theta)] R \, d\theta \qquad (h)$$

Integration of Eq. (h) gives

$$(\delta_A)_V = -2 \frac{PR^3}{EI_z}$$

Example 4-18 neglected the effects of the normal and transverse forces. The effect and order of magnitude of the contribution of these forces can be seen in the following example.

Example 4-19 For Example 4-18, determine the horizontal deflection considering the axial and transverse shear forces in addition to the bending moment.

SOLUTION From Fig. 4-27b the complete internal reactions are

$$M_z = -PR \sin \theta \qquad (a)$$

$$N = P \sin \theta \qquad (b)$$

$$V = -P \cos \theta \qquad (c)$$

Since N and V are varying, Eqs. (4-13) and (4-22) are used with $dx = R\,d\theta$. Differentiation with respect to P can be performed before integration, and the resulting deflection equation is

$$(\delta_A)_H = \frac{\partial U}{\partial P} = \frac{1}{EI_z} \int M_z \frac{\partial M_z}{\partial P} R\,d\theta + \frac{1}{EA} \int N \frac{\partial N}{\partial P} R\,d\theta + \frac{2(1+\nu)}{EA} \int V \frac{\partial V}{\partial P} R\,d\theta \qquad (d)$$

From Eqs. (a) to (c) the derivative terms are

$$\frac{\partial M_z}{\partial P} = -R \sin \theta \qquad (e)$$

$$\frac{\partial N}{\partial P} = \sin \theta \qquad (f)$$

$$\frac{\partial V}{\partial P} = -\cos \theta \qquad (g)$$

Substituting Eqs. (a) to (c) and (e) to (g) into Eq. (d) yields

$$(\delta_A)_H = \frac{1}{EI_z} \int_0^\pi (-PR \sin \theta)(-R \sin \theta) R\,d\theta + \frac{1}{EA} \int_0^\pi (P \sin \theta)(\sin \theta) R\,d\theta$$

$$+ 2\frac{1+\nu}{EA} \int_0^\pi (-P \cos \theta)(-\cos \theta) R\,d\theta$$

Integration and simplification result in

$$(\delta_A)_H = \frac{\pi}{2} \frac{PR^3}{EI_z} + \frac{\pi}{2} \frac{PR}{EA} + \pi(1+\nu) \frac{PR}{EA}$$

or

$$(\delta_A)_H = \frac{\pi}{2} \frac{PR^3}{EI_z} \left[1 + (3+2\nu) \frac{I_z}{AR^2} \right] \qquad (h)$$

It can be seen from the result of Example 4-19 that as the radius of curvature R approaches the radius of gyration $r_g = \sqrt{I_z/A}$, the axial and shear effects become quite large. In addition, as the radius of curvature approaches the radius of gyration (a thick-walled curved beam), the stress distribution due to bending changes (see Sec. 3-8) and the expression for the bending energy must be reformulated.

4-7 DEFLECTIONS OF THICK-WALLED CURVED BEAMS

The bending-stress distribution is given by

$$\sigma_\theta = -\frac{M_z(r_n - r)}{Aer} \qquad (4-36)$$

which is Eq. (3-15) repeated. Upon superposing the stress due to the axial force

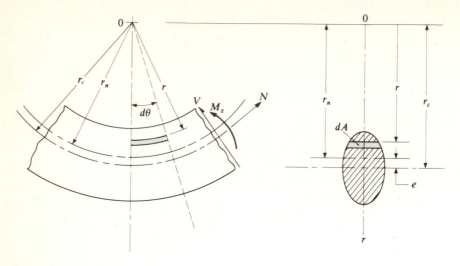

Figure 4-28

N the total normal stress across the section becomes

$$\sigma_\theta = \frac{M_z(r - r_n)}{Aer} + \frac{N}{A} \qquad (4\text{-}37)\dagger$$

In order to formulate the energy, an infinitesimal volume is necessary where the normal stress is constant. Recall Fig. 3-20, repeated here as Fig. 4-28; a volume element of $r \, d\theta \, dA$ is used. Thus the strain energy due to bending and the axial force is

$$U_b = \frac{1}{2} \int \frac{\sigma_\theta^2}{E} dV = \frac{1}{2} \int\int \frac{1}{E} \left[\frac{M_z(r - r_n)}{Aer} + \frac{N}{A} \right]^2 r \, dA \, d\theta$$

$$= \frac{1}{2} \int\int \frac{1}{EA^2} \left[\frac{M_z^2}{e^2} \frac{(r - r_n)^2}{r} + \frac{2M_z N}{e}(r - r_n) + N^2 r \right] dA \, d\theta \qquad (4\text{-}38)\ddagger$$

Integration will be performed on each term in the square brackets separately and with respect to dA first. For the first term within the brackets only $(r - r_n)^2/r$ is a function of the area A. Thus

$$\int \frac{(r - r_n)^2}{r} dA = \int \left(r - 2r_n + \frac{r_n^2}{r} \right) dA = \int r \, dA - 2r_n \int dA + r_n^2 \int \frac{dA}{r}$$

On the right-hand side of the equation, the first integral is simply $r_c A$, where r_c

---

† The strain energies due to M_z and N are not formulated independently; as will be seen later, they are coupled. This is because the centroidal axis for curved beams has a strain due to bending and thus N has an additional deflection caused by M_z, giving rise to an additional work term or strain-energy term.

‡ This neglects the energy due to the radial stress σ_r, given by Eq. (3-17). Normally this energy is extremely small compared with the bending energy.

is the location of the centroid of the cross section. The second integral is A, and the third integral, recalling Eq. (3-14), is A/r_n. Thus

$$\int \frac{(r-r_n)^2}{r}\, dA = (r_c - 2r_n + r_n)A = (r_c - r_n)A = eA$$

Factoring out the first term in the brackets of Eq. (4-38) gives

$$\frac{1}{2}\int\!\!\int \frac{1}{E}\left(\frac{M_z}{Ae}\right)^2 \frac{(r-r_n)^2}{r}\, dA\, d\theta = \frac{1}{2}\int \frac{M_z^2}{EAe}\, d\theta \qquad (4\text{-}39)$$

Returning to Eq. (4-38), we see that the second term, $r - r_n$ is a function of A. Thus

$$\int (r - r_n)\, dA = (r_c - r_n)A = eA$$

and the second term in the integral of Eq. (4-38) is

$$\frac{1}{2}\int\!\!\int \frac{2M_z N}{EA^2 e}(r - r_n)\, dA\, d\theta = \int \frac{M_z N}{EA}\, d\theta \qquad (4\text{-}40)$$

The final term in the brackets involves the integral of $r\, dA$, which is

$$\int r\, dA = r_c A$$

Thus, the last term in the integral of Eq. (4-38) is

$$\frac{1}{2}\int\!\!\int \frac{N^2 r}{EA^2}\, dA\, d\theta = \frac{1}{2}\int \frac{N^2 r_c}{EA}\, d\theta \qquad (4\text{-}41)$$

When Eqs. (4-39) to (4-41) are added, the strain energy due to bending and axial loading is

$$U_b = \frac{1}{2}\int \frac{M_z^2}{EAe}\, d\theta + \int \frac{M_z N}{EA}\, d\theta + \frac{1}{2}\int \frac{N^2 r_c}{EA}\, d\theta \qquad (4\text{-}42)$$

The second integral in Eq. (4-42) is the additional coupling term between the bending moment and the axial force.

The energy due to the shear force is approximated by

$$U_s \approx \int \frac{1+\nu}{EA} V^2 r_c\, d\theta \qquad (4\text{-}43)$$

Summing Eqs. (4-42) and (4-43) and applying Castigliano's theorem, we have

$$\delta_i = \frac{\partial U}{\partial P_i} = \int \frac{M_z}{EAe}\frac{\partial M_z}{\partial P_i}\, d\theta + \int \frac{M_z}{EA}\frac{\partial N}{\partial P_i}\, d\theta + \int \frac{N}{EA}\frac{\partial M_z}{\partial P_i}\, d\theta$$

$$+ \int \frac{N r_c}{EA}\frac{\partial N}{\partial P_i}\, d\theta + \int \frac{2(1+\nu)r_c V}{EA}\frac{\partial V}{\partial P_i}\, d\theta \qquad (4\text{-}44)\dagger$$

† Care must be taken when evaluating the sign of M_z. The sign convention for positive M_z is shown in Fig. 4-28.

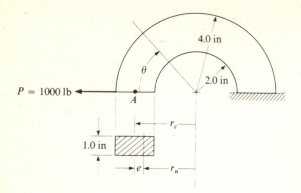

Figure 4-29

Example 4-20 Repeating Example 4-19 but for a thick-walled steel ring, as shown in Fig. 4-29, determine the deflection of point A using Eq. (4-44) and compare the results with Eq. (d) of Example 4-19.

SOLUTION Recall from Example 4-19 that the internal reactions and derivatives are

$$M_z = -PR \sin \theta \qquad \frac{\partial M_z}{\partial P} = -R \sin \theta$$

$$N = P \sin \theta \qquad \frac{\partial N}{\partial P} = \sin \theta$$

$$V = -P \cos \theta \qquad \frac{\partial V}{\partial P} = -\cos \theta$$

where $R = r_c$. Substitution of the above terms into Eq. (4-44) yields

$$\delta_A = \frac{PR^2}{EAe} \int_0^\pi \sin^2 \theta \, d\theta - \frac{PR}{EA} \int_0^\pi \sin^2 \theta \, d\theta - \frac{PR}{EA} \int_0^\pi \sin^2 \theta \, d\theta$$

$$+ \frac{PR}{EA} \int_0^\pi \sin^2 \theta \, d\theta + \frac{2(1+\nu)PR}{EA} \int_0^\pi \cos^2 \theta \, d\theta$$

Evaluating the integrals results in

$$\delta_A = \frac{\pi}{2} \frac{PR}{EA} \left(\frac{R}{e} + 1 + 2\nu \right) \qquad (a)$$

From Table 3-1, r_n is found to be

$$r_n = \frac{(2.0)(1.0)}{1.0 \ln \frac{4}{2}} = 2.885 \text{ in}$$

and from Fig. 4-29 it can be seen that $R = r_c = 3.0$ in. Thus $e = 3.000 - 2.885 = 0.115$. For steel, $E = 30 \times 10^6$ lb/in^2, and $\nu = 0.3$. Substituting the appropriate values into Eq. (a) results in

$$\delta_A = \frac{\pi}{2} \frac{(1000)(3)}{(30 \times 10^6)(2)(1)} \left[\frac{3.0}{0.115} + 1 + (2)(0.3) \right] = 0.00217 \text{ in}$$

From Eq. (d) of Example 4-19, which uses the straight-beam bending-energy formulation, the final result was given by Eq. (h).

$$\delta_A = \frac{\pi}{2} \frac{(1000)(3.0)^3}{(30 \times 10^6)(\frac{1}{12})(1)(2^3)} \left\{ 1 + [3 + (2)(0.3)] \frac{(\frac{1}{12})(1)(2^3)}{(1)(2)(3.0)^2} \right\} = 0.00240 \text{ in}$$

which is 11 percent greater than that predicted by Eq. (4-44).

4-8 CASTIGLIANO'S THEOREM APPLIED TO STATICALLY INDETERMINATE PROBLEMS

Castigliano's method can be used effectively in determining the reactions of statically indeterminate problems. The results generally arise in the form of n simultaneous equations in the unknown reactions for a structure statically indeterminate of order n. The procedure is relatively straightforward. If the structure is indeterminate of order n, then n reaction forces, or moments, can be regarded as unknown *applied* forces on the structure subject to the constraints of the deflections at the reaction points. The deflections at the supports are generally known (normally zero), and Castigliano's equation can then be used. When Castigliano's theorem is applied to statically indeterminate problems, an extremely important requirement must be understood. Once the n unknown reactions are selected to be considered as applied, the *remaining* unknown reactions must either be solved for in terms of the n unknown reactions and the actual applied forces or not used in the energy formulation. A number of examples follow to illustrate the technique.

Example 4-21 Using Castigliano's theorem, determine the reactions at points A and B of the beam shown in Fig. 4-30a.

SOLUTION First, construct the free-body diagram. Since only two equilibrium equations are applicable and there are three unknowns, the structure is indeterminate of the order $n = 1$. This means that one (and *only* one) reaction can be regarded as an applied load and the remaining reactions must be written in terms of P and the assumed applied load. Two solutions will be arrived at by separately considering R_A and M_B as applied.

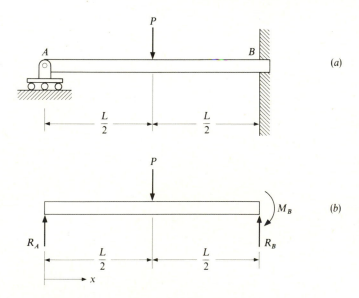

Figure 4-30

Approach 1

Consider R_A as an applied force; then

$$M_z = \begin{cases} R_A x & 0 < x < \dfrac{L}{2} \\[2ex] R_A x - P\left(x - \dfrac{L}{2}\right) & \dfrac{L}{2} < x < L \end{cases}$$

$$\frac{\partial M_z}{\partial R_A} = x \qquad 0 < x < L$$

The vertical deflection of point A is zero; thus†

$$(\delta_A)_V = \frac{\partial U}{\partial R_A} = \frac{1}{EI_z}\left\{\int_0^{L/2} (R_A x)x\,dx + \int_{L/2}^L \left[R_A x - P\left(x - \frac{L}{2}\right)\right]x\,dx\right\} = 0$$

Simplifying yields

$$\int_0^L R_A x^2\,dx - \int_{L/2}^L P\left(x - \frac{L}{2}\right)x\,dx = 0$$

Evaluating the integrals and solving for R_A yields

$$R_A = \tfrac{5}{16}P$$

Applying the equilibrium equations to Fig. 4-30b results in

$$R_B = \tfrac{11}{16}P \qquad \text{and} \qquad M_B = \tfrac{3}{16}PL$$

Approach 2

In this solution, M_B is considered to be an applied moment. In order to determine how the moment varies within the beam either R_A or R_B must be used, but in this approach P and M_B are applied. Thus, R_A or R_B must be written in terms of P and M_B. Summing moments about point B results in

$$R_A = \frac{P}{2} - \frac{M_B}{L}$$

and the bending-moment equation is ‡

$$M_z = \begin{cases} \left(\dfrac{P}{2} - \dfrac{M_B}{L}\right)x & 0 < x < \dfrac{L}{2} \\[2ex] \left(\dfrac{P}{2} - \dfrac{M_B}{L}\right)x - P\left(x - \dfrac{L}{2}\right) & \dfrac{L}{2} < x < L \end{cases}$$

$$\frac{\partial M_z}{\partial M_B} = -\frac{x}{L} \qquad 0 < x < L$$

The slope at point B is zero; thus

$$\theta_B = \frac{\partial U}{\partial M_B} = \frac{1}{EI_z}\left\{\int_0^{L/2} \left(\frac{P}{2} - \frac{M_B}{L}\right)(x)\left(-\frac{x}{L}\right)dx\right.$$

$$\left. + \int_{L/2}^L \left[\left(\frac{P}{2} - \frac{M_B}{L}\right)x - P\left(x - \frac{L}{2}\right)\right]\left(-\frac{x}{L}\right)dx\right\} = 0$$

† Note that the remaining unknown reactions R_B and M_B are not used in the energy formulation.

‡ Since in this approach M_B and P are the applied forces, the energy is not formulated using R_A and R_B.

Simplifying yields

$$-\int_0^L \left(\frac{P}{2}-\frac{M_B}{L}\right)(x)\left(\frac{x}{L}\right)dx + \int_{L/2}^L P\left(x-\frac{L}{2}\right)\left(\frac{x}{L}\right)dx = 0$$

Evaluating the integrals and solving for M_B results in

$$M_B = \tfrac{3}{16}PL$$

and the remaining unknowns, R_A and R_B, are found from equilibrium. Thus, both approaches yield the same solution, but approach 1 is slightly simpler.

Example 4-22 Determine the reactions at points C and D of the structure shown in Fig. 4-31a. All members have equal cross section A and equal modulus of elasticity E; members 1, 2, 4, and 5 are of length L, and members 3, 6, and 7 are of length $1.41L$.

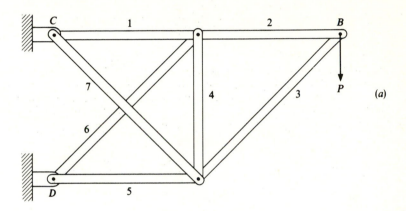

(a)

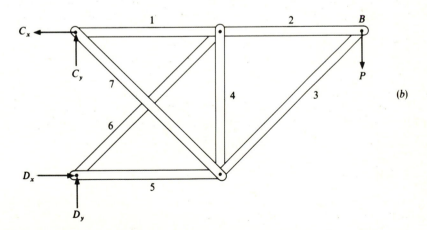

(b)

Figure 4-31

SOLUTION The free-body diagram of the structure is shown in Fig. 4-31b. Summing moments about point C yields $D_x = 2P$, and thus $C_x = 2P$. It can be seen that only one equilibrium equation remains, $C_y + D_y = P$. After using all the pertinent equilibrium equations, there remains only one equation in two unknowns. Therefore, the structure is statically indeterminate of order 1. This means that one unknown can be considered as an applied force. If D_y is selected as the applied force,

$$C_y = P - D_y$$

Solving for the axial forces in the members in terms of P and D_y only, results in

$$F_1 = P + D_y \qquad F_2 = P \qquad F_3 = -1.41P \qquad F_4 = D_y$$
$$F_5 = -(2P - D_y) \qquad F_6 = -1.41D_y \qquad F_7 = 1.41(P - D_y)$$

where positive values indicate tension and negative compression. Since the deflection of point D in the y direction is zero,

$$(\delta_D)_y = \frac{\partial U}{\partial D_y} = 0$$

where $U = U_1 + U_2 + U_3 + \cdots + U_7$. As in Example 4-9, the form of U_i is

$$U_i = \frac{1}{2}\left(\frac{F^2 L}{AE}\right)_i \qquad i = 1, 2, \ldots, 7$$

U_2 and U_3 have no terms involving D_y and hence can be ignored. Thus

$$\frac{\partial U}{\partial D_y} = \frac{1}{EA}[(P + D_y)L_1 + D_y L_4 - (2P - D_y)L_5 + (1.41)^2 D_y L_6 - (1.41)^2(P - D_y)L_7] = 0$$

and since $L_1 = L_4 = L_5 = L$ and $L_6 = L_7 = 1.41L$,

$$\frac{L}{EA}[(P + D_y) + D_y - (2P - D_y) + (1.41)^3 D_y - (1.41)^3(P - D_y)] = 0$$

Solving for D_y yields

$$D_y = 0.44P$$

and from equilibrium, $C_y = P - D_y$; then

$$C_y = 0.56P$$

Example 4-23 For the ring shown in Fig. 4-32 consider only bending and determine, (a) the bending moment as a function of θ and (b) the deflection of point B upon application of the load P.

SOLUTION (a) Since the structure is symmetric, it is necessary only to analyze a quarter segment, as shown in Fig. 4-32b. Thanks to symmetry, there cannot be any shear force at the surface at point A.† The surface isolation at point C is slightly to the right of the support force; thus symmetry at this point is lost. Since M_A and M_C are unknown, the problem is statically indeterminate. Thus, consider M_A as an applied load; thanks to symmetry there is no rotation at point A. Using Castigliano's theorem in the form of Eq. (4-33) for rotation results in

$$\theta_A = \frac{\partial U}{\partial M_A} = 0$$

Writing the bending moment as a function of θ, we can see from Fig. 4-32c that

$$M_z = -M_A + \frac{PR}{2}(1 - \cos\theta) \tag{a}$$

† If this is not clear, refer to Sec. 7-2.

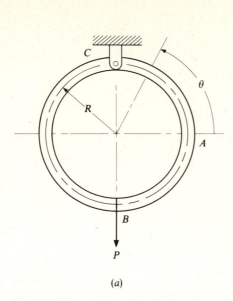

(a)

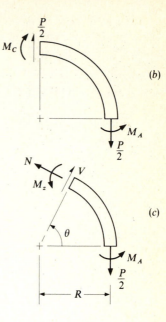

(b)

(c)

Figure 4-32

and
$$\frac{\partial M_z}{\partial M_A} = -1$$

Thus
$$\frac{\partial U}{\partial M_A} = \frac{1}{EI_z}\int_0^{\pi/2}\left[-M_A + \frac{PR}{2}(1-\cos\theta)\right](-1)R\,d\theta = 0$$

Integration yields
$$M_A = \frac{PR}{2}\left(1-\frac{2}{\pi}\right)$$

Substituting this into Eq. (a) yields the solution to part (a).

$$M_z = \frac{PR}{2}\left(\frac{2}{\pi}-\cos\theta\right)$$

M_z is greatest at $\theta = 90°$, where $M_z = M_C$. Thus

$$M_{\max} = M_C = \frac{PR}{\pi}$$

(b) Since in part (a) only one-quarter of the ring was analyzed, the total strain energy U_T is 4 times U. The deflection of point B is then

$$\delta_B = \frac{\partial U_T}{\partial P} = 4\frac{\partial U}{\partial P}$$

Since
$$M_z = \frac{PR}{2}\left(\frac{2}{\pi}-\cos\theta\right)$$

we have
$$\frac{\partial M_z}{\partial P} = \frac{R}{2}\left(\frac{2}{\pi}-\cos\theta\right) \qquad \text{and} \qquad \delta_B = \frac{4}{EI_z}\int_0^{\pi/2}P\left[\frac{R}{2}\left(\frac{2}{\pi}-\cos\theta\right)\right]^2 R\,d\theta$$

Integration yields

$$\delta_B = 0.15\frac{PR^3}{EI_z}$$

Thus, it can be seen from the examples illustrated, that Castigliano's theorem is an extremely powerful technique for performing the deflection analysis of a structure.

4-9 THE DUMMY-LOAD METHOD

The dummy-load method is directly based on the complementary-energy theorem and is thus applicable to nonlinear load-displacement problems.

The method is basically the same as that used in Example 4-3. Recall, from Sec. 4-5, that when an infinitesimal force ΔP_i is placed on a loaded structure, the complementary work, due to ΔP_i is

$$\Delta W_C = \delta_i \, \Delta P_i$$

where δ_i is the deflection of point i on the structure in the direction of ΔP_i. The increase $\Delta \Phi$ in complementary energy of the system is equal to ΔW_C. The increase in Φ of the system is equal to the sum of the increases in the complementary energy of all members. For each member, the increase in complementary energy is equal to that part of ΔP_i which performs complementary work on the member.

Example 4-24 Determine the vertical deflection of point B of the structure shown in Fig. 4-33. Members BC and BD have equal length L, area A, and modulus E.

SOLUTION Add an infinitesimal force (dummy load) ΔP to P. Thus, the complementary work due to ΔP is

$$\Delta W_C = \delta_B \, \Delta P \tag{a}$$

The forces in BC and BD are equal and are found from static analysis to be

$$F_{BC} = F_{BD} = \frac{P}{2 \cos \beta}$$

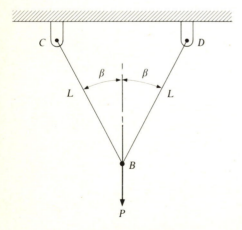

Figure 4-33

Thus the deflections of *BC* and *BD* due to the internal forces are

$$\delta_{BC} = \delta_{BD} = \frac{PL}{(2 \cos \beta)AE}$$

The forces in members *BC* and *BD* due to ΔP are

$$\Delta F_{BC} = \Delta F_{BD} = \frac{\Delta P}{2 \cos \beta}$$

Hence, the complementary work, say on member *BC*, is

$$\Delta(W_C)_{BC} = \delta_{BC} \, \Delta F_{BC}$$

which is the increase in the complementary energy $\Delta \Phi_{BC}$ of member *BC*. Since the system is symmetric, $\Delta \Phi_{BD} = \Delta \Phi_{BC}$ and the total change $\Delta \Phi$ is double that of $\Delta \Phi_{BC}$. Thus

$$\Delta \Phi = 2 \frac{PL}{(2 \cos \beta)AE} \frac{\Delta P}{2 \cos \beta} = \frac{PL}{(2 \cos^2 \beta)AE} \Delta P \qquad (b)$$

Equating Eqs. (*a*) and (*b*) results in

$$\delta_B \, \Delta P = \frac{PL}{(2 \cos^2 \beta)AE} \Delta P$$

The term ΔP cancels, and δ_B is found to be

$$\delta_B = \frac{1}{2 \cos^2 \beta} \frac{PL}{AE}$$

Axial Loading

For the general case of axial loading considering the longitudinal axis of the member to be in the *x* direction, let the deflection of an infinitesimal *dx* element due to the applied loads be $d(\delta)$. Consider ΔF to be the internal force within the element due to a *dummy* load placed on the structure at a given point; then the work performed by ΔF only is $\Delta F d(\delta)$. This is the increase in the complementary energy of the *dx* element. The total increase in complementary energy of the member is determined by integration along the entire length of the member. Thus

$$\Delta \Phi = \int_0^L \Delta F \, d(\delta) \qquad (4\text{-}45)$$

If the material is linear, the deflection of the *dx* element is given by $d(\delta) = (F \, dx)/AE$, where *F* is the internal force in the element due to the *applied* loads. Thus for a linear material Eq. (4-45) reduces to

$$\Delta \Phi = \int_0^L \Delta F \frac{F}{AE} \, dx \qquad (4\text{-}46)$$

If ΔF, *F*, *A*, and *E* are constant, Eq. (4-46) can be written

$$\Delta \Phi = \Delta F \frac{FL}{AE} \qquad (4\text{-}47)$$

Example 4-25 Repeat Example 4-24 but consider the material to be nonlinear, where the tensile load-deflection relationship for each rod is given by

$$F = k\delta^{2/3}$$

where k is a constant.

SOLUTION From Example 4-24 the internal forces in each rod due to the applied and dummy forces are, respectively,

$$F_{BC} = F_{BD} = \frac{P}{2\cos\beta} \quad \text{and} \quad \Delta F_{BC} = \Delta F_{BD} = \frac{\Delta P}{2\cos\beta}$$

Since ΔF in each rod is a constant throughout the length, from Eq. (4-45) the change in complementary energy for each rod is

$$\Delta\Phi = \Delta F \int_0^L d\delta = \Delta F \delta$$

where δ is the deflection of each rod due to the applied forces. Since for the rods, $\delta = (F/k)^{3/2}$, the deflection of each member is

$$\delta = \left[\frac{P/(2\cos\beta)}{k}\right]^{3/2}$$

and the increase in Φ for each member is

$$\Delta\Phi = \Delta F \delta = \frac{\Delta P}{2\cos\beta}\left[\frac{P/(2\cos\beta)}{k}\right]^{3/2}$$

The complementary work due to ΔP is $\delta_B \Delta P$. Equating this to the total increase in the complementary energy of the rods yields

$$W_C = \Delta\Phi_{BC} + \Delta\Phi_{BD} \qquad \delta_B \Delta P = 2\left\{\frac{\Delta P}{2\cos\beta}\left[\frac{P/(2\cos\beta)}{k}\right]^{3/2}\right\}$$

Solving for δ_B and simplifying yields

$$\delta_B = \left(\frac{P}{2k}\right)^{3/2}(\cos\beta)^{-5/2}$$

Torsional Loading

In a manner similar to that for axial loading, the increase in complementary energy for a rod of length L is

$$\Delta\Phi = \int_0^L \Delta T\, d\theta \tag{4-48}$$

where $\Delta T = $ torque being transmitted at position x in rod due to *dummy* load
$d\theta = $ angle of twist of dx element due to *applied* loads

If the material is linear, $d\theta = (1 + \nu)(T/EJ)\, dx$, where T is the torque being transmitted at x due to the *applied* loads. Thus for a linear material Eq. (4-48) reduces to

$$\Delta\Phi = \int_0^L \Delta T\,(1 + \nu)\frac{T}{EJ}\, dx \tag{4-49}$$

If ΔT, T, ν, E, and J are constant throughout the rod,

$$\Delta\Phi = \Delta T \, (1+\nu) \frac{TL}{EJ} \tag{4-50}$$

Bending

For beams in bending, the increase in the complementary energy due to the bending moment only is

$$\Delta\Phi = \int_0^L \Delta M_z \, d\theta \tag{4-51}$$

where ΔM_z = bending moment at x due to *dummy* load
$d\theta$ = rotational deflection of dx element due to *applied* loads

For a linear material, $d\theta = M_z/EI_z \, dx$, where M_z is the internal bending moment due to the *applied* loads. Thus for a linear material Eq. (4-51) reduces to

$$\Delta\Phi = \int_0^L \Delta M_z \frac{M_z}{EI_z} \, dx \tag{4-52}$$

Example 4-6 For the beam shown in Fig. 4-34a, determine the deflection and slope of point using the dummy-load method.

SOLUTION The bending moment due to the applied load is

$$M_z = -Px \qquad 0 < x < L$$

To find the deflection at point A apply a dummy force ΔP, as shown in Fig. 4-34b. Thus

$$\Delta M_z = -\Delta P x \qquad 0 < x < L$$

The work due to the dummy force is $\delta_A \, \Delta P$. The increase in complementary energy is given by

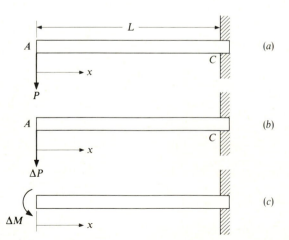

Figure 4-34

Eq. (4-52). Equating the two equations results in

$$\delta_A \, \Delta P = \int_0^L \frac{(P \, \Delta P) x^2 \, dx}{E I_z} = \frac{P \, \Delta P L^3}{3 E I_z}$$

Solving for δ_A yields

$$\delta_B = \frac{PL^3}{3EI_z}$$

To find the slope at point A, apply a dummy moment ΔM at point A, as shown in Fig. 4-34c. Thus

$$\Delta M_z = -\Delta M \qquad 0 < x < L$$

The complementary work due to ΔM is $\theta_A \, \Delta M$, where θ_A is the rotational deflection at point A in the direction of ΔM. Equating this to the increase in complementary energy given by Eq. (4-52) results in

$$\theta_A \, \Delta M = \int_0^L \frac{Px}{EI_z} \Delta M \, dx = \frac{P \, \Delta M L^2}{2 E I_z}$$

Solving for θ_A yields

$$\theta_A = \frac{PL^2}{2EI_z}$$

Deflections due to temperature changes can be handled quite effectively by the dummy-load method. Equation (4-47) can easily be modified by adding on the deflection due to a temperature change. The term FL/AE in Eq. (4-47) represents the free deflection in the member due to the internal force. Thus, when a deflection term is added to account for a temperature change, Eq. (4-47) becomes

$$\Delta \Phi = \left(\frac{FL}{AE} + \alpha \, \Delta T \, L \right) \Delta F$$

or

$$\Delta \Phi = \left(\frac{F}{AE} + \alpha \, \Delta T \right) L \, \Delta F \tag{4-53}$$

where α is the linear temperature coefficient of expansion.

Example 4-27 For the structure in Example 4-24, determine the deflection of point B if in addition to the applied load P the temperature of the structure increases ΔT. The coefficient of linear expansion of the members is α.

Solution The internal forces in each member due to the applied force are equal and are

$$F_{BC} = F_{BD} = \frac{P}{2 \cos \beta} \tag{a}$$

When a dummy load ΔP is applied at point B in the same direction as P, the resulting internal forces are simply

$$\Delta F_{BC} = \Delta F_{BD} = \frac{\Delta P}{2 \cos \beta} \tag{b}$$

Thus, from Eq. (4-53), the total change in the complementary energy is

$$\Delta \Phi = \left(\frac{P}{2 \cos \beta} + \alpha \, \Delta T \right) L \frac{\Delta P}{\cos \beta} \tag{c}$$

Equating this to the complementary work $\delta_B \, \Delta P$ yields

$$\delta_B = \left(\frac{P}{2 \cos \beta} + \alpha \, \Delta T \right) \frac{L}{\cos \beta}$$

4-10 THE DUMMY-LOAD METHOD APPLIED TO STATICALLY INDETERMINATE PROBLEMS

For statically indeterminate problems, the procedure is similar to that followed in Sec. 4-8. That is, if the structure is indeterminate of order n, then n reactions are treated as applied forces.

Example 4-28 The structure shown in Fig. 4-35a contains a cantilever beam of length L_b, area A_b, modulus E, and moment of inertia I_b. In addition, supporting the left end of the beam is a cable of length L_c, area A_c, and modulus E. Determine the reactions at points B and D.

SOLUTION The free-body diagram of the structure is shown in Fig. 4-35b. The structure is statically indeterminate of order 1. Thus, consider one unknown reaction as an applied force, say F_D at point D. The force F_D is subjected to the constraint that point D does not deflect. After isolating the two elements of the structure, the free-body diagrams are as shown in Fig. 4-35c. The bending moment in member BC can be written

$$
M_z = \begin{cases} (F_D \sin \beta)x & 0 < x < \dfrac{L_b}{2} \\[2mm] (F_D \sin \beta)x - P\left(x - \dfrac{L_b}{2}\right) & \dfrac{L_b}{2} < x < L_b \end{cases}
$$

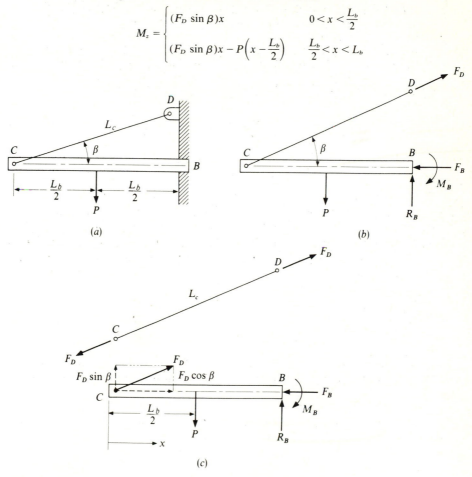

Figure 4-35

Apply a dummy force ΔF_D at point D in the direction of F_D. This is done because the equation for the deflection of point D will be written. The increase in complementary energy of member CD due to ΔF_D is found from Eq. (4-47) and is

$$\Delta\Phi_{CD} = \Delta F_D \frac{F_D L_c}{A_c E} \tag{a}$$

The bending moment in member BC due to the dummy force ΔF_D is

$$\Delta M_z = (\Delta F_D \sin \beta) x \qquad 0 < x < L_b$$

and the axial force transmitted through BC due to the dummy force is $\Delta F_D \cos \beta$. Thus the increase in complementary energy of member BC due to ΔF_D is

$$\Delta\Phi_{BC} = (\Delta F_D \cos \beta) \frac{(F_D \cos \beta) L_b}{A_b E}$$

$$+ \frac{1}{EI_z} \int_0^{L_b/2} (F_D \Delta F_D \sin^2 \beta) x^2 \, dx$$

$$+ \int_{L_b/2}^{L_b} \left[(F_D \sin \beta) x - P\left(x - \frac{L_b}{2} \right) \right] (\Delta F_D \sin \beta)(x) \, dx$$

Integration yields

$$\Delta\Phi_{BC} = F_D \Delta F_D \frac{L_b \cos^2 \beta}{A_b E} + F_D \Delta F_D \frac{L_b^3 \sin^2 \beta}{3EI_b} - \tfrac{5}{48} P \Delta F_D \frac{L_b^3 \sin \beta}{EI_b} \tag{b}$$

The complementary work done by ΔF_D is zero since the deflection of point D is zero. Thus $\Delta\Phi_{CD} + \Delta\Phi_{BC} = 0$. From this, F_D can be solved for, and is found to be

$$F_D = \frac{C_1}{C_2} P$$

where $\qquad C_1 = \dfrac{5}{48} \dfrac{L_b^3 \sin \beta}{I_b} \qquad$ and $\qquad C_2 = \dfrac{L_c}{A_c} + \dfrac{L_b \cos^2 \beta}{A_b} + \dfrac{L_b^3 \sin^2 \beta}{3I_b}$

4-11 DEFLECTIONS OF BEAMS WITH STRAINS IN THE PLASTIC RANGE AND PERMANENT SET

Consider a beam with a rectangular cross section. Recall that for a beam in bending, the deflection is given by

$$\frac{d^2 y_c}{dx^2} = -\frac{\varepsilon_x}{y} \tag{4-54}$$

When plastic behavior occurs in a section, recall that the strain is given by Eq. (3-95), repeated here:

$$\varepsilon_x = -\frac{\sigma_e}{E(c - h_p)} y \tag{4-55}$$

Recall also, that $d^2 y_c / dx^2 = M_z / EI_z$ holds true only for a beam as long as strains are in the elastic range. Thus, in the plastic range, $-\varepsilon_x / y \neq M_z / EI_z$. However, since ε_x is still linear with respect to y, it has the same appearance as that of a strain field still totally in the elastic range. Consider a fictitious bending moment M_f, which *would* be the bending moment *if* the given strain field were totally in the elastic range. Thus, $-\varepsilon_x/y = M_f/EI_z$, or

$$M_f = -EI_z \frac{\varepsilon_x}{y} \tag{4-56}$$

The moment M_f can be used in conjunction with the dummy-load method to determine deflections.†

Equations (3-87) and (3-94) are repeated here as

$$\sigma_e = \frac{M_e c}{I_z} \tag{4-57}$$

and

$$\frac{h_p}{c} = 1 - \sqrt{3 - 2\left|\frac{M_z}{M_e}\right|} \tag{4-58}$$

Substituting Eqs. (4-55), (4-57), and (4-58) into Eq. (4-56) and simplifying yields

$$M_f = \frac{M_e}{\sqrt{3 - 2\left|\dfrac{M_z}{M_e}\right|}} \tag{4-59}$$

Since M_z is the actual bending moment at a given section and a linear function of the applied loading, it can be seen that M_f is a nonlinear function of the applied loading.

Using M_f for M_z in Eq. (4-52) in the zone where plastic behavior exists, we have ‡

$$\delta_{\text{plastic}} \Delta P = \int_{\substack{\text{elastic} \\ \text{zone}}} \frac{M_z \, \Delta M_z}{EI_z} \, dx + \int_{\substack{\text{plastic} \\ \text{zone}}} \frac{M_f \, \Delta M_z}{EI_z} \, dx \tag{4-60}$$

Permanent Set

When the applied loads are released from the beam, the beam unloads elastically. Thus, the corresponding deflection for unloading can be obtained using standard elastic analysis for a beam. The residual deflection is then

$$\delta_{\text{residual}} = \delta_{\text{plastic}} - \delta_{\text{elastic}} \tag{4-61}$$

Example 4-29 A rectangular beam (Fig. 4-36a) is loaded by a vertical force P such that the bending moment at midspan is $M_z = 1.2M_e$.

(a) Determine the equation for the deflection at the midspan due to P.

(b) If P is released, determine the permanent set at midspan.

SOLUTION (a) The free-body diagram of the beam is shown in Fig. 4-36b. Total elastic behavior ceases at x_p, where

$$M_e = \frac{P}{2} x_p$$

or

$$x_p = \frac{2M_e}{P} \tag{a}$$

---

† Castigliano's method cannot be used as it will be shown that M_f is not a linear function of the applied loads, which makes the load-deflection relationship nonlinear.

‡ ΔM_z is used instead of ΔM_f in the second integral, since the dummy load is assumed to be applied before the actual loading and the dummy load causes no plastic behavior.

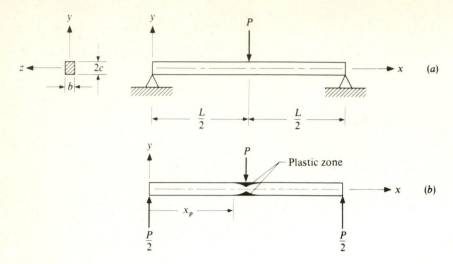

Figure 4-36

The bending moment M_z is given by

$$M_z = \tfrac{1}{2}Px \qquad\qquad (b)$$

and for a dummy load ΔP at midspan

$$\Delta M_z = \tfrac{1}{2}\Delta P\, x \qquad\qquad (c)$$

From Eq. (4-59), in the plastic zone

$$M_f = \frac{M_e}{\sqrt{3 - 2|M_z/M_e|}} = \frac{M_e}{\sqrt{3 - 2(\tfrac{1}{2}Px/M_e)}} = \frac{M_e}{\sqrt{3 - (Px/M_e)}}$$

Since the beam is symmetric, it is only necessary to deal with half of the beam and double the results. Thus from Eq. (4-60)

$$\delta_{\text{plastic}}\,\Delta P = 2\left[\int_0^{x_p} \frac{(\tfrac{1}{2}Px)(\tfrac{1}{2}\Delta P\,x)}{EI_z}\,dx + \int_{x_p}^{L/2} \frac{(M_e/\sqrt{3 - Px/M_e})(\tfrac{1}{2}\Delta P\,x)}{EI_z}\,dx\right]$$

Simplifying yields

$$\delta_{\text{plastic}} = \frac{1}{EI_z}\left(\frac{P}{2}\int_0^{x_p} x^2\,dx + M_e\int_{x_p}^{L/2} \frac{x}{\sqrt{3 - Px/M_e}}\,dx\right)$$

Integration results in

$$\delta_{\text{plastic}} = \frac{1}{EI_z}\left\{\frac{P}{6}x_p^3 - \left[\frac{2}{3}\frac{M_e}{(P/M_e)^2}\left(3 - \frac{P}{M_e}x\right)^{1/2}\left(6 + \frac{P}{M_e}x\right)\right]\Big|_{x_p}^{L/2}\right\}$$

Simplifying and using the fact that $x_p = 2M_e/P$ yields

$$\delta_{\text{plastic}} = \frac{2}{3}\frac{M_e^3}{P^2 EI_z}\left[10 - \left(3 - \frac{PL}{2M_e}\right)^{1/2}\left(6 + \frac{PL}{2M_e}\right)\right] \qquad\qquad (d)$$

At midspan, $M_z = 1.2M_e$. Thus, $PL/4 = 1.2M_e$ or $P = 4.8M_e/L$. Substituting this into Eq. (d) finally yields

$$\delta_{\text{plastic}} = 0.1011\frac{M_e L^2}{EI_z} \qquad\qquad (e)$$

(*b*) The residual deflection or permanent set after P is removed is given by Eq. (4-61). The elastic deflection is found in Appendix D, where it can be shown that

$$\delta_{\text{elastic}} = \frac{PL^3}{48EI_z} = 4.8 \frac{M_e}{L} \frac{L^3}{48EI_z} = 0.1000 \frac{M_e L^2}{EI_z} \tag{f}$$

Subtracting Eq. (*f*) from Eq. (*e*) results in

$$\delta_{\text{residual}} = 0.0011 \frac{M_e L^2}{EI_z} \tag{g}$$

When dealing with nonrectangular beams, it is a common practice to modify Eq. (4-59) by first rewriting it as

$$M_f = \frac{M_e \sqrt{\tfrac{1}{2}}}{\sqrt{\tfrac{3}{2} - |M_z/M_e|}}$$

The term $\tfrac{3}{2}$ is the shape factor SF for a rectangular beam, and the $\tfrac{1}{2}$ is SF $- 1$. Thus

$$M_f = \frac{M_e \sqrt{\text{SF} - 1}}{\sqrt{\text{SF} - |M_z/M_e|}}$$

will approximate the behavior of nonrectangular beams.

4-12 RAYLEIGH'S TECHNIQUE APPLIED TO BEAMS IN BENDING

Rayleigh's method is often employed in applied mechanics. The technique has some practical applications in deflection analyses, as will be demonstrated in Sec. 4-14. The technique assumes a deflection shape, and the strain energy is formulated based on the assumed shape and then equated to the work done by the applied forces. A problem arises, however, in the sign of the work term, and care should be exercised. When one is defining forces or force distributions, it is more consistent in terms of sign convention to define loads positive in relation to the coordinate system being used. When dealing with beams, this text has made consistent use of an *xyz* coordinate system. However, to this point applied loads have been considered to be positive in the negative *y* direction. This causes the sign problem in the work term.

Consider a load distribution $w(x)$ applied to a beam as shown in Fig. 4-37*a*, where the units of $w(x)$ are force per unit length. If a *dx* element is isolated at position *x*, the net force on the element is $w\,dx$. If the beam deflects in the *positive y* direction in a linear fashion a distance y_c, the net work performed is $-\tfrac{1}{2}wy_c\,dx$. The work is negative since the force and deflection are in opposing directions. The total work performed by the application of w is then

$$W_w = -\frac{1}{2}\int_0^L wy_c\,dx \tag{4-62}$$

Equation (4-62) does not imply that the work will be negative. In fact, for a

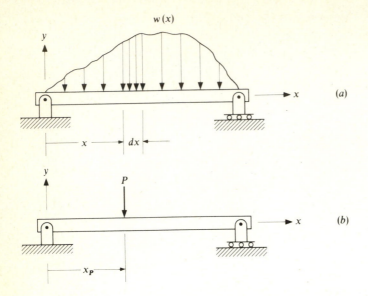

Figure 4-37

beam like that shown in Fig. 4-37a, the value of y_c will be negative everywhere. Thus, after integration of Eq. (4-62), W_w will be found to be positive.

For a concentrated force applied like that shown in Fig. 4-37b, Eq. (4-62) cannot be used to determine the work caused by the force. Again, assuming that the beam deflects in the positive y direction in a linear fashion, the force P will deflect the distance $(y_c)_{x=x_P}$. The total work is then

$$W_P = -\tfrac{1}{2}P(y_c)_{x=x_P} \tag{4-63}$$

The increase in strain energy for a beam in bending in terms of the deflection of the beam was given in Eq. (4-20) (repeated here):

$$U_b = \tfrac{1}{2}\int EI_z(y_c'')^2 \, dx \tag{4-64}$$

where $y_c'' = d^2y_c/dx^2$.

Rayleigh's technique assumes a shape for the beam deflection, defined up to one unknown constant. The shape is, in general, a function of x and must be consistent with the geometric boundary conditions of the beam. The approximate work W and strain energy U can be evaluated using Eqs. (4-62) and/or (4-63) and Eq. (4-64). Equating the two terms results in

$$W = U \tag{4-65}$$

and the value of the unknown constant can be determined from this equation. The deflection curve will then be complete; however, since the shape is assumed, the deflection will be approximate. The accuracy of the approximation depends on how closely the assumed shape matches the exact shape. After some experience, the analyst will be able to judge how good a displacement function is. In Sec. 4-14, Rayleigh's method will be applied to beams undergoing axial and lateral loading, where in many cases the exact solution is not known. It will be shown that a simple solution to the selection of an accurate function is to use the shape (from Appendix D) of the same beam minus the axial load.

Example 4-30 Determine the deflection of point A of the beam shown in Fig. 4-38. Assume that the deflection shape of the beam takes the form:†

(a)
$$y_c = - \delta\left(1 - \cos\frac{\pi x}{2L}\right)$$

(b)
$$y_c = - \delta\left(\frac{x}{L}\right)^2$$

(c)
$$y_c = \frac{\delta x^2}{2L^3}(x - 3L)$$

SOLUTION The work due to P is given by Eq. (4-63). Since for all three functions, $y_c = -\delta$ at $x = L$, the work is

$$W = -\tfrac{1}{2}P(-\delta) = \tfrac{1}{2}P\delta \qquad (a)$$

(a) The second derivative of y_c with respect to x is

$$y_c'' = -\delta\left(\frac{\pi}{2L}\right)^2 \cos\frac{\pi x}{2L}$$

Substituting this into Eq. (4-64) results in

$$U \approx U_b = \frac{EI_z}{2}\delta^2\left(\frac{\pi}{2L}\right)^4\int_0^L \cos^2\frac{\pi x}{2L}\,dx$$

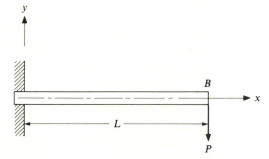

Figure 4-38

† Note that every one of the functions given satisfies the geometric boundary conditions that $y_c = dy_c/dx = 0$ at $x = 0$. Also, since the bending moment is directly related to d^2y_c/dx^2, d^2y_c/dx^2 should equal zero at $x = L$. This is not a geometric boundary condition, however, and it is usually not imposed in this type of problem. The functions of part (a) and (c) satisfy this condition, whereas part (b) does not. It will be seen that the function of part (a) is very accurate and the function of part (b) is exact. (The reader should verify the fact that this function is the same shape as given in Appendix D.) The function used in part (b) will be found to be fairly inaccurate. In addition, note that each one of the functions given was normalized to give $y_c = -\delta$ at $x = L$.

Integration yields

$$U = \frac{\pi^4 EI_z}{64L^3}\delta^2 \qquad (b)$$

Substituting Eqs. (a) and (b) into Eq. (4-65) yields

$$\tfrac{1}{2}P\delta = \frac{\pi^4}{64}\frac{EI_z}{L^3}\delta^2$$

and solving for δ results in

$$\delta = \frac{32}{\pi^4}\frac{PL^3}{EI} = 0.327\frac{PL^3}{EI_z}$$

which is approximately 2 percent less than that predicted by standard beam theory (see Appendix D). Although not asked for in the problem, the slope at point B can be found by differentiating y_c. Thus

$$\frac{dy_c}{dx} = -\frac{\pi}{2L}\delta \sin\frac{\pi x}{2L} = -0.514\frac{PL^2}{EI}\sin\frac{\pi x}{2L}$$

Therefore

$$\left(\frac{dy_c}{dx}\right)_{x=L} = -0.514\frac{PL^2}{EI_z}$$

which is approximately 3 percent greater than that predicted by standard beam theory.

(b) Again the term y_c'' is first evaluated and is $y_c'' = -2\delta/L^2$. Substituting this into Eq. (4-64) results in

$$U = \frac{2EI_z}{L^4}\delta^2\int_0^L dx = \frac{2EI_z}{L^3}\delta^2 \qquad (c)$$

Equating Eqs. (a) and (c) gives

$$\tfrac{1}{2}P\delta = \frac{2EI_z}{L^3}\delta^2$$

and solving for δ yields

$$\delta = 0.25\frac{PL^3}{EI_z}$$

which is 25 percent lower than the exact solution. Although satisfying the geometric boundary conditions, an x^2 function is not very good. If a cubic function in x is used, where $y_c = ax^3 + bx^2 + cx + d$, subject to the conditions that $y_c = y_c' = 0$ at $x = 0$ and $y_c'' = 0$ and $y_c = -\delta$ at $x = L$, the function in part (c) is arrived at.

(c) Again, y_c'' is evaluated and is found to be

$$y_c'' = 3\frac{\delta}{L^3}(x - L)$$

Substituting this into Eq. (4-64) results in

$$U = \frac{9EI_z\delta^2}{2L^6}\int_0^L (x - L)^2\, dx$$

Integration yields

$$U = \frac{3EI_z}{2L^3}\delta^2 \qquad (d)$$

Equating Eqs. (a) and (d) results in

$$\tfrac{1}{2}P\delta = \frac{3EI_z}{2L^3}\delta^2$$

Solving for δ gives

$$\delta = \frac{1}{3}\frac{PL^3}{EI_z}$$

the exact solution to the problem.

A method of improving the solution obtained by Rayleigh's technique is to select more functions, each with an unknown constant, and by optimization methods to find the best value of each constant. This procedure is most widely known as the Rayleigh-Ritz method and is covered in the following section.

4-13 THE RAYLEIGH-RITZ TECHNIQUE APPLIED TO BEAMS IN BENDING

Considering beams in bending only, assume that the deflection of the centroidal axis is given by

$$y_c = \sum_{i=1}^{n} a_i f_i(x) \tag{4-66}$$

where a_i are n unknown constants and $f_i(x)$ are n known (assumed) functions of x, where each function $f_i(x)$ satisfies the geometric boundary conditions. When y_c is substituted into Eqs. (4-62) and/or (4-63) and Eq. (4-64), the work and strain energy will be functions of a_i only. The work can then be equated to the strain energy, as is done in the Rayleigh technique. However, the resulting equation is only one equation with n unknowns a_i. The Rayleigh-Ritz technique basically performs an optimization of the a_i terms by slightly varying the a_i, keeping everything else constant. Varying the a_i causes y_c to change, thus changing the strain energy and the work. However, since $U = W$, then the changes are equal. That is,

$$dU = dW \tag{4-67}$$

Since only the a_i changes, causing a change in y_c, while the applied loads remain constant, the $\frac{1}{2}$ term in the work expression must be omitted. Thus, for a distributed load w [as defined in Eq. (4-62)] the change in work is

$$dW_w = d\left(-\int_0^L w y_c \, dx\right) \tag{4-68}$$

whereas, for a concentrated force P [as defined in Eq. (4-63)] the change in work is

$$dW_P = d[-P(y_c)_{x=x_p}] \tag{4-69}$$

The following example serves to illustrate application of the technique.

Example 4-31 In Example 4-30 consider only the geometric boundary conditions and determine the deflection of the centroidal axis, assuming that

$$y_c = a_0 + a_1 x + a_2 x^2 + a_3 x^3$$

SOLUTION In Example 4-30 the geometric boundary conditions are $y_c = y_c' = 0$ at $x = 0$. Therefore, $a_0 = a_1 = 0$, and y_c becomes

$$y_c = a_2 x^2 + a_3 x^3 \tag{a}$$

Evaluating the work term first, we have

$$(y_c)_{x=x_p} = (y_c)_{x=L} = a_2 L^2 + a_3 L^3 \tag{b}$$

Thus from Eq. (4-69)

$$dW = dW_p = d[-P(a_2L^2 + a_3L^3)]$$

or

$$dW = -PL^2(da_2 + L\, da_3) \qquad (c)$$

In order to evaluate the strain energy, y_c'' is obtained from Eq. (a) and is

$$y_c'' = 2a_2 + 6a_3x$$

Thus, the strain energy is

$$U = \frac{EI_z}{2} \int_0^L (2a_2 + 6a_3x)^2\, dx$$

Evaluating the integral results in

$$U = 2EI_zL(a_2^2 + 3a_2a_3L + 3a_3^2L^2)$$

Thus

$$dU = 2EI_zL[(2a_2 + 3La_3)\, da_2 + 3L(a_2 + 2La_3)\, da_3] \qquad (d)$$

Now, Eqs. (c) and (d) are equated, resulting in

$$-PL^2(da_2 + L\, da_3) = 2EI_zL[(2a_2 + 3La_3)\, da_2 + 3L(a_2 + 2La_3)\, da_3]$$

Equating the coefficients of da_2 and da_3 yields

$$-PL^2 = 2EI_zL(2a_2 + 3La_3) \qquad (e)$$

$$-PL^3 = 6EI_zL^2(a_2 + 2La_3) \qquad (f)$$

Solving Eqs. (e) and (f) simultaneously results in

$$a_2 = -\frac{PL}{2EI_z} \qquad a_3 = \frac{P}{6EI_z}$$

Substituting a_2 and a_3 back into Eq. (a) and simplifying algebraically yield

$$y_c = -\frac{Px^2}{6EI_z}(3L - x)$$

In Example 4-31, the solution converged to the exact solution thanks to the judicious selection of the functions. If higher-order polynomials had been used in that example, the constants associated with these higher-order polynomials would all have converged to zero.

Up to this point, the Rayleigh and the Rayleigh-Ritz techniques have been employed on simple beam problems in which the solution is either known or easily obtained by more standard methods. There are some applications in which the solution to the deflection shape is either difficult or impossible to obtain by the standard beam-deflection methods. In these cases, either the Rayleigh or the Rayleigh-Ritz technique yields relatively straightforward and accurate approximations. Some applications follow in the next section; they will be discussed using the Rayleigh method, but the Rayleigh-Ritz method is just as applicable.

4-14 STRAIGHT BEAMS UNDERGOING THE COMBINED EFFECTS OF AXIAL AND TRANSVERSE LOADING

Consider a beam loaded by a transverse load w and an axial force N, as shown in Fig. 4-39a. In an elementary strength of materials course, the deflections and

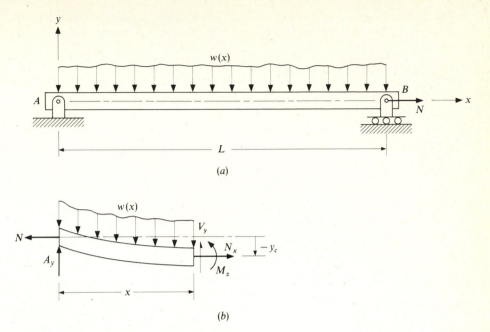

Figure 4-39

stresses caused by these loads are normally considered independently and the results added by superposition. However, if the axial force is large, this is not very accurate, as the axial force affects the lateral deflections as well as the bending stresses. If N is positive (tension), the beam "stiffens" in bending and the lateral deflections and normal stresses are less than predicted by superposition of the individual effects. Thus in terms of design applications, superposition would lead to a conservative result. However, if N is negative (compression), deflections and stresses will be larger than predicted by superposition. If the analysis is to be accurate, or if a beam is constrained axially and the axial force is unknown, the effect of the axial force on bending should be considered. This can only be accomplished by considering deflections in the force analysis.

Making a surface isolation at x, as shown in Fig. 4-39b, and accounting for the deflection due to bending, we can see that the axial force will affect the bending moment. This is where the coupling between axial and transverse loading takes place.

When the force N is known, the problem can be solved by standard beam methods. This requires the solution of a differential equation, and for discontinuous loading, the procedure can be quite involved and time-consuming. For illustration purposes, the following example with continuous loading demonstrates the exact approach, but since the loading is continuous, much of the difficulty with this approach is alleviated.

Example 4-2 For the beam shown in Fig. 4-39, determine how the vertical deflection of the centroid y_c varies as a function of x if the load distribution w is constant w_0.

SOLUTION From Fig. 4-39, w is constant, and the internal bending moment M_z is†

$$M_z = \tfrac{1}{2}w_0Lx - \tfrac{1}{2}w_0x^2 + Ny_c \qquad (a)$$

Use of the standard beam equation $M_z = EI_z\,d^2y_c/dx^2$, leads to

$$\frac{d^2y_c}{dx^2} - \frac{N}{EI_z}\,y_c = \frac{w_0x}{2EI_z}\,(L-x) \qquad (b)$$

The solution to this differential equation using the geometric conditions that $y_c = 0$ at $x = 0$ and $x = L$ is

$$y_c = \frac{w_0}{k^2N}\,(1-\cosh kx) + \frac{w_0(\cosh kL - 1)}{k^2N\sinh kL}\sinh kx - \frac{w_0x}{2N}\,(L-x) \qquad (c)$$

where

$$k = \sqrt{\frac{N}{EI_z}}$$

A large problem exists with the method used in Example 4-32 when the transverse loading is discontinuous. A differential equation must be written in each zone where M_z changes functionally. Continuity of slope and deflection at the junction of each individual zone provide the necessary conditions for evaluating the two constants of integration for each differential equation. The solution to problems where the loading is not simple is a long and tedious process that can be approximated quite easily using the techniques outlined in Secs. 4-12 and 4-13. In addition, although less serious, a problem occurs in numerically calculating deflections when the method of Example 4-32 is used. For instance, each term in Eq. (c) is very large compared with the final value of y_c. This means that each term in Eq. (c) must be calculated more accurately than usual. Using either the Rayleigh or the Rayleigh-Ritz method, a reasonably accurate approximate solution can be arrived at through a procedure which is much simpler than solving one or more differential equations. Also, in the case of axially constrained beams where the axial force is unknown, the energy approach is the only practical way to a solution.

Rayleigh's technique is used to equate the work due to the applied loading to the increase in strain energy. The work due to the transverse loading was discussed in Sec. 4-12 and is given by Eq. (4-62) or (4-63) or both. The work due to the axial force N is not as obvious as one might expect.

The work due to the axial force N is not $\tfrac{1}{2}N\delta_B$, where δ_B is the deflection of point B in the x direction. This is because δ_B is not linearly related to N, as will be seen later. Before application of the lateral loading w apply the axial force N. The resulting deflection δ_{B1} is NL/AE. The work done during this step is $\tfrac{1}{2}N\delta_{B1} = \tfrac{1}{2}N^2L/AE$. Next, with N remaining on the beam, application of w causes point B to move in the direction opposite N the distance δ_{B2} (see Fig. 4-40). The work done by N during this step is $-N\delta_{B2}$ since N is constant. Thus

† Again, due to problems in sign, $-y_c$ must be used in Fig. 4-39.

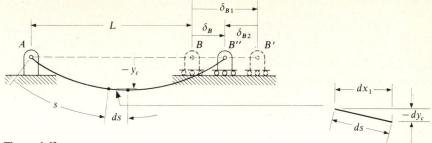

Figure 4-40

the net work due to N is

$$W_N = \frac{1}{2}\frac{N^2 L}{AE} - N\delta_{B2} \qquad (4\text{-}70)$$

Determination of δ_{B2} is rather complicated to develop, but once it is done, subsequent application utilizing the energy approach becomes straightforward.

Assuming that bending does not add any axial stress along the centroidal axis of the beam, the deflection δ_{B2} is simply the difference between the arc AB'' and the straight line AB''. Using the curvilinear variable s, one can view a segment of the beam centroid where (see Fig. 4-40)

$$ds^2 = dx_1^2 + dy_c^2$$

$$ds = dx_1\left[1 + \left(\frac{dy_c}{dx_1}\right)^2\right]^{1/2} \qquad (4\text{-}71)$$

For small slopes, $dy_c/dx_1 \ll 1$, and the term in the brackets can be expanded using the binomial expansion theorem. When terms $(dy_c/dx_1)^4$ and higher are neglected, Eq. (4-71) is approximated by

$$ds \approx dx_1\left[1 + \frac{1}{2}\left(\frac{dy_c}{dx_1}\right)^2\right]$$

Integration yields

$$\int ds \approx \int dx_1 + \frac{1}{2}\int \left(\frac{dy_c}{dx_1}\right)^2 dx_1$$

Since

$$\int ds = \text{arc } AB'' \qquad \text{and} \qquad \int dx_1 = AB''$$

we have

$$\delta_{B2} = \text{arc } AB'' - AB'' = \int ds - \int dx_1 \approx \frac{1}{2}\int \left(\frac{dy_c}{dx_1}\right)^2 dx_1$$

For small deflections $dx_1 \approx dx$; therefore,

$$\delta_{B2} \approx \frac{1}{2}\int_0^L \left(\frac{dy_c}{dx}\right)^2 dx \qquad \text{or} \qquad \delta_{B2} \approx \frac{1}{2}\int_0^L (y_c')^2 \, dx \qquad (4\text{-}72)$$

where $y_c' = dy_c/dx$.

Substituting Eq. (4-72) into Eq. (4-70) yields the net work due to N:

$$W_N \approx \frac{1}{2} \frac{N^2 L}{AE} - \frac{1}{2} N \int_0^L (y_c')^2 \, dx \qquad (4\text{-}73)$$

The increase in strain energy, neglecting shear effects, is the sum of the increase in axial and bending strain energy. Thus

$$U = U_a + U_b = \frac{1}{2} \frac{N^2 L}{AE} + \tfrac{1}{2} E I_z \int_0^L (y_c'')^2 \, dx \qquad (4\text{-}74)$$

Equating the total work from the applied loads to the total strain energy results in

$$W_w + W_P + W_N = U_a + U_b \qquad (4\text{-}75)$$

Equation (4-75) is the basic work-energy equation relating the bending-deflection curve to the combined transverse and axial loading of the beam. Any transverse load distribution and series of concentrated forces can be accommodated in Eq. (4-75).

If Rayleigh's method is used to get an approximate solution to Eq. (4-75), reasonably accurate solutions can be obtained using the beam-deflection shapes found for the same beam without an axial force (given in Appendix D) since these solutions satisfy the geometric boundary conditions.

Example 4-33 (*a*) In Example 4-32, the transverse-deflection equation was derived for the centroid of a uniformly loaded simply supported beam with an axial force. Using Eq. (*c*) from that example, determine the deflection at midspan if $L = 50$ in, $w_0 = 10$ lb/in, $I_z = 0.10$ in^4, $E = 10 \times 10^6$ lb/in^2, and $N = 900$ lb.

(*b*) Compare the above results with that obtained from Eq. (4-75) using (1) $y_c = -\delta \sin(\pi x/L)$ and (2) the shape given in Appendix D.

SOLUTION (*a*) From Example 4-32, Eq. (*c*),

$$k = \sqrt{\frac{900}{(10 \times 10^6)(0.1)}} = 3 \times 10^{-2}$$

Substituting k into Eq. (*c*) yields

$$(y_c)|_{x=L/2} = -0.662 \text{ in}\dagger$$

(*b*) For the approximate solution using Eq. (4-75) with $y_c = -\delta \sin(\pi x/L)$,

$$y_c' = -\delta \frac{\pi}{L} \cos \frac{\pi x}{L} \qquad y_c'' = \delta \left(\frac{\pi}{L}\right)^2 \sin \frac{\pi x}{L}$$

For the work, since there is no concentrated force, $W_p = 0$, and the work W_w is

$$W_w = -\frac{1}{2} \int_0^L w y_c \, dx = -\frac{1}{2} \int_0^L w_0 \left(-\delta \sin \frac{\pi x}{L}\right) dx$$

Integration results in

$$W_w = \frac{1}{\pi} w_0 L \delta \qquad (a)$$

$\dagger$ Note from Appendix D that for the same beam without the axial force the deflection at midspan is -0.814 in. Thus it can be seen that there can be quite a difference (23 percent) if N is neglected in the deflection analysis.

The work due to N is

$$W_N = \frac{1}{2}\frac{N^2L}{AE} - \frac{1}{2}N\int_0^L \left(-\delta\frac{\pi}{L}\cos\frac{\pi x}{L}\right)^2 dx$$

$$= \frac{1}{2}\frac{N^2L}{AE} - \frac{1}{2}N\left(\delta\frac{\pi}{2}\right)^2 \int_0^L \cos^2\frac{\pi x}{L} dx$$

Integration yields

$$W_N = \frac{1}{2}\frac{N^2L}{AE} - \frac{\pi^2}{4}\frac{N}{L}\delta^2 \qquad (b)$$

The strain energy is given by

$$U = \frac{1}{2}\frac{N^2L}{AE} + \frac{1}{2}\int_0^L EI_z\left[\delta\left(\frac{\pi}{L}\right)^2 \sin\frac{\pi x}{L}\right]^2 dx$$

$$= \frac{1}{2}\frac{N^2L}{AE} + \frac{1}{2}EI_z\delta^2\left(\frac{\pi}{L}\right)^4 \int_0^L \sin^2\frac{\pi x}{L} dx$$

Integration yields

$$U = \frac{1}{2}\frac{N^2L}{AE} + \frac{\pi^4}{4}\frac{EI_z}{L^3}\delta^2 \qquad (c)$$

Substituting Eqs. (a), (b), and (c) into Eq. (4-75) results in

$$\frac{1}{\pi}w_0L\delta + \frac{1}{2}\frac{N^2L}{AE} - \frac{\pi^2}{4}\frac{N}{L}\delta^2 = \frac{1}{2}\frac{N^2L}{AE} + \frac{\pi^4}{4}\frac{EI_z}{L^3}\delta^2$$

Substitution of numerical values and simplifying algebraically yields

$$\delta = \frac{(4)(10)(50^4)}{\pi^3[(900)(50^2) + \pi^2(10\times10^6)(0.1)]} = 0.665 \text{ in}$$

which is 0.3 percent higher than the result obtained by the exact technique used in part (a).

Next we see how the deflection shape of Appendix D is used. Without N, the deflection equation is given by

$$y_c = \frac{wx}{24EI_z}(2Lx^2 - x^3 - L^3)$$

Since only the shape is used, let

$$y_c = Kx(2Lx^2 - x^3 - L^3) \qquad (e)$$

where K is a constant. Thus

$$y_c' = K(6Lx^2 - 4x^3 - L^3) \qquad y_c'' = 12Kx(L-x)$$

As before, $W_P = 0$, and

$$W_w = -\frac{1}{2}\int_0^L w_0Kx(2Lx^2 - x^3 - L^3) dx$$

Integration yields

$$W_w = \frac{1}{10}w_0KL^5 \qquad (f)$$

The work W_N is

$$W_N = \frac{1}{2}\frac{N^2L}{AE} - \frac{1}{2}N\int_0^L [K(6Lx^2 - 4x^3 - L^3)]^2 dx$$

Integration yields

$$W_N = \frac{1}{2}\frac{N^2L}{AE} - \frac{17}{70}NK^2L^7 \qquad (g)$$

The strain energy is

$$U = \frac{1}{2}\frac{N^2 L}{AE} + \frac{1}{2}\int EI_z[12Kx(L-x)]^2\, dx$$

Integration yields

$$U = \frac{1}{2}\frac{N^2 L}{AE} + \frac{12}{5}EI_z K^2 L^5 \tag{h}$$

Again, combining Eqs. (f), (g), and (h) with Eq. (4-75) results in

$$K = \frac{7w_0}{17NL^2 + 168EI_z} \tag{i}$$

Substitution of numerical values yields

$$K = \frac{(7)(10)}{(17)(900)(50^2) + (168)(10 \times 10^6)(0.1)} = 3.394 \times 10^{-7}$$

Substituting K into Eq. (e) gives

$$y_c = 3.394x(2Lx^2 - x^3 - L^3) \times 10^{-7} \tag{j}$$

The deflection at midspan is maximum. Thus at $x = L/2 = 25$ in

$$(y_c)_{x=L/2} = (3.394)(25)[(2)(50)(25^2) - 25^3 - 50^3] \times 10^{-7} = -0.663 \text{ in}$$

Comparing this result with that in part (a) shows that there is less than 0.2 percent difference.

Example 4-34 Determine the maximum deflection of the cantilever beam loaded as shown in Fig. 4-41. Approximate the transverse deflection shape using the solution that neglects axial loading.

SOLUTION The deflection shape neglecting the direct compressive force F is found in Appendix D and is

$$y_c = -\frac{\delta x^2}{2L^3}(3L - x)$$

Note that y_c was manipulated so that $y_c = -\delta$ at $x = L$, where δ is the maximum deflection. The first derivative of y_c with respect to x is

$$y_c' = -\frac{3\delta x}{2L^3}(2L - x)$$

and the second derivative is

$$y_c'' = -\frac{3\delta}{L^3}(L - x)$$

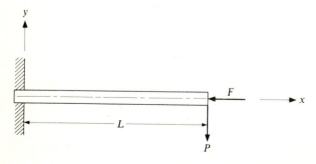

Figure 4-41

The work term W_w is zero and

$$W_P = \tfrac{1}{2}P\delta \tag{a}$$

Since $N = -F$, the work due to F is

$$W_F = \frac{1}{2}\frac{F^2L}{AE} + \tfrac{1}{2}F\int_0^L \left[-\frac{3\delta x}{2L^3}(2L - x) \right]^2 dx$$

Integration yields,

$$W_F = \frac{1}{2}\frac{F^2L}{AE} + \tfrac{3}{5}F\frac{\delta^2}{L} \tag{b}$$

The strain energy is

$$U = \frac{1}{2}\frac{F^2L}{AE} + \frac{1}{2}\int_0^L EI_z \left[-\frac{3\delta}{L^3}(L - x) \right]^2 dx$$

Integration yields

$$U = \frac{1}{2}\frac{F^2L}{AE} + \tfrac{3}{2}EI_z\frac{\delta^2}{L^3} \tag{c}$$

Substituting Eqs. (a) to (c) into Eq. (4-75) results in

$$\tfrac{1}{2}P\delta + \frac{1}{2}\frac{F^2L}{AE} + \tfrac{3}{5}F\frac{\delta^2}{L} = \frac{1}{2}\frac{F^2L}{AE} + \tfrac{3}{2}EI_z\frac{\delta^2}{L^3}$$

Simplifying yields

$$\delta = \frac{PL^3}{3EI_z(1 - 2FL^2/5EI_z)}$$

If $F = 0$, the maximum deflection agrees with beam D.1, where $\delta = PL^3/3EI_z$, since the shape used agrees with that case. Note that as the compressive force F increases in magnitude, the maximum deflection increases beyond that given by simple beam theory.

Constrained Beams

When ground supports cause an axial constraint in a beam loaded by transverse loads, an axial force develops. The problem is similar to the ones previously discussed in this section, but in this case the problem is statically indeterminate, as the axial force N cannot be found from the equilibrium equations. Consider the beam shown in Fig. 4-42. The ground constraint is modeled by a simple linear spring of spring constant k_s with units of force per unit length. If the supports are perfectly rigid, $k_s \to \infty$. Recall Fig. 4-40; the deflection of point B is

$$\delta_B = \frac{NL}{AE} - \frac{1}{2}\int_0^L (y_c')^2 \, dx \tag{4-76}$$

From Fig. 4-42, the deflection of point B is

$$\delta_B = -\frac{N}{k_s} \tag{4-77}$$

where the negative sign indicates that the deflection is in the $-x$ direction.

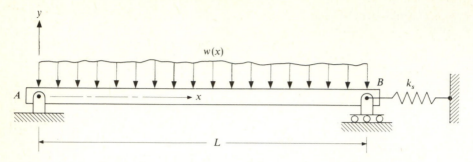

Figure 4-42

Equating Eqs. (4-76) and (4-77) and solving for N yields

$$N = \frac{k_s AE}{2(k_s L + AE)} \int_0^L (y_c')^2 \, dx \qquad (4-78)$$

Substituting Eq. (4-78) into the second term of Eq. (4-73) yields†

$$W_N = \frac{1}{2} \frac{N^2 L}{AE} - \frac{k_s AE}{4(k_s L + AE)} \left[\int_0^L (y_c')^2 \, dx \right]^2 \qquad (4-79)$$

Thus, if Eq. (4-79) is used instead of Eq. (4-73), the terms in Eq. (4-75) will not contain N. Once y_c is approximated from Eq. (4-75), N can be determined from Eq. (4-78).

> **Example 4-33** Figure 4-43 shows a uniformly loaded beam, simply supported; however, the axial movement of both supports is zero, as the foundation is very rigid. Determine the maximum deflection and approximate the axial force which develops if $L = 50$ in, $w_0 = 10$ lb/in, $E = 10 \times 10^6$ lb/in², $I_z = 0.10$ in⁴, and $A = 0.30$ in². Assume a deflection shape of $y_c = -\delta \sin(\pi x/L)$.

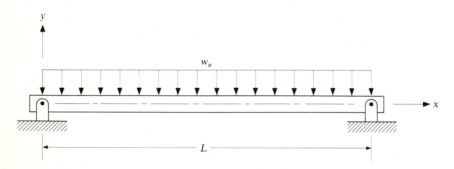

Figure 4-43

---

† Substitution of N into the first term is unnecessary as this term is canceled by the first term in the strain energy U.

SOLUTION The derivatives of y_c are

$$y_c' = -\delta \frac{\pi}{L} \cos \frac{\pi x}{L} \qquad y_c'' = \delta \left(\frac{\pi}{L}\right)^2 \sin \frac{\pi x}{L}$$

The work due to the lateral load is given by Eq. (4-62) and is,

$$W_w = -\frac{1}{2} \int_0^L w_0 \left(-\delta \sin \frac{\pi x}{L}\right) dx = \frac{1}{\pi} w_0 L \delta \tag{a}$$

The work due to N is given by Eq. (4-79). Since the supports are to be considered perfectly rigid, $k \to \infty$ and

$$W_N = \frac{1}{2} \frac{N^2 L}{AE} - \frac{AE}{4L} \left[\int_0^L \left(-\delta \frac{\pi}{L} \cos \frac{\pi x}{L}\right)^2 dx\right]^2$$

Integration yields

$$W_N = \frac{1}{2} \frac{N^2 L}{AE} - \frac{\pi^4}{16} \frac{AE}{L^3} \delta^4 \tag{b}$$

The strain energy is

$$U = \frac{1}{2} \frac{N^2 L}{AE} + \frac{1}{2} \int_0^L EI_z \left[\delta \left(\frac{\pi}{L}\right)^2 \sin \frac{\pi x}{L}\right]^2 dx$$

Integration results in

$$U = \frac{1}{2} \frac{N^2 L}{AE} + \frac{\pi^4}{4} \frac{EI_z}{L^3} \delta^2 \tag{c}$$

Substituting Eqs. (a) to (c) into Eq. (4-75) yields

$$\frac{1}{\pi} w_0 L \delta - \frac{\pi^4}{16} \frac{AE}{L^3} \delta^4 = \frac{\pi^4}{4} \frac{EI_z}{L^3} \delta^2$$

Simplifying algebraically gives

$$\delta^3 + 4 \frac{I_z}{A} \delta - \frac{16}{\pi^5} \frac{w_0 L^4}{AE} = 0 \tag{d}$$

Substituting in numerical values yields the cubic equation

$$\delta^3 + 1.333\delta - 1.089 = 0$$

Solving for δ yields one real root,

$$\delta = 0.630 \text{ in}$$

Thus the approximate deflection curve is

$$y_c = -0.630 \sin \frac{\pi x}{50} \tag{e}$$

Substituting this into Eq. (4-78) with $k_s \to \infty$ yields an approximation of the axial force N. Thus

$$N \approx \frac{1}{2} \frac{AE}{L} \int (y_c')^2 \, dx = \frac{1}{2} \frac{AE}{L} \int \left(-\delta \frac{\pi}{L} \cos \frac{\pi x}{L}\right)^2 dx = \frac{\pi^2}{4} \left(\frac{\delta}{L}\right)^2 AE$$

Substituting $\delta = 0.630$ in, $L = 50$ in, $A = 0.30$ in², and $E = 10 \times 10^6$ lb/in² yields

$$N \approx \frac{\pi^2}{4} \left(\frac{0.630}{50}\right)^2 (0.30)(10 \times 10^6) = 1175 \text{ lb}$$

Example 4-36 Compare the maximum tensile stress of Example 4-35 with that obtained when one of the supports is free to move axially. Consider the cross section of the beam to be symmetric and the height of the cross section to be 2.0 in.

SOLUTION When one support has axial freedom, the maximum bending moment occurs at midspan and is

$$(M_z)_{max} = \frac{wL^2}{8} = \frac{(10)(50)^2}{8} = 3125 \text{ lb·in}$$

and the maximum tensile stress is

$$\sigma_{max} = \frac{(M_z y)_{max}}{I_z} = \frac{(3125)(1)}{0.1} = 31{,}250 \text{ lb/in}^2$$

For the constrained beam (Example 4-35) the maximum bending moment also occurs at midspan, but the maximum bending moment is

$$(M_z)_{max} = \frac{wL^2}{8} - N\delta = 3125 - (1175)(0.63) = 2385 \text{ lb·in}$$

The maximum tensile stress is

$$\sigma_{max} = \frac{(M_z y)_{max}}{I_z} + \frac{N}{A} = \frac{(2385)(1)}{0.1} + \frac{1175}{0.3} = 27{,}800 \text{ lb/in}^2$$

By constraining this particular beam, there actually is an 11 percent reduction in the stress at the midspan. The bearing forces at the beam ends, however, can be shown to be increased by a factor of 4.8.

EXERCISES

4-1 Beam AC shown in Fig. 4-44 has a rectangular cross section 3 by 2 in and is simply supported at A and supported at C by cable DC. The cable has a cross-sectional area of 0.1 in², and both members are made of steel with $E_s = 30 \times 10^6$ lb/in². Determine the total strain energy of the system after a vertical force of 1000 lb is applied at the midpoint of the beam.

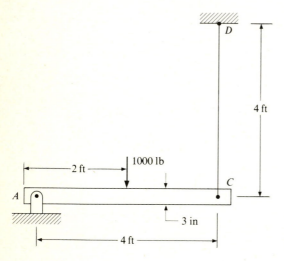

Figure 4-44

4-2 A steel shaft is transmitting a torque of 1200 N·m and a bending moment of 1500 N·m at a particular section. If the diameter is 40 mm, $E_s = 20 \times 10^{10}$ N/m², and $\nu = 0.3$, determine the maximum strain energy per unit volume.

4-3 Determine the total strain energy in the shaft of Exercise 2-5.

4-4 Using an energy or work approach, determine the vertical deflection of point C and the deflection of point D of the structure shown in Fig. 4-45. Check your results by using a geometrical approach. Each member has a modulus of elasticity E and area A; the length of members BC and CD is L.

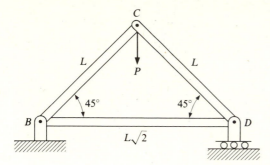

Figure 4-45

4-5 Consider each member of the structure shown in Fig. 4-45 to have a deflection-load relation of $\delta_i = k_i F_i^3$, where k_i is a constant for each member and F_i is the force transmitted in each element. For members BC and CD, $k_i = k_1$; whereas, for member BD, $k_i = k_2$. Using the complementary-energy theorem, determine the vertical deflection of point C.

4-6 For beam D.3 of Appendix D, use Castigliano's method to verify that the slope and deflection at $x = L$ are $-wL^3/6EI$ and $-wL^4/8EI$, respectively.

4-7 Determine the deflection at the midpoint of the steel step shaft loaded as shown in Fig. 4-46. $E_s = 30 \times 10^6$ lb/in².

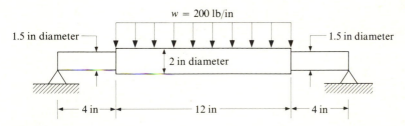

Figure 4-46

4-8 For the truss shown in Fig. 4-47 use Castigliano's method to determine (a) the vertical deflection of point C and (b) the horizontal deflection of point D. All members are of equal length L and equal cross-sectional area A and are of the same material with a modulus of elasticity E.

4-9 If the wire used in Exercise 1-3 has a diameter of $\frac{1}{2}$ in and is steel with $E_s = 30 \times 10^6$ lb/in², and $\nu = 0.29$, determine the deflection of the 100-lb force in the x direction.

4-10 For Exercise 4-1, determine the deflection of beam AC at midspan using Castigliano's method.

4-11 For Example 4-9, determine the rotation of member 3. *Hint*: A dummy concentrated moment can be placed at any point *on* member 3.

4-12 For Example 4-18, determine the rotation of the curved beam at point A.

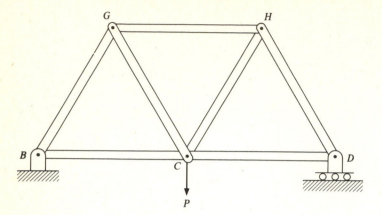

Figure 4-47

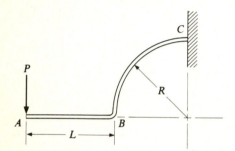

Figure 4-48

4-13 For the wire form shown in Fig. 4-48, determine the vertical deflection of point A using Castigliano's method and considering only bending. The bending rigidity of the cross section is EI.

4-14 For the wire form shown in Fig. 4-49, determine the deflection of point A in the y direction using Castigliano's method and considering only bending and torsion. The wire is steel with $E = 20 \times 10^{10} \, \text{N/m}^2$, $\nu = 0.29$, and the diameter is 5 mm. Before application of the vertical force the wire form is in the xz plane. The radius of curvature of the wire form R is 100 mm.

4-15 From Example 3-9 estimate the amount that the T-section clamp opens. Solve the problem two ways: (a) use the straight-beam approximation considering bending only, and (b) use Eq. (4-44). $E = 30 \times 10^6 \, \text{lb/in}^2$.

4-16 For the slender rod shown in Fig. 4-50, determine the reactions at points A and B and determine how the bending moment varies along the length. The bending rigidity of the cross section of the rod is EI.

4-17 The structure shown in Fig. 4-51 is fabricated by welding three I beams together. The structure is subjected to a uniformly distributed load w_0 across the horizontal beam. Determine the support reactions at A and B.

4-18 Using Castigliano's method, determine the support reactions of the beam shown in Fig. 4-52. Assume that the beam is free to slide horizontally so that no net axial force develops. Compare the results with that given in Appendix D.

4-19 For the wire form shown in Fig. 4-53, determine (a) how the bending moment varies with respect to position within the form and (b) the deflection of the load P. The rigidity of the cross section of the wire is EI.

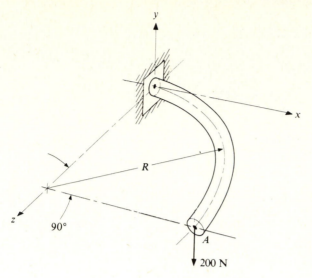

Figure 4-49

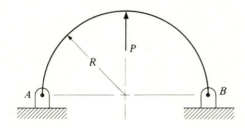

Figure 4-50

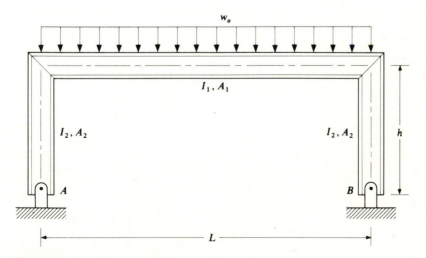

Figure 4-51

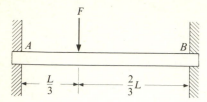

Figure 4-52

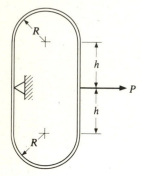

Figure 4-53

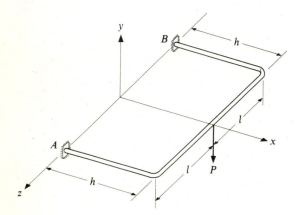

Figure 4-54

4-20 For the wire form shown in Fig. 4-54, assume that no axial force develops in the $2l$ section and determine the support reactions at points A and B. Assume that the wire is circular so that $J = 2I$.

4-21 Solve Exercise 4-6 using the dummy-load method.

4-22 Solve Exercise 4-7 using the dummy-load method.

4-23 Solve Exercise 4-8 using the dummy-load method.

4-24 Solve Exercise 4-10 using the dummy-load method.

4-25 For the truss given in Exercise 4-8, the temperature coefficient of linear expansion of each member is α. If in addition to the load P, members BG, GC, and BC experience a temperature change of ΔT, determine the vertical deflection of point C. *Note:* Symmetry cannot be employed.

4-26 Solve Exercise 4-17 using the dummy-moment method.

4-27 A uniformly loaded cantilever beam is overloaded so that the moment at the wall support reaches a value of $1.2M_e$. If the cross section of the beam is rectangular with a rigidity of EI_z and the beam length is L, determine the maximum deflection under the given load. If the load w_0 is subsequently released, determine the permanent set of the free end of the beam. *Hint*: For simplicity in the bending-moment expression, let x originate from the free end of the beam.

4-28 For the uniformly loaded cantilever beam shown in Fig. 4-55, determine the maximum deflection δ by Rayleigh's technique using the following deflection shapes:

(a)
$$y_c = -\delta\left(1 - \cos\frac{\pi x}{2L}\right)$$

(b)
$$y_c = -\frac{1}{3}\delta\frac{x^2}{L^4}(4Lx - x^2 - 6L^2)$$

Compare the results with Appendix D.

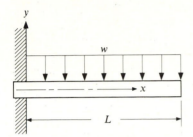

Figure 4-55

4-29 Figure 4-56 shows a cantilever beam loaded by a concentrated moment M.
   (a) Determine the work due to M in terms of the centroidal axis deflection $y_c(x)$.
   (b) Assuming a deflection shape of $y_c = \delta[1 - \cos(\pi x/2L)]$, approximate δ using Rayleigh's technique and compare the results with beam D.4 of Appendix D.

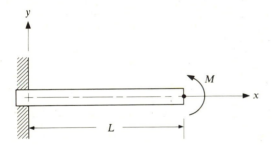

Figure 4-56

4-30 For the beam shown in Fig. 4-55, determine the deflection of the centroidal axis using the Rayleigh-Ritz technique. Assume the deflection shape to be polynomial in x. That is,

$$y_c = ax^4 + bx^3 + cx^2 + dx + e$$

For boundary conditions, employ only the geometric boundary conditions. (Uniform loads on beams lead to deflection shapes up to the fourth order in x. Thus the solution should yield exact results.)

4-31 For Example 4-34, determine $y_c(x)$ exactly through the development and solution of the appropriate differential equation. If $P = 2000$ N, $F = 3000$ N, $EI_z = 3 \times 10^4$ N·m², and $L = 1$ m, determine the maximum deflection and compare this result with that given for Example 4-34.

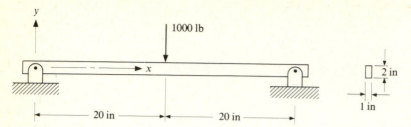

Figure 4-57

4-32 Using Rayleigh's method, estimate the maximum deflection in the steel beam shown in Fig. 4-57, the direct tensile force, and the end reactions, using (a) $y_c = -\delta \sin(\pi x/L)$ and (b) the shape given in Appendix D for a simply supported beam with a concentrated force.

STRENGTH THEORIES AND DESIGN METHODS

5-0 INTRODUCTION

The word *strength* is so overworked in the field of stress analysis that its meaning at times becomes quite vague. For example, strength of materials as the title of a course suggests that material behavior per se is the main concern. On the contrary, although material behavior is an important part of a strength of materials course, it is usually considered as a secondary topic, which simply links stress to strain (via the material properties E and ν), and provides some limits of material capability, e.g., yield strength or ultimate strength. The primary emphasis in a strength of materials course is on the behavior of *mechanical elements* under the influence of applied loads (where naturally the elements themselves are made of various materials). This is where the misuse of the word *strength* propagates. An example is a mechanical element discussed in studying bending. The element is a "beam of constant strength," which sounds confusing right from the beginning. A beam in bending, called a beam of constant strength, is actually a beam of a specific geometric configuration where the outer fibers along the entire span of the beam are of the same (or constant) *stress*. In order to avoid confusion, it is necessary to divorce the words *stress* and *strength*. Stress has already been defined as the internal-force distribution along a specific surface at a uniquely distinct point within a mechanical element. Strength denotes some kind of limit and is more difficult to define. For example, consider an aluminum rod 0.5 in in diameter loaded in tension. If the tensile force reaches a maximum value of 13,000 lb before a complete fracture, it is acceptable to say that the ultimate strength of that specific mechanical element loaded in the particular manner stated *was* 13,000 lb. However, this definition is restricted by size, type of loading, material, and the specific specimen. In an attempt to eliminate size from consideration, the force is divided by the area of the cross section, giving the maximum tensile stress reached before fracture. It is common engineering practice to calculate the maximum tensile stress based on the initial area of the cross section. Therefore, for the aluminum rod under discussion, the initial area was $A = (0.785)(0.5)^2 = 0.196$ in^2, and the maximum tensile stress before fracture was $13,000/0.196 = 66,300$ lb/in^2. Now the ultimate strength can be defined by the maximum tensile stress rather than the maximum force obtained before fracture. This value is commonly tabulated in engineering handbooks and by definition is the *ultimate tensile strength* S_u (force per unit area). To avoid confusion with stress, which is a state property, the capital letter S will be used to denote strength. For example, the value of stress when yielding of a material occurs is called the *yield strength* S_y.

It is normally assumed that once strength is defined in terms of some maximum obtainable value of stress, size effects are no longer important. Although this is normal practice, in certain cases, it is far from the truth. In general, for a given material the strength in pounds per square inch decreases as the size increases. This is most dramatic when the size is very small. For example, the ultimate strength of music wire for a wire diameter of 0.010 in is about 390,000 lb/in^2, whereas the same material with diameter of 0.250 in has an ultimate strength of 240,000 lb/in^2. However, it is important to note that this drastic difference usually levels off with increasing size. Thus, in normal situations, size effects are neglected, especially when the loading is static. Assuming this, the strength value (S_u or S_y) is some measure of the material's performance when packaged as a circular rod and loaded gradually and axially in tension. What of the various other possible geometric and loading configurations? An attempt to minimize the amount of experimentation that would be necessary for each new case comes in the form of strength theories, which try to relate a general state of stress to the strength values obtained from simple tests. The standard static tests which are performed are the tension, compression, and torsion tests. For cyclic loading, the standard test is the bending-fatigue test.

Static strength theories are discussed in the first section of this chapter. In the next section the equations of the commonly used strength theories are modified by the *design factor*, which results in the design equations typically used in engineering applications. The design factor is discussed briefly in the following section. Section 5-4 presents two methods commonly used in design to approximate the solutions of complex statically indeterminate problems. The last section of the chapter is devoted to a brief introduction to fatigue.

5-1 STRENGTH THEORIES

The stress-strain curve for a material is obtained by subjecting a bar to pure tension. When the stress in the bar goes beyond the elastic limit, the material will begin to acquire inelastic, or permanent, deformation. The value of stress at this point is called the yield strength S_y. If yielding of the material is the criterion for failure, then, all things being equal, the stress in a similar part of the same material should not be allowed to exceed the yield strength *provided* the part is loaded in pure tension.† What would be the condition for yielding of a part made of this material if the state of stress was more complicated, e.g., general biaxial or triaxial states of stress? One logical criterion might be to limit the maximum principal stress to the yield strength of the material. This might be true for one type of material and state of stress, but in general not all materials have the same mode of failure. For example, if a tensile specimen made of a ductile material is loaded to fracture, the fractures will all be at an angle of approximately 45° from the loading axis (see Fig. 5-1a). This would seem to indicate that the shear stress had more contributing effect to the failure than the tensile stress since the maximum shear stress occurs at 45° (see Example 2-14). To further substantiate this, ductile specimens tend to yield at the same normal stress, whether in tension or compression, because the shear stress is the major cause. Thus, for ductile materials the maximum-shear-stress theory is often used. On the other hand, a tensile specimen made of a brittle material does tend to fail in tension, as can be seen by the fracture shown in Fig. 5-1b.

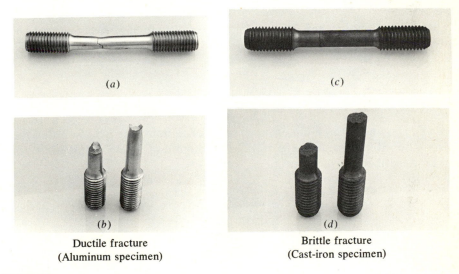

(a)

(c)

(b)

(d)

Ductile fracture
(Aluminum specimen)

Brittle fracture
(Cast-iron specimen)

Figure 5-1

† This is neglecting statistical variations in the material as well as size effects and loading conditions.

There are many theories of static failure of which the consequences can be seen in the tensile test. When the specimen begins to yield, the following events occur:

1. *The maximum-principal-stress theory*: the maximum principal stress reaches the tensile yield strength S_y.
2. *The maximum-shear-stress theory*: the maximum shear stress reaches the shear yield strength $S_y/2$.
3. *The maximum-principal-strain theory*: the maximum principal strain reaches the yield strain S_y/E.
4. *The maximum-strain-energy theory*: the potential energy (called the strain energy) per unit volume reaches a maximum of $\frac{1}{2}S_y^2/E$ (see Sec. 4-2).
5. *The maximum-distortion-energy theory* (sometimes called von Mises' theory): the energy causing a change in shape reaches a maximum of $[(1+\nu)/3E]S_y^2$.†

Theories 1, 3, and 4 will not be discussed in this section since their application to real materials is quite limited. The maximum-shear-stress theory or the maximum-distortion-energy theory is generally applied when the structural material is ductile. The distortion-energy theory generally predicts failure more accurately, but the maximum-shear-stress theory is often used in design because it is more conservative. For brittle materials, the Coulomb-Mohr failure theory is often used in design. The theory is similar to the maximum-shear-stress theory and is conservative. There are more accurate theories for specific brittle materials, but because of their specialized nature, they will not be discussed here.

Before continuing, consider a general triaxial state of stress at a point given by

$$\sigma = \begin{bmatrix} \sigma_x & \tau_{xy} & \tau_{zx} \\ \tau_{xy} & \sigma_y & \tau_{yz} \\ \tau_{zx} & \tau_{yz} & \sigma_z \end{bmatrix}$$

It can be shown, by three-dimensional transformations (see Sec. 6-1), that there exists a coordinate system $x'y'z'$ where the state of stress at the same point can be described by the matrix

$$\sigma = \begin{bmatrix} \sigma_{x'} & 0 & 0 \\ 0 & \sigma_{y'} & 0 \\ 0 & 0 & \sigma_{z'} \end{bmatrix}$$

The stresses $\sigma_{x'}$, $\sigma_{y'}$, $\sigma_{z'}$ are the principal stresses at the point, and are denoted by σ_1, σ_2, and σ_3, where arbitrarily, $\sigma_1 > \sigma_2 > \sigma_3$. The principal stresses are the

---

† The maximum-octahedral-stress theory yields the same results as the distortion-energy theory and will not be discussed. See Exercise 6-4.

three roots of the equation [see Eq. (6-14)]

$$\sigma^3 - (\sigma_x + \sigma_y + \sigma_z)\sigma^2 + (\sigma_x\sigma_y + \sigma_y\sigma_z + \sigma_z\sigma_x - \tau_{xy}^2 - \tau_{yz}^2 - \tau_{zx}^2)\sigma$$
$$- (\sigma_x\sigma_y\sigma_z + 2\tau_{xy}\tau_{yz}\tau_{zx} - \sigma_x\tau_{yz}^2 - \sigma_y\tau_{zx}^2 - \sigma_z\tau_{xy}^2) = 0 \qquad (5\text{-}1)$$

For a biaxial state of stress, $\sigma_z = \tau_{yz} = \tau_{zx} = 0$ and Eq. (5-1) reduces to

$$\sigma^3 - (\sigma_x + \sigma_y)\sigma^2 + (\sigma_x\sigma_y - \tau_{xy}^2)\sigma = 0 \qquad (5\text{-}2)$$

of which the *three* roots are

$$\sigma = 0 \qquad \sigma = \frac{\sigma_x + \sigma_y}{2} \pm \sqrt{\left(\frac{\sigma_x - \sigma_y}{2}\right)^2 + \tau_{xy}^2} \qquad (5\text{-}3)$$

Once the three principal stresses are determined, three Mohr's circles can be constructed; the three circles intersect σ_1, σ_2, and σ_3 on the σ axis as shown in Fig. 5-2. It can be shown that all possible values of σ and τ for any surface intersecting the point in question are on or within the boundaries of the three circles. The possible values of σ and τ are shown in the shaded areas of Fig. 5-2*b*. Thus, the maximum shear stress is given by

$$\tau_{max} = \frac{\sigma_1 - \sigma_3}{2} \qquad (5\text{-}4)$$

Example 5-1 Determine the maximum shear stress on the outer surface of a closed thin-walled cylinder internally pressurized at 1000 lb/in². The inner diameter of the cylinder is 20 in, and the wall thickness is 0.5 in.

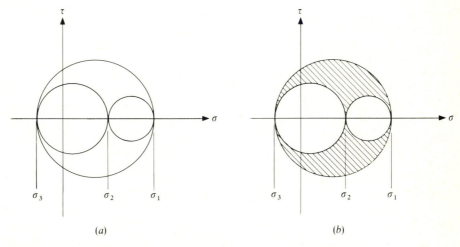

(a)   (b)

Figure 5-2

SOLUTION The stresses for a thin-walled cylinder are given in Chap. 2, where the hoop stress σ_θ is

$$\sigma_\theta = \frac{pD}{2t} = \frac{(1000)(20)}{(2)(0.5)} = 20{,}000 \text{ lb/in}^2$$

and the longitudinal stress σ_z is

$$\sigma_z = \frac{pD}{4t} = \frac{(1000)(20)}{(4)(0.5)} = 10{,}000 \text{ lb/in}^2$$

The stress perpendicular to the outer surface of the cylinder is zero because it is exposed to the atmosphere (free surface). Since there are no shear stresses on the radial, tangential, and longitudinal surfaces, the principal stresses are

$$\sigma_1 = 20{,}000 \text{ lb/in}^2 \qquad \sigma_2 = 10{,}000 \text{ lb/in}^2 \qquad \sigma_3 = 0$$

and the maximum shear stress is

$$\tau_{max} = \frac{\sigma_1 - \sigma_3}{2} = \frac{20{,}000 - 0}{2} = 10{,}000 \text{ lb/in}^2$$

Note that if the analysis were restricted to a biaxial analysis considering only the hoop and longitudinal stresses, the maximum shear stress would have been only 5000 lb/in^2, which is much smaller than the actual maximum. Thus one must be very careful when performing a two-dimensional analysis.

Maximum-Shear-Stress Theory

Since the maximum shear stress in an element loaded in pure tension is one-half the maximum tensile stress, the shear yield strength is $S_y/2$. Thus, the yield criterion for the maximum-shear-stress theory is

$$\frac{\sigma_1 - \sigma_3}{2} = \frac{S_y}{2} \qquad \text{or} \qquad \sigma_1 - \sigma_3 = S_y \tag{5-5}$$

The Maximum-Energy-of-Distortion Theory

This theory relates the distortional energy of a point under a general state of stress to that of the tensile specimen at yielding. A hydrostatic state of stress occurs when all three principal stresses are equal and no distortion occurs. Any deviation from this state will cause distortion. A general state of stress can be thought of as a pure hydrostatic state plus a distortion state. To obtain each state, the average normal stress σ_{av} is determined, where

$$\sigma_{av} = \frac{\sigma_x + \sigma_y + \sigma_z}{3} \tag{5-6}$$

Thus, the general state of stress can be written

$$\sigma = \begin{bmatrix} \sigma_{av} & 0 & 0 \\ 0 & \sigma_{av} & 0 \\ 0 & 0 & \sigma_{av} \end{bmatrix} + \begin{bmatrix} \sigma_x - \sigma_{av} & \tau_{xy} & \tau_{zx} \\ \tau_{xy} & \sigma_y - \sigma_{av} & \tau_{yz} \\ \tau_{zx} & \tau_{yz} & \sigma_z - \sigma_{av} \end{bmatrix}$$

The first matrix is the hydrostatic part of the stress, and the second is the distortional part of the stress. Another way of expressing this using the principal stresses is

$$
\sigma = \begin{bmatrix} \sigma_{av} & 0 & 0 \\ 0 & \sigma_{av} & 0 \\ 0 & 0 & \sigma_{av} \end{bmatrix} + \begin{bmatrix} \sigma_1 - \sigma_{av} & 0 & 0 \\ 0 & \sigma_2 - \sigma_{av} & 0 \\ 0 & 0 & \sigma_3 - \sigma_{av} \end{bmatrix}
$$

where $\sigma_{av} = (\sigma_1 + \sigma_2 + \sigma_3)/3$.† The energy per unit volume of a stressed element is derived in Sec. 4-2. The energy per unit volume of an element without shear stresses is given by Eq. (4-6) and is

$$
u = \tfrac{1}{2}[\sigma_x^2 + \sigma_y^2 + \sigma_z^2 - 2\nu(\sigma_x\sigma_y + \sigma_y\sigma_z + \sigma_z\sigma_x)] \tag{5-7}
$$

To determine the distortion energy, simply substitute $\sigma_1 - \sigma_{av}$, $\sigma_2 - \sigma_{av}$, and $\sigma_3 - \sigma_{av}$, for σ_x, σ_y, and σ_z respectively; when the relationship $\sigma_{av} = (\sigma_1 + \sigma_2 + \sigma_3)/3$ is used, the energy of distortion per unit volume reduces to

$$
u_d = \frac{1+\nu}{6E}[(\sigma_1 - \sigma_2)^2 + (\sigma_2 - \sigma_3)^2 + (\sigma_3 - \sigma_1)^2] \tag{5-8}
$$

For the tensile test, the state of stress at yielding is $\sigma_1 = S_y$, $\sigma_2 = \sigma_3 = 0$. Thus at yielding, the energy of distortion is

$$
u_d = \frac{1+\nu}{3E}S_y^2 \tag{5-9}
$$

Thus, equating Eqs. (5-8) and (5-9), we see that the criterion for yielding of an element under a general state of stress is

$$
\sqrt{\tfrac{1}{2}[(\sigma_1 - \sigma_2)^2 + (\sigma_2 - \sigma_3)^2 + (\sigma_3 - \sigma_1)^2]} = S_y \tag{5-10}
$$

Example 5-2 Estimate the torque on a 10-mm-diameter steel shaft when yielding begins using (a) the maximum-shear-stress theory and (b) the maximum-distortion-energy theory. The yield strength of the steel is 140 MN/m².

SOLUTION (a) The shear yield strength is one-half the tensile yield strength. The maximum shear stress in torsion is thus equated to the shear yield strength, resulting in

$$
\tau_{max} = \frac{16T}{\pi d^3} = \tfrac{1}{2}S_y
$$

Solving for the torque yields

$$
T = \frac{\pi d^3}{32}S_y = \frac{(\pi)(0.010)^3}{32} \, 140 \times 10^6 = 13.74 \text{ N·m}
$$

(b) The principal stresses for an element with one set of shear stresses, say $\tau_{xy} = \tau$, are

---

† In Chap. 6 it is shown that the sum of the normal stresses on any three mutually perpendicular surfaces at a point is constant. Thus, $\sigma_x + \sigma_y + \sigma_z = \sigma_1 + \sigma_2 + \sigma_3$.

found using Eqs. (5-3) to be

$$\sigma_1 = \tau \qquad \sigma_2 = 0 \qquad \sigma_3 = -\tau$$

Substituting this into Eq. (5-10) yields

$$3\tau^2 = S_y^2$$

or

$$\tau = 0.577 S_y$$

Substituting $\tau = 16T/\pi d^3$ and solving for the torque results in

$$T = \frac{(0.577)\pi d^3}{16} S_y = \frac{(0.577)\pi (0.010)^3}{16} 140 \times 10^6 = 15.86 \text{ N·m}$$

Thus, it can be seen that for yielding in pure torsion, the distortion theory predicts a torque which is 15 percent greater than the prediction of the maximum-shear-stress theory. Tests on ductile materials have shown that the distortion theory is much more accurate for predicting yield, but in design work the more conservative answer predicted by shear stress is commonly used.

For biaxial stress,

$$\sigma = \begin{bmatrix} \sigma_x & \tau_{xy} \\ \tau_{xy} & \sigma_y \end{bmatrix}$$

The principal stress, in the z direction, is zero. Consider the principal stresses in the xy plane to be $\hat\sigma_1$ and $\hat\sigma_2$; they are calculated from Eq. (2-35) or (5-3). To understand the difference between the two theories, a graph of $\hat\sigma_1$ vs. $\hat\sigma_2$ is drawn. First, the maximum-shear-stress theory will be developed for the following cases.

Case 1. $\hat\sigma_1 > \hat\sigma_2 > 0$ Since the third principal stress $\sigma_3 = 0$,

$$\tau_{max} = \frac{\hat\sigma_1 - 0}{2} = \frac{S_y}{2} \qquad \text{or} \qquad \hat\sigma_1 = S_y \qquad (5\text{-}11a)$$

Case 2. $\hat\sigma_1 > 0 > \hat\sigma_2$ For this case, $\sigma_1 = \hat\sigma_1$ and $\sigma_3 = \hat\sigma_2$, and the maximum shear stress is

$$\tau_{max} = \frac{\hat\sigma_1 - \hat\sigma_2}{2} = \frac{S_y}{2} \qquad \text{or} \qquad \hat\sigma_1 - \hat\sigma_2 = S_y \qquad (5\text{-}11b)$$

Case 3. $0 > \hat\sigma_1 > \hat\sigma_2$ For this case, $\sigma_1 = 0$ and $\sigma_3 = \hat\sigma_2$, and the maximum shear stress is

$$\tau_{max} = \frac{0 - \hat\sigma_2}{2} = \frac{S_y}{2} \qquad \text{or} \qquad \hat\sigma_2 = -S_y \qquad (5\text{-}11c)$$

There are three other cases similar to these, corresponding to $\hat\sigma_2 > \hat\sigma_1$, these relationships are plotted in Fig. 5-3a.

For the distortion-energy-theory, substitution of $\hat\sigma_1$, $\hat\sigma_2$, and 0 for the three

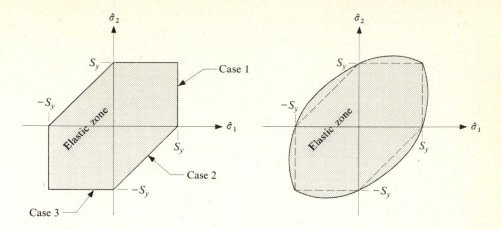

(a) Maximum shear stress theory   (b) Maximum distortion energy theory

Figure 5-3

principal stresses in Eq. (5-10) yields

$$\sqrt{\hat{\sigma}_1^2 + \hat{\sigma}_2^2 - \hat{\sigma}_1\hat{\sigma}_2} = S_y \tag{5-12}$$

A graph of this equation is presented in Fig. 5-3b.

For simplification, an equivalent single normal stress which is as damaging as the case given in Eq. (5-12) is defined as the von Mises' stress σ_{vM}, where

$$\sigma_{vM} = \sqrt{\hat{\sigma}_1^2 + \hat{\sigma}_2^2 - \hat{\sigma}_1\hat{\sigma}_2} \tag{5-13}$$

Thus, according to the maximum-distortion-energy theory, yielding occurs when

$$\sigma_{vM} = S_y \tag{5-14}$$

Brittle materials exhibit no distinct point of yielding, but in terms of ultimate strength, brittle materials tend to fail in a tension mode when loaded in tension; when loaded in compression, they tend to fail in a shear mode at a higher compressive load than when in tension. Thus, two failure points are generally specified, the ultimate strength in tension S_{ut} and the ultimate strength in compression S_{uc}, where $|S_{uc}| > S_{ut}$. To accommodate this, the maximum-shear theory can be modified to the *Coulomb-Mohr theory of failure.* For a biaxial state of stress, the three cases of Eqs. (5-11) are modified so that

Case 1, $\hat{\sigma}_1 > \hat{\sigma}_2 > 0$:

$$\hat{\sigma}_1 = S_{ut} \tag{5-15a}$$

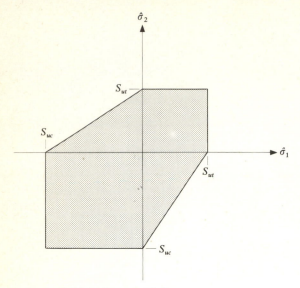

Coulomb-Mohr theory of failure   **Figure 5-4**

Case 2, $\hat{\sigma}_1 > 0 > \hat{\sigma}_2$:

$$\frac{\hat{\sigma}_1}{S_{ut}} + \frac{\hat{\sigma}_2}{S_{uc}} = 1 \qquad (5\text{-}15b)$$

(Note that S_{uc} is a negative number.)

Case 3, $0 > \hat{\sigma}_1 > \hat{\sigma}_2$:

$$\hat{\sigma}_2 = S_{uc} \qquad (5\text{-}15c)$$

The three cases and the corresponding ones for $\hat{\sigma}_2 > \hat{\sigma}_1$ are plotted in Fig. 5-4.

5-2 DESIGN EQUATIONS

The equations obtained in the previous section can be modified for design purposes. To obtain a design equation, the strength is reduced by a design factor n, occasionally referred to as the factor of safety. Modifying the commonly used strength theories for biaxial problems results in the following design equations.

Ductile Materials

Maximum-distortion-energy-theory:

$$\sqrt{\hat{\sigma}_1^2 + \hat{\sigma}_2^2 - \hat{\sigma}_1\hat{\sigma}_2} = \frac{S_y}{n} \qquad (5\text{-}16)$$

Maximum-shear-stress theory:

Case 1, $\hat{\sigma}_1 > \hat{\sigma}_2 > 0$:

$$\hat{\sigma}_1 = \frac{S_y}{n} \tag{5-17a}$$

Case 2, $\hat{\sigma}_1 > 0 > \hat{\sigma}_2$:

$$\hat{\sigma}_1 - \hat{\sigma}_2 = \frac{S_y}{n} \tag{5-17b}$$

Case 3, $0 > \hat{\sigma}_1 > \hat{\sigma}_2$:

$$\hat{\sigma}_2 = -\frac{S_y}{n} \tag{5-17c}$$

Brittle Materials

Coulomb-Mohr theory of failure:

Case 1, $\hat{\sigma}_1 > \hat{\sigma}_2 > 0$:

$$\hat{\sigma}_1 = \frac{S_{ut}}{n} \tag{5-18a}$$

Case 2, $\hat{\sigma}_1 > 0 > \hat{\sigma}_2$:

$$\frac{\hat{\sigma}_1}{S_{ut}} + \frac{\hat{\sigma}_2}{S_{uc}} = \frac{1}{n} \tag{5-18b}$$

Case 3, $0 > \hat{\sigma}_1 > \hat{\sigma}_2$:

$$\hat{\sigma}_2 = \frac{S_{uc}}{n} \tag{5-18c}$$

Example 5-3 Figure 5-5a shows a round shaft of diameter 1.5 in loaded by a bending moment $M_z = 5000$ lb·in, a torque $T = 8000$ lb·in, and an axial tensile force $N = 6000$ lb. If the material is ductile with a yield strength $S_y = 40,000$ lb/in², determine the design factor corresponding to yield using (a) the maximum-shear-stress theory and (b) the maximum distortion-energy theory.

SOLUTION Initially, consider each loading state separately. The tensile force N creates a tensile stress σ_x, which is constant over the cross section:

$$(\sigma_x)_N = \frac{N}{A} = \frac{6000}{(0.785)(1.5)^2} = 3400 \text{ lb/in}^2$$

The bending moment M gives a normal stress distribution which is linear with respect to y, and the maximum tensile stress due to bending alone occurs at the bottom of the shaft, where

$$(\sigma_x)_M = \frac{Mc}{I} = \frac{(5000)(1.5/2)}{(\pi/64)(1.5)^4} = 15,090 \text{ lb/in}^2$$

Shear stresses τ arise from the torque T. These stresses are maximum at the outer surface, where

$$\tau_T = \frac{Tr_o}{J} = \frac{(8000)(1.5/2)}{(\pi/32)(1.5)^4} = 12,070 \text{ lb/in}^2$$

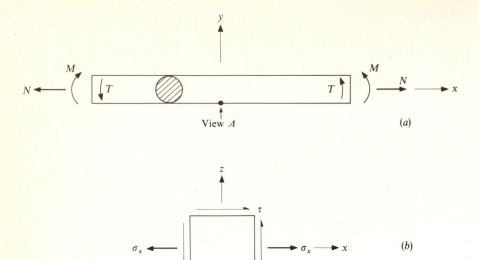

View A

(a)

View A

(b)

Figure 5-5

Combining the stresses using superposition, one can see that the worst case will occur on an element located on the bottom of the shaft, where (see Fig. 5-5b)

$$\sigma_x = (\sigma_x)_N + (\sigma_x)_{Mz} = 3400 + 15{,}090 = 18{,}490 \text{ lb/in}^2$$

$$\tau = \tau_T = 12{,}070 \text{ lb/in}^2$$

The state of stress is biaxial. Considering only surfaces perpendicular to the plane of the page, the principal stresses are given by Eqs. (2-35), where

$$\hat{\sigma}_1 = \frac{18{,}490 + 0}{2} + \sqrt{\left(\frac{18{,}490 - 0}{2}\right)^2 + (12{,}070)^2} = 24{,}450 \text{ lb/in}^2$$

$$\hat{\sigma}_2 = \frac{18{,}490 + 0}{2} - \sqrt{\left(\frac{18{,}490 - 0}{2}\right)^2 + (12{,}070)^2} = -5960 \text{ lb/in}^2$$

(*a*) For the maximum-shear-stress theory this is case 2, given by Eq. (5-17b), and so

$$24{,}450 - (-5960) = \frac{40{,}000}{n}$$

Solving for n yields $n = 1.32$.

(*b*) For the maximum-distortion-energy theory, Eq. (5-16) applies, where

$$\sqrt{24{,}450^2 + (-5960)^2 - (24{,}450)(-5960)} = \frac{40{,}000}{n}$$

Solving for n yields $n = 1.43$.

Since the maximum-distortion-energy theory is less conservative, it makes sense for a higher factor of safety to be obtained in part (*b*).

5-3 THE DESIGN FACTOR

The design factor n is used to account for uncertainties in the determination of the strength of the part as well as uncertainties in the evaluation of the stresses in the part (see Ref. 1).

The uncertainties relating to part strength are:

1. The uncertainty of the exact properties of the base material of the part
2. The uncertainty of the size effect, discussed earlier
3. The uncertainty in the manufacturing of the part; material forming (casting, forging, drawing, etc.), machining, welding, heat treatment, and surface treatment (such as plating)—all of which have an effect on the part-strength uncertainty
4. The uncertainty of the effect of the operating environment, e.g., temperature effects and corrosion.

Define a part-strength uncertainty factor f_s such that the "true" minimum strength of the part $S_{\min}$ is given by

$$S_{\min} = \frac{S}{f_s} \tag{5-19}$$

where S is the material strength of a tensile specimen of a given material obtained from direct test, published data, or manufacturer's specifications. The value of S may not reflect the true strength of the actual design part and is corrected by the uncertainty factor f_s. The strength uncertainty factor is the product of each individual uncertainty of items 1 to 4 above. In general, $f_s > 1$.

The uncertainties associated with the part stresses are:

1. The uncertainties in the type of calculations made for dynamic forces and moments.
2. The uncertainties of stress calculations due to manufacturing inaccuracies. Because of manufacturing tolerances, stresses calculated using nominal dimensions may be in error due to cross-sectional dimension tolerances and true load positions relative to the part.
3. The uncertainties in assembly operations. Initial stresses may arise from methods of fastening one part to another (bolting, riveting, welding, press fitting, etc.), where human error may also be involved. Alignment also affects stress calculations. If self-alignment devices are used, the uncertainty in stress is reduced.
4. The uncertainties in the internal stresses before assembly. Residual stresses may be present because of material forming or fabrication techniques.
5. The uncertainties in the analysis of the stresses. This depends on how good the analysis was in terms of modeling and the particular stress equations used. For example, if a pressurized cylinder was analyzed using the thin-walled equations, some error will exist, as the thick-walled equations

are more exact. The thin-walled equations are limited to $D \geq 20t$, where D is the inner diameter and t is the wall thickness. When $D = 20t$, the thin-walled equations predict stresses which are approximately 5 percent less than those given by the thick-walled equations.

Another area of uncertainty in the stress analysis occurs in establishing the stress concentration K_t, if appropriate. Although many cases have been determined analytically and experimentally, no general analysis is available.

Define a stress uncertainty factor f_σ such that the "true" maximum stress σ_{max} is given by

$$\sigma_{max} = \frac{\sigma}{f_\sigma} \tag{5-20}$$

where σ is the stress determined by analysis. The stress-uncertainty factor is the product of the factors of items 1 to 5 above. In general, $f_\sigma < 1$.

Combining Eqs. (5-19) and (5-20) gives

$$\frac{S}{\sigma} = \frac{f_s S_{min}}{f_\sigma \sigma_{max}}$$

A strength-reduction factor would not be necessary if S_{min} and σ_{max} were known. Thus, let $S_{min}/\sigma_{max} = 1$, and since the ratio S/σ is the strength-reduction factor n,

$$n = \frac{f_s}{f_\sigma} \tag{5-21}$$

There is even a good deal of uncertainty associated with obtaining the strength- and stress-uncertainty factors themselves. In many cases, codes dictate the appropriate value of n. Where codes do not apply, the designer must often make an intelligent guess for n based either on intuitive reasoning or past experience. Most of the uncertainty factors must be arrived at through statistical studies, especially those dealing with strength.

Making a guesstimate of the design factor may be the only alternative open to the designer. However, the decision is not to be taken lightly. The following example attempts to illustrate how subtle the selection of a factor of safety can be.

Example 5-4 It is necessary to make a plate of the geometry shown in Fig. 5-6. The material to be used has a published yield strength of 50,000 lb/in². The designer is to specify the maximum acceptable load so that absolutely no yielding of the material will occur. For sake of illustration, the following will be assumed.

1. Everything is ideal, except for the tolerances on the cross section; i.e., the published data on the yield strength are exact, and the material is homogeneous and isotropic; the load is positioned perfectly so that the loading causes no bending; the loading is gradually applied; etc.
2. The designer is unaware of stress-concentration effects.
3. No codes for the establishment of a factor of safety apply to the given design.

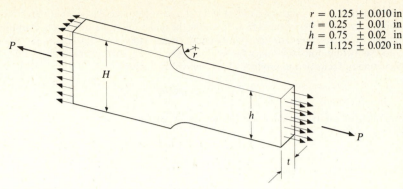

$$r = 0.125 \pm 0.010 \text{ in}$$
$$t = 0.25 \pm 0.01 \text{ in}$$
$$h = 0.75 \pm 0.02 \text{ in}$$
$$H = 1.125 \pm 0.020 \text{ in}$$

Figure 5-6

DESIGNER'S SOLUTION It is unfortunate, but assumption 2 applies to many designers. Our designer will probably use nominal dimensions, the normal procedure, finding a stress

$$\sigma = \frac{P_{max}}{A} = \frac{P_{max}}{(\frac{3}{4})(\frac{1}{4})} = \tfrac{16}{3} P_{max}$$

He will now equate this to the allowable yield strength divided by the factor of safety. According to the given assumptions, the only place he feels that he is in error is in the determination of the cross-sectional area. The stress-uncertainty factor based on dimensional tolerances is

$$f_\sigma = \frac{\sigma}{\sigma_{max}} = \frac{P_{max}/A_{nom}}{P_{max}/A_{min}} = \frac{A_{min}}{A_{nom}} = \frac{(0.73)(0.24)}{(0.75)(0.25)} = 0.9344$$

Since the strength value is assumed to be exact, $f_s = 1$ and the design factor is

$$n = \frac{f_s}{f_\sigma} = \frac{1}{0.9344} = 1.07$$

To be on the "safe" side, the designer decides to use a design factor of 1.1. To complete the analysis, the stress is equated to the strength divided by the design factor and the result is

$$\tfrac{16}{3} P_{max} = \frac{50,000}{1.1}$$

$$P_{max} = 8523 \text{ lb}$$

EXACT SOLUTION The designer was ignorant of the fact that the true stress distribution is not uniform and that the stress near the fillets is larger than the nominal stress he calculated. From Appendix G, the stress-concentration factor is $K_t = 1.8$. Thus, the maximum stress is actually

$$\sigma_{max} = K_t \frac{P_{max}}{A}$$

If the nominal cross-sectional dimensions are used to calculate the area A, then, as before, a factor of safety of 1.07 should be used to reduce the working stress:

$$1.8 \frac{P_{max}}{(\frac{3}{4})(\frac{1}{4})} = \frac{50,000}{1.07}$$

$$P_{max} = 4870 \text{ lb}$$

In the above analysis the factor of safety $n = 1.07$ is based on an uncertainty in strength, $f_s = 1$, and in stress, $f_\sigma = 1/1.07$. If in the exact analysis the nominal stress equation is used (the same one used by the designer) a different factor of safety should be used. That is, the uncertainty factor in stress changes to account for the omission of the stress concentration in the nominal stress equation. The uncertainty factor for the stress concentration is $f_\sigma = 1/1.8 = 0.556$, and, as before, the uncertainty factor due to the dimensional tolerances is $1/1.07$. The total uncertainty factor is the product of the individual factors. Thus,

$$f_\sigma = \frac{1}{1.8}\frac{1}{1.07} = 0.5192$$

and the factor of safety using the nominal stress should be

$$n = \frac{f_s}{f_\sigma} = \frac{1}{0.5192} = 1.926 \approx 1.93$$

Thus, a designer who had used this value of n would have obtained the exact solution, i.e.,

$$\tfrac{16}{3}P_{\text{max}} = \frac{50,000}{1.926}$$

$$P_{\text{max}} = 4870 \text{ lb}$$

In general, the exact solution is not known as in the overidealized problem just presented. The purpose of the example was to illustrate that a factor of safety is based on how close the analysis reflects the given problem. There will be many cases where a guess is necessary, but one should not be too casual about it.

5-4 ENGINEERING APPROXIMATIONS USED IN STATICALLY INDETERMINATE PROBLEMS

Although much time is spent on analyzing statically determinate problems, in reality, structures are normally designed to be highly statically indeterminate. To a large extent this is due to the engineer's basic tendency toward conservatism. Statically indeterminate structures are supported in a redundant fashion. Thus, more than one point will provide safety against collapse. In addition, statically indeterminate structures generally distribute stresses better. Therefore, in a given situation, certain stresses can be reduced by adding more restraint. Another basic reason for highly statically indeterminate structures is to reduce the large elastic deflections and rotations which generally arise in statically determinate structures. Although more desirable structurally, statically indeterminate problems are more difficult to analyze because the deflections of the structure must be included in the analysis. As the structure becomes more redundant and the number of mechanical elements increases, the exact solution becomes more difficult to obtain. Energy techniques provide the easiest approach to complex statically indeterminate problems; however, there are many times when it is neither possible nor necessary to perform an elaborate and exact analysis of a structure. There are two basic techniques used to make engineering approximations where a statically indeterminate problem is reduced to a statically determinate one. The first is the method of neglecting elastic deformations of members much more rigid than others. This technique is used quite extensively with fasteners like bolts and rivets. The second method of approximation is called *limit analysis* (sometimes referred to as limit design or ultimate analysis). Limit analysis allows for a given section under high stress to go completely into the plastic range. Since in this circumstance the stress distribution is known, a deflection equation is no longer necessary and the number of necessary deflection equations for an exact analysis can be reduced to zero.

Considering Deflections of Flexible Elements Only

Consider the riveted connection shown in Fig. 5-7*a*. If a break is made isolating the right-hand member by slicing through the rivets, the resulting free-body diagram is as shown in Fig. 5-7*b*.

Thanks to symmetry, the force on the top of each rivet equals the force on the bottom. Summing forces yields $2F_1 + 2F_2 = P$. Since the system is statically indeterminate, the exact values of F_1 and F_2 would have to be determined by deflection theory, but this is where an approximation is often made. If the deflections of the plates are ignored, the shear deflections of the rivets will be the same. Thus the shear stresses are identical in each rivet. If $A_1 = A_2$, $F_1 = F_2$. Consequently, applying the equilibrium equation results in $F_1 = F_2 = P/4$.

If the rivets are located as shown in Fig. 5-8, twisting of the plates might occur. To avoid twisting, the shear deflections of each rivet should be equal (assuming again that the plates are basically rigid). Since shear deflections are

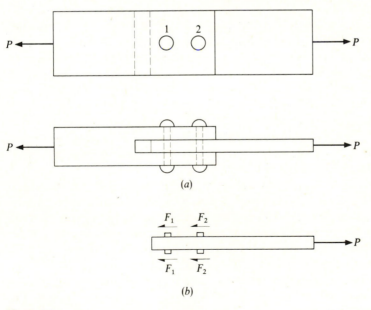

(a)

(b)

Figure 5-7

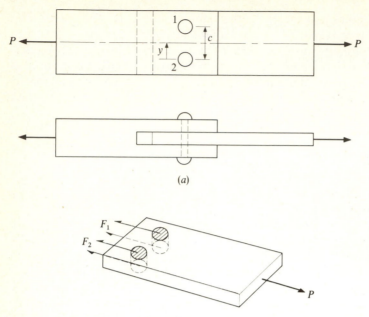

(a)

Figure 5-8

proportional to shear stresses, for no twisting the shear stresses in each rivet should be equal. Thus, $F_1/A_1 = F_2/A_2$, and using the fact that $2F_1 + 2F_2 = P$ results in

$$F_1 = \frac{A_1}{A_1 + A_2}\frac{P}{2} \qquad F_2 = \frac{A_2}{A_1 + A_2}\frac{P}{2}$$

Summing moments about the center of rivet 2 (keeping in mind that a net force of $2F_1$ is acting on rivet 1, top and bottom surfaces), gives the necessary line of load application for no twist as

$$\bar{y} = \frac{cA_1}{A_1 + A_2}$$

Note that this is the location of the centroid of the shear areas of the rivets. In general, then, loading through the centroid of the shear areas eliminates twisting of the plates.

If the plates are loaded in torsion, as in shaft couplings, or if the load is not through the rivet centroid, twisting will occur and will be centered about the rivet centroid. Considering only the forces which arise from twisting, we see that the line of action of the forces between the rivets or bolts and the plates will be perpendicular to a line drawn between the given rivet and the centroid. An example of the forces applied by a set of rivets to a plate is shown in Fig. 5-9, where a torque is being applied to the plate. The forces which arise from the torque must be such that the torque balances for equilibrium and the net force is zero.

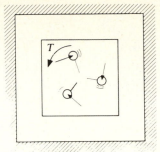

Figure 5-9

Example 5-5 The plate shown in Fig. 5-10 is mounted on a wall with four bolts. Bolts 1 and 2 are $\frac{1}{2}$ in in diameter, and bolts 3 and 4 are $\frac{3}{4}$ in in diameter. Approximate the shear stress in each rivet if a force of 1000 lb is applied to the plate as shown. Neglect any out-of-plane bending.

SOLUTION First, the centroid of the bolt areas is found. Thanks to symmetry, $\bar{y} = 1$ in. The horizontal centroid is[†]

$$\bar{x} = \frac{(2)(2)(\pi/4)(\frac{3}{4})^2}{2[(\pi/4)(\frac{1}{2})^2 + (\pi/4)(\frac{3}{4})^2]} = 1.385 \text{ in}$$

The net torque about the bolt-area centroid is $(4.615)(1000) = 4615$ lb·in. The direct force of 1000 lb is distributed to each bolt through equal stresses (no twisting for the direct force). Since the areas are equal for bolts 1 and 2, the forces will be equal. That is, $F_2' = F_1'$. Likewise for bolts 3 and 4, $F_4' = F_3'$. From the geometry shown in Fig. 5-10c, the relationships for the forces due to the direct 1000-lb load are

$$1.385F_1' = 0.615F_3' \tag{a}$$

and

$$2F_1' + 2F_3' = 1000 \tag{b}$$

Solving Eqs. (a) and (b) results in $F_1' = 154$ lb and $F_3' = 346$ lb.

The next step is to find the forces that arise from the torque. Considering only torque, we see that the forces in bolts 1 and 2 are equal, as are the forces in bolts 3 and 4, because each respective set of rivets has equal area and is an equal distance from the centroid. Since twisting is about the centroid, shear deflections and consequently shear stresses are proportional to the distance from the centroid. Since $r_1 = r_2 = \sqrt{1^2 + (1.385)^2} = 1.71$ in and $r_3 = r_4 = \sqrt{1^2 + (0.615)^2} = 1.17$ in,

$$\frac{F_1''/(\pi/4)(\frac{1}{2})^2}{1.71} = \frac{F_3''/(\pi/4)(\frac{3}{4})^2}{1.17}$$

or

$$1.539F_1'' = F_3'' \tag{c}$$

Balancing the torque results in

$$T = 2F_1''r_1 + 2F_3''r_3$$

or

$$4615 = 2F_1''(1.71) + 2F_3''(1.17) \tag{d}$$

Substitution of Eq. (c) into (d) yields

$$F_1'' = 657 \text{ lb} \qquad \text{and} \qquad F_3'' = 1012 \text{ lb}$$

The total force on each bolt is found by vector addition. The net results are shown in Fig. 5-10c. The shear stresses on bolts 1, 2, 3, and 4 are

$$\tau_1 = \tau_2 = \frac{436}{(\pi/4)(\frac{1}{2})^2} = 2220 \text{ lb/in}^2 \qquad \tau_3 = \tau_4 = \frac{1232}{(\pi/4)(\frac{3}{4})^2} = 2789 \text{ lb/in}^2$$

[†]The $\pi/4$ can be omitted in these calculations and the centroid of d^2 can be found.

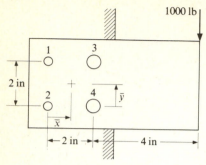

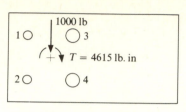

(a) Actual loading

(b) Equivalent loading

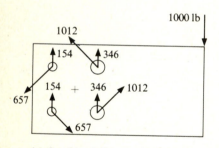

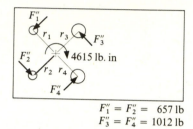

(c) Direct force

(d) Torsional load

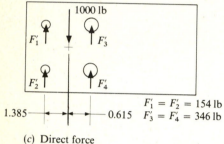

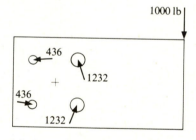

(e) Combined direct and torsional loading

(f) Complete free-body diagram

Figure 5-10

When all the bolts or rivets have equal areas, the problem becomes much simpler, as the forces rather than the stresses can be dealt with directly.

Example 5-6 Repeat Example 5-5 but consider all the bolts to be 1/2 in in diameter.

SOLUTION Since the areas of the bolts are equal, $\bar{x} = 1$ in and $\bar{y} = 1$ in. The reactions from the direct force of 1000 lb divide equally. Thus (see Fig. 5-11c)

$$F_1' = F_2' = F_3' = F_4' = \frac{1000}{4} = 250 \text{ lb}$$

Considering the torque, again, we see that the forces are equal as the bolts are of equal area

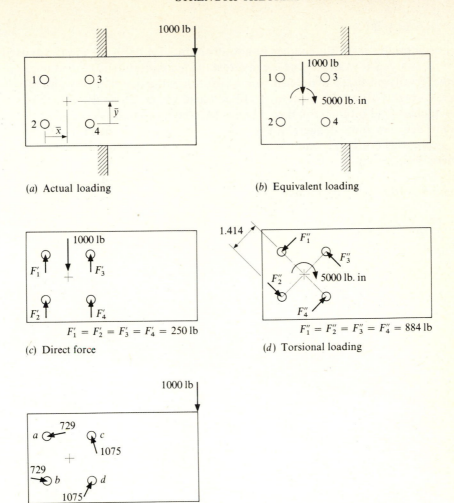

(*a*) Actual loading

(*b*) Equivalent loading

$F_1' = F_2' = F_3' = F_4' = 250\,\text{lb}$

(*c*) Direct force

$F_1'' = F_2'' = F_3'' = F_4'' = 884\,\text{lb}$

(*d*) Torsional loading

(*e*) Combined direct and torsional loading

Figure 5-11

and spacing. Thus $F_1'' = F_2'' = F_3'' = F_4'' = F''$, and for a balance of torque (see Fig. 5-11*d*)

$$4F''(1.414) = 5000$$

$$F'' = 884\,\text{lb}$$

Adding the force vectors results in the forces shown in Fig. 5-11*e*. Thus, for bolts 1 to 4,

$$\tau_1 = \tau_2 = \frac{729}{(\pi/4)(\tfrac{1}{2})^2} = 3713\,\text{lb/in}^2 \qquad \tau_3 = \tau_4 = \frac{1075}{(\pi/4)(\tfrac{1}{2})^2} = 5475\,\text{lb/in}^2$$

Comparing the results of Examples 5-5 and 5-6 shows that increasing the diameter of bolts 3 and 4 reduces the stresses in all four bolts.

Limit Analysis†

In considering a statically indeterminate structure of order n the approach used in limit analysis is to allow n of the greatest loaded restraints to reach their limit value (totally plastic) and determine the case when the $(n + 1)$st restraint totally yields. At this point, the structure will collapse without bounds. The value of the applied load on the structure when total collapse occurs is called the limit load. There are some dangers in using limit analysis which the designer should be aware of. The analysis should be used for statically loaded ductile materials‡ where the maximum forces are known with confidence. If the structure is undergoing large variation cyclic loading, where the danger of a fatigue failure is present, the analysis should not be used.

Example 5-7 Determine the limit load P_L which will cause the beam shown in Fig. 5-12a to collapse. The material is EPP with an elastic limit of $\sigma_e = 40,000$ lb/in².

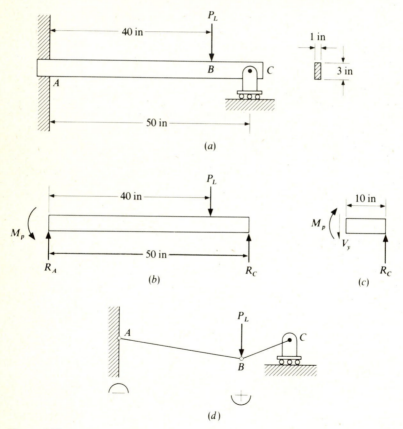

Figure 5-12

† See Ref. 2.
‡ In this section, discussion will be limited to EPP material, as defined in Sec. 3-15.

SOLUTION The structure is statically indeterminate of the first order. Thus, when two restraints reach their limit, the beam will collapse. Since the bending stresses are much greater than the transverse shear stresses, only the bending stresses will be considered. It can be seen that the bending moment will reach maximum values at *two* points, A and B.† Thus, when the moment reaches the limit moment M_p at A and B, the structure will collapse. Isolating the beam at A results in the free-body diagram shown in Fig. 5-12b. Summing moments at A yields

$$M_p + 50R_C - 40P_L = 0. \tag{a}$$

Isolating section BC just to the right of P_L yields the free-body diagram shown in Fig. 5-12c. Summing moments at B yields

$$R_C = \tfrac{1}{10}M_p \tag{b}$$

Substituting this into Eq. (a) results in

$$P_L = 0.15M_p \tag{c}$$

From Eq. (3-88), $M_p = \sigma_e bc^2$. Thus

$$P_L = (0.15)(40,000)(1)(1.5)^2 = 13,500 \text{ lb}‡$$

Note that the directions of the limit moments are found by considering hinges forming at the limit points and determining whether the beam is concave up or down at each point. This is illustrated in Fig. 5-12d.

In some cases, more than $n + 1$ maximum load points may be possible, and more than one collapse mode must be considered.

Example 5-8 Two forces equal in magnitude are applied to the beam shown in Fig. 5-13a. Determine the limit value of P_L in terms of M_p.

SOLUTION Again the structure is indeterminate of the first order, and when two points reach their limit, the structure will collapse. However, there are three positions in which the moment may reach a maximum, points A, B, and C. Thus there are three possible modes of failure shown in Fig. 5-13b, c, and d.

The analysis is the same as Example 5-7. That is, isolations are made where the moment is assumed to be M_p. For example, the free-body diagrams for mode 2 are shown in Fig. 5-13e and f. From Fig. 5-13e, summing moments at A for equilibrium yields

$$M_p + R_D L - \frac{L}{3}P_L - \frac{2L}{3}P_L = 0 \tag{a}$$

† In limit analysis, it is not important which point reaches the limit first. For example, from Appendix D, an exact analysis would show that point B reaches the limit moment M_p first. When this occurs, it is said that a *hinge* forms at point B. The load can still be increased beyond this point, as section AB still has bending capacity. Since the material is EPP, the moment remains constant at the hinge at point B at a value of M_p. Thus the additional bending from the increasing load goes into section AB until the moment at point A reaches the limit M_p. At this point the structure has completely lost its bending capacity and collapses. In the elastic analysis, if the limit moment occurred at point A first, by the same argument, collapse would not occur until the moment at B reached M_p. Thus, in either case the same limit is reached, and consequently the order of the process is immaterial and the results from the elastic analysis are not needed.

‡ Note that if an elastic analysis is performed on the beam using the equations for beam D.11 of Appendix D, when the outer fibers of the beam at point B reach σ_e, the applied force is 8520 lb. Any force greater than this will cause some permanent deformation. However, if this can be tolerated, the beam has much more load capacity than the force found in the elastic analysis.

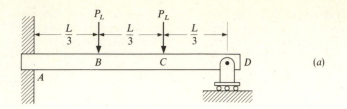

(a)

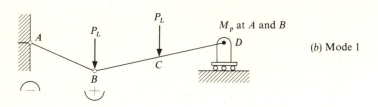

(b) Mode 1

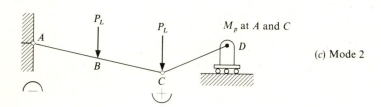

(c) Mode 2

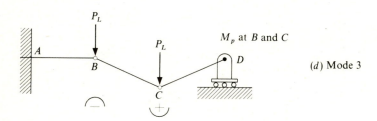

(d) Mode 3

Free-body diagrams
for mode 2

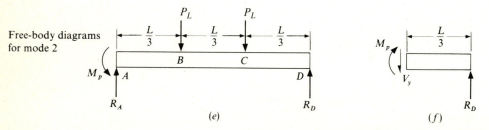

(e)

(f)

Figure 5-13

From Fig. 5-13f, summing moments at C results in

$$R_D = \frac{3}{L} M_p$$

Substituting R_D into Eq. (a) and solving for P_L yields

$$P_L = \frac{4}{L} M_p \qquad \text{mode 2}$$

In a similar manner, modes 1 and 3 can be examined, and the final results are

$$P_L = \begin{cases} \dfrac{5}{L} M_p & \text{mode 1} \\[2mm] \dfrac{9}{L} M_p & \text{mode 3} \end{cases}$$

Since the limit force for collapse mode 2 is the lowest, the limiting value for the structure is $P_L = 4M_P/L$.

More than one mode of collapse can also occur in a multielement structure, and, as in the previous example, each mode should be examined.

Example 5-9 The beam shown in Fig. 5-14a has a rectangular cross section 25 mm wide by 50 mm deep and is completely fixed at point B and supported by a cable at point D. The net tensile area of the cable is $A = 12.5$ mm², and both the beam and cable are made of an EPP material with an elastic limit of $\sigma_e = 240$ MN/m². A vertical force is applied at midspan of the beam. Determine the limiting value of the load P_L.

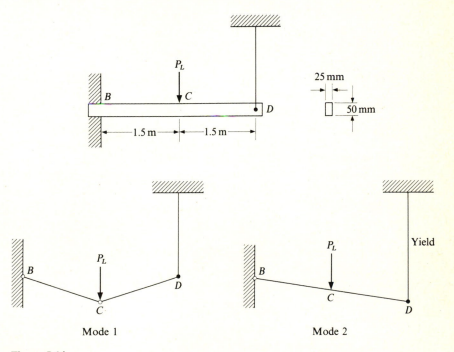

Mode 1

Mode 2

Figure 5-14

SOLUTION Since the structure is statically indeterminate of the first order, collapse will occur when two separate sections reach their yield condition. The two possible collapse modes are shown in Fig. 5-14b. In mode 1, plastic hinges have formed at the two points where the bending moments are maximum. For mode 2, a hinge has formed at the wall, and the stress in the cable has reached the limit value. A third mode may seem possible, where a hinge forms at the midspan and the cable yields. It can be shown that this is not possible but the reader should verify through physical reasoning that if this mode occurred, the left-hand side of the beam would still have structural integrity and therefore the structure would not collapse in this mode.

Mode 1 is similar to the case treated in Example 5-7. Thus, in a similar manner, the limit load for mode 1 is found to be

$$P_L = \frac{6}{L} M_p = \frac{6}{L} bc^2\sigma_e = \frac{6}{3}(0.025)(0.025)^2(240 \times 10^6)$$

$$= 7500 \text{ N} \qquad \text{mode 1}$$

For mode 2, the force in the cable is $\sigma_e A$. Thus, summing moments at point B yields

$$(\sigma_e A)L - P_L\frac{L}{2} + M_p = 0$$

Solving for P_L results in

$$P_L = 2\left(\sigma_e A + \frac{1}{L} M_p\right)$$

Since the limit moment in bending of a rectangular beam is $M_p = bc^2\sigma_e$, we have

$$P_L = 2\sigma_e\left(A + \frac{bc^2}{L}\right) = 2(240 \times 10^6)\left[12.5 \times 10^{-6} + \frac{(0.025)(0.025)^2}{3}\right] = 8500 \text{ N} \qquad \text{mode 2}$$

Thus, mode 1 will be the collapse mode, and the limit force is $P_L = 7500$ N.

5-5 FATIGUE ANALYSIS†

Cyclic loading of machine elements occurs often in mechanical systems, and the designer should be aware of some of the characteristics that influence fatigue failures. The governing factor in static design is either the yield or ultimate strength in tension or compression. These limiting values for a given material are obtained from a gradual axial loading of a standard test specimen. For cyclic loading, a different test specimen is used, and the normal mode of loading is pure bending. Figure 5-15 shows a schematic representation of the standard fatigue test. The rotating specimen has a polished surface to minimize premature failures due to surface finish and is loaded in pure bending by applying a weight symmetrically with respect to the simple supports. The shear force V_y between the load application bearings is zero, and the bending moment is constant at a value of $M = WL_1/2$. The diameter of the specimen is gradually varying to avoid any stress concentration at the center of the specimen, where the maximum bending stress is

$$\sigma_{max} = \frac{32M}{\pi d^3} = \frac{32WL_1/2}{\pi(0.30)^3} = 188.6WL_1 \qquad (5\text{-}22)$$

† For more advanced treatment of fatigue see Refs. 3 and 4.

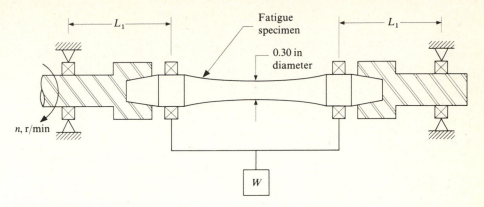

Figure 5-15

Since the load is stationary and the shaft is rotating, at any given point on the surface of the specimen the stress undergoes a variation of $\sigma = \pm\sigma_{max}$ for each cycle (see Fig. 5-16).

The specimen is loaded with a known weight W and cycled until the specimen fractures, and the number of cycles N to fracture is recorded. The value of σ_{max} corresponding to fracture at a specific number of cycles is normally referred to as S_f, the fatigue strength. The test is usually terminated at 10^6 to 10^7 cycles if a fracture does not occur. A number of specimens of a given material are tested using different values of W, and the results of S_f vs. N are normally plotted on log-log or semilog paper. The test data are generally highly scattered, and either statistical considerations or a minimum base curve should be used. For most steels the S–N diagram has the appearance illustrated in Fig. 5-17. There is a sharp knee in the curve, which occurs at approximately 10^6 cycles; beyond the knee, failure will not occur no matter how great the number of cycles. The strength at this point is called the endurance limit S_e.

For most steels the relationship between $\log S_f$ and $\log N$ is linear from 10^3 to about 10^6 cycles, whereas from about 10^6 and greater, S_f remains constant at a value S_e. Usually, under 10^3 cycles, the loading is considered to be static and at 10^3 cycles the failure occurs at about 0.9 times the ultimate strength of the

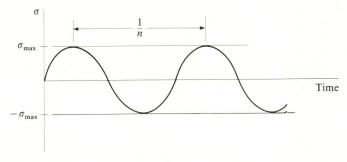

Figure 5-16

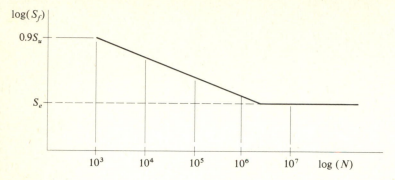

Figure 5-17

material S_u. Under these conditions the relationship between S_f and N for 10^3 to 10^6 cycles is

$$S_f = \frac{(0.9S_u)^2}{S_e} N^{-(1/3)\log(0.9S_u/S_e)} \tag{5-23}$$

or (if solving for N)

$$N = \left[\frac{S_f S_e}{(0.9S_u)^2}\right]^{-3/[\log(0.9S_u/S_e)]} \tag{5-24}$$

The endurance limit for most steels is approximately one half the ultimate strength. For analysis work, however, it is recommended that the endurance limit be obtained from testing. The data will be scattered, and considerable variance will require the analyst to use a statistically acceptable value.

Example 5-10 Estimate the life of a fatigue specimen with a force $W = 186$ lb and $L_1 = 2.0$ in (see Fig. 5-15). The material has an ultimate strength of 100,000 lb/in² and an endurance limit of 48,000 lb/in².

SOLUTION

$$\sigma_{max} = 188.6\,WL_1 = (188.6)(186)(2.0) = 70,160 \text{ lb/in}^2$$

Letting $S_f = 70,160$ and substituting into Eq. (5-24) results in

$$N = \left[\frac{(70,160)(48,000)}{[(0.9)(100,000)]^2}\right]^{-3/\{\log[(0.9)(100,000)/48,000]\}} = 15,430 \text{ cycles}$$

Quite often a machine element may have fluctuating stresses which are not purely reversing. For example, if a constant tensile load is placed on the fatigue specimen, the stress curve appears as shown in Fig. 5-18.

This case is not purely static, where σ_{max} cannot be directly compared with either S_y or S_u, and it is not purely reversing, where σ_{max} cannot be compared with S_e. There is no exact analytical representation for this case, and the designer is forced to use an empirical approach. Defining the average or mean stress σ_m (analogous to the static stress component) and the fluctuating stress

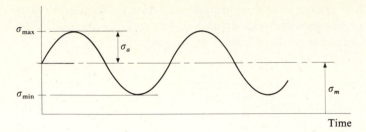

Figure 5-18

amplitude σ_a (analogous to the reversing stress component), we have

$$\sigma_m = \tfrac{1}{2}(\sigma_{\max} + \sigma_{\min}) \qquad \sigma_a = \tfrac{1}{2}(\sigma_{\max} - \sigma_{\min}) \qquad (5\text{-}25)$$

Up to this point conditions of failure are known for only two cases of combined values of σ_m and σ_a. That is, when $\sigma_a = 0$, $\sigma_m = S_y$ or $\sigma_m = S_u$, and when $\sigma_m = 0$, $\sigma_a = S_e$. When σ_a and σ_m both have values, some other type of criterion must be developed.

Possible criteria for failure are shown in Fig. 5-19. The Gerber line and the

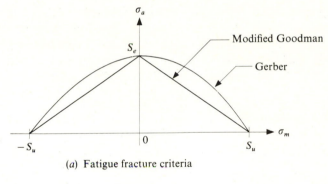

(a) Fatigue fracture criteria

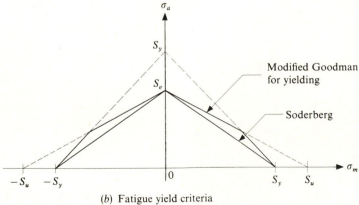

(b) Fatigue yield criteria

Figure 5-19

modified Goodman line are used for fracture criteria.† For yield criteria either the Soderberg line or the modified Goodman line for yielding are used. The equations for these lines for positive values of σ_m are:

Fracture:

Modified Goodman:
$$\frac{\sigma_m}{S_u} + \frac{\sigma_a}{S_e} = 1 \tag{5-26}$$

Gerber:
$$\left(\frac{\sigma_m}{S_u}\right)^2 + \frac{\sigma_a}{S_e} = 1 \tag{5-27}$$

Yield:

Soderberg:
$$\frac{\sigma_m}{S_y} + \frac{\sigma_a}{S_e} = 1 \tag{5-28}$$

Modified Goodman:
$$\frac{\sigma_m}{S_u} + \frac{\sigma_a}{S_e} = 1 \qquad \frac{\sigma_a}{\sigma_m} \geq \frac{S_e(S_u - S_y)}{S_u(S_y - S_e)}$$

$$\sigma_a + \sigma_m = S_y \qquad \frac{\sigma_a}{\sigma_m} \leq \frac{S_e(S_u - S_y)}{S_u(S_y - S_e)} \tag{5-29}$$

All other factors being equal, it has been found that a mean tensile stress is more detrimental than a mean compressive stress. Thus, all the previously mentioned diagrams are generally much too conservative for negative values of σ_m. Considering only the Goodman yield criteria, the range of permissible values of σ_m and σ_a when σ_m is negative can be increased as shown in Fig. 5-20.

Example 5-11 An additional constant tensile load of 2475 lb is to be applied to a rotating fatigue specimen, where the fluctuating bending stress is 40,000 lb/in². The material values for

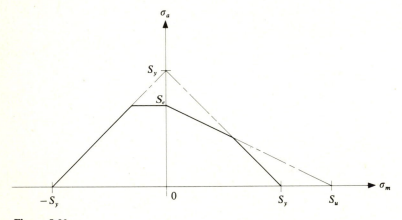

Figure 5-20

† The modified Goodman line is normally used in practice, as it is more conservative than the Gerber line.

the specimen are $S_y = 80,000 \text{ lb/in}^2$, $S_u = 100,000 \text{ lb/in}^2$, and $S_e = 50,000 \text{ lb/in}^2$. Determine whether or not the design criterion of Fig. 5-20 is exceeded.

SOLUTION The axial stress is

$$\sigma_{\text{axial}} = \frac{2475}{(\pi/4)(0.3)^2} = 35,000 \text{ lb/in}^2$$

Thus

$$\sigma_{\text{max}} = 35,000 + 40,000 = 75,000 \text{ lb/in}^2$$

and

$$\sigma_{\text{min}} = 35,000 - 40,000 = -5000 \text{ lb/in}^2$$

From Eqs. (5-25)

$$\sigma_m = \tfrac{1}{2}[75,000 + (-5000)] = 35,000 \text{ lb/in}^2$$

and

$$\sigma_a = \tfrac{1}{2}[75,000 - (-5000)] = 40,000 \text{ lb/in}^2$$

Plotting the given data with respect to the modified Goodman diagram gives Fig. 5-21. It can be seen that the state of stress fails to meet the criterion of the modified Goodman diagram, and a failure may occur.

Many empirical approaches are used in fatigue analysis, and the results must be interpreted with some degree of scepticism. So far we have been dealing with the geometric configuration of the fatigue specimen. Differences between the actual machine element and specimen in size, surface finish, rapid reduction in cross section, temperature, residual stresses, corrosion, and plating, all contribute to earlier fatigue failures. In addition, the state of stress may be more complicated than the fatigue specimen, thus possibly calling for an approach to compare a combined state of stress with the material values of S_y, S_u, and S_e. A third complication is that the cyclic loading of the machine element may change periodically or even randomly. For the remainder of this section, some of the major points will be discussed, and the reader is urged to consult some of the advanced design books in this area for further discussion.

Many designers account for the physical differences between the machine part and the fatigue specimen of the same material by reducing the allowable

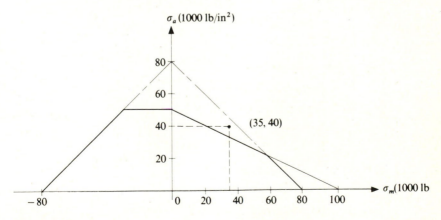

Figure 5-21

endurance limit through the use of various modifying factors. The *main* factors, which are the only ones discussed here, are surface-finish imperfections and stress concentrations due to a drastic change in cross section.† Figure 5-22 illustrates the reduction factor due to surface effects for steel.

Considering stress concentrations due to a rapid reduction in cross section, the reduction factor is $1/K_t$ provided the transition radius of the two differing cross sections is large (which it should be when anticipating a fatigue problem). K_t is the static stress-concentration factor. For a small transition radius (called the *notch radius*), the condition is generally not as severe as the static stress concentration depicts. In these cases a fatigue stress-concentration factor K_f is defined, where $K_f < K_t$; K_f is dependent on K_t, material, and notch size. For conservative analysis, a reduction of the endurance limit by multiplying by $1/K_t$ is reasonable.

Example 5-12 The machined steel shaft shown in Fig. 5-23 is in purely reversed bending. The endurance limit of the material, as predicted from the standard fatigue test, is 50,000 lb/in². The ultimate tensile strength is 100,000 lb/in². Determine the maximum value of the bending moment for which the cyclic life of the shaft is indefinite.

SOLUTION Neglecting the stress concentration, we see that the maximum stress at section A-A is

$$\sigma_{max} = \frac{32M}{\pi(1^3)} = 10.2M$$

The surface-reduction factor from Fig. 5-22 is 0.73. The stress-concentration factor is obtained from Appendix G, and for $D/d = 1.5$ and $r/d = 0.05$, $K_t = 1.8$. Thus, the endurance

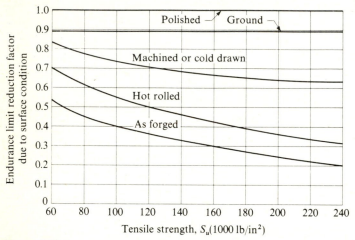

Figure 5-22

† Many other factors contribute toward reducing the endurance limit of a machine part, e.g., size effect, reliability (the published data on endurance limits are usually based on a 50 percent reliability), temperature, etc. It is a standard approach to consider all these effects as separate and independent and reduce the endurance limit by each factor. This is not accurate, but this procedure is used since it is conservative.

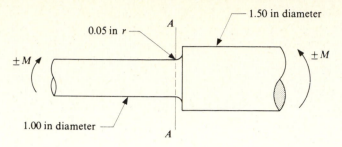

Figure 5-23

limit for the part is

$$(S_e)_{part} \approx (0.73)\frac{1}{1.8}(50,000) = 20,280 \text{ lb/in}^2$$

Thus

$$10.2M \approx 20,280 \qquad M \approx 1990 \text{ lb·in}$$

Biaxial State of Stress

Under the conditions of a general biaxial state of stress which is cyclic, it is common practice to modify the static failure theories for analysis purposes.

If the material is ductile, either the maximum-shear or the maximum-distortion-energy theory is used. One technique is first to separate the state of stress into its average and reversing components, then solve for the respective principal stresses, σ_{1m}, σ_{2m} and σ_{1a}, σ_{2a}. The equivalent von Mises stresses are

$$(\sigma_m)_{vM} = \sqrt{\sigma_{1m}^2 + \sigma_{2m}^2 - \sigma_{1m}\sigma_{2m}} \qquad (5\text{-}30a)$$

$$(\sigma_a)_{vM} = \sqrt{\sigma_{1a}^2 + \sigma_{2a}^2 - \sigma_{1a}\sigma_{2a}} \qquad (5\text{-}30b)$$

The final step is to construct the Goodman diagram and plot the von Mises stresses determined from Eqs. (5-30).

Example 5-13 For Example 5-12 assume that $S_y = 80,000 \text{ lb/in}^2$, and a purely reversing moment of 1000 lb·in and a constant torque of 1200 lb·in are applied. Is the criterion established by the Goodman diagram exceeded?

SOLUTION The mean stress is due to the torsion, where

$$\tau = \frac{16T}{\pi d^3} = \frac{(16)(1200)}{\pi(1^3)} = 6100 \text{ lb/in}^2$$

The principal stresses for this case are

$$\sigma_{1m} = 6100 \text{ lb/in}^2 \qquad \sigma_{2m} = -6100 \text{ lb/in}^2$$

and from Eq. (5-30a)

$$(\sigma_m)_{vM} = \sqrt{6100^2 + (-6100)^2 - (6100)(-6100)} = 10,560 \text{ lb/in}^2$$

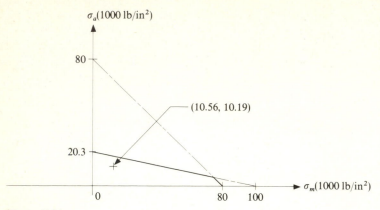

Figure 5-24

The reversing stress is due to bending, and

$$\sigma = \pm\frac{32M}{\pi d^3} = \pm\frac{(32)1000}{\pi(1^3)} = \pm10,190 \text{ lb/in}^2$$

Thus $\qquad\qquad \sigma_{1a} = 10,190 \text{ lb/in}^2 \qquad \sigma_{2a} = 0$

and from Eq. (5-30b)

$$(\sigma_a)_{vM} = 10,190 \text{ lb/in}^2$$

The final step is to construct the Goodman diagram; the state of stress can be seen to be within the Goodman diagram shown in Fig. 5-24.

Estimating Life for Nonreversing Stress Cycles

If the cyclic stress is not purely reversing, that is, $\sigma_m \neq 0$, and a finite life is predicted from the Goodman diagram, the S–N diagram cannot be used directly as the diagram applies only to a purely reversing state of stress. However, an estimate of the life can be obtained by first determining an *equivalent reversing stress*, which is as damaging as the actual cyclic stress conditions. Consider the case where σ_m and σ_a are outside the Goodman diagram, as shown in Fig. 5-25. Under these conditions a finite life is predicted.

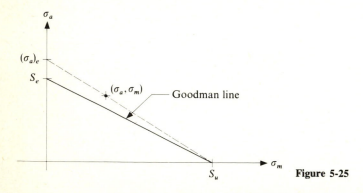

Figure 5-25

A line drawn connecting the points (σ_a, σ_m) and $(0, S_u)$ will also predict failure, as does the Goodman line (see Fig. 5-25). The Goodman line predicts a failure at 10^6 cycles, whereas the line constructed outside the Goodman line predicts a failure at a specific number of cycles less than 10^6. The point at which this line (shown dotted in Fig. 5-25) intersects the σ_a axis corresponds to a purely reversing stress as damaging as the actual stress conditions (σ_a, σ_m). Thus, this point is called the equivalent reversing stress $(\sigma_a)_e$.

The equation for this line is

$$\frac{\sigma_a}{(\sigma_a)_e} + \frac{\sigma_m}{S_u} = 1$$

Solving for $(\sigma_a)_e$ results in

$$(\sigma_a)_e = \frac{\sigma_a}{1 - \sigma_m/S_u} \tag{5-31}$$

To obtain an estimate of the number of cycles to failure, $(\sigma_a)_e$ is substituted into Eq. (5-24) in place of S_f.

Example 5-14 Since in Example 5-11, the values of σ_a and σ_m fell outside the modified Goodman line, estimate the number of cycles to failure.

SOLUTION From Example 5-11

$$\sigma_m = 35,000 \text{ lb/in}^2 \qquad \sigma_a = 40,000 \text{ lb/in}^2$$

and

$$S_e = 50,000 \text{ lb/in}^2 \qquad S_u = 100,000 \text{ lb/in}^2$$

From Eq. (5-31) the equivalent reversing stress is

$$(\sigma_a)_e = \frac{40,000}{1 - 35,000/100,000} = 61,500 \text{ lb/in}^2$$

Substituting this into Eq. (5-24) in place of S_f results in

$$N = \left[\frac{(61,500)(50,000)}{[(0.9)(100,000)]^2}\right]^{-3/\{\log [(0.9)(100,000)/50,000]\}} = 87,800 \text{ cycles}$$

If the cyclic loading of the machine element varies as shown in Fig. 5-26, the damaging effect of each individual stress cycle accumulates. A technique which takes this cumulative effect into consideration is Miner's method (see Ref. 5).

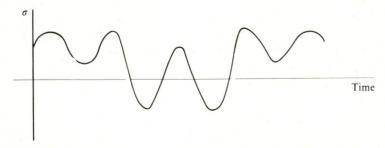

Figure 5-26

The basic equation is

$$\frac{n_1}{N_1} + \frac{n_2}{N_2} + \cdots + \frac{n_i}{N_i} \le 1 \tag{5-32}$$

where n_i is the number of cycles of stress $(\sigma_a)_i$ applied to the specimen and N_i is the number of cycles for infinite life at $(\sigma_a)_i$. Here equivalent reversing stresses can be used as well.

Example 5-15 A fatigue specimen with an endurance limit of 350 MN/m² and an ultimate strength of 700 MN/m² is cycled 20 percent of the time with $(\sigma_a)_1 = 490$ MN/m², 50 percent with $(\sigma_a)_2 = 385$ MN/m², and 30 percent with $(\sigma_a)_3 = 280$ MN/m². Estimate the number of cycles to failure.

SOLUTION The number of cycles to failure at $(\sigma_a)_1 = 490$ MN/m² is found from Eq. (5-24) and is $N_1 = 19,170$ cycles. For $(\sigma_a)_2 = 385$ MN/m², $N_2 = 326,250$ cycles. For $(\sigma_a)_3 = 280$ MN/m², $N_3 = \infty$;

$$\frac{0.20N}{19,170} + \frac{0.50N}{326,250} + 0 = 1$$

Solving for N yields

$$N = 83,600 \text{ cycles}$$

The discussion on fatigue in this section is far from complete, but the purpose was to expose the reader to some of the concepts and approaches used in fatigue analysis. The results and approaches are always subject to question and argument since there is no universally accepted analytical approach to the exact solution of fatigue problems. For this reason, if the design is critical, reliability tests must be extensively pursued in the final analysis.

EXERCISES

5-1 The various cases of biaxial stress given in Exercise 2-18 are repeated here. For each case (1) draw the *three* Mohr's circles corresponding to the three principal stresses, (2) determine the surface location and magnitude of the maximum shear stress, and (3) evaluate the magnitude of the equivalent von Mises' stress.

(a) $\sigma_x = 10,000$ lb/in², $\sigma_y = \tau_{xy} = 0$ lb/in²

(b) $\tau_{xy} = 5000$ lb/in², $\sigma_x = \sigma_y = 0$ lb/in²

(c) $\sigma_x = 10,000$ lb/in², $\sigma_y = 5000$ lb/in², $\tau_{xy} = 0$ lb/in²

(d) $\sigma_x = \sigma_y = 5000$ lb/in², $\tau_{xy} = 0$ lb/in²

(e) $\sigma_x = 8000$ lb/in², $\sigma_y = 0$ lb/in², $\tau_{xy} = -3000$ lb/in²

(f) $\sigma_x = 10,000$ lb/in², $\sigma_y = -2000$ lb/in², $\tau_{xy} = -8000$ lb/in²

5-2 In Exercise 2-11, determine the maximum shear stress and the maximum von Mises stress.

5-3 In Exercise 3-4, determine the maximum shear stress and the maximum von Mises stress.

5-4 In Exercise 3-15, determine the maximum shear stress and the maximum von Mises stress.

5-5 Repeat Exercise 3-17 but let the design criterion be that the maximum shear stress should not exceed 24,000 lb/in² *after* the internal pressure of 5000 lb/in² is applied.

(*a*) Determine the maximum acceptable radial interference and the corresponding interface pressure before application of the internal pressure of 5000 lb/in².

(*b*) Under the conditions of part (*a*) plot the stress distribution in each member.

(*c*) Under the conditions of part (*a*) the assembly is subjected to an internal pressure of 5000 psi. Determine the resulting stress distributions in both members.

5-6 A 20-mm-diameter rod made of a ductile material with a yield strength of 350 MN/m² is subjected to a torque of 100 N·m and a bending moment of 150 N·m. An axial tensile force is then gradually applied. What is the value of the axial force when yielding of the rod occurs? Solve the problem two ways using (*a*) the maximum-shear-stress theory and (*b*) the maximum-distortional-energy theory.

5-7 A thin-walled pressure vessel is made of a ductile material with a yield strength of 53,000 lb/in². The cylindrical vessel has an outside diameter of 3 in and a wall thickness of 0.0938 in. What maximum internal pressure will the vessel be capable of withstanding before yielding occurs?

5-8 In Chap. 3 the locations and values of the maximum radial and tangential stresses were given for a rotating disk. Determine the location and magnitude of the maximum shear stress.

5-9 A cast-iron flywheel is in the form of a disk of constant thickness of 1.0 in with an outside diameter of 20 in and an inside diameter of 4 in. The tensile strength of the material is 30,000 lb/in², and the compressive strength is 100,000 lb/in². Estimate the rotational speed, in revolutions per minute, for which fracture will occur. The weight density of the material is 0.26 lb/in³, and Poisson's ratio is 0.21.

5-10 A ductile steel has a yield strength of 40,000 lb/in². Find the factors of safety corresponding to yielding according to the maximum-shear-stress theory and the distortion-energy theory, respectively, for the following biaxial stress states:

(*a*) $\sigma_x = \sigma_y = 20,000$ lb/in², $\qquad \tau_{xy} = 0$

(*b*) $\sigma_x = \sigma_y = \tau_{xy} = 105$ MN/m²

(*c*) $\sigma_x = \sigma_y = 0$, $\qquad \tau_{xy} = 20,000$ lb/in²

(*d*) $\sigma_x = -2000$ lb/in², $\qquad \sigma_y = 10,000$ lb/in², $\qquad \tau_{xy} = -8000$ lb/in²

5-11 A brittle material with $S_{ut} = 20,000$ lb/in² and $S_{uc} = -80,000$ lb/in² is subjected to the following biaxial stress states:

(*a*) $\sigma_x = 12,000$ lb/in², $\qquad \sigma_y = 0$, $\qquad \tau_{xy} = -8000$ lb/in²

(*b*) $\sigma_x = -39,000$ lb/in², $\qquad \sigma_y = -31,000$ lb/in², $\qquad \tau_{xy} = 3000$ lb/in²

(*c*) $\sigma_x = 120$ MN/m², $\qquad \sigma_y = 70$ MN/m², $\qquad \tau_{xy} = 0$ MN/m²

For each case, determine the factor of safety corresponding to failure.

5-12 In a particular mechanism, a connecting rod is used where a tensile force is transmitted through pins located at both ends of the rod. The main portion of the rod will have a circular cross section of diameter *d*. From past experience it has been found that the following tolerances can be maintained:

1. The diameter can be machined to within a tolerance of ±0.2 percent.
2. The centerline of the pins can be maintained to within a tolerance of 0.004*d* with respect to the centroidal axis of the link.
3. The maximum axial force transmitted through the link can be predicted within an accuracy of ±10 percent.
4. The strength of the material is known within the limits of ±1 percent.

For analysis purposes the designer will use nominal dimensions, simple axial stress formulation, nominal maximum force value, and the nominal material strength value. Determine (*a*) the stress and strength uncertainties associated with items 1 to 4 and (*b*) the net factor of safety necessary for the design based on the type of analysis performed by the designer.

5-13 Four rivets each with a cross-sectional area of 250 mm² hold the plate as shown in Fig. 5-27. Estimate the magnitude of the maximum rivet shear stress.

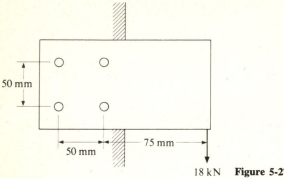

50 mm

50 mm   75 mm

18 kN   **Figure 5-27**

5-14 A plate is mounted on two walls, as shown in Fig. 5-28. Considering only the moment and vertical force reactions at the wall, estimate the magnitude of the maximum rivet shear stress if the rivets have a cross-sectional area of 0.4 in².

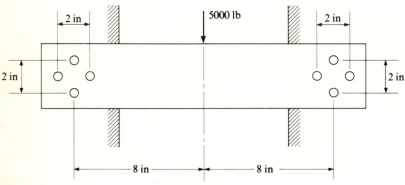

2 in   5000 lb   2 in

2 in   2 in

8 in   8 in

Figure 5-28

5-15 A gusset plate is welded to a mounting plate as shown in Fig. 5-29. The mounting plate is secured to a base by four bolts, each with an effective cross-sectional area of 0.2 in². Before application of the 1000-lb force, the bolts were torqued down so that a preload of 5000 lb tension was developed in each bolt. Determine the maximum bolt tensile stress upon the application of the 1000-lb force. (Assume that the plate tends to rotate about corner *A*.)

5-16 For the structure shown in Fig. 5-30, three steel wires of the same diameter are to be used to support a static load of 100,000 lb. The horizontal member is assumed to be rigid. Assuming that the wire material is EPP with $\sigma_e = 40,000$ lb/in², determine the minimum size of wire that can be used based on (*a*) an elastic analysis; (*b*) a limit analysis.

5-17 For beam D.12 of Appendix D determine the ultimate value of *w* for collapse if the internal moment is limited to M_p.

5-18 A beam of rectangular cross section *b* by 2*c* is cantilevered at one end and simply supported at the other end. Three equal forces *P* are placed at equal intervals of *L*/4, where *L* is the span length of the beam. If the elastic limit of the material is σ_e, determine the limiting value of *P* and show the resulting collapse mode.

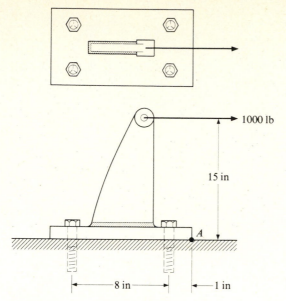

1000 lb

15 in

A

8 in   1 in

Figure 5-29

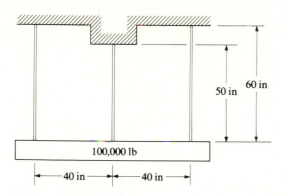

60 in
50 in

100,000 lb

40 in   40 in

Figure 5-30

5-19 A clamp of rectangular cross section is loaded as shown in Fig. 5-31. The elastic limit of the material is 40,000 lb/in², and the material is EPP. Determine the value of P when (a) the highest stressed element at section a-a reaches the elastic limit; (b) the ultimate limit of the cross section is reached. *Note*: the neutral axis due to the combined axial and bending load across section a-a *shifts* when plastic behavior occurs. Keep in mind that the stress distribution must always satisfy the state of equilibrium of the structure.

5-20 The bar shown in Fig. 5-32 is machined from steel with $S_y = 420$ MN/m² and $S_u = 560$ MN/m². The axial force F is completely reversing. Estimate the value of the force amplitude which will cause a failure at 100,000 cycles.

5-21 If the axial force in Exercise 5-20 is cycling from zero to a maximum value F_{max}, estimate the value of F_{max} for a life of 100,000 cycles.

5-22 The rotating shaft of Exercise 2-22 is machined from a steel, where $S_y = 60,000$ lb/in² and $S_u = 80,000$ lb/in². Neglecting any stress concentration near the pulley, determine the factor of safety of the design for an indefinite life based on (a) a static-yield analysis and (b) a fatigue-yield analysis.

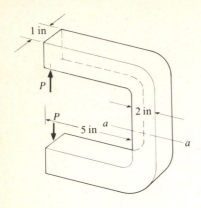

Figure 5-31

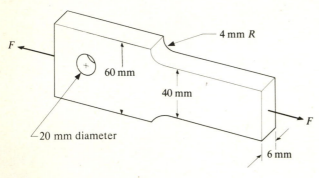

Figure 5-32

5-23 A mechanical part is made of steel with properties $S_u = 90,000$ lb/in^2, $S_y = 70,000$ lb/in^2, and $S_e = 30,000$ lb/in^2. It is subjected to a bending stress which alternates between 6000 and 14,000 lb/in^2. Determine the equivalent reversing stress and evaluate the factor of safety corresponding to a life of 500,000 cycles.

REFERENCES

1. Johnson, R. C.: Predicting Part Failures, *Mach. Des.*, vol. 37, no. 1, pp. 137–142, January 1965; no. 2, pp. 157–162, January 1965.
2. Van den Broek, J. A.: "Theory of Limit Design," Wiley, New York, 1948.
3. Juvinall, R. C.: "Stress Strain and Strength," McGraw-Hill, New York, 1967.
4. Osgood, C. C.: "Fatigue Design," Wiley, New York, 1970.
5. Miner, M. A.: Cumulative Damage in Fatigue, *J. Appl. Mech.*, vol. 12, pp. A159–A164, September 1945.

CONCEPTS FROM THE THEORY OF ELASTICITY†

6-0 INTRODUCTION

The mathematical theory of elasticity is an elegant and organized representation of the elastic behavior of continuous solids and is quite different from the approaches used in basic strength of materials, where the governing equations are arrived at through physical reasoning and incomplete, although fairly accurate, techniques of solid-body mechanics.

Some of the basic concepts and formulations developed in the theory of elasticity will be introduced in this chapter. Indicial or tensor notation is commonly used since they greatly simplify the lengthy equations of general three-dimensional elasticity. The stress matrix given in Chap. 1 is actually a stress tensor, since a tensor has the property of being transformable from one coordinate system to another. The strain matrix, on the other hand, if defined as

$$\varepsilon = \begin{bmatrix} \varepsilon_x & \gamma_{xy} & \gamma_{xz} \\ \gamma_{yx} & \varepsilon_y & \gamma_{yz} \\ \gamma_{zx} & \gamma_{zy} & \varepsilon_z \end{bmatrix}$$

is not a tensor since it cannot be transformed from one coordinate system to another. This was demonstrated in Sec. 2-8 for the two-dimensional Mohr's circle of strain. If $\gamma/2$ is used to define shear strain, the strain matrix will

† See Refs. 1 and 2.

transform and hence be a tensor. Thus, if the strain matrix is defined as

$$\varepsilon = \begin{bmatrix} \varepsilon_x & \varepsilon_{xy} & \varepsilon_{xz} \\ \varepsilon_{yx} & \varepsilon_y & \varepsilon_{yz} \\ \varepsilon_{zx} & \varepsilon_{zy} & \varepsilon_z \end{bmatrix} \tag{6-1}$$

where $\varepsilon_{xy} = \gamma_{xy}/2$, $\varepsilon_{xz} = \gamma_{xz}/2$, etc., the strain will transform.

A vector is another example of a tensor since it can be represented by any coordinate system. Tensor notation will not be used in this chapter since it sometimes decreases physical understanding and the notation already developed in the text is not suitable for tensor notation, but the reader is urged to read Appendix K on matrices and tensor notation for some additional insight.

6-1 TRANSFORMATIONS OF STRESS IN THREE DIMENSIONS

Stress Transformations

In Chap. 2 transformations for the biaxial state of stress were presented, where Mohr's circle was used quite effectively. For a general three-dimensional state of stress, transformations are more complicated.

Consider a general state of stress at a point where the state of stress is described by the three orthogonal planes perpendicular to the x, y, and z directions (see Fig. 6-1). If an arbitrary plane is passed through the solid so that the plane intersects the three mutually perpendicular reference planes, an element about the point will be isolated as shown in Fig. 6-2a, where the x' axis is perpendicular to the oblique plane and the y' and z' axes are tangent to the plane. The orientation of the perpendicular to the oblique surface can be established by the angles $\theta_{x'x}$, $\theta_{x'y}$, and $\theta_{x'z}$, which relate the x' axis to the x, y, and z axes respectively, as shown in Fig. 6-2b. Let the area of the oblique surface be A. Then the areas of the remaining surfaces can be shown to be

$$A_1 = A \cos \theta_{x'x} \tag{6-2a}$$

$$A_2 = A \cos \theta_{x'y} \tag{6-2b}$$

$$A_3 = A \cos \theta_{x'z} \tag{6-2c}$$

The terms $\cos \theta_{x'x}$, $\cos \theta_{x'y}$, and $\cos \theta_{x'z}$ are referred to as the *directional cosines* of the x' axis of the oblique surface relative to the xyz coordinate system. Since the directional cosines will be repeated often in this section, it is convenient to define them as

$$n_{x'x} = \cos \theta_{x'x} \tag{6-3a}$$

$$n_{x'y} = \cos \theta_{x'y} \tag{6-3b}$$

$$n_{x'z} = \cos \theta_{x'z} \tag{6-3c}$$

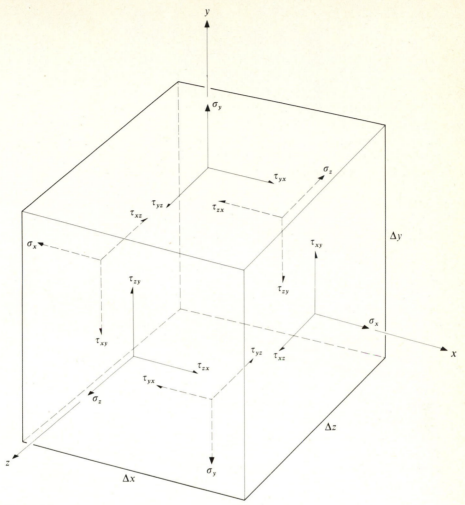

Figure 6-1

Thus, Eqs. (6-2) can be rewritten in the form

$$A_1 = An_{x'x} \tag{6-4a}$$

$$A_2 = An_{x'y} \tag{6-4b}$$

$$A_3 = An_{x'z} \tag{6-4c}$$

To determine the normal stress $\sigma_{x'}$ on the oblique surface, summation of forces in the x' direction is performed. The force due to each stress is obtained by multiplying the stress by the area over which the stress acts. Next, the component of the force in the x' direction is found. For example, referring back to Fig. 6-2a, the force due to σ_x is $\sigma_x A_1 = \sigma_x An_{x'x}$. The component of this

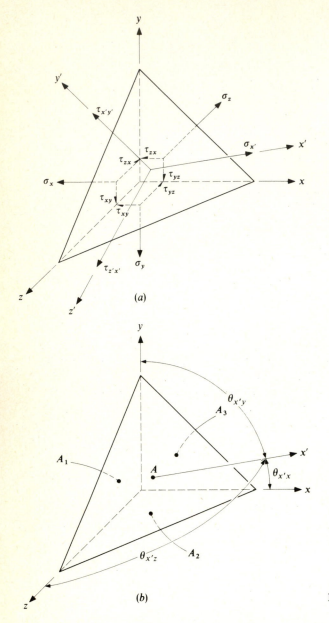

(a)

(b)

Figure 6-2

force in the x' direction is $-(\sigma_x A n_{x'x}) n_{x'x}$. Summing all forces in the x' direction for equilibrium yields

$$\sigma_{x'} A - (\sigma_x A n_{x'x}) n_{x'x} - (\sigma_y A n_{x'y}) n_{x'y} - (\sigma_z A n_{x'z}) n_{x'z}$$
$$- (\tau_{xy} A n_{x'x}) n_{x'y} - (\tau_{zx} A n_{x'x}) n_{x'z} - (\tau_{xy} A n_{x'y}) n_{x'x}$$
$$- (\tau_{yz} A n_{x'y}) n_{x'z} - (\tau_{zx} A n_{x'z}) n_{x'x} - (\tau_{yz} A n_{x'z}) n_{x'y} = 0$$

Factoring A and solving for $\sigma_{x'}$ results in

$$\sigma_{x'} = \sigma_x n_{x'x}^2 + \sigma_y n_{x'y}^2 + \sigma_z n_{x'z}^2 + 2\tau_{xy}n_{x'x}n_{x'y} + 2\tau_{yz}n_{x'y}n_{x'z} + 2\tau_{zx}n_{x'z}n_{x'x} \tag{6-5a}$$

For a complete transformation of stresses with respect to the arbitrary oblique surface, it is necessary to determine the shear stresses $\tau_{x'y'}$ and $\tau_{x'z'}$. If we define the directional cosines of the y' axis relative to the xyz coordinate system as $n_{y'x}$, $n_{y'y}$, and $n_{y'z}$, a summation of forces in the y' direction for equilibrium yields

$$\tau_{x'y'}A - (\sigma_x A n_{x'x})n_{y'x} - (\sigma_y A n_{x'y})n_{y'y} - (\sigma_z A n_{x'z})n_{y'z} - (\tau_{xy}A n_{x'x})n_{y'y}$$
$$- (\tau_{zx}A n_{x'x})n_{y'z} - (\tau_{xy}A n_{x'y})n_{y'x} - (\tau_{yz}A n_{x'y})n_{y'z} - (\tau_{zx}A n_{x'z})n_{y'x}$$
$$- (\tau_{yz}A n_{x'z})n_{y'y} = 0$$

Factoring A and solving for $\tau_{x'y'}$ results in

$$\tau_{x'y'} = \sigma_x n_{x'x}n_{y'x} + \sigma_y n_{x'y}n_{y'y} + \sigma_z n_{x'z}n_{y'z} + \tau_{xy}(n_{x'x}n_{y'y} + n_{x'y}n_{y'x})$$
$$+ \tau_{yz}(n_{x'y}n_{y'z} + n_{x'z}n_{y'y}) + \tau_{zx}(n_{x'x}n_{y'z} + n_{x'z}n_{y'x}) \tag{6-5b}$$

The final shear stress on the oblique surface $\tau_{z'x'}$ is found in a similar fashion. Defining the directional cosines of the z' axis relative to the xyz coordinate system as $n_{z'x}$, $n_{z'y}$, and $n_{z'z}$, we find the shear stress to be

$$\tau_{z'x'} = \sigma_x n_{x'x}n_{z'x} + \sigma_y n_{x'y}n_{z'y} + \sigma_z n_{x'z}n_{z'z} + \tau_{xy}(n_{x'x}n_{z'y} + n_{x'y}n_{z'x})$$
$$+ \tau_{yz}(n_{x'y}n_{z'z} + n_{x'z}n_{z'y}) + \tau_{zx}(n_{x'x}n_{z'z} + n_{x'z}n_{z'x}) \tag{6-5c}$$

Equations (6-5a), (6-5b), and (6-5c) are completely sufficient for the determination of the state of stress on any internal surface in which an arbitrarily selected tangential set of coordinates is used (in this case the $y'z'$ coordinates). Most textbooks treating this subject generally go beyond this point for mathematical conciseness and include the remaining two orthogonal surfaces, i.e., the surfaces perpendicular to the y' and z' axes, respectively, containing $\sigma_{y'}$, $\tau_{y'z'}$, and $\tau_{y'x'}$ and $\sigma_{z'}$, $\tau_{z'y'}$, and $\tau_{z'x'}$. Assuming that body moments do not exist, $\tau_{y'x'} = \tau_{x'y'}$, $\tau_{z'x'} = \tau_{x'z'}$, and $\tau_{z'y'} = \tau_{y'z'}$, and there remain only three additional stresses to evaluate, $\sigma_{y'}$, $\sigma_{z'}$, and $\tau_{y'z'}$. The analyses for the surfaces containing these stresses are identical to the previous development except that the surface normals are different. For the surface perpendicular to the y' axis, the two additional stresses are

$$\sigma_{y'} = \sigma_x n_{y'x}^2 + \sigma_y n_{y'y}^2 + \sigma_z n_{y'z}^2 + 2\tau_{xy}n_{y'x}n_{y'y} + 2\tau_{yz}n_{y'y}n_{y'z} + 2\tau_{zx}n_{y'z}n_{y'x} \tag{6-5d}$$

$$\tau_{y'z'} = \sigma_x n_{y'x}n_{z'x} + \sigma_y n_{y'y}n_{z'y} + \sigma_z n_{y'z}n_{z'z} + \tau_{xy}(n_{y'x}n_{z'y} + n_{y'y}n_{z'x})$$
$$+ \tau_{yz}(n_{y'y}n_{z'z} + n_{y'z}n_{z'y}) + \tau_{zx}(n_{y'x}n_{z'z} + n_{y'z}n_{z'x}) \tag{6-5e}$$

Finally, the normal stress $\sigma_{z'}$ is

$$\sigma_{z'} = \sigma_x n_{z'x}^2 + \sigma_y n_{z'y}^2 + \sigma_z n_{z'z}^2 + 2\tau_{xy}n_{z'x}n_{z'y} + 2\tau_{yz}n_{z'y}n_{z'z} + 2\tau_{zx}n_{z'z}n_{z'x} \tag{6-5f}$$

Biaxial State of Stress

For the biaxial state of stress, as discussed in Sec. 2-7, $\sigma_z = \tau_{yz} = \tau_{zx} = 0$, and

$$\theta_{x'x} = \theta \qquad \theta_{x'y} = 90° - \theta \qquad \theta_{x'z} = 90°$$

$$\theta_{y'x} = -(90° - \theta) \qquad \theta_{y'y} = \theta \qquad \theta_{y'z} = 90°$$

$$\theta_{z'x} = 90° \qquad \theta_{z'y} = 90° \qquad \theta_{z'z} = 0$$

Thus

$$n_{x'x} = \cos\theta \qquad n_{x'y} = \sin\theta \qquad n_{x'z} = 0$$

$$n_{y'x} = -\sin\theta \qquad n_{y'y} = \cos\theta \qquad n_{y'z} = 0$$

$$n_{z'x} = 0 \qquad n_{z'y} = 0 \qquad n_{z'z} = 1$$

Substituting these conditions into Eqs. (6-5a) to (6-5f) results in

$$\sigma_{x'} = \sigma_x \cos^2\theta + \sigma_y \sin^2\theta + 2\tau_{xy} \cos\theta \sin\theta \qquad (a)$$

$$\sigma_{y'} = \sigma_x \sin^2\theta + \sigma_y \cos^2\theta - 2\tau_{xy} \sin\theta \cos\theta \qquad (b)$$

$$\tau_{x'y'} = -\sigma_x \cos\theta \sin\theta + \sigma_y \sin\theta \cos\theta + \tau_{xy}(\cos^2\theta - \sin^2\theta)$$

$$= -(\sigma_x - \sigma_y)\sin\theta \cos\theta + \tau_{xy}(\cos^2\theta - \sin^2\theta) \qquad (c)$$

and $\sigma_{z'} = \tau_{z'x'} = \tau_{y'z'} = 0$. It can be seen that Eqs. (a) and (c) are identical to Eqs. (2-29) and (2-30) respectively. Equation (b) can be shown to be the same as Eq. (2-33).

Example 6-1 The state of stress at a point with respect to the xyz system is

$$\sigma = \begin{bmatrix} 3 & 2 & -2 \\ 2 & 0 & -1 \\ -2 & -1 & 1 \end{bmatrix} (1000) \text{ lb/in}^2$$

Determine the stress matrix relative to the $x'y'z'$ coordinate system shown in Fig. 6-3.

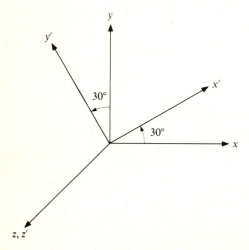

Figure 6-3

SOLUTION From Fig. 6-3 the directional cosines are

$$n_{x'x} = \frac{\sqrt{3}}{2} \qquad n_{x'y} = \tfrac{1}{2} \qquad n_{x'z} = 0$$

$$n_{y'x} = -\tfrac{1}{2} \qquad n_{y'y} = \frac{\sqrt{3}}{2} \qquad n_{y'z} = 0$$

$$n_{z'x} = 0 \qquad n_{z'y} = 0 \qquad n_{z'z} = 1$$

From the stress matrix given, $\sigma_x = 3000$ lb/in², $\tau_{xy} = 2000$ lb/in², $\tau_{zx} = -2000$ lb/in², $\sigma_y = 0$, $\tau_{yz} = -1000$ lb/in², and $\sigma_z = 1000$ lb/in². Substitution of the above in Eqs. (6-5) yields

$$\sigma_{x'} = \left[(3)\left(\frac{\sqrt{3}}{2}\right)^2 + (0)(\tfrac{1}{2})^2 + (1)(0^2) + (2)(2)\left(\frac{\sqrt{3}}{2}\right)(\tfrac{1}{2}) + (2)(-1)(\tfrac{1}{2})(0) + (2)(-2)(0)\left(\frac{\sqrt{3}}{2}\right)\right](1000)$$

$$= \left(\frac{9}{4} + \sqrt{3}\right)(1000) \text{ lb/in}^2$$

$$\tau_{x'y'} = \left\{(3)\left(\frac{\sqrt{3}}{2}\right)(-\tfrac{1}{2}) + (0)(\tfrac{1}{2})\left(\frac{\sqrt{3}}{2}\right) + (1)(0)(0) + (2)\left[\left(\frac{\sqrt{3}}{2}\right)\left(\frac{\sqrt{3}}{2}\right) + (\tfrac{1}{2})(-\tfrac{1}{2})\right]\right.$$

$$\left. + (-1)\left[(\tfrac{1}{2})(0) + (0)\left(\frac{\sqrt{3}}{2}\right)\right] + (-2)\left[\left(\frac{\sqrt{3}}{2}\right)(0) + (0)(-\tfrac{1}{2})\right]\right\}(1000)$$

$$= \left(1 - \frac{3}{4}\sqrt{3}\right)(1000) \text{ lb/in}^2$$

$$\tau_{z'x'} = \left\{(3)\left(\frac{\sqrt{3}}{2}\right)(0) + (0)(\tfrac{1}{2})(0) + (1)(0)(1) + (2)\left[\left(\frac{\sqrt{3}}{2}\right)(0) + (\tfrac{1}{2})(0)\right]\right.$$

$$\left. + (-1)[(\tfrac{1}{2})(1) + (0)(0)] + (-2)\left[\left(\frac{\sqrt{3}}{2}\right)(1) + (0)(0)\right]\right\}(1000)$$

$$= -\left(\frac{1}{2} + \sqrt{3}\right)(1000) \text{ lb/in}^2$$

$$\sigma_{y'} = \left[(3)(-\tfrac{1}{2})^2 + (0)\left(\frac{\sqrt{3}}{2}\right)^2 + (1)(0^2) + (2)(2)(-\tfrac{1}{2})\left(\frac{\sqrt{3}}{2}\right) + (2)(-1)\left(\frac{\sqrt{3}}{2}\right)(0) + (2)(-2)(0)(-\tfrac{1}{2})\right](1000)$$

$$= \left(\frac{3}{4} - \sqrt{3}\right)(1000) \text{ lb/in}^2$$

$$\tau_{y'z'} = \left\{(3)(-\tfrac{1}{2})(0) + (0)\left(\frac{\sqrt{3}}{2}\right)(0) + (1)(0)(1) + (2)\left[(-\tfrac{1}{2})(0) + \left(\frac{\sqrt{3}}{2}\right)(0)\right]\right.$$

$$\left. + (-1)\left[\left(\frac{\sqrt{3}}{2}\right)(1) + (0)(0)\right] + (-2)\left[(-\tfrac{1}{2})(1) + (0)(0)\right]\right\}(1000)$$

$$= \left(1 - \frac{1}{2}\sqrt{3}\right)(1000) \text{ lb/in}^2$$

$$\sigma_{z'} = [(3)(0^2) + (0)(0^2) + (1)(1^2) + (2)(2)(0)(0) + (2)(-1)(0)(1) + (2)(-2)(1)(0)](1000) = 1000 \text{ lb/in}^2†$$

The stress matrix for the $x'y'z'$ coordinate system is therefore

$$\sigma = \begin{bmatrix} \frac{9}{4} + \sqrt{3} & 1 - \frac{3}{4}\sqrt{3} & -\frac{1}{2} - \sqrt{3} \\ 1 - \frac{3}{4}\sqrt{3} & \frac{3}{4} - \sqrt{3} & 1 - \frac{1}{2}\sqrt{3} \\ -\frac{1}{2} - \sqrt{3} & 1 - \frac{1}{2}\sqrt{3} & 1 \end{bmatrix}(1000) \text{ lb/in}^2$$

† Note that the z and z' axes are identical. Thus, it should be obvious that $\sigma_{z'} = \sigma_z$.

Transformation Simplified

It is important to keep in mind that only Eqs. (6-5a), (6-5b), and (6-5c) are necessary to obtain the state of stress at any one given surface. Furthermore, Eqs. (6-5b) and (6-5c) provide the components of the total shear stress on the surface with respect to a particular set of orthogonal axes tangential to the surface. However, the *total* shear stress on the surface depends only on the particular surface as defined by the surface normal. If we redefine the surface normal by the n axis, as shown in Fig. 6-4, where the directional cosines of the normal are

$$n_x = \cos \theta_x \tag{6-6a}$$

$$n_y = \cos \theta_y \tag{6-6b}$$

$$n_z = \cos \theta_z \tag{6-6c}$$

the normal stress σ is given by Eq. (6-5a) and is

$$\sigma = \sigma_x n_x^2 + \sigma_y n_y^2 + \sigma_z n_z^2 + 2\tau_{xy}n_x n_y + 2\tau_{yz}n_y n_z + 2\tau_{zx}n_z n_x \tag{6-7}$$

The total shear stress τ is now the only unknown in Fig. 6-4. Thus τ can be found from a summation of forces. The total force in the x direction due to the stresses on the orthogonal surfaces is

$$F_x = -\sigma_x A_1 - \tau_{xy}A_2 - \tau_{zx}A_3$$

However, from Eqs. (6-4), $A_1 = An_x$, $A_2 = An_y$, and $A_3 = An_z$. Thus, F_x can be rewritten as

$$F_x = -A(\sigma_x n_x + \tau_{xy}n_y + \tau_{zx}n_z) \tag{6-8a}$$

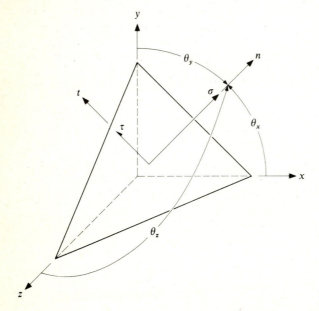

Figure 6-4

Likewise, the forces in the y and z directions due to the stresses on the orthogonal surfaces are

$$F_y = - A (\tau_{xy} n_x + \sigma_y n_y + \tau_{yz} n_z) \tag{6-8b}$$

$$F_z = - A (\tau_{zx} n_x + \tau_{yz} n_y + \sigma_z n_z) \tag{6-8c}$$

The magnitude of τ can easily be found since for equilibrium the magnitude of the forces on the oblique surface $A (\sigma^2 + \tau^2)^{1/2}$ must balance the magnitude of forces on the orthogonal surfaces $(F_x^2 + F_y^2 + F_z^2)^{1/2}$. Equating the two and solving for τ results in

$$\tau = [(\sigma_x n_x + \tau_{xy} n_y + \tau_{zx} n_z)^2 + (\tau_{xy} n_x + \sigma_y n_y + \tau_{yz} n_z)^2$$
$$+ (\tau_{zx} n_x + \tau_{yz} n_y + \sigma_z n_z)^2 - \sigma^2]^{1/2} \tag{6-9}$$

The directional cosines establishing the direction of τ can be defined as t_x, t_y, and t_z. For equilibrium of forces in the x direction,

$$\tau A t_x + \sigma A n_x + F_x = 0$$

Solving for t_x yields

$$t_x = - \frac{1}{\tau} \left(\frac{F_x}{A} + \sigma n_x \right)$$

Substituting F_x from Eq. (6-8a) results in

$$t_x = \frac{1}{\tau} [(\sigma_x - \sigma) n_x + \tau_{xy} n_y + \tau_{zx} n_z] \tag{6-10a}$$

Likewise, t_y and t_z are

$$t_y = \frac{1}{\tau} [\tau_{xy} n_x + (\sigma_y - \sigma) n_y + \tau_{yz} n_z] \tag{6-10b}$$

$$t_z = \frac{1}{\tau} [\tau_{zx} n_x + \tau_{yz} n_y + (\sigma_z - \sigma) n_z] \tag{6-10c}$$

Example 6-2 The state of stress at a particular point relative to the xyz coordinate system is given by the stress matrix

$$\sigma = \begin{bmatrix} 14 & 7 & -7 \\ 7 & 10 & 0 \\ -7 & 0 & 35 \end{bmatrix} \text{MN/m}^2$$

Determine the normal stress and the magnitude and direction of the shear stress on a surface intersecting the point and parallel to the plane given by the equation

$$2x - y + 3z = 9$$

SOLUTION The perpendicular of the surface is established by the direction numbers of the plane d_x, d_y, and d_z. These are simply the respective coefficients of the x, y, and z terms of the equation of the plane. Thus

$$d_x = 2 \qquad d_y = -1 \qquad d_z = 3$$

The directional cosines are found by simply normalizing the direction numbers so that the sum of their squares is unity. Therefore,

$$n_x = \frac{2}{\sqrt{2^2 + (-1^2) + 3^2}} = \frac{2}{\sqrt{14}}$$

Likewise

$$n_y = \frac{-1}{\sqrt{14}} \qquad n_z = \frac{3}{\sqrt{14}}$$

From the stress matrix, $\sigma_x = 14 \, \text{MN/m}^2$, $\tau_{xy} = 7 \, \text{MN/m}^2$, $\tau_{zx} = -7 \, \text{MN/m}^2$, $\sigma_y = 10 \, \text{MN/m}^2$, $\tau_{yz} = 0 \, \text{MN/m}^2$, and $\sigma_z = 35 \, \text{MN/m}^2$. Substituting the stresses and direction cosines in Eq. (6-7) yields

$$\sigma = 14\left(\frac{2}{\sqrt{14}}\right)^2 + 10\left(\frac{-1}{\sqrt{14}}\right)^2 + 35\left(\frac{3}{\sqrt{14}}\right)^2 + (2)(7)\frac{2}{\sqrt{14}}\frac{-1}{\sqrt{14}}$$

$$+ (2)(0)\frac{-1}{\sqrt{14}}\frac{3}{\sqrt{14}} + (2)(-7)\frac{3}{\sqrt{14}}\frac{2}{\sqrt{14}} = 19.21 \, \text{MN/m}^2$$

The shear stress is found using Eq. (6-9) and is

$$\tau = \left[\left(14\frac{2}{\sqrt{14}} + 7\frac{-1}{\sqrt{14}} + (-7)\frac{3}{\sqrt{14}}\right)^2 + \left(7\frac{2}{\sqrt{14}} + 10\frac{-1}{\sqrt{14}} + 0\frac{3}{\sqrt{14}}\right)^2\right.$$

$$\left. + \left(-7\frac{2}{\sqrt{14}} + 0\frac{-1}{\sqrt{14}} + 35\frac{3}{\sqrt{14}}\right)^2 - (19.2)^2\right]^{1/2} = 14.95 \, \text{MN/m}^2$$

From Eqs. (6-10), the directional cosines for the direction of τ are

$$t_x = \frac{1}{14.95}\left[(14 - 19.21)\frac{2}{\sqrt{14}} + 7\frac{-1}{\sqrt{14}} + (-7)\frac{3}{\sqrt{14}}\right] = -0.687$$

$$t_y = \frac{1}{14.95}\left[7\frac{2}{\sqrt{14}} + (10 - 19.21)\frac{-1}{\sqrt{14}} + 0\frac{3}{\sqrt{14}}\right] = 0.415$$

$$t_z = \frac{1}{14.95}\left[(-7)\frac{2}{\sqrt{14}} + 0\frac{-1}{\sqrt{14}} + (35 - 19.21)\frac{3}{\sqrt{14}}\right] = 0.596$$

There are two simple checks that can be made on the directional cosines of τ:

1. The sum of the squares of the directional cosines is unity. Thus

$$t_x^2 + t_y^2 + t_z^2 = 1$$

2. The directions of σ and τ are perpendicular. Thus

$$n_x t_x + n_y t_y + n_z t_z = 0$$

Principal Stresses

For the general three-dimensional state of stress there exist at least three mutually perpendicular surfaces in which there are no shear stresses. The normal stresses on these surfaces are denoted σ_1, σ_2, and σ_3, where usually the ordering is such that $\sigma_1 > \sigma_2 > \sigma_3$. If two of the principal stresses are equal, there will exist an infinite set of surfaces containing these principal stresses, where the normals to these surfaces are perpendicular to the third principal stress. If all three principal stresses are equal, a hydrostatic state of stress exists, where all surfaces contain the same principal stress and no shear stress exists on any surface. Returning to Fig. 6-4, assume that on the oblique surface the shear stress τ equals zero. Then σ will be a principal stress denoted by σ_i. The components of the force on the oblique surface in the x, y, and z directions are

$$F'_x = \sigma_i A n_x \tag{6-11a}$$

$$F'_y = \sigma_i A n_y \tag{6-11b}$$

$$F'_z = \sigma_i A n_z \tag{6-11c}$$

Recall Eqs. (6-8), which give the forces in the x, y, and z directions due to the stresses on the orthogonal surfaces. For equilibrium, $F_x + F'_x = 0$, etc. This results in

$$(\sigma_x - \sigma_i)n_x + \tau_{xy}n_y + \tau_{zx}n_z = 0 \tag{6-12a}$$

$$\tau_{xy}n_x + (\sigma_y - \sigma_i)n_y + \tau_{yz}n_z = 0 \tag{6-12b}$$

$$\tau_{zx}n_x + \tau_{yz}n_y + (\sigma_z - \sigma_i)n_z = 0 \tag{6-12c}$$

One possible solution to Eqs. (6-12) is $n_x = n_y = n_z = 0$. This cannot occur since

$$n_x^2 + n_y^2 + n_z^2 = 1 \tag{6-13}$$

In order to avoid the zero solution of the directional cosines the determinant of the coefficients of n_x, n_y, and n_z of Eqs. (6-12) is equated to zero. This makes the solution of the directional cosines indeterminate from Eqs. (6-12). Thus

$$\begin{vmatrix} \sigma_x - \sigma_i & \tau_{xy} & \tau_{zx} \\ \tau_{xy} & \sigma_y - \sigma_i & \tau_{yz} \\ \tau_{zx} & \tau_{yz} & \sigma_z - \sigma_i \end{vmatrix} = 0$$

Evaluating the determinant results in

$$\sigma_i^3 - (\sigma_x + \sigma_y + \sigma_z)\sigma_i^2 + (\sigma_x\sigma_y + \sigma_y\sigma_z + \sigma_z\sigma_x - \tau_{yz}^2 - \tau_{zx}^2 - \tau_{xy}^2)\sigma_i$$
$$- (\sigma_x\sigma_y\sigma_z + 2\tau_{yz}\tau_{zx}\tau_{xy} - \sigma_x\tau_{yz}^2 - \sigma_y\tau_{zx}^2 - \sigma_z\tau_{xy}^2) = 0 \tag{6-14}$$

Equation (6-14) is a cubic equation in the unknown, σ_i, where three solutions result, the principal stresses σ_1, σ_2, and σ_3.

If the values of the directional cosines for a principal stress are wanted, the principal stress is substituted into Eqs. (6-12); then only *two* independent equations in the remaining unknowns, n_x, n_y, and n_z, result. These together with Eq. (6-13) are sufficient for determining the directional cosines for the surface containing the principal stress. However, instead of solving two linear and one second-order equation simultaneously, there is a simplified method demonstrated in the following example.

Example 6-3 For the stress matrix given below, determine the principal stresses and the directional cosines associated with the normals to the surfaces of each principal stress.

$$\sigma = \begin{bmatrix} 3 & 1 & 1 \\ 1 & 0 & 2 \\ 1 & 2 & 0 \end{bmatrix} (1000) \text{ lb/in}^2$$

SOLUTION The stresses are $\sigma_x = 3 \times 10^3 \text{ lb/in}^2$, $\tau_{xy} = 1 \times 10^3 \text{ lb/in}^2$, $\tau_{zx} = 1 \times 10^3 \text{ lb/in}^2$, $\sigma_y = 0$,

$\tau_{yz} = 2 \times 10^3$ lb/in^2, and $\sigma_z = 0$. Substituting this into Eq. (6-14) yields

$$\sigma_i^3 - (3 + 0 + 0)\sigma_i^2 + [(3)(0) + (0)(0) + (0)(3) - 2^2 - 1^2 - 1^2]\sigma_i$$
$$- [(3)(0)(0) + (2)(2)(1)(1) - (3)(2^2) - (0)(1^2) - (0)(1^2)] = 0$$

which simplifies to

$$\sigma_i^3 - 3\sigma_i^2 - 6\sigma_i + 8 = 0 \qquad (a)$$

The solutions to the cubic equation are $\sigma_i = 4 \times 10^3$, 1×10^3, and -2×10^3 lb/in^2. Thus

$$\sigma_1 = 4000 \text{ lb/in}^2 \qquad \sigma_2 = 1000 \text{ lb/in}^2 \qquad \sigma_3 = -2000 \text{ lb/in}^2$$

The directional cosines associated with each principal stress are determined independently. First, consider σ_1 and substitute $\sigma_i = 4000$ lb/in^2 into Eqs. (6-12). This results in

$$-n_x + n_y + n_z = 0 \qquad (b)$$
$$n_x - 4n_y + 2n_z = 0 \qquad (c)$$
$$n_x + 2n_y - 4n_z = 0 \qquad (d)$$

Solving Eqs. (b), (c), and (d) simultaneously will not yield a solution, as the three equations are not independent. For example, operating on the equations, one finds $2(b) + (c) = -(d)$. Thus only two independent equations can be used. In this example any two can be used, and a third equation comes from Eq. (6-13). However, instead of solving the three equations simultaneously, let

$$n_x = ac_x \qquad n_y = ac_y \qquad n_z = ac_z$$

where a is an unknown constant. Next let $c_x = 1$ arbitrarily. Substituting this into Eqs. (b) and (c) results in

$$-a + ac_y + ac_z = 0 \qquad \text{and} \qquad a - 4ac_y + 2ac_z = 0$$

or

$$c_y + c_z = 1 \qquad (e)$$

and

$$-4c_y + 2c_z = -1 \qquad (f)$$

Solving Eqs. (e) and (f) simultaneously yields $c_y = \frac{1}{2}$ and $c_z = \frac{1}{2}$.† Now using Eq. (6-13),

$$a^2 + (\tfrac{1}{2}a)^2 + (\tfrac{1}{2}a)^2 = 1$$

results in $a = 2/\sqrt{6}$. Therefore

$$n_x = ac_x = \frac{2}{\sqrt{6}}(1) = \frac{2}{\sqrt{6}}$$

$$n_y = ac_y = \frac{2}{\sqrt{6}}\frac{1}{2} = \frac{1}{\sqrt{6}}$$

$$n_z = ac_z = \frac{2}{\sqrt{6}}\left(\frac{1}{2}\right) = \frac{1}{\sqrt{6}}$$

Since these directional cosines are associated with the principal stress σ_1, the subscript 1 is used. Thus, the directional cosines for the normal to the surface containing the principal stress $\sigma_1 = 4000$ lb/in^2 are

$$(n_x)_1 = \frac{2}{\sqrt{6}} \qquad (n_y)_1 = \frac{1}{\sqrt{6}} \qquad (n_z)_1 = \frac{1}{\sqrt{6}}$$

† This technique has one flaw. If n_x had been zero in this example, then an erroneous solution would have resulted since this would have required a to be zero, thus causing $n_x = n_y = n_z = 0$ and Eq. (6-13) could not have been satisfied. If an erroneous solution results when this technique is used, simply repeat the solution letting c_y or c_z equal unity.

Repeating the same procedure for $\sigma_2 = 1000$ lb/in^2, we find the directional cosines to be

$$(n_x)_2 = \frac{1}{\sqrt{3}} \qquad (n_y)_2 = \frac{-1}{\sqrt{3}} \qquad (n_z)_2 = \frac{-1}{\sqrt{3}}$$

and for $\sigma_3 = -2000$ lb/in^2

$$(n_x)_3 = 0 \qquad (n_y)_3 = \frac{1}{\sqrt{2}} \qquad (n_z)_3 = \frac{-1}{\sqrt{2}}$$

Stress Invariants

Let us return to Eq. (6-14). The solutions of σ_i are independent of the coordinate system used to define the coefficients of the cubic equation for σ_i. The coefficients of σ_i in Eq. (6-14) are constant and are normally referred to as the stress invariants. Thus

$$\sigma_x + \sigma_y + \sigma_z = I_1 \tag{6-15a}$$

$$\sigma_x\sigma_y + \sigma_y\sigma_z + \sigma_z\sigma_x - \tau_{yz}^2 - \tau_{zx}^2 - \tau_{xy}^2 = I_2 \tag{6-15b}$$

$$\sigma_x\sigma_y\sigma_z + 2\tau_{yz}\tau_{zx}\tau_{xy} - \sigma_x\tau_{yz}^2 - \sigma_y\tau_{zx}^2 - \sigma_z\tau_{xy}^2 = I_3 \tag{6-15c}$$

The constants I_1, I_2, and I_3 are referred to as the first, second, and third stress invariants, respectively. Equations (6-15) are particularly helpful in checking the results of a stress transformation, as illustrated in the following example.

Example 6-4 Returning to Example 6-1, verify that the stresses obtained for the $x'y'z'$ coordinate system satisfy Eqs. (6-15).

SOLUTION From Eq. (6-15a)

$$I_1 = \sigma_x + \sigma_y + \sigma_z = 3000 + 0 + 1000 = 4000 \text{ lb/in}^2$$

Relative to the $x'y'z'$ system the first invariant is

$$I_1 = \sigma_{x'} + \sigma_{y'} + \sigma_{z'} = (\tfrac{9}{4} + \sqrt{3} + \tfrac{3}{4} - \sqrt{3} + 1)(1000) = 4000 \text{ lb/in}^2$$

which checks. The second invariant, using the stresses relative to the xyz coordinate system, is

$$I_2 = \sigma_x\sigma_y + \sigma_y\sigma_z + \sigma_z\sigma_x - \tau_{xz}^2 - \tau_{zx}^2 - \tau_{xy}^2$$

$$= [(3)(0) + (0)(1) + (1)(3) - (-1^2) - (-2^2) - 2^2](10^3)^2 = -6 \times 10^6 (\text{lb/in}^2)^2$$

and relative to the $x'y'z'$ system

$$I_2 = \sigma_{x'}\sigma_{y'} + \sigma_{y'}\sigma_{z'} + \sigma_{z'}\sigma_{x'} - \tau_{y'z'}^2 - \tau_{z'x'}^2 - \tau_{x'y'}^2$$

$$= [(\tfrac{9}{4} + \sqrt{3})(\tfrac{3}{4} - \sqrt{3}) + (\tfrac{3}{4} - \sqrt{3})(1) + (1)(\tfrac{9}{4} + \sqrt{3}) - (1 - \tfrac{1}{2}\sqrt{3})^2$$

$$-(-\tfrac{1}{2} - \sqrt{3})^2 - (1 - \tfrac{3}{4}\sqrt{3})^2](10^3)^2 = -6 \times 10^6 (\text{lb/in}^2)^2$$

The third invariant is

$$I_3 = \sigma_x\sigma_y\sigma_z + 2\tau_{yz}\tau_{zx}\tau_{xy} - \sigma_x\tau_{yz}^2 - \sigma_y\tau_{zx}^2 - \sigma_z\tau_{xy}^2$$

$$= [(3)(0)(1) + (2)(-1)(-2)(2) - (3)(-1^2) - (0)(-2^2) - (1)(2^2)](10^3)^3 = 1 \times 10^9 (\text{lb/in}^2)^3$$

and relative to the $x'y'z'$ system is

$$I_3 = \sigma_{x'}\sigma_{y'}\sigma_{z'} + 2\tau_{y'z'}\tau_{z'x'}\tau_{x'y'} - \sigma_{x'}\tau_{y'z'}^2 - \sigma_{y'}\tau_{z'x'}^2 - \sigma_{z'}\tau_{x'y'}^2$$

$$= [(\tfrac{9}{4} + \sqrt{3})(\tfrac{3}{4} - \sqrt{3})(1) + (2)(1 - -\tfrac{1}{2}\sqrt{3})(-\tfrac{1}{2} - \sqrt{3})(1 - \tfrac{3}{4}\sqrt{3}) - (\tfrac{9}{4} + \sqrt{3})(1 - \tfrac{1}{2}\sqrt{3})^2$$

$$- (\tfrac{3}{4} - \sqrt{3})(-\tfrac{1}{2} - \sqrt{3})^2 - (1)(1 - \tfrac{3}{4}\sqrt{3})^2](10^3)^3 = 1 \times 10^9 (\text{lb/in}^2)^3$$

Thus, all three stress invariants check out for both coordinate systems.

Strain Transformations

Recall from Chap. 2 for the biaxial state of strain, that the strain transformation equations are identical to the stress equations, where σ is substituted by ε and τ by $\gamma/2$. Since this applies to transformations in three dimensions as well, the strain transformation equations will not be repeated here. Simply make the substitution of $\varepsilon_x \leftrightarrow \sigma_x \; \varepsilon_y \leftrightarrow \sigma_{y'}, \; \gamma_{xy}/2 \leftrightarrow \tau_{xy}, \; \gamma_{y'z'}/2 \leftrightarrow \tau_{y'z'}$ etc., in Eqs. (6-5).

6-2 THE EQUILIBRIUM EQUATIONS

In Chap. 1, the stresses and body forces on an infinitesimal element were shown in Fig. 1-22, repeated here as Fig. 6-5. In general, the stresses vary with respect to position, and the changes in the stresses are depicted by $\Delta\sigma_x$, $\Delta\sigma_y$, etc. If these changes are represented in terms of a Taylor's series expansion and only the linear terms retained, we have

$$\Delta\sigma_x = \frac{\partial\sigma_x}{\partial x}\,\Delta x \qquad \Delta\sigma_y = \frac{\partial\sigma_y}{\partial y}\,\Delta y \qquad \text{etc.}$$

$$\Delta\tau_{xy} = \frac{\partial\tau_{xy}}{\partial x}\,\Delta x \qquad \Delta\tau_{yx} = \frac{\partial\tau_{yx}}{\partial y}\,\Delta y = \frac{\partial\tau_{xy}}{\partial y}\,\Delta y \qquad \text{etc.}$$

If we recall that the body forces, $\bar{F}_x$, $\bar{F}_y$, and $\bar{F}_z$ are forces per unit volume, we can now write the equilibrium equations. For equilibrium, the sum of the forces in the x direction equals zero. Thus

$$\left(\sigma_x + \frac{\partial\sigma_x}{\partial x}\,\Delta x\right)\Delta y\,\Delta z + \left(\tau_{xy} + \frac{\partial\tau_{xy}}{\partial y}\,\Delta y\right)\Delta z\,\Delta x + \left(\tau_{zx} + \frac{\partial\tau_{zx}}{\partial z}\,\Delta z\right)\Delta x\,\Delta y$$

$$- \sigma_x\,\Delta y\,\Delta z - \tau_{xy}\,\Delta z\,\Delta x - \tau_{zx}\,\Delta x\,\Delta y + \bar{F}_x\,\Delta x\,\Delta y\,\Delta z = 0$$

Simplification of this expression results in

$$\frac{\partial\sigma_x}{\partial x} + \frac{\partial\tau_{xy}}{\partial y} + \frac{\partial\tau_{zx}}{\partial z} + \bar{F}_x = 0 \tag{6-16a}$$

Similarly, summing forces in the y and z directions respectively yields

$$\frac{\partial\tau_{xy}}{\partial x} + \frac{\partial\sigma_y}{\partial y} + \frac{\partial\tau_{yz}}{\partial z} + \bar{F}_y = 0 \tag{6-16b}$$

$$\frac{\partial\tau_{zx}}{\partial x} + \frac{\partial\tau_{yz}}{\partial y} + \frac{\partial\sigma_z}{\partial z} + \bar{F}_z = 0 \tag{6-16c}$$

For the biaxial state of stress, where $\sigma_z = \tau_{zx} = \tau_{yz} = 0$, the equilibrium equations reduce to

$$\frac{\partial\sigma_x}{\partial x} + \frac{\partial\tau_{xy}}{\partial y} + \bar{F}_x = 0 \tag{6-17a}$$

$$\frac{\partial\tau_{xy}}{\partial x} + \frac{\partial\sigma_y}{\partial y} + \bar{F}_y = 0 \tag{6-17b}$$

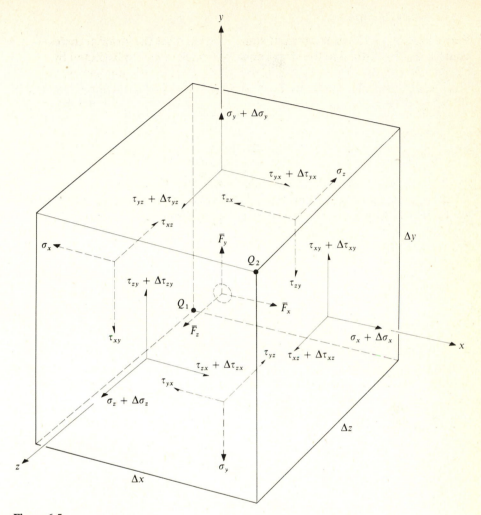

Figure 6-5

Example 6-5 For the beam shown in Fig. 6-6, if the weight is neglected, the bending moment as a function of x is $M_z = -\frac{1}{2}wx^2$. The bending stress is given by $\sigma_x = -M_z y/I_z$, and thus, for this example,

$$\sigma_x = \frac{w}{2I_z} x^2 y \tag{a}$$

Starting with Eq. (a), use Eqs. (6-17)† to determine how τ_{xy} and σ_y vary as functions of x and y.

---

† Using Eqs. (6-17) assumes a biaxial stress field. For a thin rectangular beam, this is a relatively valid assumption since the side faces are free surfaces, i.e., are stress-free, and the stresses in the z direction are zero. If the beam is thin, the stresses in the z direction will not have a chance to become appreciable.

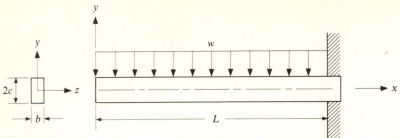

Figure 6-6

SOLUTION If the weight of the beam is neglected, the body forces are zero. Thus $\bar{F}_x = \bar{F}_y = 0$. Substituting Eq. (*a*) into Eq. (6-17*a*) results in

$$\frac{\partial \tau_{xy}}{\partial y} = -\frac{w}{I_z}xy$$

Keeping in mind that the derivative in the above equation is a partial derivative, we see that integration of both sides of the equation with respect to y results in

$$\tau_{xy} = -\frac{w}{2I_z}xy^2 + f(x) \tag{b}$$

Instead of a constant of integration, a function of integration is added, where $f(x)$ is a function of x alone [note that $f(x)$ can still contain a constant]. The function is found from the boundary or surface conditions. From Fig. 6-6 it can be seen that there is no imposed shear force on the top and bottom surfaces of the beam. For the top surface, the condition is that $\tau_{xy} = 0$ at $y = c$. Solving for $f(x)$ gives

$$f(x) = \frac{wc^2}{2I_z}x$$

Substituting $f(x)$ into Eq. (*b*) yields

$$\tau_{xy} = \frac{w}{2I_z}x(c^2 - y^2) \tag{c}$$

Note that the condition that $\tau_{xy} = 0$ at $y = -c$ is automatically satisfied in Eq. (*c*).

Using Eq. (2-17), the reader should verify that Eq. (*c*) gives the same result as that of $\tau_{xy} = V_y Q / I_z b$.

Next, Eq. (*c*) is substituted into Eq. (6-17*b*), resulting in

$$\frac{\partial \sigma_y}{\partial y} = -\frac{w}{2I_z}(c^2 - y^2)$$

Again integrating with respect to y (note that a function of integration must be added) yields

$$\sigma_y = -\frac{w}{2I_z}y(c^2 - \tfrac{1}{3}y^2) + g(x) \tag{d}$$

where $g(x)$ is the function of integration.

A boundary condition from Fig. 6-6 is $\sigma_y = -w/b$ at $y = c$. Thus, Eq. (*d*) is written

$$-\frac{w}{b} = -\frac{w}{2I_z}c[c^2 - \tfrac{1}{3}(c^2)] + g(x)$$

When we recall that $b = 3I_z/2c^3$, the above equation simplifies to

$$g(x) = -\frac{wc^3}{3I_z}$$

and Eq. (*d*) becomes

$$\sigma_y = \frac{-w}{6I_z}(2c^3 + 3c^2 y - y^3) \tag{e}$$

Note that Eq. (*e*) also satisfies the condition that $\sigma_y = 0$ at $y = -c$.

6-3 GENERALIZED HOOKE'S LAW

The stress-strain equations given in Chap. 1 only apply to homogeneous and isotropic materials. Homogeneous refers to the fact that the elastic properties do not change from point to point in the body, and isotropic means that the properties are not varying with respect to direction. Many engineering materials, however, are neither homogeneous nor isotropic, e.g., rolled or drawn metals, wood, laminates, and fiber-filled epoxy materials. These materials are not isotropic since their elastic properties depend on direction.

In general, each strain is dependent on each stress. For example, the strain ε_x written as a linear function of each stress is

$$\varepsilon_x = C_{11}\sigma_x + C_{12}\sigma_y + C_{13}\sigma_z + C_{14}\tau_{xy} + C_{15}\tau_{yz} + C_{16}\tau_{zx} + C_{17}\tau_{xz} + C_{18}\tau_{zy} + C_{19}\tau_{yx}$$

Each of the remaining eight strains can be written in a similar fashion. If no symmetry is assumed, $9^2 = 81$ independent constants would result. That is, in matrix notation the stress-strain relations would be

$$
\begin{Bmatrix} \varepsilon_x \\ \varepsilon_y \\ \varepsilon_z \\ \gamma_{xy} \\ \gamma_{yz} \\ \gamma_{zx} \\ \gamma_{xz} \\ \gamma_{zy} \\ \gamma_{yx} \end{Bmatrix} =
\begin{bmatrix}
C_{11} & C_{12} & C_{13} & C_{14} & C_{15} & C_{16} & C_{17} & C_{18} & C_{19} \\
C_{21} & C_{22} & C_{23} & C_{24} & C_{25} & C_{26} & C_{27} & C_{28} & C_{29} \\
 & & & & & & & & \\
\cdot & \cdot & \cdot & \cdot & \cdot & \cdot & \cdot & \cdot & \cdot \\
 & & & & & & & & \\
C_{91} & C_{92} & C_{93} & C_{94} & C_{95} & C_{96} & C_{97} & C_{98} & C_{99}
\end{bmatrix}
\begin{Bmatrix} \sigma_x \\ \sigma_y \\ \sigma_z \\ \tau_{xy} \\ \tau_{yz} \\ \tau_{zx} \\ \tau_{xz} \\ \tau_{zy} \\ \tau_{yx} \end{Bmatrix} \tag{6-18}
$$

However, shear is normally symmetric, where $\tau_{yx} = \tau_{xy}$, etc., and $\gamma_{yx} = \gamma_{xy}$, etc. This reduces the number of stress and strain terms to six each. Thus, there are only $6^2 = 36$ independent elastic constants, i.e.,

$$
\begin{Bmatrix} \varepsilon_x \\ \varepsilon_y \\ \varepsilon_z \\ \gamma_{xy} \\ \gamma_{yz} \\ \gamma_{zx} \end{Bmatrix} = \begin{bmatrix} K_{11} & K_{12} & K_{13} & K_{14} & K_{15} & K_{16} \\ K_{21} & K_{22} & K_{23} & K_{24} & K_{25} & K_{26} \\ \cdot & \cdot & \cdot & \cdot & \cdot & \cdot \\ & & & & & \\ & & & & & \\ K_{61} & \cdot & \cdot & \cdot & & K_{66} \end{bmatrix} \begin{Bmatrix} \sigma_x \\ \sigma_y \\ \sigma_z \\ \tau_{xy} \\ \tau_{yz} \\ \tau_{zx} \end{Bmatrix} \tag{6-19}
$$

The $[K]$ matrix can be shown to be symmetric. That is, $K_{12} = K_{21}$, $K_{13} = K_{31}$, etc. This can be verified by looking at the net work due to the stresses, which should be independent of the order of application. Thus, in general (except when cross-shear stresses are unequal) there are at most 21 possible independent elastic constants.

If a material has one plane of symmetry, there can be no interaction between the out-of-plane shear stresses and the remaining strains. If the xy plane is assumed to be a plane of symmetry, Eq. (6-19) reduces to

$$
\begin{Bmatrix} \varepsilon_x \\ \varepsilon_y \\ \varepsilon_z \\ \gamma_{xy} \\ \gamma_{yz} \\ \gamma_{zx} \end{Bmatrix} = \begin{bmatrix} K_{11} & K_{12} & K_{13} & K_{14} & 0 & 0 \\ K_{21} & K_{22} & K_{23} & K_{24} & 0 & 0 \\ K_{31} & K_{32} & K_{33} & K_{34} & 0 & 0 \\ K_{41} & K_{42} & K_{43} & K_{44} & 0 & 0 \\ 0 & 0 & 0 & 0 & K_{55} & K_{56} \\ 0 & 0 & 0 & 0 & K_{65} & K_{66} \end{bmatrix} \begin{Bmatrix} \sigma_x \\ \sigma_y \\ \sigma_z \\ \tau_{xy} \\ \tau_{yz} \\ \tau_{zx} \end{Bmatrix} \tag{6-20}
$$

where $K_{ij} = K_{ji}$.

A material which exhibits symmetry with respect to three mutually orthogonal planes is called an *orthotropic material*. If the xy, yz, and zx planes are considered planes of symmetry, Eq. (6-19) reduces to

$$
\begin{Bmatrix} \varepsilon_x \\ \varepsilon_y \\ \varepsilon_z \\ \gamma_{xy} \\ \gamma_{yz} \\ \gamma_{zx} \end{Bmatrix} = \begin{bmatrix} K_{11} & K_{12} & K_{13} & 0 & 0 & 0 \\ K_{21} & K_{22} & K_{23} & 0 & 0 & 0 \\ K_{31} & K_{32} & K_{33} & 0 & 0 & 0 \\ 0 & 0 & 0 & K_{44} & 0 & 0 \\ 0 & 0 & 0 & 0 & K_{55} & 0 \\ 0 & 0 & 0 & 0 & 0 & K_{66} \end{bmatrix} \begin{Bmatrix} \sigma_x \\ \sigma_y \\ \sigma_z \\ \tau_{xy} \\ \tau_{yz} \\ \tau_{zx} \end{Bmatrix} \tag{6-21}
$$

where $K_{ij} = K_{ji}$.

If the properties of an orthotropic material are identical in all three directions, the material is said to have a *cubic structure*. Thus, if we let

$$
K_{11} = K_{22} = K_{33} = \frac{1}{E}
$$

$$
K_{12} = K_{13} = K_{23} = K_{21} = K_{31} = K_{32} = -\frac{\nu}{E}
$$

$$
K_{44} = K_{55} = K_{66} = \frac{1}{G}
$$

where E, ν, and G are the material constants normally used in engineering applications, for an orthotropic material with a cubic structure, Eq. (6-21) reduces to

$$
\begin{Bmatrix} \varepsilon_x \\ \varepsilon_y \\ \varepsilon_z \\ \gamma_{xy} \\ \gamma_{yz} \\ \gamma_{zx} \end{Bmatrix} = \frac{1}{E}
\begin{bmatrix}
1 & -\nu & -\nu & 0 & 0 & 0 \\
-\nu & 1 & -\nu & 0 & 0 & 0 \\
-\nu & -\nu & 1 & 0 & 0 & 0 \\
0 & 0 & 0 & \dfrac{E}{G} & 0 & 0 \\
0 & 0 & 0 & 0 & \dfrac{E}{G} & 0 \\
0 & 0 & 0 & 0 & 0 & \dfrac{E}{G}
\end{bmatrix}
\begin{Bmatrix} \sigma_x \\ \sigma_y \\ \sigma_z \\ \tau_{xy} \\ \tau_{yz} \\ \tau_{zx} \end{Bmatrix}
\tag{6-22}
$$

If the material properties of a cubic structure are the same for any arbitrary coordinate system, the material is truly isotropic. The derivation given in Appendix A utilized this fact, and the relationship between E, ν, and G resulted, where

$$
G = \frac{E}{2(1+\nu)}
$$

Substituting this into Eq. (6-22) gives

$$
\begin{Bmatrix} \varepsilon_x \\ \varepsilon_y \\ \varepsilon_z \\ \gamma_{xy} \\ \gamma_{yz} \\ \gamma_{zx} \end{Bmatrix} = \frac{1}{E}
\begin{bmatrix}
1 & -\nu & -\nu & 0 & 0 & 0 \\
-\nu & 1 & -\nu & 0 & 0 & 0 \\
-\nu & -\nu & 1 & 0 & 0 & 0 \\
0 & 0 & 0 & 2(1+\nu) & 0 & 0 \\
0 & 0 & 0 & 0 & 2(1+\nu) & 0 \\
0 & 0 & 0 & 0 & 0 & 2(1+\nu)
\end{bmatrix}
\begin{Bmatrix} \sigma_x \\ \sigma_y \\ \sigma_z \\ \tau_{xy} \\ \tau_{yz} \\ \tau_{zx} \end{Bmatrix}
\tag{6-23}
$$

Lamé Constants

In the mathematical theory of elasticity, the material constants may be expressed in terms of Lamé's constants μ and λ rather than the engineering constants E and ν. In indicial notation, the stress-strain relations can be more conveniently written in terms of Lamé's constants. When indicial notation is used, the stress-strain relation is

$$
\sigma_{ij} = 2\mu\varepsilon_{ij} + \lambda\delta_{ij}\varepsilon_{kk} \qquad i, j = 1, 2, 3
\tag{6-24}
$$

where δ_{ij} is called the *Kronecker delta* and

$$
\delta_{ij} = \begin{cases} 1 & \text{when } i = j \\ 0 & \text{when } i \neq j \end{cases}
$$

The term ε_{kk} represents a summation term given by

$$
\varepsilon_{kk} = \varepsilon_{11} + \varepsilon_{22} + \varepsilon_{33}
$$

The σ_{ij} and ε_{ij} terms are†

$$\sigma_{11} = \sigma_x \qquad\qquad \sigma_{12} = \tau_{xy} \qquad\qquad \sigma_{13} = \tau_{xz} = \tau_{zx}$$

$$\sigma_{21} = \tau_{yx} = \tau_{xy} \qquad \sigma_{22} = \sigma_y \qquad\qquad \sigma_{23} = \tau_{yz}$$

$$\sigma_{31} = \tau_{zx} \qquad\qquad \sigma_{32} = \tau_{zy} = \tau_{yz} \qquad \sigma_{33} = \sigma_z$$

$$\varepsilon_{11} = \varepsilon_x \qquad\qquad \varepsilon_{12} = \tfrac{1}{2}\gamma_{xy} \qquad\qquad \varepsilon_{13} = \tfrac{1}{2}\gamma_{xz} = \tfrac{1}{2}\gamma_{zx}$$

$$\varepsilon_{21} = \tfrac{1}{2}\gamma_{yx} = \tfrac{1}{2}\gamma_{xy} \quad\; \varepsilon_{22} = \varepsilon_y \qquad\qquad \varepsilon_{23} = \tfrac{1}{2}\gamma_{yz}$$

$$\varepsilon_{31} = \tfrac{1}{2}\gamma_{zx} \qquad\qquad \varepsilon_{32} = \tfrac{1}{2}\gamma_{zy} = \tfrac{1}{2}\gamma_{yz} \quad\; \varepsilon_{33} = \varepsilon_z$$

where the shear strains are as defined in Eq. (6-1).

If, for example, we want the equation for τ_{xy}, let $i = 1$ and $j = 2$ in Eq. (6-24). Thus, $\delta_{12} = 0$, and from Eq. (6-24)

$$\sigma_{12} = 2\mu\varepsilon_{12}$$

or

$$\tau_{xy} = 2\mu\frac{\gamma_{xy}}{2} = \mu\gamma_{xy}$$

It can be seen that the Lamé constant μ is equal to the shear modulus G, or, in terms of E and ν, it is

$$\mu = \frac{E}{2(1 + \nu)} \tag{6-25}$$

If σ_y is wanted from Eq. (6-24), let $i = j = 2$. Thus, $\delta_{22} = 1$, and

$$\sigma_{22} = 2\mu\varepsilon_{22} + \lambda(\varepsilon_{11} + \varepsilon_{22} + \varepsilon_{33}) = (2\mu + \lambda)\varepsilon_{22} + \lambda(\varepsilon_{11} + \varepsilon_{33})$$

or

$$\sigma_y = (2\mu + \lambda)\varepsilon_y + \lambda(\varepsilon_x + \varepsilon_z)$$

If this is compared with Eq. (1-22b), the second Lamé constant λ is found to be

$$\lambda = \frac{E\nu}{(1 + \nu)(1 - 2\nu)} \tag{6-26}$$

6-4 LARGE STRAINS

In Chap. 1, the strain-deflection equations were developed for small strains. Recall Fig. 1-26 (repeated here as Fig. 6-7); the extension of line QD was approximated by $\partial u / \partial x\, \Delta x$. If $Q'D'$ is first considered to be in the xy plane only, the actual length is

$$Q'D' = \left[\left(\Delta x + \frac{\partial u}{\partial x}\Delta x\right)^2 + \left(\frac{\partial v}{\partial x}\Delta x\right)^2\right]^{1/2}$$

$$= \Delta x\left[1 + 2\frac{\partial u}{\partial x} + \left(\frac{\partial u}{\partial x}\right)^2 + \left(\frac{\partial v}{\partial x}\right)^2\right]^{1/2}$$

† Note that the index numbers 1, 2, 3 represent the xyz coordinate system.

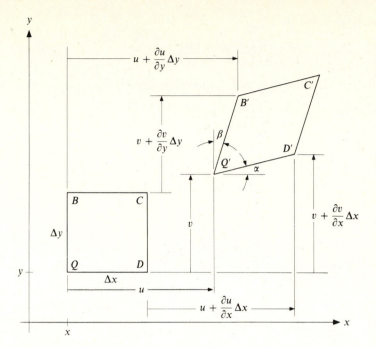

Figure 6-7

Using the binomial expansion theorem retaining *second*-order terms, we can reduce the above to†

$$Q'D' \approx \Delta x\left[1 + \frac{\partial u}{\partial x} + \frac{1}{2}\left(\frac{\partial v}{\partial x}\right)^2\right]$$

The normal strain in the x direction is given by

$$\varepsilon_x = \frac{Q'D' - QD}{QD}$$

where $QD = \Delta x$. Thus

$$\varepsilon_x \approx \frac{\partial u}{\partial x} + \frac{1}{2}\left(\frac{\partial v}{\partial x}\right)^2$$

In general, $Q'D'$ is not in the xy plane, and D' will be displaced in the z direction relative to Q' by $\partial w/\partial x\ \Delta x$, where w is displacement in the z direction. This additional term can be accounted for in the overall length of

† Provided

$$2\frac{\partial u}{\partial x} + \left(\frac{\partial u}{\partial x}\right)^2 + \left(\frac{\partial v}{\partial x}\right)^2 < 1$$

$Q'D'$, and the resulting strain is

$$\varepsilon_x = \frac{\partial u}{\partial x} + \frac{1}{2}\left[\left(\frac{\partial v}{\partial x}\right)^2 + \left(\frac{\partial w}{\partial x}\right)^2\right] \tag{6-27a}$$

Similarly the normal strains in the y and z direction are

$$\varepsilon_y = \frac{\partial v}{\partial y} + \frac{1}{2}\left[\left(\frac{\partial u}{\partial y}\right)^2 + \left(\frac{\partial w}{\partial y}\right)^2\right] \tag{6-27b}$$

$$\varepsilon_z = \frac{\partial w}{\partial z} + \frac{1}{2}\left[\left(\frac{\partial u}{\partial z}\right)^2 + \left(\frac{\partial v}{\partial z}\right)^2\right] \tag{6-27c}$$

Note that the second terms on the right-hand side of Eqs. (6-27) are the second-order correction terms to the normal strains.

The shear strain γ_{xy} is found from $\cos \angle B'Q'D'$, since

$$\cos \angle B'Q'D' = \cos\left(\frac{\pi}{2} - \gamma_{xy}\right) = \sin \gamma_{xy} \approx \gamma_{xy}$$

The cosine is best found through the use of vectors, where the scalar product of the vectors $\overline{B'Q'}$ and $\overline{Q'D'}$ is divided by the product of their magnitudes. This results in

$$\gamma_{xy} \approx \cos \angle B'Q'D'$$

$$= \frac{\left(1+\dfrac{\partial u}{\partial x}\right)\dfrac{\partial u}{\partial y} + \dfrac{\partial v}{\partial x}\left(1+\dfrac{\partial v}{\partial y}\right) + \dfrac{\partial w}{\partial x}\dfrac{\partial w}{\partial y}}{\left[\left(1+\dfrac{\partial u}{\partial x}\right)^2 + \left(\dfrac{\partial v}{\partial x}\right)^2 + \left(\dfrac{\partial w}{\partial x}\right)^2\right]^{1/2}\left[\left(\dfrac{\partial u}{\partial y}\right)^2 + \left(1+\dfrac{\partial v}{\partial y}\right)^2 + \left(\dfrac{\partial w}{\partial y}\right)^2\right]^{1/2}}$$

The terms in the denominator are expanded using the binomial expansion theorem. Only the first-order terms in strain are retained since these terms are multiplied with the numerator, which already carries terms in strain no lower than the first order. This results in

$$\gamma_{xy} \approx \left[\left(1+\frac{\partial u}{\partial x}\right)\frac{\partial u}{\partial y} + \frac{\partial v}{\partial x}\left(1+\frac{\partial v}{\partial y}\right) + \frac{\partial w}{\partial x}\frac{\partial w}{\partial y}\right]\left(1-\frac{\partial u}{\partial x}\right)\left(1-\frac{\partial v}{\partial y}\right)$$

Multiplying the terms out and neglecting terms higher than second order yields

$$\gamma_{xy} = \frac{\partial v}{\partial x} + \frac{\partial u}{\partial y} - \frac{\partial u}{\partial x}\frac{\partial u}{\partial y} - \frac{\partial v}{\partial x}\frac{\partial v}{\partial y} + \frac{\partial w}{\partial x}\frac{\partial w}{\partial y} \tag{6-28a}$$

In a similar manner, the remaining shear strains are found to be

$$\gamma_{yz} = \frac{\partial w}{\partial y} + \frac{\partial v}{\partial z} + \frac{\partial u}{\partial y}\frac{\partial u}{\partial z} - \frac{\partial v}{\partial y}\frac{\partial v}{\partial z} - \frac{\partial w}{\partial y}\frac{\partial w}{\partial z} \tag{6-28b}$$

$$\gamma_{zx} = \frac{\partial u}{\partial z} + \frac{\partial w}{\partial x} - \frac{\partial u}{\partial z}\frac{\partial u}{\partial x} + \frac{\partial v}{\partial z}\frac{\partial v}{\partial x} - \frac{\partial w}{\partial z}\frac{\partial w}{\partial x} \tag{6-28c}$$

The majority of the references dealing with elasticity express the large strain-deflection equations somewhat differently because the strain is defined so that no approximations are necessary. The definition of strain, however, is not the same as that normally used in engineering. For the sake of illustration only, consider the normal strain in the x direction. In the theory of elasticity the large strain is defined in terms of the square of the element length instead of the length itself. Thus, the normal strain in the x direction is defined as E_x, where

$$E_x = \frac{(Q'D')^2 - QD^2}{2QD^2} \tag{6-29}$$

Since

$$(Q'D')^2 = \left[\left(1 + \frac{\partial u}{\partial x}\right)^2 + \left(\frac{\partial v}{\partial x}\right)^2 + \left(\frac{\partial w}{\partial x}\right)^2\right](\Delta x)^2$$

we have

$$E_x = \frac{\left[1 + 2\frac{\partial u}{\partial x} + \left(\frac{\partial u}{\partial x}\right)^2 + \left(\frac{\partial v}{\partial x}\right)^2 + \left(\frac{\partial w}{\partial x}\right)^2\right](\Delta x)^2 - (\Delta x)^2}{2(\Delta x)^2}$$

which reduces to

$$E_x = \frac{\partial u}{\partial x} + \frac{1}{2}\left[\left(\frac{\partial u}{\partial x}\right)^2 + \left(\frac{\partial v}{\partial x}\right)^2 + \left(\frac{\partial w}{\partial x}\right)^2\right] \tag{6-30}$$

Equation (6-30) is similar to Eq. (6-27a) except for the additional $\frac{1}{2}(\partial u/\partial x)^2$ term. For small strains, where second-order terms are neglected, $E_x = \varepsilon_x$.

6-5 COMPATIBILITY

When one is seeking a solution to the stress distribution in a body, the dynamic state of the body must be satisfied. For example, if the body is in equilibrium, any segment of the body together with its corresponding internal-force distribution must maintain the segment in static equilibrium. At any given section it is possible to find many stress distributions which will ensure equilibrium. An *acceptable* stress distribution is one which ensures a piecewise-continuous-deformation distribution of the body. This is the essential characteristic of *compatibility*; i.e., the stress distribution and the resulting deflection distribution must be compatible with boundary conditions and a continuous distribution of deformations so that no "holes" or overlapping of specific points in the body occur. Recall the simple stress distributions given in Chap. 2. For axial loading, torsion, and bending, the stress distributions are initially based on intuitive reasoning of the strain or deflection fields. For example, consider the axial loading shown in Fig. 6-8. For an isolation made at section c-c, Fig. 6-8b shows a uniform stress distribution, whereas Fig. 6-8c shows a parabolic distribution. Both distributions will satisfy equilibrium provided $\int \sigma_x\, dA = P$. As a matter of fact, any stress distribution symmetric with respect to the centroidal axis can ensure equilibrium. Then why is the

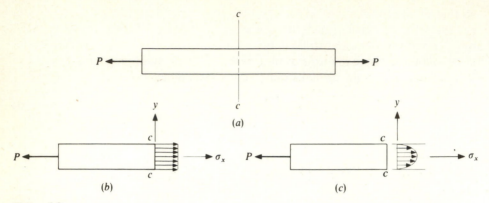

Figure 6-8

uniform distribution the correct one? This is where the concept of compatibility becomes important. The uniform stress distribution of Fig. 6-8a not only ensures equilibrium (provided $\sigma_x = P/A$), it is also compatible with a continuous strain and displacement field consistent with the boundary conditions of the axially loaded member.

If a stress field is assumed, and if a continuous displacement field consistent with the boundary conditions can be found, the stress field is acceptable. The case of axial loading is left as an exercise at the end of the chapter. Consider instead Example 1-4. Recall that the displacement field was given in this example, from which the strain and stress fields were determined. However, in a problem like this, one normally begins with the standard stress formulations which satisfy equilibrium. If a continuous displacement field can then be found which satisfies the geometric boundary conditions of the problem, the stress field is compatible.

Example 6-6 For the beam shown in Fig. 6-9, determine the displacement field due to bending only. Consider the cross section of the beam to be rectangular and thin so that deflections are not functions of z. Check the solution against that given in Example 1-4. The stiffness of the beam is EI_z, and Poisson's ratio is ν.

SOLUTION Since only the bending stresses are being considered, only the bending moment as a function of x is necessary. This is

$$M_z = -P(L - x) \qquad 0 < x < L \tag{a}$$

Thus the stress field is

$$\sigma_x = -\frac{M_z y}{I_z} = \frac{P}{I_z} y(L - x) \tag{b}$$

$$\sigma_y = 0 \tag{c}$$

$$\tau_{xy} = 0 \tag{d}$$

From Hooke's law, the strain field is

$$\varepsilon_x = \frac{1}{E}(\sigma_x - \nu\sigma_y) = \frac{P}{EI_z} y(L - x) \tag{e}$$

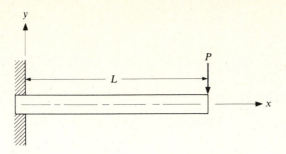

Figure 6-9

$$\varepsilon_y = \frac{1}{E}(\sigma_y - \nu\sigma_x) = -\frac{\nu P}{EI_z} y(L-x) \tag{f}$$

and
$$\gamma_{xy} = \frac{2(1+\nu)}{E} \tau_{xy} = 0 \tag{g}$$

Next the normal strains ε_x and ε_y are substituted into the strain-displacement equations, resulting in

$$\frac{\partial u}{\partial x} = \varepsilon_x = \frac{P}{EI_z} y(L-x) \tag{h}$$

and
$$\frac{\partial v}{\partial y} = \varepsilon_y = -\frac{\nu P}{EI_z} y(L-x) \tag{i}$$

Equations (h) and (i) can be integrated, but it must be kept in mind that since these equations contain partial derivatives, appropriate *functions* of integration must be added. Thus

$$u = \frac{P}{2EI_z} xy(2L-x) + f(y) \tag{j}$$

$$v = -\frac{\nu P}{2EI_z} y^2(L-x) + g(x) \tag{k}$$

The functions of integration, $f(y)$ and $g(x)$, are pure functions of y and x, respectively, where both functions possibly contain constants. The next step is where the concept of compatibility is used. If Eqs. (j) and (k) are acceptable, they can be substituted into the shear strain-displacement equation, $\gamma_{xy} = \partial v/\partial x + \partial u/\partial y$, and if Eq. ($g$) can be satisfied, a compatible solution will emerge. As will be seen, Eq. (g) can be satisfied only if $f(y)$ and $g(x)$ can be separated from the equation. From Eqs. (j) and (k)

$$\gamma_{xy} = \frac{\partial v}{\partial x} + \frac{\partial u}{\partial y} = \frac{\nu P}{2EI_z} y^2 + \frac{\partial g(x)}{\partial x} + \frac{P}{2EI_z} x(2L-x) + \frac{\partial f(y)}{\partial y}$$

From Eq. (g), the above equation is equated to zero. Rearranging the resulting equation yields

$$\frac{\partial g(x)}{\partial x} + \frac{P}{2EI_z} x(2L-x) = -\frac{\partial f(y)}{\partial y} - \frac{\nu P}{2EI_z} y^2 \tag{l}$$

Since the beam is not a line in space, a pure function of x [the left-hand side of Eq. (l)] can equal a pure function of y [the right-hand side of Eq. (l)] for a continuous two-dimensional body only if both sides of Eq. (l) equal a constant, say C_1. Setting both sides of Eq. (l) equal to C_1 separates the functions $f(y)$ and $g(x)$, and a compatible solution will result. Thus†

$$\frac{dg(x)}{dx} + \frac{P}{2EI_z} x(2L-x) = C_1 \tag{m}$$

† Once the equations are separated, each equation is written in terms of one independent variable only. This means that the partial derivative can be replaced by an ordinary derivative.

$$\frac{df(y)}{dy} + \frac{\nu P y^2}{2EI_z} = -C_1 \qquad (n)$$

Solving Eqs. (m) and (n) results in

$$g(x) = -\frac{P}{6EI_z} x^2(3L - x) + C_1 x + C_2 \qquad (o)$$

$$f(y) = -\frac{\nu P y^3}{6EI_z} - C_1 y + C_3 \qquad (p)$$

where C_2 and C_3 are constants of integration. Substituting Eqs. (o) and (p) into Eqs. (j) and (k) gives

$$u = \frac{P}{2EI_z} xy(2L - x) - \frac{\nu P}{6EI_z} y^3 - C_1 y + C_3 \qquad (q)$$

$$v = -\frac{\nu P}{2EI_z} y^2(L - x) - \frac{P}{6EI_z} x^2(3L - x) + C_1 x + C_2 \qquad (r)$$

The constants C_1, C_2, and C_3 are evaluated using the geometric boundary conditions of the particular problem. In this example, there is an infinite set of boundary conditions since the beam cross section is completely fixed at the wall at $x = 0$. Thus, all conditions except three must be relaxed, and some error at the wall must be tolerated. This is acceptable since even the stress distribution [Eq. (b)] is in error there. Probably the most accepted conditions are that of $u = v = \theta_{xy} = 0$ at $x = y = 0$. [Recall from Chap. 1, that θ_{xy} is the rotation given by $\theta_{xy} = \frac{1}{2}(\partial v/\partial x - \partial u/\partial y)$.] Substituting these conditions into Eqs. (q) and (r) yields $C_1 = C_2 = C_3 = 0$. Thus the final results for the complete deflection field are

$$u = \frac{P}{2EI_z} xy(2L - x) - \frac{\nu P}{6EI_z} y^3 \qquad (s)$$

and

$$v = -\frac{\nu P}{2EI_z} y^2(L - x) - \frac{P}{6EI_z} x^2(3L - x) \qquad (t)$$

Referring back to Example 1-4, we can see that Eqs. (s) and (t) agree with what was given there.

Observations on Example 6-6

1. It can be seen from Eqs. (s) and (t) that the left surface of the beam is not rigidly held for $x = 0$ and $y \neq 0$, as there are nonzero values for u and v. This is due to the fact that the actual stress distribution at the wall does not behave like that originally given in the problem if the entire left surface is clamped. However, the actual correction would be quite small.

2. The lateral deflection of the neutral axis $v(x, 0)$ is normally discussed in a basic strength of materials course using a purely geometric approach, where for an end-loaded cantilever beam the deflection is given as (see Appendix D)

$$y_c = v(x, 0) = -\frac{Px^2}{6EI}(3L - x)$$

The same result is obtained from Eq. (t).

3. The slope or angle that the deflected neutral axis has with respect to the horizontal is given by $\theta_{xy}(x, 0)$. From Eqs. (s) and (t) the rotation of the

neutral axis is

$$\frac{dy_c}{dx} = \theta_{xy}(x, 0) = \frac{1}{2}\left(\frac{\partial v}{\partial x} - \frac{\partial u}{\partial y}\right)\bigg|_{y=0} = -\frac{Px}{2EI}(2L - x)$$

which again agrees with the results obtained from simple deflection theory (see Appendix D).

4. A surface of the beam cross section for any particular value of x does not remain planar. The surfaces warp, as can be seen in Eqs. (s) and (t). If the surfaces remained planar, u would be a linear function of y and v would not be a function of y.

Sometimes stress distributions are used in the solution of a given problem where the stress distributions are not compatible. In these cases, the complete deflection field will not be solvable, and depending on the severity of the noncompatibility, the accuracy of the stress distribution may be questionable.

Example 6-7 In Example 6-5 the equilibrium equations are used on a uniformly loaded cantilever beam to find the τ_{xy} and σ_y stress distributions using the conventional strength of material equation for σ_x. Each one of these stress distributions ensures equilibrium, but do they provide a compatible displacement field? Using the method in Example 6-6, determine whether or not a compatible displacement field can be found for the beam shown in Fig. 6-6; use the following stress fields:

$$\sigma_x = \frac{w}{2I_z}x^2y \tag{a}$$

$$\sigma_y = -\frac{w}{6I_z}(2c^3 + 3c^2y - y^3) \tag{b}$$

$$\tau_{xy} = \frac{w}{2I_z}x(c^2 - y^2) \tag{c}$$

SOLUTION As in Example 6-6, the stresses are substituted into the strain-stress equations,

$$\varepsilon_x = \frac{1}{E}(\sigma_x - \nu\sigma_y) = \frac{w}{6EI_z}[3x^2y + \nu(2c^3 + 3c^2y - y^3)] \tag{d}$$

$$\varepsilon_y = \frac{1}{E}(\sigma_y - \nu\sigma_x) = -\frac{w}{6EI_z}(2c^3 + 3c^2y - y^3 + 3\nu x^2y) \tag{e}$$

$$\gamma_{xy} = \frac{2(1+\nu)}{E}\tau_{xy} = \frac{(1+\nu)w}{EI_z}x(c^2 - y^2) \tag{f}$$

Since $\partial u/\partial x = \varepsilon_x$ and $\partial v/\partial y = \varepsilon_y$, u and v are found by integrating Eqs. (d) and (e), resulting in

$$u = \frac{w}{6EI_z}x[x^2y + \nu(2c^3 + 3c^2y - y^3)] + f(y) \tag{g}$$

$$v = -\frac{w}{6EI_z}y(2c^3 + \frac{3}{2}c^2y - \frac{1}{4}y^3 + \frac{3}{2}\nu x^2y) + g(x) \tag{h}$$

Next, Eqs. (g) and (h) are substituted into $\gamma_{xy} = \partial v/\partial x + \partial u/\partial y$ and equated to Eq. (f). After simplification this yields

$$\frac{\partial f(y)}{\partial y} + \frac{\partial g(x)}{\partial x} + \frac{w}{6EI_z}[x^3 - 3c^2(2 + \nu)x] + \frac{w}{EI_z}xy^2 = 0 \tag{i}$$

The last term in Eq. (i) contains an xy^2 term which is a pure function of neither x nor y. Thus, Eq. (i) is an invalid equation, and the stress field given by Eqs. (a) to (c) is *not* compatible.

The stress distribution given by Eqs. (a) to (c) of Example 6-8 are fairly accurate, but they are not capable of providing an acceptable displacement field. A correction can be made to the stress distribution so that a displacement field can be determined. One technique is simply to modify the stress equations so as to cancel any undesirable terms.

Example 6-8 Determine the simplest correction which will cancel the unwanted xy^2 term obtained in Example 6-7.

SOLUTION If in Example 6-7, σ_x in Eq. (a) and consequently ε_x contained a y^3 term, u would contain an xy^3 term. This would then provide an xy^2 term in $\partial u/\partial y$. The addition of y^3 to σ_x would yield no additional term to $\partial v/\partial x$. Thus as a trial, let

$$\sigma_x = F(x)y + K_1 y^3 \qquad (j)$$

where K_1 is a constant and $F(x)$ is an unknown function of x.† The moment induced by this stress is M_z. Thus

$$\int_{-c}^{+c} y\sigma_x (b\, dy) = -M_z = -\frac{w}{2}x^2$$

Substition of Eq. (j) and integration yield

$$2b\left[F(x)\frac{c^3}{3} + K_1 \frac{c^5}{5}\right] = \frac{w}{2}x^2$$

or

$$\tfrac{2}{3}bc^3[F(x) + \tfrac{3}{5}K_1 c^2] = \frac{w}{2}x^2$$

The moment of inertia is $I_z = \tfrac{2}{3}bc^3$. Thus

$$F(x) + \tfrac{3}{5}K_1 c^2 = \frac{w}{2I_z}x^2 \qquad (k)$$

The additional $K_1 y^3$ term in σ_x creates an additional $K_1 y^3/E$ in ε_x and $K_1 xy^3/E$ in u. Thus, the additional term in $\partial u/\partial y$ is $3K_1 xy^2/E$. Equating this to $-(w/EI)xy^2$ will eliminate the unwanted xy^2 term in Eq. (i) of Example 6-7. Thus, K_1 is obtained, where $K_1 = -(w/3I_z)$. Substituting this into Eq. (k) yields

$$F(x) = \frac{wx^2}{2I_z} + \frac{wc^2}{5I_z} = \frac{w}{10I_z}(5x^2 + 2c^2)$$

The modified stress distribution of Eq. (j) becomes

$$\sigma_x = \frac{w}{10I_z}(5x^2 + 2c^2)y - \frac{w}{3I_z}y^3 \qquad (l)$$

Thus the stress distribution given by the above equation and Eqs. (b) and (c) of Example 6-7 will provide a solution to the displacement field of the beam. One point to note, however, is that throughout the beam the stress distribution gives equilibrium but at $x = 0$ there is a distribution of the normal stress σ_x. This means that the distribution is still basically incorrect. However, the error is small. Further improvement can be made by assuming series solutions for the stresses. However, these improvements would prove negligible.

---

† This is fortunate. The additional term contains no function of x. Thus, Eqs. (b) and (c) for σ_y and τ_{xy} will be unaffected by the additional term, and equilibrium is still ensured. Also, the additional y^3 term in σ_x gives no net axial force across the section.

Considering the xy plane only, we can define compatibility rigorously as follows. Recall that the strain-displacement equations are given by

$$\varepsilon_x = \frac{\partial u}{\partial x} \qquad \varepsilon_y = \frac{\partial v}{\partial y} \qquad \gamma_{xy} = \frac{\partial v}{\partial x} + \frac{\partial u}{\partial y}$$

Differentiating γ_{xy} with respect to x and then with respect to y yields

$$\frac{\partial^2 \gamma_{xy}}{\partial x \, \partial y} = \frac{\partial^3 u}{\partial x \, \partial y^2} + \frac{\partial^3 v}{\partial x^2 \, \partial y}$$

Noting that

$$\frac{\partial^3 u}{\partial x \, \partial y^2} = \frac{\partial^2 \varepsilon_x}{\partial y^2} \qquad \text{and} \qquad \frac{\partial^3 v}{\partial x^2 \, \partial y} = \frac{\partial^2 \varepsilon_y}{\partial x^2}$$

we have

$$\frac{\partial^2 \gamma_{xy}}{\partial x \, \partial y} = \frac{\partial^2 \varepsilon_x}{\partial y^2} + \frac{\partial^2 \varepsilon_y}{\partial x^2} \tag{6-31}$$

Equation (6-31), called the *compatibility equation* neglecting z dependence, provides a check on whether a given strain field is compatible in the xy plane.

Example 6-9 Show that the strains given by Eqs. (d) to (f) of Example 6-7 do not satisfy the compatibility equation.

SOLUTION Differentiation of the strains given in Example 6-7 results in

$$\frac{\partial^2 \gamma_{xy}}{\partial x \, \partial y} = -\frac{2w(1+\nu)}{EI} y$$

$$\frac{\partial^2 \varepsilon_x}{\partial y^2} = -\frac{w\nu}{EI} y$$

$$\frac{\partial^2 \varepsilon_y}{\partial x^2} = -\frac{w\nu}{EI} y$$

Substituting these into Eq. (6-31) yields

$$\frac{-2w(1+\nu)}{EI} y = -\frac{w\nu}{EI} y - \frac{w\nu}{EI} y$$

Simplifying results in

$$(1+\nu)y = \nu y$$

Thus compatibility is valid only at $y = 0$, that is, along the centroidal axis. However, as was found in Example 6-7, this strain field is *not compatible* in general. Note that since the solution is compatible along the centroidal axis, the solution for deflection using the elementary strength of materials approach is accurate.

There are six compatibility equations, given by

$$\frac{\partial^2 \gamma_{xy}}{\partial x \, \partial y} = \frac{\partial^2 \varepsilon_x}{\partial y^2} + \frac{\partial^2 \varepsilon_y}{\partial x^2} \tag{6-32a}$$

$$\frac{\partial^2 \gamma_{yz}}{\partial y \, \partial z} = \frac{\partial^2 \varepsilon_y}{\partial z^2} + \frac{\partial^2 \varepsilon_z}{\partial y^2} \tag{6-32b}$$

$$\frac{\partial^2 \gamma_{zx}}{\partial z\, \partial x} = \frac{\partial^2 \varepsilon_z}{\partial x^2} + \frac{\partial^2 \varepsilon_x}{\partial z^2} \tag{6-32c}$$

$$2\frac{\partial^2 \varepsilon_x}{\partial y\, \partial z} = \frac{\partial}{\partial x}\left(-\frac{\partial \gamma_{yz}}{\partial x} + \frac{\partial \gamma_{zx}}{\partial y} + \frac{\partial \gamma_{xy}}{\partial z} \right) \tag{6-32d}$$

$$2\frac{\partial^2 \varepsilon_y}{\partial z\, \partial x} = \frac{\partial}{\partial y}\left(\frac{\partial \gamma_{yz}}{\partial x} - \frac{\partial \gamma_{zx}}{\partial y} + \frac{\partial \gamma_{xy}}{\partial z} \right) \tag{6-32e}$$

$$2\frac{\partial^2 \varepsilon_z}{\partial x\, \partial y} = \frac{\partial}{\partial z}\left(\frac{\partial \gamma_{yz}}{\partial x} + \frac{\partial \gamma_{zx}}{\partial y} - \frac{\partial \gamma_{xy}}{\partial z} \right) \tag{6-32f}$$

The first three equations are the in-plane dependence, like Eq. (6-31). The last three equations are the out-of-plane dependence and are left for the reader to verify as an exercise at the end of the chapter.

6-6 PLANE ELASTIC PROBLEMS

In the theory of elasticity there is a special class of problems known as plane elastic problems which can be solved fairly readily. To be classified as a plane elastic problem, it must have the following characteristics.

Geometry (see Fig. 6-10) A plane body consists of a region of uniform thickness (say in the z direction) bounded by two parallel planes and by any closed surface. If the thickness t is very large, the problem is considered to be a plane-strain problem, and if the thickness is small compared with the lateral dimensions, the problem is considered to be a plane-stress problem.

Loading Body forces, if they exist, cannot vary in the direction of the body thickness and cannot have components in the z direction. The applied surface loads cannot have components in the z direction and must be uniformly distributed across the thickness, i.e., constant in the z direction; no loads can be applied on the parallel planes bounding the top and bottom surfaces.

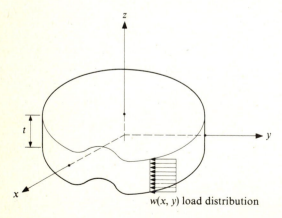

$w(x, y)$ load distribution   **Figure 6-10** Plane elastic problem.

The assumption for plane-stress problems is that since the stresses are zero on the parallel planes, and since the body is very thin, the stresses in the z direction will not grow to any appreciable values. Thus, for plane stress $\sigma_z = \tau_{yz} = \tau_{zx} = 0$.

For plane-strain problems where the model is very thick in the z direction, the assumption is that the strains in the z direction are zero. That is, $\varepsilon_z = \gamma_{yz} = \gamma_{zx} = 0$. Also the strains in the x and y directions are not functions of z.

A simple example which illustrates the difference between plane-stress and plane-strain problems is that of narrow-beam theory vs. wide-beam theory. If a rectangular beam is narrow, the problem is a plane-stress problem. If a rectangular beam is very wide, the problem is a plane-strain problem.

6-7 THE AIRY STRESS FUNCTION

For plane elastic problems, a function of x and y can be developed which automatically satisfies the equilibrium equations and compatibility. In the plane-stress case where $\sigma_z = \tau_{yz} = \tau_{zx} = 0$, the strain-stress equations are

$$\varepsilon_x = \frac{1}{E}(\sigma_x - \nu\sigma_y) \qquad \varepsilon_y = \frac{1}{E}(\sigma_y - \nu\sigma_x) \qquad \gamma_{xy} = \frac{2(1+\nu)}{E}\tau_{xy} \qquad (6\text{-}33)$$

Substituting Eqs. (6-33) into the compatibility equation (6-32a) yields

$$\frac{\partial^2}{\partial y^2}(\sigma_x - \nu\sigma_y) + \frac{\partial^2}{\partial x^2}(\sigma_y - \nu\sigma_x) = 2(1+\nu)\frac{\partial^2\tau_{xy}}{\partial x\,\partial y} \qquad (6\text{-}34)$$

The equilibrium equations in two dimensions are

$$\frac{\partial\sigma_x}{\partial x} + \frac{\partial\tau_{xy}}{\partial y} + \bar{F}_x = 0 \qquad \text{and} \qquad \frac{\partial\tau_{xy}}{\partial x} + \frac{\partial\sigma_y}{\partial y} + \bar{F}_y = 0$$

or
$$\frac{\partial\tau_{xy}}{\partial y} = -\frac{\partial\sigma_x}{\partial x} - \bar{F}_x \qquad \text{and} \qquad \frac{\partial\tau_{xy}}{\partial x} = -\frac{\partial\sigma_y}{\partial y} - \bar{F}_y$$

Differentiating the first equation by x and the second by y yields

$$\frac{\partial^2\tau_{xy}}{\partial y\,\partial x} = -\frac{\partial^2\sigma_x}{\partial x^2} - \frac{\partial\bar{F}_x}{\partial x} \qquad (6\text{-}35a)$$

$$\frac{\partial^2\tau_{xy}}{\partial x\,\partial y} = -\frac{\partial^2\sigma_y}{\partial y^2} - \frac{\partial\bar{F}_y}{\partial y} \qquad (6\text{-}35b)$$

Assuming τ_{xy} to be continuous in x and y, we have

$$\frac{\partial^2\tau_{xy}}{\partial y\,\partial x} = \frac{\partial^2\tau_{xy}}{\partial x\,\partial y}$$

Adding Eqs. (6-35a) and (6-35b) yields

$$\frac{\partial^2 \tau_{xy}}{\partial x\,\partial y} = -\frac{1}{2}\left(\frac{\partial^2 \sigma_x}{\partial x^2} + \frac{\partial^2 \sigma_y}{\partial y^2} + \frac{\partial \bar{F}_x}{\partial x} + \frac{\partial \bar{F}_y}{\partial y}\right)$$

Substituting this into Eq. (6-34) results in

$$\frac{\partial^2}{\partial y^2}(\sigma_x - \nu\sigma_y) + \frac{\partial^2}{\partial x^2}(\sigma_y - \nu\sigma_x) = -(1+\nu)\left(\frac{\partial^2 \sigma_x}{\partial x^2} + \frac{\partial^2 \sigma_y}{\partial y^2} + \frac{\partial \bar{F}_x}{\partial x} + \frac{\partial \bar{F}_y}{\partial y}\right)$$

Simplification yields

$$\frac{\partial^2}{\partial x^2}(\sigma_x + \sigma_y) + \frac{\partial^2}{\partial y^2}(\sigma_x + \sigma_y) = -(1+\nu)\left(\frac{\partial \bar{F}_x}{\partial x} + \frac{\partial \bar{F}_y}{\partial y}\right) \tag{6-36}$$

In operational notation, $\partial^2/\partial x^2 + \partial^2/\partial y^2 = \nabla^2$. Thus, the left-hand side of Eq. (6-36) is $\nabla^2(\sigma_x + \sigma_y)$, and the equation can be rewritten

$$\nabla^2(\sigma_x + \sigma_y) = -(1+\nu)\left(\frac{\partial \bar{F}_x}{\partial x} + \frac{\partial \bar{F}_y}{\partial y}\right) \tag{6-37}$$

For problems in which the body forces are zero, Eq. (6-37) reduces to

$$\nabla^2(\sigma_x + \sigma_y) = 0 \tag{6-38}$$

Assume that there exists a function $\phi(x, y)$ of x and y such that

$$\sigma_x = \frac{\partial^2 \phi}{\partial y^2} \tag{6-39a}$$

$$\sigma_y = \frac{\partial^2 \phi}{\partial x^2} \tag{6-39b}$$

$$\tau_{xy} = -\frac{\partial^2 \phi}{\partial x\,\partial y} \tag{6-39c}$$

Substituting Eqs. (6-39) into the equilibrium equations with $\bar{F}_x = \bar{F}_y = 0$ demonstrates that equilibrium is satisfied. For compatibility, substitution of Eqs. (6-39) into Eq. (6-38) yields

$$\frac{\partial^4 \phi}{\partial x^4} + 2\frac{\partial^2 \phi}{\partial x^2\,\partial y^2} + \frac{\partial^4 \phi}{\partial y^4} = 0$$

Using the notation that $(\partial^2/\partial x^2 + \partial^2/\partial y^2)^2 = \nabla^4$, we can rewrite the above equation as

$$\nabla^4 \phi = 0 \tag{6-40}$$

This is called the *biharmonic equation*, and the function ϕ is referred to as the *Airy stress function*.

The advantage of the stress function is that if polynomial series are used for ϕ, then from Eq. (6-39) the resulting stresses will automatically satisfy equilibrium; if Eq. (6-40) is satisfied, compatibility is ensured and it is only necessary to look at the boundary conditions for a complete solution. The disadvantage of the stress function, however, is that trials of the function must be made to determine whether it will provide an appropriate stress field which satisfies the boundary conditions. After some experience, the analyst will begin to understand the types of stress fields which arise from the various functions.

Example 6-10 Determine the stress fields that arise from the following stress functions:

(a) $$\phi = Cy^2$$

(b) $$\phi = Ax^2 + Bxy + Cy^2$$

(c) $$\phi = Ax^3 + Bx^2y + Cxy^2 + Dy^3$$

where A, B, C, and D are constants.

SOLUTION (a)
$$\sigma_x = \frac{\partial^2 \phi}{\partial y^2} = 2C$$

$$\sigma_y = \frac{\partial^2 \phi}{\partial x^2} = 0$$

$$\tau_{xy} = -\frac{\partial^2 \phi}{\partial x\, dy} = 0$$

This function is suitable for a bar or plate in a uniform tensile or compressive state of stress.

(b)
$$\sigma_x = \frac{\partial^2 \phi}{\partial y^2} = 2C$$

$$\sigma_y = \frac{\partial^2 \phi}{\partial x^2} = 2A$$

$$\tau_{xy} = \frac{-\partial^2 \phi}{\partial x\, \partial y} = -B$$

This function is suitable for a general uniform stress field over the entire body.

(c)
$$\sigma_x = \frac{\partial^2 \phi}{\partial y^2} = 2Cx + 6Dy$$

$$\sigma_y = \frac{\partial^2 \phi}{\partial x^2} = 6Ax + 2By$$

$$\tau_{xy} = \frac{-\partial^2 \phi}{\partial x\, \partial y} = -(2Bx + 2Cy)$$

All stresses vary linearly with respect to x and y. Note that if $A = B = C = 0$, the state of stress would correspond to a beam in pure bending. Thus, the function $\phi = Dy^3$ can be used for bending. Also, note that compatibility for the above cases is automatically satisfied since $\nabla^4 \phi = 0$.

Stress functions with constants and linear terms of x and y will give zero stresses, so that only functions of xy, x^2, y^2, and higher are usable. To see how the Airy stress function is generated and used we return to the familiar problem of the uniformly loaded cantilever beam of Example 6-5.

Example 6-11 Determine a stress function for the stress field of Fig. 6-6 and evaluate the stress field.

SOLUTION Constructing the free-body diagram and evaluating the reactions results in Fig. 6-11. The conditions of stress at the boundaries are:

At $x = 0$: $$\sigma_x = 0 \tag{a}$$

$$\tau_{xy} = 0 \tag{b}$$

At $x = L$: $$\int_{-c}^{c} \tau_{xy} b \, dy = wL \tag{c}$$

$$\int_{-c}^{c} \sigma_x b \, dy = 0 \tag{d}$$

$$\int_{-c}^{c} \sigma_x yb \, dy = \tfrac{1}{2} wL^2 \tag{e}$$

At $y = c$: $$\sigma_y = -\frac{w}{b} \tag{f}$$

$$\tau_{xy} = 0 \tag{g}$$

At $y = -c$ $$\sigma_y = 0 \tag{h}$$

$$\tau_{xy} = 0 \tag{i}$$

It is impossible to satisfy condition (a) with a simple stress function. Recall Example 6-8, where it was found that the stress distribution was not zero at the free end. Thus condition (a) will be ignored.†

Some observations should be made before attempting to establish the stress function. Since the cross section is symmetric about the y axis and $\sigma_y = -w/b$ at $y = c$ and $\sigma_y = 0$ at $y = -c$, σ_y should be an odd function of y. Thus, the stress function should contain odd functions of y. This is substantiated by the fact that σ_x should also contain odd functions of y since the net axial force is zero [this automatically satisfies condition (d)]. Second, since σ_y is constant as a function of x on the top and bottom faces, the stress function should not contain powers of x greater than x^2. Thus, the following stress function will be tried:

$$\phi = Axy + Bx^2 + Cx^2y + Dy^3 + Exy^3 + Fx^2y^3 + Gy^5 \tag{j}‡$$

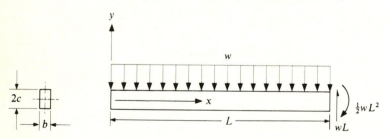

Figure 6-11

† However, at $x = 0$, the conditions that $\int_{-c}^{c} \sigma_x b \, dx = 0$ and $\int_{-c}^{c} \sigma_x yb \, dy = 0$ must hold for equilibrium.

‡ Note that with the omission of condition (a), there are eight conditions and one equation remaining, Eq. (6-40). Thus it would seem that nine constants are in order. However, by the selection of odd functions of y, condition (d) is automatically satisfied. In addition, τ_{xy} will be found to contain even functions of y, making τ_{xy} symmetric with respect to y. Thus, only one of conditions (g) or (i) can be used. This explains the selection of only seven unknown constants.

Since $\nabla^4\phi = 0$,

$$\nabla^4\phi = 24Fy + 120Gy = 0$$

Thus
$$F = -5G$$

Substituting this into Eq. (j) and evaluating the stress field using Eqs. (6-39) results in

$$\sigma_x = \frac{\partial^2\phi}{\partial y^2} = 6Dy + 6Exy - 30Gx^2y + 20Gy^3 \tag{k}$$

$$\sigma_y = \frac{\partial^2\phi}{\partial x^2} = 2B + 2Cy - 10Gy^3 \tag{l}$$

$$\tau_{xy} = -\frac{\partial^2\phi}{\partial x\,\partial y} = -(A + 2Cx + 3Ey^2 - 30Gxy^2) \tag{m}$$

Dealing with τ_{xy} first, we see that condition (b) at $x = 0$ requires that

$$A + 3Ey^2 = 0$$

The only way this can be true for all values of y is

$$A = E = 0$$

When either one of conditions (g) or (i) is used at $y = \pm c$, Eq. (m) becomes

$$0 = -(2Cx - 30Gc^2x)$$

or
$$C = 15Gc^2 \tag{n}$$

Substituting this into Eq. (l) yields

$$\sigma_y = 2B + 30Gc^2y - 10Gy^3 \tag{o}$$

From conditions (f) and (h) at $y = c$ and $y = -c$, respectively, Eq. (o) becomes

$$-\frac{w}{b} = 2B + 30Gc^3 - 10Gc^3 = 2B + 20Gc^3$$

and
$$0 = 2B - 30Gc^3 + 10Gc^3 = 2B - 20Gc^3$$

Solving the two simultaneous equations yields

$$B = -\frac{w}{4b} \qquad G = -\frac{w}{40bc^3}$$

The area moment of inertia is $I_z = \frac{2}{3}bc^3$. Thus, the above terms can be rewritten as

$$B = \frac{wc^3}{6I_z} \qquad G = -\frac{w}{60I_z}$$

Then, from Eq. (n)

$$C = 15Gc^2 = -\frac{wc^2}{4I_z}$$

Substituting E and G into Eq. (k) results in

$$\sigma_x = 6Dy + \frac{w}{2I_z}x^2y - \frac{w}{3I_z}y^3 \tag{p}$$

From condition (e) at $x = L$, Eq. (p) becomes

$$\int_{-c}^{c}\left(6Dy + \frac{w}{2I_z}L^2y - \frac{w}{3I_z}y^3\right)yb\,dx = \tfrac{1}{2}wL^2$$

Solving for D yields

$$D = \frac{wc^2}{30I_z}$$

Substituting the constants back into the equations for the stresses gives

$$\sigma_x = \frac{w}{10I_z}(5x^2 + 2c^2)y - \frac{w}{3I_z}y^3 \tag{q}$$

$$\sigma_y = -\frac{w}{6I_z}(2c^3 + 3c^2y - y^3) \tag{r}$$

$$\tau_{xy} = \frac{w}{2I_z}x(c^2 - y^2) \tag{s}$$

Comparing these results with Eq. (*l*) of Example 6-8 and Eqs. (*b*) and (*c*) of Example 6-7 shows exact agreement.

Conversion between Plane-Strain and Plane-Stress Problems

If a solution is obtained for a plane-stress problem where the object under analysis is thin in the z direction, there is a straightforward method for converting this solution into the same problem where the object is very wide in the z direction. That is, a conversion of a plane-stress solution into a plane-strain solution. Likewise, if a solution is obtained based on the plane-strain assumption, the solution can be changed to apply to the identical problem (but based on the plane-stress assumption).

For the plane-stress assumption, $\sigma_z = \tau_{yz} = \tau_{xz} = 0$, and Hooke's law for the normal strains ε_x and ε_y is

$$\varepsilon_x = \frac{1}{E}(\sigma_x - \nu\sigma_y) \tag{6-41a}$$

and

$$\varepsilon_y = \frac{1}{E}(\sigma_y - \nu\sigma_x) \tag{6-41b}$$

Assuming plane-strain, $\varepsilon_z = \gamma_{yz} = \gamma_{zx} = 0$. Since $\varepsilon_z = [\sigma_z - \nu(\sigma_x + \sigma_y)]/E = 0$, then $\sigma_z = \nu(\sigma_x + \sigma_y)$. Thus the strain in the x direction is

$$\varepsilon_x = \frac{1}{E}\{\sigma_x - \nu[\sigma_y + \nu(\sigma_x + \sigma_y)]\} = \frac{1-\nu^2}{E}\left(\sigma_x - \frac{\nu}{1-\nu}\sigma_y\right) \tag{6-42a}$$

Likewise, in the y direction

$$\varepsilon_y = \frac{1-\nu^2}{E}\left(\sigma_y - \frac{\nu}{1-\nu}\sigma_x\right) \tag{6-42b}$$

Now if a solution for σ_x and σ_y is obtained for a plane-stress problem, Eqs. (6-41) are appropriate. However, if instead of using E in Eqs. (6-41) the term $E/(1-\nu^2)$ is used and instead of ν the term $\nu/(1-\nu)$ is used, the strains will behave as in Eqs. (6-42), which apply to the plane-strain problem. The conversion rules are shown in Table 6-1.

For example, the equation $d^2y_c/dx^2 = M_z/EI_z$ applies to a narrow beam (plane stress). For a wide beam (plane strain) substitute $E/(1-\nu^2)$ for E, which results in $d^2y_c/dx^2 = (1-\nu^2)M_z/EI_z$. Recall from Sec. 3-7, that this equation applies for wide beams.

Table 6-1

Solution	Convert to:	Substitute:	For:
Plane stress	Plane strain	$\dfrac{E}{1-\nu^2}, \dfrac{\nu}{1-\nu}$	E, ν
Plane strain	Plane stress	$\dfrac{1+2\nu}{(1+\nu)^2}E, \dfrac{\nu}{1+\nu}$	E, ν

Polar Coordinates

There are many practical plane elastic problems where polar coordinates are used. If Airy's stress function is defined as a function $\phi(r, \theta)$ of r and θ, the biharmonic equation can be shown to be†

$$\nabla^4\phi = \left(\frac{\partial^2}{\partial r^2} + \frac{1}{r}\frac{\partial}{\partial r} + \frac{1}{r^2}\frac{\partial^2}{\partial\theta^2}\right)^2 \phi = 0 \tag{6-43}$$

and the stresses are given by

$$\sigma_r = \frac{1}{r}\frac{\partial\phi}{\partial r} + \frac{1}{r^2}\frac{\partial^2\phi}{\partial\theta^2} \tag{6-44a}$$

$$\sigma_\theta = \frac{\partial^2\phi}{\partial r^2} \tag{6-44b}$$

$$\tau_{r\theta} = \frac{1}{r^2}\frac{\partial\phi}{\partial\theta} - \frac{1}{r}\frac{\partial^2\phi}{\partial r\,\partial\theta} \tag{6-44c}$$

Example 6-12 Determine the stress in a semi-infinite plate due to a normal load, w force/unit length, acting on its edge, as shown in Fig. 6-12. Try the stress function

$$\phi = A\theta + Br^2\theta + Cr\theta \sin \theta + Dr\theta \cos \theta \tag{a}$$

where A, B, C, and D are constants.

SOLUTION The reader should verify that Eq. (a) does indeed satisfy Eq. (6-43) and compatability is ensured. For equilibrium the stresses from Eqs. (6-44) are

$$\sigma_r = 2B\theta + \frac{2}{r}(C \cos \theta - D \sin \theta) \tag{b}$$

$$\sigma_\theta = 2B\theta \tag{c}$$

$$\tau_{r\theta} = \frac{A}{r^2} - B \tag{d}$$

The stresses σ_r and σ_θ both contain a linear term in θ. The stress must be single-valued. Thus,

---

† The superscript 2 on the expression in parentheses is an operational notation and does not imply squaring. That is,

$$\nabla^4\phi = \left(\frac{\partial^2}{\partial r^2} + \frac{1}{r}\frac{\partial}{\partial r} + \frac{1}{r^2}\frac{\partial^2}{\partial\theta^2}\right)\left(\frac{\partial^2\phi}{\partial r^2} + \frac{1}{r}\frac{\partial\phi}{\partial r} + \frac{1}{r^2}\frac{\partial^2\phi}{\partial\theta^2}\right) = 0$$

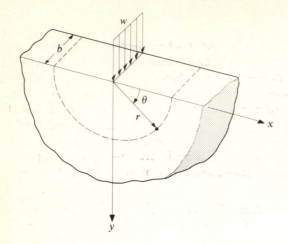

Figure 6-12

$B = 0$, and consequently

$$\sigma_r = \frac{2}{r}(C \cos \theta - D \sin \theta) \tag{e}$$

$$\sigma_\theta = 0 \tag{f}$$

$$\tau_{r\theta} = \frac{A}{r^2} \tag{g}$$

The sine function in σ_r gives symmetry with respect to the y axis, whereas the cosine function does not. Thus, thanks to symmetry, $C = 0$. Thus

$$\sigma_r = -\frac{2D}{r}\sin \theta \tag{h}$$

$$\sigma_\theta = 0 \tag{i}$$

$$\tau_{r\theta} = \frac{A}{r^2} \tag{j}$$

The boundary conditions from Fig. 6-12 are

$$\tau_{r\theta} = 0 \qquad \theta = 0, \pi \tag{k}$$

and

$$\int_0^\pi \sigma_r \sin \theta br\, d\theta = -wb \tag{l}$$

Applying condition (k) to Eq. (j) yields $A = 0$. Substituting Eq. (h) into condition (l) results in

$$-2Db \int_0^\pi \sin^2 \theta\, d\theta = -wb$$

Integrating and solving for D yields, $D = w/\pi$. Thus, the compatible stress distribution is

$$\sigma_r = -\frac{2w}{\pi r}\sin \theta \tag{m}$$

$$\sigma_\theta = 0 \tag{n}$$

$$\tau_{r\theta} = 0 \tag{o}$$

note that Eq. (m) agrees with Eq. (3-64).

6-8 THE TORSIONAL STRESS FUNCTION

The stress-function approach is also useful when dealing with torsion on a mechanical element with a noncircular cross-section. Airy's stress function cannot be used since the general torsion problem does not fall in the category of a plane elastic problem. Thus, the equations of equilibrium and compatibility must be recalled for the case of pure torsion. Consider a general cross-section undergoing torsion as shown in Fig. 6-13. A surface isolation perpendicular to the rod axis is also shown, where it can be seen that the shear stresses τ_{xy} and τ_{zx} act over a $dy\,dz$ element. It will be assumed here that σ_x is negligible.

For equilibrium, Eq. (6-16a), neglecting the body force $\bar{F}_x$ and the normal stress σ_x, is given by

$$\frac{\partial \tau_{xy}}{\partial y} + \frac{\partial \tau_{zx}}{\partial z} = 0 \tag{6-45}$$

If a stress function $\phi_t(y, z)$† exists such that

$$\tau_{xy} = \frac{\partial \phi_t}{\partial z} \tag{6-46a}$$

$$\tau_{zx} = -\frac{\partial \phi_t}{\partial y} \tag{6-46b}$$

Eq. (6-45) (equilibrium) is satisfied.

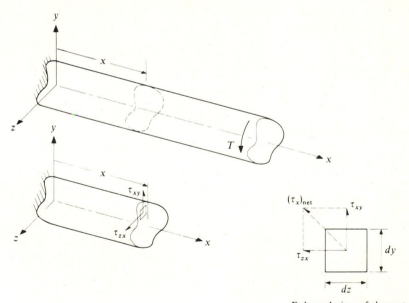

Figure 6-13

Enlarged view of element

† ϕ_t is called the torsional stress function, differing from Airy's stress function.

When deflections in terms of the angle of twist θ are considered, the deflection of a point on the isolated surface is as shown in Fig. 6-14, from which it can be seen that

$$w = \theta y \qquad v = -\theta z$$

Assuming that the rod is fixed at the origin of xyz (see Fig. 6-13), we can represent the angle of twist by the angle of twist per unit length θ', where $\theta = \theta'x$, and

$$w = \theta'xy \tag{6-47a}$$

$$v = -\theta'xz \tag{6-47b}$$

For noncircular cross sections, the surface warps, and in general the point will also deflect in the x direction u. However, this is independent of x and $u = u(y, z)$. The strains γ_{xy} and γ_{zx} are

$$\gamma_{xy} = \frac{\partial v}{\partial x} + \frac{\partial u}{\partial y} = -\theta'z + \frac{\partial u}{\partial y} \qquad \gamma_{zx} = \frac{\partial w}{\partial x} + \frac{\partial u}{\partial z} = \theta'y + \frac{\partial u}{\partial z}$$

Since $\tau = [E/2(1 + v)]\gamma$, we have

$$\tau_{xy} = \frac{E}{2(1 + v)}\left(-\theta'z + \frac{\partial u}{\partial y}\right) \tag{a}$$

$$\tau_{zx} = \frac{E}{2(1 + v)}\left(\theta'y + \frac{\partial u}{\partial z}\right) \tag{b}$$

Writing the expression $\partial\tau_{zx}/\partial y - \partial\tau_{xy}/\partial z$ using Eqs. (a) and (b) yields

$$\frac{\partial\tau_{zx}}{\partial y} - \frac{\partial\tau_{xy}}{\partial z} = \frac{E}{1 + v}\theta'$$

Substitution of Eqs. (6-46) results in

$$\frac{\partial^2\phi_t}{\partial y^2} + \frac{\partial^2\phi_t}{\partial z^2} = \frac{-E}{1 + v}\theta' \tag{6-48}$$

Equation (6-48) is the governing equation for the torsional stress function ϕ_t.

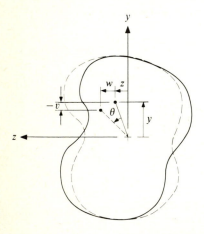

Figure 6-14

At the boundary, the net shear stress must be tangent to the boundary. Thus, from Fig. 6-13

$$\frac{\tau_{xy}}{\tau_{zx}} = \frac{dy}{dz} \qquad \text{or} \qquad \tau_{zx}\, dy - \tau_{xy}\, dz = 0$$

Substituting in Eqs. (6-46) results in

$$\frac{\partial \phi_t}{\partial y}\, dy + \frac{\partial \phi_t}{\partial z}\, dz = 0$$

However, since $\phi_t = \phi_t(y, z)$, the above can be written

$$d\phi_t = 0 \qquad\qquad (6\text{-}49)$$

Since Eq. (6-49) applies to the boundary, ϕ_t is constant along the boundary of the cross section. The value of this constant is arbitrary and is normally chosen to be zero. If the boundary of the cross section is a well-behaved function of y and z such as a circle, ellipse, etc., the equation of the boundary becomes an excellent stress function. This is demonstrated in Example 6-13, below.

To relate the stress function to the transmitted torque T return to Fig. 6-13. The net torque about the x axis due to the stresses on the $dy\,dz$ element is $(y\tau_{zx} - z\tau_{xy})\,dy\,dz$. Thus the total torque is

$$T = \iint (y\tau_{zx} - z\tau_{xy})\, dy\, dz$$

Substitution of Eqs. (6-46) and integrating each term by parts letting $\phi_t = 0$ on the boundary results in

$$T = 2 \iint \phi_t\, dy\, dz \qquad\qquad (6\text{-}50)$$

Example 6-13 A circular shaft of radius r_o is transmitting a torque T. Determine the corresponding shear-stress distribution.

SOLUTION The equation for the boundary of a circle of radius r_o in the xz plane is $y^2 + z^2 = r_o^2$. Try the following stress function:

$$\phi_t = k(y^2 + z^2 - r_o^2) \qquad\qquad (a)$$

where k is a constant. Note that $\phi_t = 0$ along the entire boundary. To establish the value of k, Eq. (6-50) is used. From Eq. (a) it can be seen that polar coordinates are more suitable to the problem. Let $r^2 = y^2 + z^2$, where r is a variable radial position. The infinitesimal area $dy\,dz$ can be replaced by $2\pi r\,dr$ since at any given position r the stress function is constant. Thus, the double integral of Eq. (6-50) reduces to a single integral, written

$$T = 4\pi k \int_0^{r_o} (r^2 - r_o^2)r\, dr$$

Integrating and solving for k results in

$$k = -\frac{T}{\pi r_o^4}$$

The polar moment of inertia of a circular cross section is $J = (\pi/2)r_o^4$. Thus

$$k = -\frac{T}{2J}$$

Upon substituting k into Eq. (a), the stress function is found to be

$$\phi_t = -\frac{T}{2J}(y^2 + z^2 - r_o^2) \tag{b}$$

The stresses are found from Eqs. (6-46):

$$\tau_{xy} = -\frac{Tz}{J} \tag{c}$$

and

$$\tau_{zx} = \frac{Ty}{J} \tag{d}$$

Note that at any given point, the net shear stress is given by $(\tau_x)_{net} = \sqrt{\tau_{xy}^2 + \tau_{zx}^2}$ (see Fig. 6-13). Thus, from Eqs. (c) and (d),

$$(\tau_x)_{net} = \frac{T}{J}\sqrt{z^2 + y^2} = \frac{Tr}{J} \tag{e}$$

which is identical to that used in elementary strength of materials.

For the angle of twist, substitution of Eq. (b) into Eq. (6-48) yields

$$-\frac{T}{J} - \frac{T}{J} = -\frac{E}{1+\nu}\theta'$$

or

$$\theta' = 2(1+\nu)\frac{T}{EJ}$$

If the total length of the bar is L, the angle of twist across the entire length is $\theta = \theta'L$, and thus

$$\theta = 2(1+\nu)\frac{TL}{EJ}$$

which again agrees with the elementary strength of materials solution.

Determining the stress functions for cross sections whose boundaries are not well defined by continuous functions is a more complicated matter. When this occurs, the stress functions are generally used in the form of an infinite series. Such a case is a rectangular cross section in torsion. Using a series solution, we find that the shear stress is maximum at the midpoint of the longest side (points A and B of Fig. 6-15), where the maximum stress is given by (see Ref. 3, p. 309)

$$\tau_{max} = k_1 \frac{T}{hb^2} \tag{6-51}$$

and the angle of twist is given by

$$\theta = k_2 \frac{(1+\nu)TL}{Ehb^3} \tag{6-52}$$

where the constants k_1 and k_2 are tabulated in Table 6-2 in terms of the ratio of h/b.

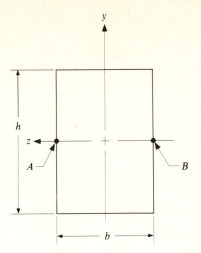

Figure 6-15

Table 6-2

h/b	k_1	k_2	h/b	k_1	k_2
1	4.80	14.2	4	3.54	7.12
1.5	4.33	10.2	5	3.44	6.87
2.0	4.06	8.73	10	3.21	6.41
2.5	3.87	8.03	∞	3.00	6.00
3	3.74	7.60			

Thin Rectangular Shapes

It can be seen that when $h \gg b$, $k_1 \approx 3$ and $k_2 \approx 6$. Thus for a thin rectangular cross section, Eqs. (6-51) and (6-52) reduce to

$$\tau_{\max} \approx \frac{3T}{hb^2} \tag{6-53}$$

$$\theta \approx \frac{6(1+\nu)TL}{Ehb^3} \tag{6-54}$$

Equations (6-53) and (6-54) can also be used to approximate the shear stress in "open sections," like that shown in Fig. 6-16, where the wall thickness is b and the total extended length of the section is h.

Equation (6-53) can be modified to handle sections in which the thickness varies. Rewrite Eq. (6-53) in the form

$$\tau_{\max} = \frac{3T}{hb^2} = \frac{Tb}{\frac{1}{3}hb^3}$$

The term in the denominator is called the effective polar moment of inertia of

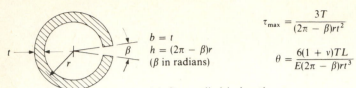

$$\tau_{max} = \frac{3T}{(2\pi - \beta)rt^2}$$

$b = t$
$h = (2\pi - \beta)r$
(β in radians)

$$\theta = \frac{6(1 + v)TL}{E(2\pi - \beta)rt^3}$$

(a) Open cylindrical section

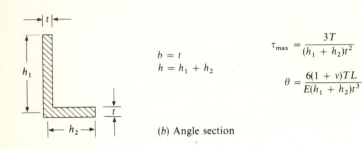

$b = t$
$h = h_1 + h_2$

$$\tau_{max} = \frac{3T}{(h_1 + h_2)t^2}$$

$$\theta = \frac{6(1 + v)TL}{E(h_1 + h_2)t^3}$$

(b) Angle section

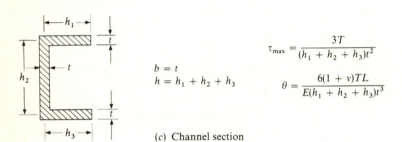

$b = t$
$h = h_1 + h_2 + h_3$

$$\tau_{max} = \frac{3T}{(h_1 + h_2 + h_3)t^2}$$

$$\theta = \frac{6(1 + v)TL}{E(h_1 + h_2 + h_3)t^3}$$

(c) Channel section

Figure 6-16

the cross section J_e. Thus†

$$\tau_{max} = \frac{Tb}{J_e} \tag{6-55}$$

If a section has varying thickness,

$$J_e = \sum \tfrac{1}{3}hb^3 \tag{6-56}$$

and b in Eq. (6-55) corresponds to the thickness of the section where the shear stress is being evaluated.

Example 6-14 Estimate the maximum shear stress on the section shown in Fig. 6-17 if a torque of 1000 lb·in is applied to the section.

SOLUTION The moment of inertia of the horizontal leg is

$$J_{e1} = (\tfrac{1}{3})(3)(0.5)^3 = 0.125 \text{ in}^4$$

†Note the similarity between Eqs. (6-55) and (2-10a).

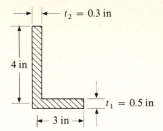

Figure 6-17

and in the vertical leg

$$J_{e2} = (\tfrac{1}{3})(4)(0.3)^3 = 0.036 \text{ in}^4$$

Thus

$$J_e = 0.125 + 0.036 = 0.161 \text{ in}^4$$

The maximum shear occurs on the lower leg and is

$$\tau_{\max} \approx \frac{Tb}{J} = \frac{(1000)(0.5)}{0.161} = 3106 \text{ lb/in}^2$$

6-9 DISCUSSION

The treatment of the topics of elasticity in this chapter could certainly continue, but beyond this point, the mathematical capabilities of the reader must be considerably greater than the level intended for this book. Keep in mind, however, that the subject of elasticity is invaluable in continuing study and understanding in the field of stress analysis. As the reader's mathematical background increases, he or she will discover a vast warehouse of advanced problems which have been solved by the various mathematical techniques in the theory of elasticity.

EXERCISES

6-1 The state of stress at a point relative to an xyz coordinate system is

$$\sigma = \begin{bmatrix} 4 & 1 & -1 \\ 1 & 0 & 2 \\ -1 & 2 & 0 \end{bmatrix} (1000) \text{ lb/in}^2$$

Determine the complete state of stress relative to an $x'y'z'$ coordinate system if

$$n_{x'x} = \frac{\sqrt{3}}{2} \qquad n_{x'y} = \tfrac{1}{2} \qquad n_{x'z} = 0$$

$$n_{y'x} = -\frac{\sqrt{3}}{4} \qquad n_{y'y} = \tfrac{3}{4} \qquad n_{y'z} = \tfrac{1}{2}$$

$$n_{z'x} = \tfrac{1}{4} \qquad n_{z'y} = -\frac{\sqrt{3}}{4} \qquad n_{z'z} = \frac{\sqrt{3}}{2}$$

6-2 In Exercise 6-1, determine the total shear stress on the surface perpendicular to the x' axis. Solve the problem two ways: (a) using Eqs. (6-5b) and (6-5c), determine $\tau_{x'y'}$ and $\tau_{z'x'}$ and add the results using vector addition; (b) use Eqs. (6-7) and (6-9)

6-3 At point Q in a body the state of stress relative to an xyz coordinate system is

$$\sigma = \begin{bmatrix} 5 & 2 & -2 \\ 2 & 0 & 4 \\ -2 & 4 & 3 \end{bmatrix} \times 10^8 \text{ N/m}^2$$

Using the cube defined in Fig. 6-18, determine the normal and shear stress at point Q for surfaces parallel to the following planes: (a) BCGF, (b) ABEF, (c) BGE.

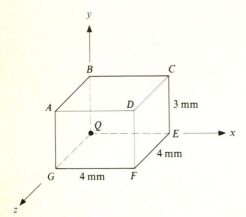

Figure 6-18

6-4 In the maximum-distortion-energy theory (see Sec. 5-1) it is stated that the state of stress can be divided into a hydrostatic part and a distortional part. In the theory, this distortional part is assumed to be the mechanism for failure. The hydrostatic stress is simply the average of the normal stresses; i.e.,

$$\sigma_{av} = \tfrac{1}{3}(\sigma_1 + \sigma_2 + \sigma_3)$$

There will be eight faces (forming an octahedron) which have a normal stress of σ_{av}. The normals to these surfaces are symmetric to the principal axes; this makes each of the directional cosines for each "octahedral" surface equal to $\pm\sqrt{\tfrac{1}{3}}$ with respect to each of the principal axes. The shear stresses on these surfaces cause the distortion and are called the *octahedral shear stresses*.

(a) Show that the octahedral shear stress on each surface is

$$\tau_{oct} = \tfrac{1}{3}\sqrt{(\sigma_1 - \sigma_2)^2 + (\sigma_2 - \sigma_3)^2 + (\sigma_3 - \sigma_1)^2} \qquad (a)$$

(b) Show that the maximum-octahedral-shear-stress theory gives the same yield criteria as the maximum-distortion-energy theory by comparing Eq. (a) with the maximum octahedral shear stress in a tensile-test specimen when the tensile stress reaches the yield strength S_y.

6-5 For the stress matrix given in Exercise 6-1, determine the principal stresses and the directional cosines of the normals to each surface containing a principal stress. As a check, verify that the three normals are mutually perpendicular.

6-6 The state of stress at a point relative to an xyz coordinate system is

$$\sigma = \begin{bmatrix} 2 & 1 & 1 \\ 1 & 2 & 1 \\ 1 & 1 & 2 \end{bmatrix} \times 10^8 \text{ N/m}^2$$

Evaluate the principal stresses and the directional cosines of the normals to each surface containing a principal stress.

6-7 Evaluate the Lamé constants for the following materials:
  (a) Steel: $E_s = 29 \times 10^6$ lb/in², $\nu_s = 0.285$
  (b) Aluminum: $E_a = 10.3 \times 10^6$ lb/in², $\nu_a = 0.33$
  (c) CR-39 (a plastic material used in photoelasticity) $E_{CR-39} = 0.3 \times 10^6$ lb/in², $\nu_{CR-39} = 0.42$.

6-8 Figure 6-19 shows two cantilever beams. The material of each beam is Douglas fir, which is an orthotropic material. In Fig. 6-19a the axes of symmetry of the material properties align with the geometric axes xyz, whereas, in Fig. 6-19b the wood was planed so that the axes of symmetry of the material properties xyz do not align with the geometric axes $x'y'z'$. For both beams determine the normal strain in the direction of the geometric longitudinal axis of the beam at point A and compare the results. With respect to the xyz coordinate system used in either case, the material property relating stress to strain is

$$K_{ij} = \begin{bmatrix} 6 & -2.5 & -3 & 0 & 0 & 0 \\ -2.5 & 90 & -4 & 0 & 0 & 0 \\ -3 & -4 & 120 & 0 & 0 & 0 \\ 0 & 0 & 0 & 96 & 0 & 0 \\ 0 & 0 & 0 & 0 & 84 & 0 \\ 0 & 0 & 0 & 0 & 0 & 78 \end{bmatrix} \times 10^{-7} \text{ in}^2/\text{lb}$$

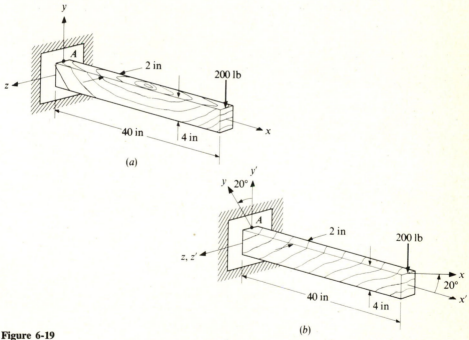

Figure 6-19

6-9 The stress field for the cantilever beam shown in Fig. 6-9 considering bending and transverse shear effects is

$$\sigma_x = \frac{P}{I_z} y(L - x) \qquad \tau_{xy} = -\frac{P}{2I_z}(c^2 - y^2) \qquad \sigma_y = 0$$

Using Eqs. (6-17), verify that this stress field satisfies equilibrium.

6-10 For the uniformly loaded cantilever beam of Examples 6-7 and 6-8, the compatible stress field was found to be

$$\sigma_x = \frac{w}{10I_z}(5x^2 + 2c^2)y - \frac{w}{3I_z}y^3 \tag{a}$$

$$\tau_{xy} = \frac{w}{2I_z}x(c^2 - y^2) \tag{b}$$

$$\sigma_y = -\frac{w}{6I_z}(2c^3 + 3c^2y - y^3) \tag{c}$$

Using Eqs. (6-17) verify that this stress field satisfies equilibrium.

6-11 Figure 2-63 of Exercise 2-20 shows a tapered plate of constant thickness t in tension. Assuming that the axial stress formulation is valid, at section $a\text{-}a$

$$\sigma_x = \frac{P}{ht} \tag{a}$$

where h is the width of the plate at $a\text{-}a$ and therefore a function of x. Using the equilibrium Eq. (6-17a), prove that at section $a\text{-}a$ the shear stress varies as

$$\tau_{xy} = 2\frac{y}{h}\sigma_x \tan\theta \tag{b}$$

Hint: The boundary condition that $\tau_{xy} = 0$ at $y = 0$ can be used since the x axis is an axis of symmetry.

6-12 Extending Exercise 6-11, use Eq. (b) and the equilibrium Eq. (6-17b) to prove that

$$\sigma_y = 4\frac{y^2}{h^2}\sigma_x \tan^2\theta \tag{c}$$

Hint: The boundary condition for σ_y can be found in the solution of Exercise 2-20. That is, σ_y can be evaluated at $y = \pm h/2$ using Mohr's circle at the free surface.

6-13 Figure 6-20 shows a tapered beam of constant width b. Assume that the bending stress can be expressed by the elementary formula

$$\sigma_x = -\frac{M_z}{I_z}y = \frac{3}{2}\frac{P}{bc^3}xy \tag{a}$$

where $2c$ is the depth of the beam at x and thus is a function of x. Applying the equilibrium equations to Eq. (a), show that

$$\tau_{xy} = \frac{P}{2I_z}\left[c^2 - y^2 + \frac{x}{c}(3y^2 - c^2)\tan\theta\right] \tag{b}$$

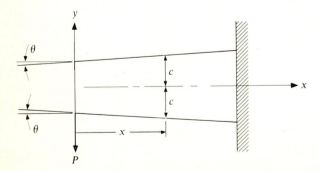

Figure 6-20

and
$$\sigma_y = \frac{P}{I_z}\frac{y}{c}\tan\theta\left[c^2 - y^2 + \frac{x}{c}(2y^2 - c^2)\tan\theta\right] \tag{c}$$

Hint: The boundary conditions are found by determining τ_{xy} and σ_y at $y = \pm c$ using Mohr's circle.

6-14 Show that Eq. (*a*) from Exercise 6-11 and (*c*) from Exercise 6-12 do not, in general, satisfy compatibility. Use Eq. (6-38), $\nabla^2(\sigma_x + \sigma_y) = 0$. Is there any specific location in which compatibility is satisfied? Since Exercise 6-11 started with the assumption that σ_x was uniform, suggest a possible method of improving the assumption using intuitive reasoning and outline the steps that should be taken to obtain a more accurate solution.

6-15 Consider a long rectangular bar of width w and thickness t uniformly loaded in tension at both ends by P/wt (force per unit area). Using an Airy stress function, obtain the stress field which provides compatibility and equilibrium and satisfies the boundary conditions.

6-16 Prove that the stress function given in Example 6-12, Eq. (*a*), satisfies the biharmonic equation in polar coordinates.

6-17 Verify that Eqs. (6-32*d*) to (6-32*f*) are valid.

6-18 For a curved beam in pure bending, polar coordinates are used [see Eqs. (3-15) and (3-17)]. If the cross section of the beam is rectangular, the following stress function is applicable:

$$\Phi = A + B \ln r + Cr^2 + Dr^2 \ln r \tag{a}$$

where A, B, C, and D are constants.

(*a*) Prove that Φ satisfies Eq. (6-43).

(*b*) If the inner and outer radii of the beam are r_i and r_o, respectively, and the width of the cross section is b, determine the stress field from Φ and compare the results with Eqs. (3-15) and (3-17) as applied to a rectangular curved beam.

6-19 Figure 1-35 shows a plate hanging under its own weight, where the weight density is D_p. Considering the weight density as a body force $\bar{F}_x$, determine σ_x using Eq. (6-17*a*). Verify your results using the elementary surface-isolation technique.

6-20 For a thin plate with no body forces, a biaxial state of stress exists where

$$\sigma_x = ay^3 + bx^2 y - cx \qquad \sigma_y = dy^3 - e \qquad \tau_{xy} = fxy^2 + gx^2 y - h$$

where a, b, c, d, e, f, g, and h are constants. What are the constraints on the constants such that the stress field satisfies both equilibrium and compatibility?

6-21 Using the elementary strength of material stress formulations for σ_x, σ_y, and τ_{xy}, derive the expressions for $u(x, y)$ and $v(x, y)$ of Exercise 1-10 using the technique outlined in Example 6-6.

6-22 Repeat the analysis of Example 6-6 to determine $u(x, y)$ and $v(x, y)$ but consider the beam to be rectangular of depth $2c$ and include the effects of the transverse shear stress. Verify that the transverse-shear-stress field is given by

$$\tau_{xy} = -\frac{P}{2I_z}(c^2 - y^2)$$

6-23 A thin plate of thickness t is immersed in water as shown in Fig. 6-21. The weight density of the plate is D_p and of the water is D_w. Determine the displacement field of the plate neglecting z dependence on displacement but do *not* neglect the stress in the z direction.

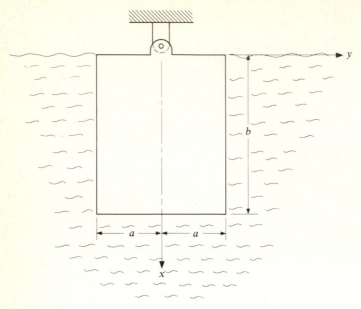

Figure 6-21

6-24 Determine the deflection field of the uniformly loaded cantilever beam of Fig. 6-6 using Eqs. (*b*) and (*c*) of Example 6-7 and Eq. (*l*) of Example 6-8.

6-25 Equation (3-53) represents the interface pressure for two thin cylinders press-fitted together. Convert this equation so that the pressure represents the press fit of two long cylinders.

6-26 Using the stress function of Example 6-11, Eq. (*j*), obtain the stress field for the simply supported beam shown in Fig. 6-22. Compare the results with the elementary strength of materials formulations.

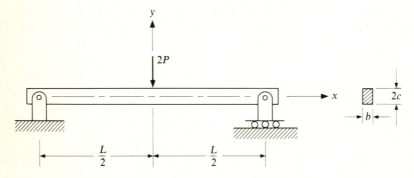

Figure 6-22

6-27 Determine the stresses in a semi-infinite plate due to a shear load as shown in Fig. 6-23. The load *w* (force per unit length) is to be considered to be applied along the thickness of the plate at $x = 0$. *Hint*: Use the stress function of Example 6-12, Eq. (*a*).

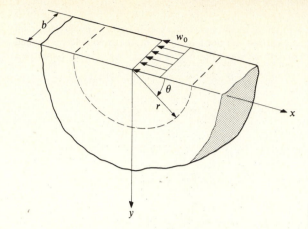

Figure 6-23

6-28 A steel shaft with an elliptical cross section as shown in Fig. 6-24 is undergoing a net torque of 500 N·m. Determine the value and location of the maximum shear stress on the section and the angle of twist per unit length ($E = 20 \times 10^{10}$ N/m², $\nu = 0.3$).

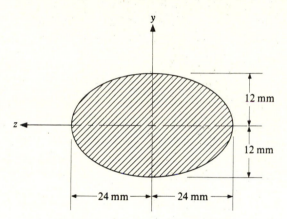

Figure 6-24

6-29 Each of the sections shown in Fig. 6-25 is transmitting a torque of 200 lb·in. Estimate the maximum shear stress in each section and the angle of twist per unit length.

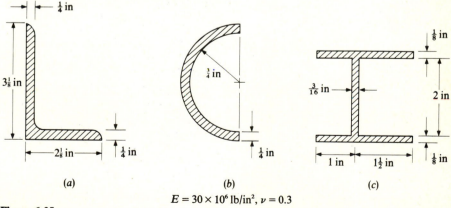

(a) (b) (c)

$E = 30 \times 10^6$ lb/in², $\nu = 0.3$

Figure 6-25

6-30 In Chap. 3, shear flow in beams with thin sections was discussed. In order to avoid additional shear stresses due to torsion it was shown that the net loading of the beam should be applied through the shear center. The channel section shown in Fig. 6-26 is undergoing bending about the z

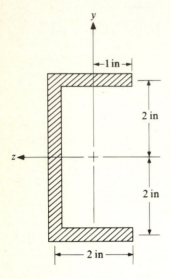

Figure 6-26

axis, where the net transverse shear force V_y is in the positive y direction. If the net effect of the applied load is on the top flange and coincides with the y axis, determine the magnitude and location of the resulting maximum shear stress due to V_y and the torsion on the section if $V_y = 1000$ lb. The thickness of the web and flanges is $\frac{3}{16}$ in.

REFERENCES

1. Sokolnikoff, I. S.: "Mathematical Theory of Elasticity," McGraw-Hill, New York, 1956.
2. Love, A. E. H.: "A Treatise on the Mathematical Theory of Elasticity," 4th ed., Dover, New York, 1944.
3. Timoshenko, S. P., and J. N. Goodier: "Theory of Elasticity," 3d ed., McGraw-Hill, New York, 1970.

EXPERIMENTAL STRESS ANALYSIS

7-0 INTRODUCTION

When dealing with a complex stress-analysis problem in which a complete theoretical solution may prove impractical with respect to time, cost, or degree of difficulty, experimental techniques are often employed. In the design phase, scale models can be analyzed using experimental methods, and in many cases design changes can be scrutinized quite effectively. If the structure is already in existence and its effectiveness or its behavior under a change in loading specifications is to be determined, the structure can easily be analyzed by experimental methods. Even if a given problem is solved by a theoretical approach, the results must generally be verified, in which case, some degree of experimentation is necessary. Thus, a knowledge of experimental methods is an essential part of an analyst's background.

In this chapter, the three basic conventional experimental methods will be described. It will be necessary in some cases, to revert to mathematical theory, but the intention is to minimize theory and concentrate on physical understanding. The three basic topics are:

1. Strain gages and instrumentation
2. Photoelastic methods
3. Brittle-coating methods

Although there are many other topics in this field (see Refs. 1 and 2), the purpose here is to introduce the fundamental concepts and applications of the most conventional methods used in practice. Before discussing the above topics, dimensional analysis and analysis techniques are discussed.

7-1 DIMENSIONAL ANALYSIS

When attempting to predict the stresses (or any state property) of a prototype design, it sometimes is necessary to construct and test a scale model, which may or may not be of the same material. The prototype may be so large that cost and practicality make testing out of the question. Conversely, the prototype may be so small that test data are difficult to obtain. In addition, the prototype may be planned to be of a costly or unmanageable material. In either case, a different material for the model would be advantageous. Furthermore the method of testing may require special materials for the model, as in transmission photoelasticity, a change in materials between the prototype and model is usually necessary.

Thus, the questions arise: How to design the model and how does the prototype behave with respect to the results obtained from the testing of the model? These questions are resolved through the method of dimensional analysis.

For a structure which obeys Hooke's law it is only necessary to prescribe one state property, the stress matrix σ, and two material properties, E and ν. With this, strain and displacements are derivable. The independent spatial variables are x, y, and z. The loads can be prescribed by one of the applied loads, say F, and the other forces by simply determining their ratios to F. Likewise, dimensions can be related by ratios to one of the dimensions of the structure, say L. Thus, if all the ratios are known, the fundamental quantities are σ, E, ν, x, y, z, F, and L. The units of all of these quantities can be obtained from the units of force and length. These quantities can also be put in dimensionless form using F and L:

$$\frac{\sigma}{F/L^2} \qquad \frac{E}{F/L^2} \qquad \nu \qquad \frac{x}{L} \qquad \frac{y}{L} \qquad \frac{z}{L} \qquad \frac{F}{F} \qquad \frac{L}{L}$$

Since the last two quantities are obviously trivial, the independent dimensionless quantities are

$$\frac{L^2\sigma}{F} \qquad \frac{L^2E}{F} \qquad \nu \qquad \frac{x}{L} \qquad \frac{y}{L} \qquad \frac{z}{L}$$

For a given prototype, each quantity will have some value based on geometry, material, and the applied loading. For example, consider the quantity

$$\frac{L_p^2\sigma_p}{F_p} = K$$

where the subscript p denotes the prototype values and K is some dimensionless constant. In order to maintain similarity between the prototype and model, this quantity should be the same for the model:

$$\frac{L_m^2\sigma_m}{F_m} = K = \frac{L_p^2\sigma_p}{F_p}$$

where the subscript m denotes the model values. Thus, in order to predict the prototype stress values from the results obtained from the model,

$$\sigma_p = \left(\frac{L_m}{L_p}\right)^2\left(\frac{F_p}{F_m}\right)\sigma_m \tag{7-1}$$

Let σ be the stress scale factor; then

$$\sigma_p = \sigma\sigma_m$$

or

$$\sigma = \frac{\sigma_p}{\sigma_m} \tag{7-2}$$

Likewise, let the force scale factor be

$$F = \frac{F_p}{F_m} \tag{7-3}$$

and the length scale factor be

$$L = \frac{L_p}{L_m} \tag{7-4}$$

Thus, it can be seen from Eqs. (7-1) to (7-4) that the stress factor is related to the force and length scale factors

$$\sigma = \frac{F}{L^2} \tag{7-5}$$

A number of scale factors can be created which (except for ν) can be described by the scale factors F and L; they are presented in Table 7-1.

Example 7-1 A half-scale plastic model is made to represent a prototype design. A single force of 2000 lb will be applied to the proposed prototype.

(a) The model is loaded to 200 lb, and the maximum stress is found to be 1000 lb/in². Determine the maximum stress in the prototype when loaded to 2000 lb.

(b) If the modulus of elasticity of the model is 0.5×10^6 lb/in² and the modulus of the prototype is 10×10^6 lb/in², estimate the total deflection of the prototype if the total deflection of the model is 0.20 in.

SOLUTION (a)

$$L = \frac{L_p}{L_m} = \frac{L_p}{\frac{1}{2}L_p} = 2 \qquad F = \frac{F_p}{F_m} = \frac{2000}{200} = 10$$

The stress scale is

$$\sigma = \frac{F}{L^2} = \frac{10}{2^2} = 2.5$$

Table 7-1

Scale factor	Prototype model	Force and length dependence
Force scale F	$F = \dfrac{F_p}{F_m}$	$F = F$
Length scale L	$L = \dfrac{L_p}{L_m}$	$L = L$
Stress scale σ	$\sigma = \dfrac{\sigma_p}{\sigma_m}$	$\sigma = \dfrac{F}{L^2}$
Modulus scale E	$E = \dfrac{E_p}{E_m}$	Function of materials
Moment scale M	$M = \dfrac{M_p}{M_m}$	$M = FL$
Pressure scale P	$P = \dfrac{P_p}{P_m}$	$P = \dfrac{F}{L^2}$
Strain scale ε	$\varepsilon = \dfrac{\varepsilon_p}{\varepsilon_m}$	$\varepsilon = \dfrac{\sigma}{E} = \dfrac{F}{L^2 E}$
Displacement scale δ	$\delta = \dfrac{\delta_p}{\delta_m}$	$\delta = \varepsilon L = \dfrac{F}{LE}$

and since $\sigma = \sigma_p / \sigma_m$, we have

$$\sigma_p = (2.5)(1000) = 2500 \text{ lb/in}^2$$

(*b*)

$$E = \frac{E_p}{E_m} = \frac{10 \times 10^6}{0.5 \times 10^6} = 20$$

$$\delta = \frac{F}{LE} = \frac{10}{(2)(20)} = 0.25 \qquad \delta = \frac{\delta_p}{\delta_m}$$

or

$$\delta_p = \delta \delta_m = (0.25)(0.20) = 0.05 \text{ in}$$

7-2 ANALYSIS TECHNIQUES

Stress in a solid cannot be measured directly and must be deduced from strain measurements and Hooke's law. In general, to determine the complete state of stress or strain at a point, six quantities must be known. In free-surface measurements where the z axis is perpendicular to the surface, $\sigma_z = \tau_{yz} = \tau_{zx} = 0$, the state of stress is biaxial. In this case only *three* measurements are necessary to establish the state at a given point on the surface. Important problems and experimental techniques are not restricted to the biaxial state of stress, but most problems of interest involve the measurement of biaxial states of stress since to a large extent maximum-stress states tend to develop at free surfaces.

Considering only biaxial cases from this point on, a further reduction in the number of measured quantities necessary to establish the state of stress can be made if some additional information is known. If the directions of the principal axes are known, it is only necessary to determine the two principal strains where Hooke's law is used to find the principal stresses. Then Mohr's circle can be used to establish the state of stress in any direction. The direction of the principal axes can be found:

1. Along lines of symmetry
2. When $\sigma_z = \tau_{yz} = \tau_{zx} = 0$ (where the part is thin in the z direction) and the y direction is perpendicular to a free surface
3. From brittle-coating or photoelastic methods

Symmetry

The shear stress in the direction of a line of symmetry is zero, and thus the line and its perpendicular establish the principal axes. For example, consider the ring loaded as shown in Fig. 7-1a. The lines a-a and b-b are lines of symmetry. Thus $\tau_{xy} = 0$ along lines a-a and b-b.

To understand why this is true, separate the structure along line a-a. If a shear stress is present on the top element at a given point on line a-a, then thanks to symmetry the shear stress will be present on the bottom element and in the same direction (see Fig. 7-1b). This disobeys Newton's law of action and reaction unless $\tau = 0$.

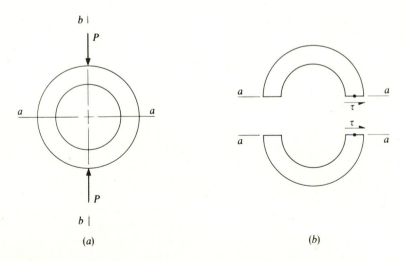

(a)   (b)

Figure 7-1

The same argument holds for line b-b, but because of the loading along the line, a sharp shear-stress gradient occurs and shear stresses develop quickly from line b-b.

When $\sigma_z = \tau_{yz} = \tau_{zx} = 0$ and the y Direction Is Perpendicular to a Free Surface or a Surface with a Normal-Force Distribution

This occurs in plane-stress elastic problems (see Chap. 6) where thin platelike structures undergo in-plane loading. Consider Fig. 7-1 again; if the thickness is small, $\sigma_z = \tau_{yz} = \tau_{zx} = 0$ throughout. When an xy localized coordinate system is established at points on a free surface, such as points A and B shown in Fig. 7-2, there is no shear force (or stress) on the free side of the element corresponding to points A and B. Thus x_A, y_A and x_B, y_B are principal axes for points A and B, respectively. There is an additional piece of information under these conditions; i.e., since we are dealing with a stress-free surface, $(\sigma_y)_A$ and $(\sigma_y)_B = 0$. Thus, it is only necessary to determine $(\sigma_x)_A$ and $(\sigma_x)_B$. Therefore only one piece of information is needed at each point.

If an external surface is exposed to a known pressure distribution, again only one piece of information is needed. For example, assume that the inner

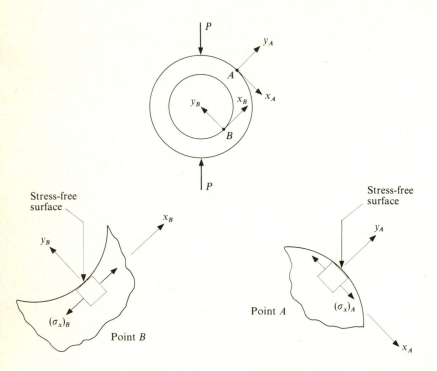

Figure 7-2

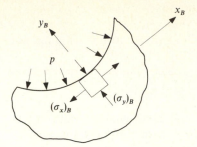

Figure 7-3

surface of the ring is experiencing a uniform normal pressure along the circumference of a value p. Thus $(\sigma_y)_B = -p$, and the element appears as shown in Fig. 7-3; since there is no shear stress on the inner surface, x_B and y_B remain principal axes. In order to determine the state of stress at point B, again only one additional piece of information is needed, $(\sigma_x)_B$.

Brittle-Coating and Photoelastic Methods

These methods enable the analyst to locate the principal axes. With brittle coating, the location is approximate and usually applicable only in areas undergoing high principal tensile stresses. The photoelastic method, however, provides more accurate information. The techniques will be discussed in more detail later.

Thus it can be seen that there are a number of ways of reducing the amount of information necessary to completely specify the state of stress at a point by experimental methods.

7-3 STRAIN GAGES, GENERAL

The use of electrical-resistance strain gages is probably the most common method of measurement in experimental stress analysis. In addition, strain-gage technology is quite important in the design of transducer instrumentation for the measurement of force, torque, pressure, etc.

Electrical-resistance strain gages are based on the principle that the resistance R of a conductor changes as a function of normal strain ε. The change in resistance of a strain gage is given by

$$\Delta R = S_A R \varepsilon \qquad (7\text{-}6)$$

where S_A = sensitivity of conductor material
ε = average normal strain impressed on conductor

The most common material for strain gages is a copper-nickel alloy known as Advance. The strain sensitivity for Advance is about 2.1. The primary advantages of this material are:

1. The strain sensitivity S_A is linear over a wide range of strain and does not change as the Advance goes plastic.
2. The thermal stability is excellent and is not greatly influenced by temperature changes when used on common structural materials.

Other materials with higher sensitivities are used when sensitivity is necessary (as in dynamic measurements). However, care must be exercised, as these materials do not behave as well as Advance. Semiconductor gages, for example, reach sensitivities as high as 145, accounted for by large changes in resistivity with strain. However, care must be exercised with respect to the thermal stability of these piezoresistive gages.

Most gages have a nominal resistance of 120 Ω. This minimizes gage current and overloading of the power supply. To obtain an accuracy of ±5 μin/in using Advance it is necessary to measure a change in resistance of approximately ±1.2 mΩ.

To measure these small changes in resistance accurately two basic circuits are used. The Wheatstone bridge circuit is used for static analysis and temperature-compensated dynamic analysis, and the potentiometer circuit is used for dynamic analysis when temperature compensation is not required. The advantage of the potentiometer circuit is that the readout is direct, whereas the Wheatstone bridge circuit is a null-balance technique requiring an initial balance. The potentiometer circuit utilizes a common ground for the circuit, amplifier, and recording instrument, which reduces the electronic noise. However, since a null-balance strain indicator is simple to operate and can be used with or without temperature compensation, it is generally preferred for static applications.

Metallic electrical-resistance strain gages used in experimental stress analysis come in two basic types, bonded-wire and bonded-foil (see Fig. 7-4).

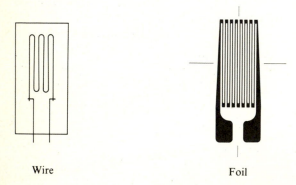

Wire   Foil   **Figure 7-4**

The resistivity of Advance is approximately 49 $\mu\Omega\cdot$cm. Thus, if a strain gage is to be fabricated using a wire 0.001 in in diameter and is to have a resistance of 120 Ω, the gage would require a wire almost 5 in long. To shorten the gage length it is necessary to construct the gage as shown in Fig. 7-4a.

For normal applications, wire gages are sandwiched between two sheets of thin paper; foil gages are generally mounted on a thin epoxy carrier or paper or sandwiched (encapsulated) between two thin sheets of epoxy; this improves the temperature range, fatigue life, and chemical and mechanical protection of the sensing grid but at additional cost. Other carrier materials are available for special applications.

The most popular adhesive used for the normal installation of a gage is Eastman 910, or for mounting a paper-backed gage cellulose nitrate cement such as Duco cement is used. Extreme care must be exercised when installing a gage as a good bond and insulated gage are necessary. The installation procedures can be obtained from technical instruction bulletins supplied by the manufacturer. The fundamental steps are given in Appendix H. An installation of a foil gage with and without strain-relief loops is shown in Fig. 7-5.

Once the gages are correctly mounted, wired, tested for resistance, and waterproofed, they are ready for instrumentation and testing. Certain fundamental characteristics of the gages should be pointed out. The following primary effects should be considered:

1. Linearity of the grid material
2. Temperature effects
3. Strain gradient
4. Cross sensitivity of the gage
5. Zero shift and hysteresis effects
6. Dynamic response
7. Gage-current heating effects

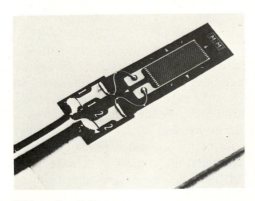

With strain relief loops   Without strain relief loops

Figure 7-5 Examples of strain gage installations. *(Courtesy of Micro-Measurements Division of Vishay Intertechnology, Inc.)*

Linearity of the Grid Material

As mentioned before, Advance is linear over a wide range of strain. However, if higher sensitivity is desired, as in dynamic measurements, grid materials such as isoelastic or Stabiloy are commonly used. These materials do not remain linear. For example, isoelastic remains linear at a value of $S_A = 3.6$ up to about 7500 μin/in. Beyond this point S_A drops from 3.6 to about 2.5. This must be accounted for in the instrumentation if testing goes beyond 7500 μin/in.

Temperature Effects

When the ambient temperature changes, the gage material tends to change in length by an amount different from the change in the specimen. This difference will strain the gage by the amount

$$\varepsilon = (\alpha_s - \alpha_g)\,\Delta T$$

where α_s and α_g are the thermal coefficients of expansion of the specimen and gage, respectively. The corresponding change in resistance is

$$\left(\frac{\Delta R}{R}\right)_1 = (\alpha_s - \alpha_g)S_g\,\Delta T$$

where S_g is the calibrated strain sensitivity of the gage (known as the gage factor).

In addition, a temperature change will cause a change in resistance of the gage material

$$\left(\frac{\Delta R}{R}\right)_2 = \gamma\,\Delta T$$

where γ is the temperature coefficient of resistivity of the gage material. The combined effect of these factors will produce an overall change in resistance of

$$\left(\frac{\Delta R}{R}\right)_{\Delta T} = (\alpha_s - \alpha_g)S_g\,\Delta T + \gamma\,\Delta T \tag{7-7}$$

This effect can be canceled or compensated for by two different methods. The first method involves compensation within the gage itself. This can be accomplished if the net effects of the factors in Eq. (7-7) can be arranged so that $(\Delta R/R)_{\Delta T} \approx 0$. Thus, if $(\alpha_s - \alpha_g)S_g \approx -\gamma$, the gage will be self-temperature-compensating. This is possible in a restricted temperature range either by a manufacturer's modification of the gage or by sampling each lot of alloy material to determine the thermal behavior of the lot and then matching the thermal characteristics to a particular specimen material. The manufacturer will supply the thermal characteristics of the self-temperature-compensating gages as applied to the intended specimen material. An example of the thermal characteristics of a specific BLH gage is shown in Fig. 7-6. Note that the apparent strain is zero at 75 and 125°F and relatively stable within this range.

The second method of temperature compensation involves the instrumen-

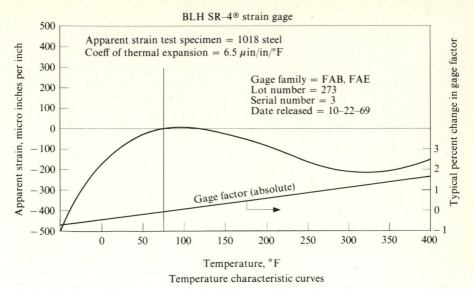

Figure 7-6 Strain gage temperature characteristics. *(Courtesy of Baldwin-Lima-Hamilton Corporation.)*

tation circuit and the use of an additional gage which can be actively or passively used in the circuit. This method will be described in detail in the section dealing with instrumentation.

Strain Gradient

Since the gage is of finite length, the change in resistance is due to the average strain along the gage and not the center strain in general. If the strain along the gage is constant or linearly varying, the average strain is the center strain. However, for any other case, the average strain will differ from the center strain. If the strain gradient or change in strain along the gage is small, the error will be small, but if there is a large change in strain over a small distance, the gage should be as small as possible to minimize the error.

Cross Sensitivity of the Gage

The strain sensitivity of a single straight uniform length of a conductor is defined as

$$S_A = \frac{\Delta R/R}{\varepsilon} \tag{7-8}$$

where ε is a uniform strain along, and in the direction of, the conductor. However, due to the configuration of the gage, cross strains (transverse strains) will effect a change in resistance also. This is not desirable, as only the strain in the direction of the gage is expected. In addition, the gages are tested and

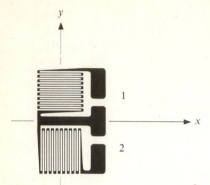

Figure 7-7

calibrated by the manufacturer by mounting the gage in the axial direction of a tensile specimen. The resulting strain measured is assumed due to axial strain alone. However, there is a transverse strain present in a uniaxial stress field due to the Poisson effect. The gage factor, or sensitivity, provided by the manufacturer is thus

$$S_g = \frac{\Delta R / R}{\varepsilon_x} \qquad \text{with } \varepsilon_y = -\nu\varepsilon_x \qquad (7\text{-}9)$$

(ν is usually 0.285), where the x axis is in the axial direction of the tensile specimen and the y axis is perpendicular to the x axis. If the gage is used under conditions where the transverse strain is such that $\varepsilon_y = -\nu\varepsilon_x$, then

$$\varepsilon_x = \frac{\Delta R / R}{S_g}$$

would yield exact results. If $\varepsilon_y \neq -\nu\varepsilon_x$, some error will be induced. Usually this error is small; however, as the value of the ratio of $\varepsilon_y / \varepsilon_x$ deviates from $-\nu$, the errors may become appreciable.

Part of the data supplied by the manufacturer is a value of the transverse sensitivity K_t (sometimes denoted by k or K). Technical bulletins are available from the manufacturer which describe the correction equations for standard gage configurations. For the two-element 90° rosette shown in Fig. 7-7 the correction equations are

$$\varepsilon_x = \frac{1 - \nu K_t}{1 - K_t^2}(\hat{\varepsilon}_x - K_t\hat{\varepsilon}_y) \qquad \varepsilon_y = \frac{1 - \nu K_t}{1 - K_t^2}(\hat{\varepsilon}_y - K_t\hat{\varepsilon}_x) \qquad (7\text{-}10)$$

where $\hat{\varepsilon}_x$ and $\hat{\varepsilon}_y$ are the indicated strains from gages 1 and 2, respectively, and ε_x and ε_y are the corrected strains (for further details, see Appendix J).

Zero Shift and Hysteresis Effects

When a new gage goes through the first reversal-strain cycle, the resistance will not return to its initial value, thereby causing a zero shift as well as a deviation in linearity, or hysteresis. These factors can cause significant errors in interpreting the results. This is primarily due to cold working of the gage, which causes changes in resistivity. These effects are reduced significantly if the gage is cycled about five times before taking data.

Dynamic Response

In dynamic-strain measurements there is a small lag time before the gage responds to a strain wave; this is due to the intermediate material, namely, the adhesive and carrier. The response time is of the order of 0.1 μs, but this valve has not been verified because of the inherent difficulty in measuring this extremely small response time.

A second, and more important, characteristic of dynamic-strain measurement is the fact that since a gage is measuring the average strain across the grid, there are measurement distortions of the actual strain wave. To illustrate this, consider a strain pulse traveling through a specimen as shown in Fig. 7-8a. The magnitude of the pulse is ε_p with time duration t_p, and it is propagating with a velocity c. Thus, the effective length of the pulse is ct_p.

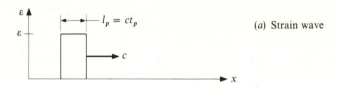

(a) Strain wave

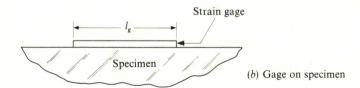

(b) Gage on specimen

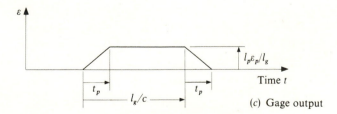

(c) Gage output

Figure 7-8

As the wave enters the gage area, the gage output is zero. It begins to increase linearly (since averaging is basically an integrating method) until the full wave is within the gage length, where the average strain would be $(l_p/l_g)\varepsilon_p$. The time this takes is t_p. The average strain across the gage remains constant at this value until the wave begins to leave the gage area. The total time this takes is l_g/c. As the wave leaves the gage area, the gage output again drops to zero in a linear fashion at the same rate of entry. Thus, the output shape, magnitude, and time duration are distorted due to the wavelength vs. gage length. The smaller the gage, the less distortion in the output will be observed.

Gage-Current Heating Effects

Although this effect is primarily related to the instrumentation circuitry, the gage current alters the gage behavior. The gage is a resistor R, and since the gage current I produces heat according to the I^2R law, there is a temperature rise on the gage which depends on the gage current and the heat-dissipation characteristics of the installation. Since the gage temperature affects the gage output, a zero shift of the gage will occur if there is a temperature change. This temperature rise reaches an equilibrium point, and the gage will then remain stable. However, during the warm-up time, large errors can occur. If the gage current is 25 mA or less, the warm-up time, in general, is quite small and the zero shift is negligible.

Other factors influence the behavior of gages, but if the gage is installed correctly, they are not as critical as those outlined above.

7-4 STRAIN-GAGE CONFIGURATIONS

Both in wire or foil gages many configurations and sizes are available. In wire form, gage lengths ranging from $\frac{1}{16}$ to 8 in can be obtained, whereas foil gages range from 0.008 to 4 in. Strain gages come in many forms for transducer or stress-analysis applications. The fundamental configurations for stress-analysis work are shown in Fig. 7-9.

As mentioned in Sec. 7-2, to find the complete state of stress or strain on a free surface, three pieces of information are necessary. Thus, if the state of stress is completely unknown at a point on a free surface, it is necessary to use a three-element rectangular or delta rosette since *each element* provides only one piece of information, the *normal strain* in one particular direction at the point.

The rectangular rosette provides normal-strain components in three directions spaced at angles of 45°, as shown in Fig. 7-10a. If an xy coordinate system is assumed to coincide with gages A and C, then $\varepsilon_x = \varepsilon_A$ and $\varepsilon_y = \varepsilon_C$. Gage B in conjunction with gages A and C provides information necessary to determine γ_{xy}. Using the strain-transformation equation (2-41a) with $\theta = 45°$, we can show

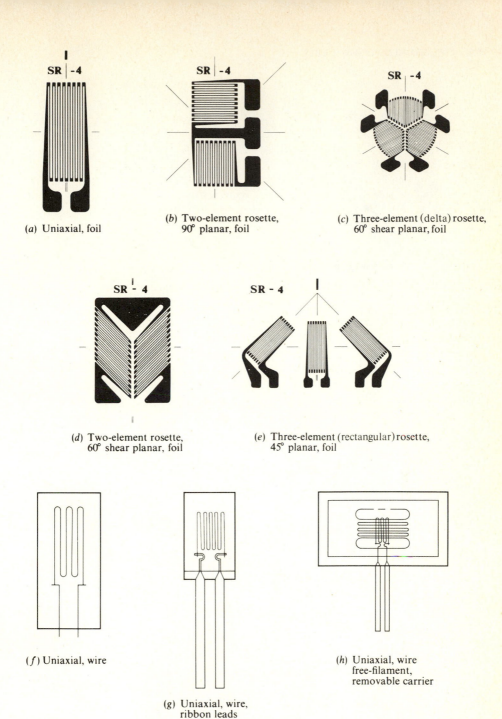

(a) Uniaxial, foil

(b) Two-element rosette,
90° planar, foil

(c) Three-element (delta) rosette,
60° shear planar, foil

(d) Two-element rosette,
60° shear planar, foil

(e) Three-element (rectangular) rosette,
45° planar, foil

(f) Uniaxial, wire

(g) Uniaxial, wire,
ribbon leads

(h) Uniaxial, wire
free-filament,
removable carrier

Figure 7-9 Examples of strain gage configurations. *(Courtesy of Baldwin-Lima-Hamilton Corporation.)*

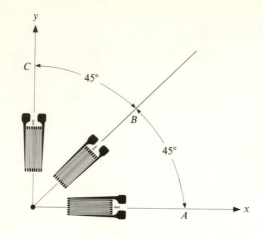

(a) Rectangular rosette

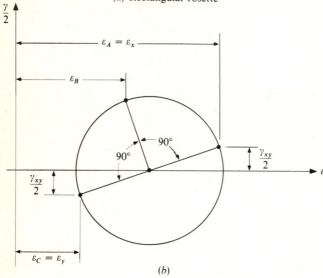

(b)

Figure 7-10

that

$$\varepsilon_B = \tfrac{1}{2}(\varepsilon_x + \varepsilon_y + \gamma_{xy}) = \tfrac{1}{2}(\varepsilon_A + \varepsilon_C + \gamma_{xy})$$

Thus $\qquad \gamma_{xy} = 2\varepsilon_B - \varepsilon_A - \varepsilon_C$

Once ε_x, ε_y, and γ_{xy} are known, we can use Hooke's law [Eqs. (1-24)] to determine the stresses σ_x, σ_y, and τ_{xy} from the stress-strain relations.

The relationship between ε_A, ε_B, and ε_C can be seen from the Mohr's circle of strain corresponding to the strain state at the point under investigation (see Fig. 7-10b).

Example 7-2 A three-element rectangular rosette is mounted on a steel specimen, and at a particular state of loading of the structure the strains are

$$\varepsilon_A = 200 \ \mu\text{in/in} \qquad \varepsilon_B = 900 \ \mu\text{in/in} \qquad \varepsilon_C = 1000 \ \mu\text{in/in}$$

Determine the values and location of the principal stresses and the value of the maximum shear stress at the point (assume the cross sensitivity of the gages to be negligible); $E = 29 \times 10^6 \ \text{lb/in}^2$, and $\nu = 0.285$.

SOLUTION

$$\varepsilon_x = \varepsilon_A = 200 \ \mu\text{in/in} \qquad \varepsilon_y = \varepsilon_C = 1000 \ \mu\text{in/in}$$

$$\gamma_{xy} = 2\varepsilon_B - \varepsilon_A - \varepsilon_C = (2)(900) - 200 - 1000 = 600 \ \mu\text{in/in}$$

The strains ε_x, ε_y, γ_{xy} can be converted into stresses σ_x, σ_y, γ_{xy} using Hooke's law, and the principal stresses can then be found using Mohr's circle, or the principal strains can be found using Mohr's strain circle, and then the principal stresses can be determined using Hooke's law.

Using the second approach, we plot Mohr's strain circle. Since γ_{xy} is positive, the shear strain is plotted below the ε axis for the x face and above for the y face (see Fig. 7-11). From the diagram, the center distance is found to be $600 \ \mu\text{in/in}$; thus

$$\varepsilon_1 = 600 + \sqrt{400^2 + 300^2} = 1100 \ \mu\text{in/in}$$

$$\varepsilon_2 = 600 - \sqrt{400^2 + 300^2} = 100 \ \mu\text{in/in}$$

$$\frac{\gamma_{\text{max}}}{2} = \sqrt{400^2 + 300^2} = 500 \qquad \gamma_{\text{max}} = 1000 \ \mu\text{in/in}$$

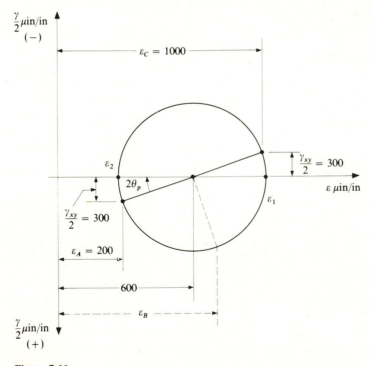

Figure 7-11

From Hooke's Law [Eqs. (1-24)] with

$$E = 29 \times 10^6 \text{ lb/in}^2 \quad \text{and} \quad \nu = 0.285$$

the principal stresses are

$$\sigma_1 = \frac{29 \times 10^6}{1 - (0.285)^2}[1100 \times 10^{-6} + (0.285)(100 \times 10^{-6})] = 35,620 \text{ lb/in}^2$$

$$\sigma_2 = \frac{29 \times 10^6}{1 - (0.285)^2}[100 \times 10^{-6} + (0.285)(1100 \times 10^{-6})] = 13,050 \text{ lb/in}^2$$

Since the principal stresses are of like sign (see Sec. 5-1), the maximum shear stress is

$$\tau_{max} = \frac{\sigma_1}{2} = 17,810 \text{ lb/in}^2$$

and does not occur on a surface perpendicular to the xy plane. The location of the principal axes is found from Fig. 7-11, where

$$2\theta_p = \tan^{-1} \tfrac{300}{400}$$

Thus, $\theta_p = 18.4°$ clockwise from gages A and C.

If in the previous example, the cross sensitivity of the gages was $K_t = 4$ percent (0.04), a correction would be necessary. From Appendix J, Eq. (J-11), the corrected strain on gage A would be

$$\varepsilon_A = \frac{1 - (0.285)(0.04)}{1 - (0.04)^2}[200 - (0.04)(1000)] = 158 \ \mu\text{in/in}$$

Likewise, for gages B and C using Eqs. (J-12) and (J-13), respectively, gives

$$\varepsilon_B = 880 \ \mu\text{in/in} \qquad \varepsilon_C = 980 \ \mu\text{in/in}$$

which in turn would result in principal strains of

$$\varepsilon_1 = 1079 \ \mu\text{in/in} \qquad \varepsilon_2 = 59 \ \mu\text{in/in}$$

and
$$\sigma_1 = 34,800 \text{ lb/in}^2 \qquad \sigma_2 = 11,600 \text{ lb/in}^2$$

The equations which relate three-element gage outputs to the principal strains and stresses are given in Appendix I.

As stated in Sec. 7-2, if the state of stress at a point is not completely unknown, less gage information is required for quantification. For example, if the directions of the principal axes are known, it is only necessary to use a two-element rectangular rosette installed along the principal axes. The gage outputs would be the principal strains, and a complete Mohr's circle analysis could be made using only two gages.

If only one piece of information is necessary to describe the complete state of stress at a point, only one single-element gage is necessary. Examples of when a single-element gage is sufficient are rods in axial loading, bending and torsion, and free surfaces whose normals lie in the plane of a plane-stress problem (thin structures).

Stress Gage

A gage which gives an output directly proportional to the stress is called a *stress gage*. For a biaxial stress field the equation for stress in any direction, say the x direction, is given by

$$\sigma_x = \frac{E}{1 - \nu^2}(\varepsilon_x + \nu\varepsilon_y) \tag{7-11}$$

Assume a gage oriented as shown in Fig. 7-12. The x axis is the bisector of the angle 2β, where the A half of the gage is oriented at an angle of $+\beta$ from the x axis and the B half at $-\beta$. The strains on the two halves are found using the transformation equation

$$\varepsilon_{x'} = \varepsilon_x \cos^2 \theta + \varepsilon_y \sin^2 \theta + \gamma_{xy} \sin \theta \cos \theta$$

Therefore
$$\varepsilon_A = \varepsilon_x \cos^2 \beta + \varepsilon_y \sin^2 \beta + \gamma_{xy} \sin \beta \cos \beta$$

$$\varepsilon_B = \varepsilon_x \cos^2 \beta + \varepsilon_y \sin^2 \beta - \gamma_{xy} \sin \beta \cos \beta$$

The gage reads the average strain, $\varepsilon_{av} = \frac{1}{2}(\varepsilon_A + \varepsilon_B)$, and so

$$\varepsilon_{av} = \varepsilon_x \cos^2 \beta + \varepsilon_y \sin^2 \beta \tag{7-12}$$

If β is such that $\beta = \tan^{-1} \sqrt{\nu}$, we have

$$\sin^2 \beta = \frac{\nu}{1 + \nu} \quad \text{and} \quad \cos^2 \beta = \frac{1}{1 + \nu}$$

Thus
$$\varepsilon_{av} = \frac{1}{1 + \nu}(\varepsilon_x + \nu\varepsilon_y) \tag{7-13}$$

Since $1 - \nu^2 = (1 - \nu)(1 + \nu)$, substitution of Eq. (7-13) into Eq. (7-11) yields

$$\sigma_x = \frac{E}{1 - \nu}\varepsilon_{av} \tag{7-14}$$

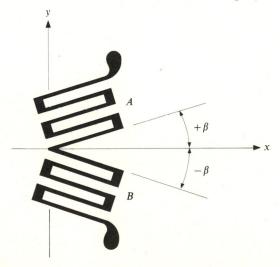

Figure 7-12

Thus it can be seen that the output ε_{av} of the stress gage is directly proportional to the stress in the direction of the angle bisector of the gage. To obtain the stress, simply multiply the output of the gage by $E/(1-\nu)$.

If the x direction is in the direction of a principal stress, $\gamma_{xy} = 0$ and only one leg of the gage, A or B, is necessary to obtain a stress gage. Thus, a single-element gage can be used in this case, where the gage is oriented at an angle of $\beta = \tan^{-1}\sqrt{\nu}$ from the principal axis.

7-5 STRAIN-GAGE INSTRUMENTATION

As mentioned earlier, the Wheatstone bridge circuit is the primary circuit used with strain gages, whereas the potentiometer circuit is used in special dynamic measurements. Only the basic applications using the Wheatstone bridge will be discussed in this section. The reader is urged to consult the references cited at the end of this chapter for further detail on instrumentation.

The Wheatstone bridge, a circuit sensitive to small resistance changes, is shown in Fig. 7-13. A dc voltage V is applied across contacts ac. It can be shown that the resulting voltage across contacts bd is given by

$$E = \frac{R_1R_3 - R_2R_4}{(R_1 + R_2)(R_3 + R_4)} V \tag{7-15}$$

When measuring changes in the output voltage or current, the procedure is to balance the initial resistances so that the nominal output voltage E is zero. Thus, for initial balance

$$R_1R_3 = R_2R_4 \tag{7-16}$$

With the circuit balanced, small changes in the resistances will cause a corresponding change in the output voltage. Substituting $R_1 + \Delta R_1$, $R_2 + \Delta R_2$,

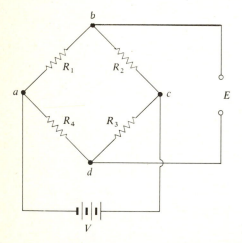

Figure 7-13

etc., into Eq. (7-15) will yield the change in voltage ΔE. After some algebraic manipulation if all but the first-order terms in $\Delta R/R$ are neglected (since $\Delta R/R$ is quite small), it can be shown that (see Exercise 7-11)

$$\Delta E \approx Vr\left(\frac{\Delta R_1}{R_1} - \frac{\Delta R_2}{R_2} + \frac{\Delta R_3}{R_3} - \frac{\Delta R_4}{R_4}\right) \tag{7-17}$$

where

$$r = \frac{R_1 R_2}{(R_1 + R_2)^2} = \frac{R_3 R_4}{(R_3 + R_4)^2}$$

For specific applications where maximum sensitivity is desired, there is some flexibility in adjusting the ratios of R_1/R_2, and the analyst can design a specific Wheatstone bridge circuit. Again, the reader is urged to consult a more advanced reference for further information.

For most analysis work, a null-balance system is used. The null-balance system is usually more accurate than the direct readout and is less expensive. The earlier inexpensive commercial strain indicators, which may still be available, use a reference-bridge circuit to provide the null-balance system (see Fig. 7-14). Through calibrated resistance changes of the non-strain-gage resistors of the circuit, the voltage E can be reduced to its null, or zero, value. The amount of calibrated resistance change is then related to the strain-gage resistance change. With commercial strain indicators, the adjustment resistance gives a direct readout in strain. A commercial null-balance system is shown in Fig. 7-15, the Baldwin-Lima-Hamilton (BLH) model 120C strain indicator with a range of 0 to 60,000 μin/in. This model is available for either ac or battery operation. Also shown in Fig. 7-15 is a Vishay's instruments model P-350A strain indicator, which operates with an ac source or battery.

Switch and balance units are available for multiple gage installations (see Fig. 7-16). Usually these units come with 10 independent strain-gage channels, which can be individually balanced, and the output of any channel can then be directed to the strain indicator via a manual or automatic channel-selection switch. Several switch and balance units can be interconnected for installations involving more than 10 gages.

Direct-digital-readout strain indicators are also available (see Fig. 7-17). Although these units are more costly, they are much more flexible and easier to use. The output can be connected to an oscilloscope for dynamic applications, the units can also be coupled with an automatic switching and balancing unit with a scan controller, and the indicator output can be automatically recorded onto a printer (see Fig. 7-17c); or the output can be directly fed into digital processing equipment.

From Eq. (7-17) it can be seen that the output voltage is proportional to resistance change in the bridge circuit. If only one gage of resistance R_g is used in the bridge so that $R_1 = R_g$, then $\Delta R_1 = \Delta R_g$ and $\Delta R_2 = \Delta R_3 = \Delta R_4 = 0$. Thus the voltage change is given by

$$\Delta E = Vr\frac{\Delta R_g}{R_g}$$

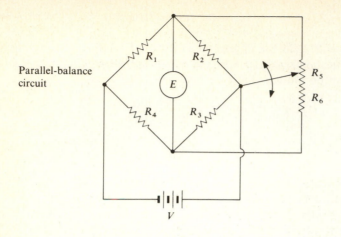

Parallel-balance circuit

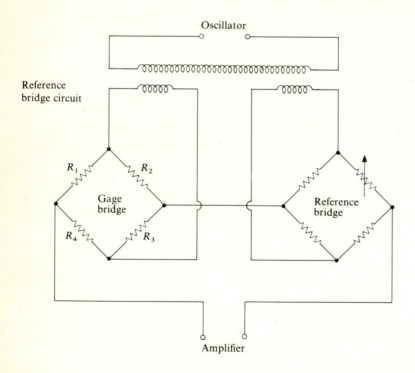

Figure 7-14 Null-balance systems.

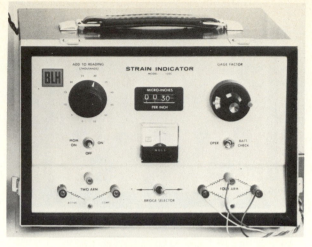

(a)

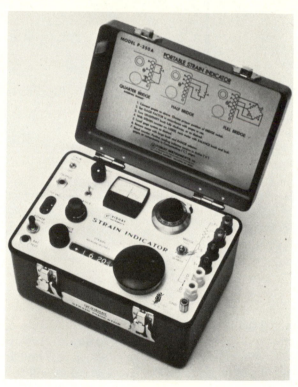

(b)

Figure 7-15 (a) Strain indicator. BLH model 120C. (*Courtesy of Baldwin-Lima-Hamilton Corporation.*) (b) Strain indicator. Vishay model P-350A. (*Courtesy of Vishay Instruments Division of Vishay Intertechnology, Inc.*)

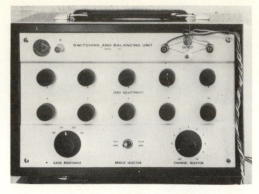

(a)

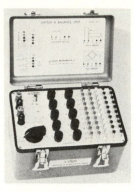

(b)

Figure 7-16 (a) Switching and balancing unit. BLH model 225. (*Courtesy of Baldwin-Lima-Hamilton Corporation.*) (b) Switch and balance unit. Vishay model SB-1. (*Courtesy of Vishay Instruments Division of Vishay Intertechnology, Inc.*)

Since

$$\frac{\Delta R_g}{R_g} = S_g \varepsilon$$

it follows that

$$\Delta E = VrS_g\varepsilon \qquad (7\text{-}18)$$

Thus the output voltage is directly proportional to the strain on the gage, and the strain indicator can be calibrated directly to strain. However, if the strain gage is not a self-temperature-compensated gage, ambient-temperature changes will cause a gage-resistance change. The result will be a change in the output voltage, which then gives a false indication of strain. As discussed in Sec. 7-3, this effect can be compensated for by using a self-temperature-compensating gage. An alternative method is to compensate for temperature changes by using a second gage in the circuit, called the *compensating gage*. The compensating gage has the same resistance and gage factor as the active gage and is normally mounted on any unloaded part made of the same material as the test specimen. If the compensating gage is the resistor R_2, then as the test

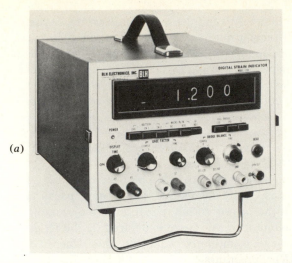

(a)

(b)

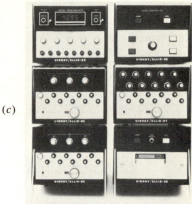

(c)

Figure 7-17 (a) Digital strain indicator. BLH model 1200. (*Courtesy of Baldwin-Lima-Hamilton Corporation.*) (b) Digital strain indicator. Vishay/Ellis model 20. (*Courtesy of Vishay Instruments Division of Vishay Intertechnology, Inc.*) (c) Digital strain indicator with scan controller, switch and balance units, and printer. (*Courtesy of Vishay Instruments Division of Vishay Intertechnology, Inc.*)

specimen is loaded and a temperature change ΔT occurs, the resulting resistance changes are

$$\frac{\Delta R_1}{R_1} = S_g \varepsilon + \frac{(\Delta R_g)_{\Delta T}}{R_g} \qquad \text{and} \qquad \frac{\Delta R_2}{R_2} = \frac{(\Delta R_g)_{\Delta T}}{R_g}$$

where $(\Delta R_g)_{\Delta T}$ is the resistance change due to ΔT. Substitution of the resistance changes into Eq. (7-17) results in

$$\Delta E = V r S_g \varepsilon$$

which is identical to the result obtained when the active gage is used alone [Eq. (7-18)]. Thus, it can be seen that temperature effects are cancelled.

Commercial strain indicators can usually be operated in three modes.

1. One-arm mode (one-fourth bridge). One strain gage is used, and three

resistors are internal to the device. Temperature compensation is possible only if a self-temperature-compensating strain gage is used.

2. Two-arm mode (one-half bridge). In this mode, two resistors are internal to the device, and the remaining two resistors can either be a strain gage and a precision resistor for manipulation of the circuit or two strain gages for temperature compensation or transducer applications. One arm of this bridge is commonly labeled the *active arm*, which yields positive readings for positive strain. The other arm is labeled the *compensating arm*, which yields negative readings for positive strain. Thus, temperature compensation is possible, as already discussed.

   The compensating gage can also be used to obtain a "gain" factor in the reading. This is especially useful when using gages as transducers. For example, say that a rod within a machine is undergoing a changing axial force and a monitor of this force is desired. If the R_1 gage is the axial gage and the R_2 gage is mounted laterally, then due to the Poisson effect, $\Delta R_2 = -\nu \Delta R_1$. Substituting ΔR_1 and ΔR_2 into Eq. (7-17) results in

$$\Delta E = (1 + \nu)Vr\frac{\Delta R_1}{R_1} = (1 + \nu)VrS_g\varepsilon$$

   where ε is the axial strain on the R_1 gage. Thus, approximately a 30 percent gain is realized if the compensating gage can be used in an active manner.

3. Four-arm mode (full bridge). No internal resistors are used in this mode. The analyst can now design the entire bridge to obtain larger gain factors either by the arrangement of nominal resistance values or by using four active strain gages.

Output sensitivity can be increased by using more active gages. This is quite important in transducer applications. For example, if two strain gages are installed where the strain on each gage will be the same and the gages are the R_1 and R_3 resistors, then

$$\frac{\Delta R_1}{R_1} = \frac{\Delta R_3}{R_3} = S_g\varepsilon$$

and substitution into Eq. (7-17) yields

$$\Delta E = 2VrS_g\varepsilon$$

Thus the output is doubled, and since the indicator reads strain directly, the actual strain is one-half the indicated strain. Temperature compensation can be achieved by using dummy gages for the R_2 and R_4 resistors. Further increases in sensitivity can be achieved by various means. Since this is primarily used for transducer applications, further development of these techniques is omitted here; it can be investigated in the references.

The methods used in strain-gage instrumentation constitute a large subject and were barely introduced in this section. For further information the reader is urged to consult the references and to obtain the copious literature available from the manufacturers of instrumentation supplies and equipment.

7-6 THE THEORY OF PHOTOELASTICITY

Photoelastic analysis is a very powerful tool both for educational purposes and applications. Beyond understanding the basic photoelastic phenomenon the method of analysis is quite straightforward. However, one can become so fully engrossed in the phenomenon when first introducing the theory that the methods of actually using the technique become unclear and often misunderstood. To avoid becoming bogged down in a complex mathematical development of the phenomenon, a rather loose (but physically accurate) presentation of the photoelastic effect will be given in this section.

To begin to understand photoelasticity, it is necessary to review the properties of light, as based on Maxwell's electromagnetic wave theory, and the polarization and refraction of light.

A monochromatic light source emits light rays of one particular wavelength λ, which propagates at the speed of light c. Assuming the general case, this emission of light rays propagates in many directions from the source. Consider each ray to be made up of a series of waves, where each wave can be thought of as a vector varying sinusoidally with time and position. One such wave, shown in Fig. 7-18, has wavelength λ and amplitude A_1 and is propagating at speed c in the z direction. This wave can be described by

$$a_1(t, z) = A_1 \sin \frac{2\pi}{\lambda}(z - ct) \tag{7-19}$$

If the observation of the wave is fixed at some point along the z axis, the wave can be described as

$$a_1(t) = A_1 \sin \frac{2\pi}{\lambda} ct \tag{7-20}$$

Thus the light amplitude at some point along the z axis can be thought of as a vector whose magnitude varies with time between the values of $\pm A_1$, as shown

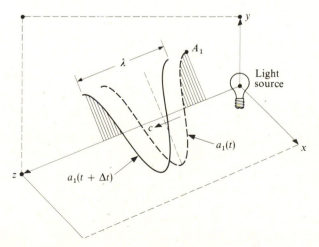

Figure 7-18

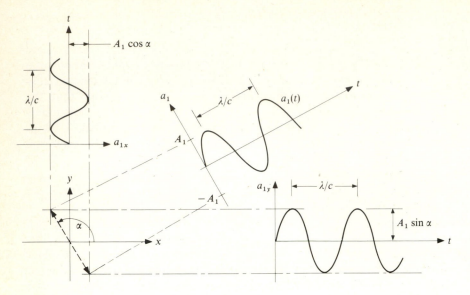

Figure 7-19

in Fig. 7-19. Since the period of a complete cycle is λ/c, the frequency of the wave is c/λ.

Because each wave, a_1, a_2, etc., can be thought of as a vector, each wave can be divided into two perpendicular components. For example, the wave a_1 of Fig. 7-19 has components in the x and y directions, which are, respectively,

$$a_{1x}(t) = A_1 \cos \alpha \, \sin \frac{2\pi}{\lambda} ct \qquad (7\text{-}21a)$$

$$a_{1y}(t) = A_1 \sin \alpha \, \sin \frac{2\pi}{\lambda} ct \qquad (7\text{-}21b)$$

Thus, the wave $a_1(t)$ can be considered as the vector sum of two waves oscillating in phase in the x and y directions propagating in the z direction.

Polarization

Consider an arbitrary wave $a(t)$. The components of $a(t)$ in the x and y direction can be written

$$a_x(t) = A_x \, \sin \frac{2\pi}{\lambda} ct \qquad (7\text{-}22a)$$

$$a_y(t) = A_y \, \sin \frac{2\pi}{\lambda} ct \qquad (7\text{-}22b)$$

where A_x and A_y are the magnitudes of the components of the wave in the x and y directions respectively (see Fig. 7-20).

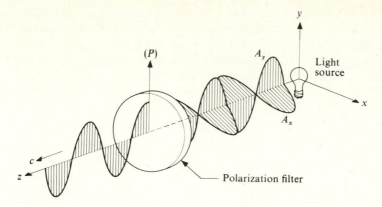

Figure 7-20

One of the resultant waves in Fig. 7-20 can be canceled by placing a polarization filter in the light path. A polarization filter can almost completely absorb a light wave perpendicular to the polarization axis of the filter. Thus if the polarization axis is arbitrarily aligned with the y direction, most of $A_x(t)$ will be absorbed and most of $A_y(t)$ will be transmitted.

The filter which polarizes the light is called the polarizer P. If a second polarization filter is placed in the light path, the light polarized by the first filter will be completely absorbed when the polarization axes of the two filters are perpendicular to each other (crossed), as shown in Fig. 7-21.

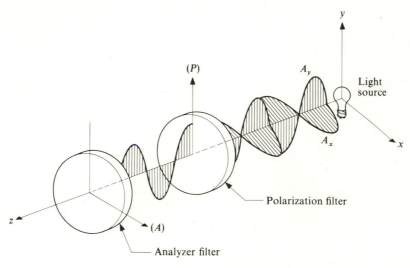

Figure 7-21

The second filter is called the analyzer A. If the analyzer axis is at an angle less than 90° with respect to the polarizer axis, some light will be transmitted. That is, the component of the polarized wave along the analyzer axis will be transmitted. If the analyzer and polarizer axes are aligned (parallel), most of the polarized light will be transmitted through the analyzer.

Refraction

When light travels from one medium into another, the speed of wave propagation changes. The relative index of refraction n_r of a substance is defined as the index of refraction of the substance relative to a surrounding medium, e.g., air, and is the ratio of the speed c_a in the surrounding medium to the speed c_s in the substance. Thus

$$n_r = \frac{c_a}{c_s} \tag{7-23}$$

When a wave travels from air into a transparent solid, the wave slows down. However, when the wave reenters the air from the solid, the wave regains its original speed in air. This can be seen dramatically when light enters the substance at an oblique angle, e.g., a pencil in a glass of water. On the other hand, if the angle of light incidence is zero, there is no visible change although the wave still slows down. Figure 7-22a shows a wave traveling through air, and Fig. 7-22b shows the same wave with a transparent solid in the light path. The slowing down of the wave causes a reduction in the wavelength as the light passes through the substance. This then causes a shift in phase Δ. The time t for the wave of Fig. 7-22a to travel a distance $d + \Delta$ is equal to the time for the wave of Fig. 7-22b to travel through the thickness of the solid the distance d. Thus

$$t = \frac{d + \Delta}{c_a} = \frac{d}{c_s} \quad \text{or} \quad \Delta = \left(\frac{c_a}{c_s} - 1\right)d$$

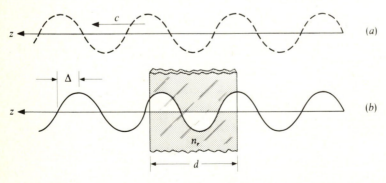

Figure 7-22

Since, $n_r = c_a/c_s$,

$$\Delta = (n_r - 1)d \tag{7-24}$$

Although there is this retardation of the light wave through the solid for normal incidence, the effect can not be detected visibly.

Birefringence

Certain natural and synthetic materials refract light differently with respect to two distinct perpendicular axes. Materials which behave in this manner are called doubly refracting or birefringent. Figure 7-23 shows a disk made of a birefringent material, where the two refraction axes are labeled 1 and 2 and n_1 and n_2 are the relative indices of refraction and c_1 and c_2 are the speeds of light

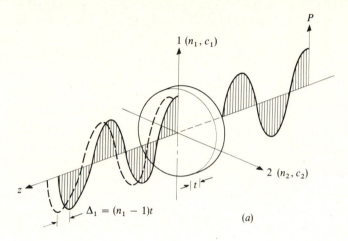

(a)

$1 (n_1, c_1)$

$2 (n_2, c_2)$

$\Delta_1 = (n_1 - 1)t$

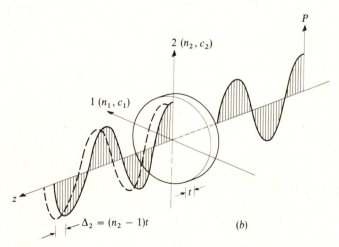

(b)

$2 (n_2, c_2)$

$1 (n_1, c_1)$

$\Delta_2 = (n_2 - 1)t$

Figure 7-23

transmission along these axes, respectively. If $n_2 > n_1$, then $c_1 > c_2$ and axis 1 is referred to as the *fast axis*, whereas axis 2 is called the *slow axis*. If light enters a birefringent plate when the light is polarized in the direction of either the fast or slow axis, as shown in Fig. 7-23, the light will emerge from the plate polarized as before but retarded by either $(n_1 - 1)t$ or $(n_2 - 1)t$, where t is now used to represent the thickness of the plate. If $n_2 > n_1$, the phase shifts are such that $\Delta_2 > \Delta_1$. If the polarization axis of the entering light is in the direction of neither the fast nor the slow axes, the form of the emerging wave is complex, since part of the wave is retarded along the fast axis and part along the slow axis.

As a special example, assume that the birefringent plate is such that $\Delta_2 - \Delta_1 = \lambda/4$. This means that there is a quarter of a wavelength difference in retardation between the slow and fast axes, and the plate is called a *quarter-wave plate*. Also assume that the wave is polarized at an angle $\beta = 45°$ with respect to the fast and slow axes. The light entering the plate is illustrated in Fig. 7-24.

As before, the wave of magnitude A can be divided into components A_1 and A_2 along axes 1 and 2, respectively. This results in two waves each of magnitude $0.707A$ and each traveling in phase. As the waves enter the plate, they both slow down. However, since axis 2 is the slow axis, the wave along axis 2 slows down more than the wave along axis 1. When the two waves emerge from the plate, they regain their original and identical speed c but the wave along axis 1 is ahead of the wave along axis 2 by the amount $\Delta = \Delta_2 - \Delta_1 = \lambda/4$, as represented in Fig. 7-25. To establish the characteristics of the emerging wave, the next step is to recombine the two component waves by vector addition. Figure 7-25 establishes the waves at a specific time and at an

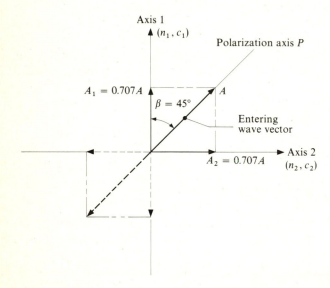

Figure 7-24

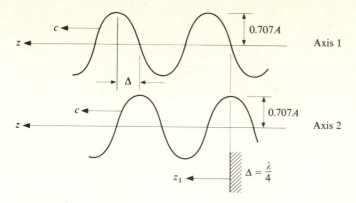

Figure 7-25

arbitrary point of reference z_1 along the light path. Let the angle that the final amplitude vector makes with axis 1 be α. The final amplitude of the wave at position z_1 is $A_f = \sqrt{a_1^2 + a_2^2}$, and the wave will be oriented with respect to axis 1 at an angle of $\alpha = \tan(a_2/a_1)$, where a_1 and a_2 are the amplitudes at position z_1 of the wave components along axes 1 and 2, respectively. Table 7-2 presents the results for A_f and α for specific values of z_1. For the points selected it can be seen that A_f is constant at a value of $0.707A$. It can be shown that for any value of z_1 the final amplitude A_f remains constant at $0.707A$. Thus it can be seen that the wave has a constant amplitude but is rotating along the z axis, making one complete revolution in one wavelength. Thus, the complete picture of the behavior of the quarter-wave plate can be seen in Fig. 7-26. The light emerging from the plate is called *circularly polarized light*.

If the analyzer filter A is placed in the light path with its polarization axis perpendicular to the polarizer P, the components of the circularly polarized light will be transmitted and observed through the analyzer. Without the quarter-wave plate in the light path, virtually no light is observed through the analyzer when it is perpendicular to the polarizer. Thus, this special example of

Table 7-2 Amplitude and orientation of wave emerging from quarter-wave plate

z_1	A_f	α
0	$0.707A$	0°
$\lambda/4$	$0.707A$	90°
$\lambda/2$	$0.707A$	180°
$3\lambda/4$	$0.707A$	270°
λ	$0.707A$	360°

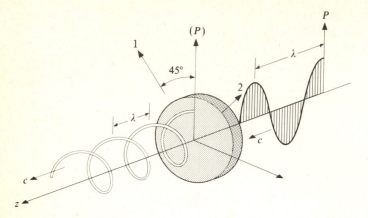

Figure 7-26

a doubly refracting material illustrates how light can be transmitted through crossed polarization filters.

If the relative wave shift of a birefringent plate Δ is an integral multiple of λ, that is, $\Delta = \lambda, 2\lambda, 3\lambda, \ldots$, the component waves leaving the plate will be in phase regardless of the value of β and their vector combination will produce a wave identical to the wave entering the birefringent plates. Thus, in this case, with the polarization filters crossed, no light will be observed since the birefringent plate does not change the light characteristics. If Δ is a half order of λ, that is, $\Delta = \lambda/2, \frac{3}{2}\lambda, \frac{5}{2}\lambda, \ldots$, it can be shown that the emerging wave is identical to the entering wave except that it is perpendicular to the polarization axis. Then if the analyzer is perpendicular to the polarizer, the full light wave will be transmitted through the analyzer (minus transmission losses).

Returning to the quarter-wave plate, if the angle β between the polarizer axis and axis 1 of the quarter-wave plate is different from 45°, the waves along axis 1 and 2 will have different amplitudes A_1 and A_2. The emerging wave is rotating as before, but instead of being circular the wave will be elliptical with a value of A_1 along axis 1 and A_2 along axis 2, where $A_1 \neq A_2$. If the angle β between the polarizer and axis 1 is 0 or 90°, the wave will not have components along one of the retardation axes regardless of the value of Δ. Hence, the emerging wave remains polarized in the same direction as when it entered, and if the analyzer is crossed with the polarizer, again no light will be transmitted. Thus, when $\beta = 0$ or 90° or when $\Delta = N\lambda$ ($N = 0, 1, 2, 3, \ldots$), no light will be transmitted through the analyzer. These two cases are the most important concepts in understanding photoelasticity.

Stress and Birefringence

When they are stressed, certain transparent materials behave in a birefractive manner, where the values of n_1 and n_2 are directly related to the principal stresses σ_1 and σ_2. Actually, the behavior is related to strain, but since in most cases the principal stresses are directly proportional to the principal strains, it is acceptable to relate the refraction behavior to the stresses. The relationship between the indices of refraction along the axes containing the principal stresses is

$$n_1 - n_2 = k(\sigma_1 - \sigma_2) \qquad (7\text{-}25)$$

where axes 1 and 2 refer to the principal-stress axes and k is called the relative stress-optic coefficient, a material property. If t is the thickness of a specimen made of this type of material, then from Eq. (7-24)

$$n_1 = \frac{\Delta_1}{t} + 1 \qquad (7\text{-}26a)$$

and

$$n_2 = \frac{\Delta_2}{t} + 1 \qquad (7\text{-}26b)$$

Therefore

$$n_1 - n_2 = \frac{\Delta_1 - \Delta_2}{t} = \frac{\Delta}{t} \qquad (7\text{-}27)$$

where Δ is the relative shift in phase between the waves emerging from the specimen along axes 1 and 2. Combining Eqs. (7-25) and (7-27) yields

$$\sigma_1 - \sigma_2 = \frac{\Delta}{kt} \qquad (7\text{-}28)$$

When $\Delta = N\lambda$, where $N = 0, 1, 2, 3, \ldots$, there is a full-wave shift which theoretically results in no net change in the wave. Thus, when the model is between two crossed polarization filters (called a *plane polariscope*, see Fig. 7-27) and $\Delta = N\lambda$ for integer values of N, no light is transmitted. Therefore one

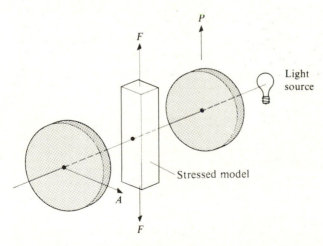

Plane polariscope

Figure 7-27

condition for the extinction of light is

$$\sigma_1 - \sigma_2 = \frac{N\lambda}{kt} \qquad N = 0, 1, 2, 3, \ldots \tag{7-29}$$

Equation (7-29) is the basic photoelastic equation which enables one to quantify the stress difference $\sigma_1 - \sigma_2$ at points where light is not transmitted. Notice that the extinction of light is also a function of the wavelength λ except when $N = 0$. This explains why whenever white light is directed through a photoelastic model, various colored bands or fringes will be seen. Each specific color represents the extinction of a particular wavelength. Due to the relative-retardation effect, only the $N = 0$ fringe will be black since it is independent of λ (all wavelengths extinguish for $N = 0$). The remainder of these fringes will be like colors in the rainbow. Thus, the entire collection of fringes (including $N = 0$) is called *isochromatic fringes*. The technique for counting fringe orders will be discussed shortly.

There is a second condition for the extinction of light. Recall from the discussion of the quarter-wave plate that when axes 1 or 2 align with the polarizer and the analyzer is crossed, no light is transmitted. This applies to the stressed material as well. Thus, when the principal stresses *align* with the polarizer and the analyzer is crossed, light is extinguished. Again, bands or fringes appear where light is extinguished. These bands, however, are always black regardless of whether or not a white light source is used. In addition, the sensitivity of this effect is relatively small, so that the fringes are usually wider than the isochromatic fringes. The fringes which develop due to the alignment of either principal stress with the polarizer are called *isoclinic fringes* since all points on a specific fringe have principal stresses in the same direction of the polarizer. If the polarizer and analyzer are rotated together relative to the stressed model and their crossed orientation is maintained, the isoclinic fringes change since only points whose stresses align with the polarizer extinguish the light. If one knows the location of the polarization axis, it is an easy task to determine the orientation of the principal axes for any point on a given model undergoing a general biaxial state of stress.

To recapitulate; when the polarizer and analyzer filters are crossed and a thin stressed birefringent material is inserted between the filters, two sets of fringes will be observed:

1. *Isochromatic fringes* These are dependent on the stress-optic effect. Whenever a zero or multiple of a full-wave shift of a particular wavelength occurs, the light wave is extinguished. This behavior is according to Eq. (7-29). If white light is directed through the model, only the zero fringe ($N = 0$) is black and the remaining isochromatic fringes are colored. Orientation of the crossed filter set is independent of this effect.
2. *Isoclinic fringes* These fringes depend on the orientation of the crossed filter set. For a particular angular position of the crossed filter set, a series of wide black fringes appear on the model. The principal stresses for every

point within the fringes are roughly in the same direction as the polarizer and analyzer. This is approximate, as there is no sharp cutoff of light. Normally the center of the band is used when recording an isoclinic fringe. These fringes are always black, regardless of the wavelength of the light source. The fringes move as the filter set is rotated relative to the model. If all fringes go through the same point regardless of the orientation of the filters, the point is an *isotropic point*. That is, Mohr's circle at this location is a point, and the state of stress in any direction is the same, totally lacking shear stress *in the plane* of investigation.

Since the isoclinic fringes are wide and black, they tend to mask the isochromatic fringes, making it difficult to analyze them. There is a method using *two quarter-wave plates* which eliminates the isoclinic fringes. Before describing it, however, the analysis of the isoclinic fringes will be discussed.

Isoclinic-Fringe Analysis

Consider a disk loaded in compression by two diametrically opposing nearly concentrated forces located on the top and bottom of the disk. The corresponding fringe patterns are shown in Fig. 7-28. The faint fringes are the isochromatic fringes and will be discussed later.† The wide, dark fringes are the isoclinic fringes. The white dotted lines were hand-drawn on the photographs to show the location of the centerline of each fringe. Each photograph shows a different angular orientation of the filter set relative to the model. It can be seen that a different set of isoclinics occur for each orientation. The cases of 0 and 90° are actually the same.

Figure 7-29 illustrates the relative positions of all the isoclinics shown in Fig. 7-28. Knowledge of the location of the principal axes will make it easier to analyze the isochromatic fringes, as will be seen later. The isoclinic data alone can be quite helpful as from this the *stress trajectories* can be constructed graphically. Stress trajectories depict the "flow" of the principal stresses (recall the hydrodynamic analogy described in Sec. 3-14) and can also help in design.

One method of constructing the stress trajectories is shown in Fig. 7-30a. For illustration, consider one quarter of the disk. Mark off a uniform grid along the horizontal diameter. At each point in the grid (here, for illustration, point a) draw a vertical line ab to the next isoclinic. Since the horizontal line is a 0-90° isoclinic, one of the principal stresses at point a is in the vertical direction. Divide line ab in half at point c. From point c, draw a line 15° in the direction that the filter set rotated when the 15° isoclinic fringe formed. The 15° line intersects the 15° isoclinic at point d and the 30° isoclinic at point e. Again,

---

† The photographs were made by directing white light through the model, and a time exposure longer than optimum was used to wash out the isochromatics. For this model, the black zero isochromatic fringe occurs only along the outer edge. Thus, the black line completely encircling the outer edge of the disk is not an isoclinic fringe.

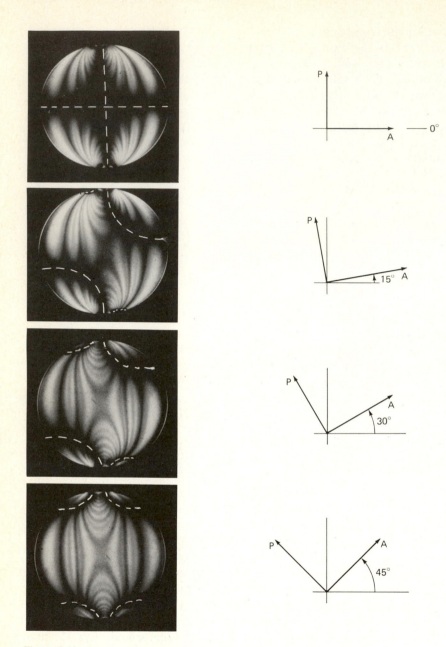

Figure 7-28

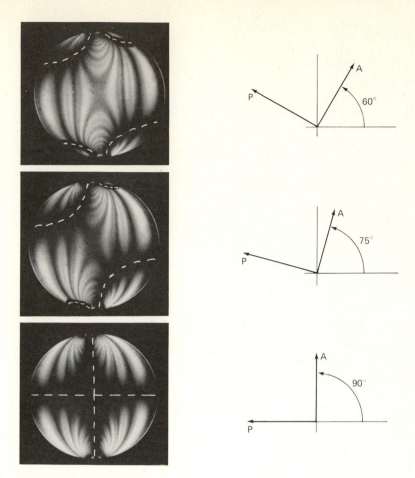

Figure 7-28 (*Continued*)

divide line *de* in half, now at point *f*. From point *f* draw a line 30° from the vertical. Continue the process as shown until the 90° isoclinic is reached. Then using a french curve, draw a continuous line tangent to each of the construction lines. This presents a good approximation of one stress trajectory. The process is continued for each initial horizontal grid point. Thanks to the symmetry of this problem, the remaining three quarters can be drawn. Thus, the first set of the stress trajectories (primary) is constructed as shown in Fig. 7-31*a*. In a similar fashion, the secondary stress trajectories can be constructed, but in this case, the construction is begun along the vertical-diameter line, as shown in Fig. 7-31*b*. It can be seen from Fig. 7-31*b* that the primary principal stresses provide an indication of how the flow of stress is transmitting through the disk.

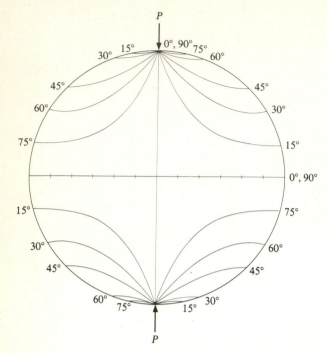

Figure 7-29

Isochromatic Fringe Analysis

There is an optical method which completely eliminates the isoclinic fringes when an analysis of the isochromatic fringes is performed. Recall that when a quarter-wave plate is positioned after the polarizer and the fast and slow axes of the plate are at 45° with respect to the polarizer axis, the wave which emerges from the plate is circularly polarized. If the model is inserted in the path of this circular polarized light, the relative refractions still take place, but since the light is spiraling through the model, the light vector aligns only with the principal stress axes at infinitesimal distances. Thus, the isoclinics will not form. As it leaves the model however, the spiraling effect induced by the quarter-wave plate must be canceled, otherwise the isochromatic fringes will be affected. This is done by placing a second quarter-wave plate between the model and analyzer filter, as shown in Fig. 7-32. This arrangement of the polarization filters and the quarter-wave plates is referred to as a *circular polariscope.*

If the fast or slow axes of both quarter-wave plates are rotated 45° so that the axes align with the polarizer, the wave plates become ineffective, the polariscope reverts to a plane polariscope, and consequently the isoclinic fringes reappear. Polariscopes are generally constructed so that the quarter-wave plates can be rotated to form either a plane or a circular polariscope.

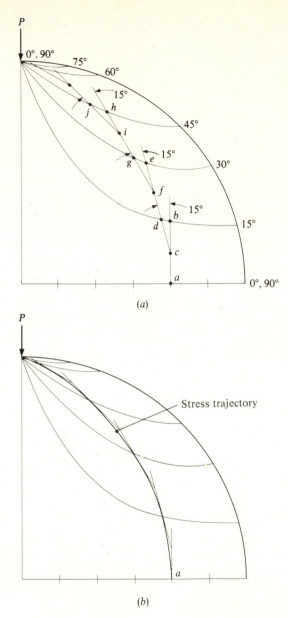

(a)

(b)

Stress trajectory

a

Figure 7-30

Consider the compressed disk again. If the disk is placed in a circular polariscope, only the isochromatic fringes will be present. A schematic representation of the fringes and their color ordering is presented in Fig. 7-33 (because of high cost a color photograph cannot be reproduced). In white light, the black isochromatic fringe is the zero-order fringe ($N = 0$). On the compressed disk, a zero-order fringe occurs everywhere along the outer edge except

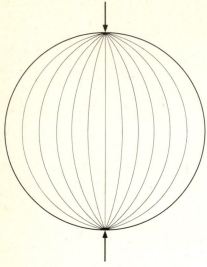

(a) Primary stress trajectories

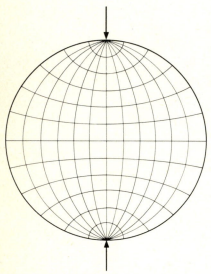

(b) Primary and secondary stress trajectories   **Figure 7-31**

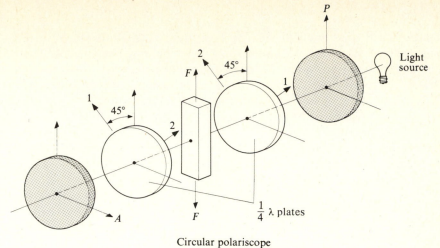

Circular polariscope

Figure 7-32

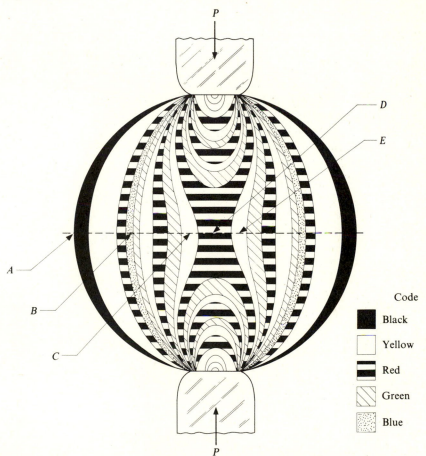

Code

■	Black
□	Yellow
▤	Red
▨	Green
▦	Blue

Figure 7-33 Isochromatic fringes using a white light source in a circular polariscope.

near the points of load application. The horizontal dotted line shown in Fig. 7-33 was drawn to illustrate the method of fringe counting. At point A, $N = 0$. Now, as one moves from point A to the right and along the dotted line, the color makes the following changes: black to yellow, yellow to orange, orange to red, red to blue, blue to green, and finally green back to yellow. This is due to the fact that each wavelength of the light refracts differently. From Eq. (7-29), $\sigma_1 - \sigma_2 = N\lambda/kt$, it can be seen that lower wavelengths extinguish first since less stress difference is required. Since the wavelength of blue or violet is the shortest in the visible spectrum, it extinguishes first, leaving a yellowish light. As the higher wavelengths extinguish, the transmission of light goes from yellow to red to green (the progression of a traffic light). The first fringe zone begins from the zero-order black fringe starting with the yellowish area and extending to the end of green. This band of colors constitutes the first fringe. Quantitative work, however, should be done using monochromatic light since Eq. (7-29) pertains to a particular wavelength. Also, the fringes will be narrow and distinct for a given wavelength.

As the fringe order increases, the order or progression of yellow, red, green continues. The order of progression changes if the fringe order decreases. For example, if we observe the color order going from point C to point E in the model shown in Fig. 7-33, we see that the order begins to reverse at point D, where the order changes to green, red, yellow, indicating that the fringe order is decreasing. This makes sense since for the model being discussed the loading is symmetric. Using white light initially makes it easier to count the fringes later with monochromatic light since the ascending and descending orders can be distinguished with white light. If the disk is now observed using monochromatic light, only black fringes will be observed as shown in Fig. 7-34. However, through the use of the white light, the orders are determined quite easily.

Using white light is not the only method of obtaining the fringe order. If the formation of the fringes is observed, one can keep account of the fringes as they develop. Another method is to use a *Babinet compensator*, covered in detail in Sec. 7-7.

Another characteristic of isochromatic fringes in white light is that a blue zone occurs between the red and green zones and is very predominant in the first-order fringe ($N = 1$); in higher-order fringes the blue zone is extremely fine and almost indiscernible. The wavelength corresponding to this color is 22.7 μin. Thus, a monochromatic filter of this wavelength is preferred for producing narrow fringes. For approximate work using white light, this red-to-green transition (called the *tint of passage*) can be used to establish fringe orders.

If a black isochromatic fringe is present ($N = 0$), fringe counting should start at this point. If no black fringe is present, a fringe present with a distinct blue between red and green is a first-order fringe ($N = 1$) and counting should start from this point. If neither the zero nor first-order fringes are present in the model, fringe development up to the final loading should be watched carefully, or the Babinet compensator should be used.

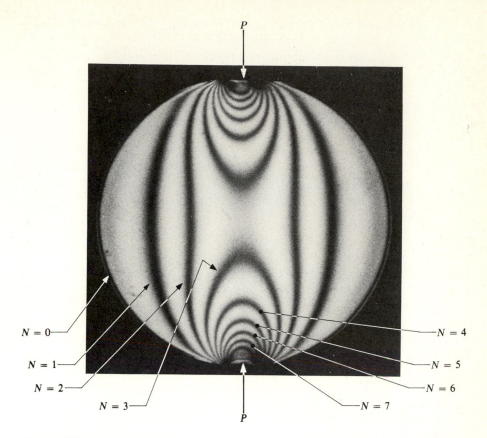

Figure 7-34

7-7 TECHNIQUES USED IN PHOTOELASTIC APPLICATIONS

Photoelastic techniques are primarily applied in practice to:

1. Two-dimensional models using transmission photoelasticity
2. Photoelastic coatings on actual components using reflection photoelastic techniques
3. Three-dimensional models using the stress-freezing method and transmission photoelasticity†

Since the scope of this chapter is introductory in nature, only a discussion of

---

† Thanks to the advent of the laser, the scattered-light technique for three-dimensional photoelasticity has become more applicable than in the past. For a description of this technique, consult Ref. 1.

the basic procedures used in the above techniques is presented; further development can be found in the references.

As shown in the previous section, two types of fringe patterns develop in a photoelastic model, isoclinic and isochromatic fringes. The isoclinic fringes provide full-field information on the directions of the principal stress axes, and the isochromatic fringes give full-field information on the principal stress difference, $\sigma_1 - \sigma_2$, according to Eq. (7-29).

The material constant k in Eq. (7-29) can be determined from a tensile test where the state of stress is known. In practice, the specimen test and tensile test are performed for a given λ (whether the test is performed in mono-chromatic light or using the tint of passage under white light) and thus λ/k is constant and is evaluated from the tensile test. If we let $\lambda/k = C$, Eq. (7-29) can be written

$$\sigma_1 - \sigma_2 = \frac{CN}{t} \tag{7-30}$$

For the tensile test, $\sigma_1 = P/A$ and $\sigma_2 = 0$, and if the thickness of the tensile specimen is t and width w, $A = wt$. Thus

$$\sigma_1 - \sigma_2 = \frac{P}{wt}$$

and from Eq. (7-30) $CN/t = P/wt$, or

$$\frac{P}{N} = Cw \tag{7-31}$$

The procedure is to load the specimen until $N = 1$ in the area where P/A is valid and record the load P. Continuing this for $N = 2, 3$, etc. and plotting P vs N yields a curve, as shown in Fig. 7-35. From Eq. (7-31), the slope of the curve is Cw. Table 7-3 summarizes some of the mechanical properties and the photoelastic constants of various photoelastic materials.

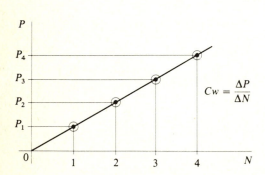
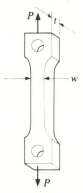

$$Cw = \frac{\Delta P}{\Delta N}$$

Figure 7-35

Table 7-3

Material	Tensile strength, lb/in²	E, lb/in²	ν	C, lb/(in) (fringe)†
CR-39	3000	300,000	0.42	100
PSM-5 epoxy‡	—	450,000	0.36	60
PSM-1 polyester‡	—	340,000	0.38	40
PSM-4 polyurethane‡	—	1,000	0.50	3–5

† $\lambda = 22.7\ \mu\text{in}.$
‡ Supplied by Photolastic Inc.

For a complete quantification of the value of the principal stresses at a given point in the model two questions remain unanswered: If the point does not fall on an isochromatic fringe, how is the fractional fringe order established? Since Eq. (7-30) represents only one equation in two unknowns, σ_1 and σ_2, how does one obtain the individual values of σ_1 and σ_2?

Fractional Fringe Orders

In the previous section it was shown how to determine the fringe orders at points in the model where light extinction (isochromatic fringes) occurs for a particular wavelength. The fringe orders for these points have integer values, that is, $N = 0, 1, 2, 3, \ldots$. Points not on the fringe lines have fractional values of N. There are two basic methods for determining the fractional fringe order at a point, the Tardy method and the Babinet-Soleil method of compensation. The Tardy method can be used with a standard polariscope, but the second requires an additional piece of equipment called a Babinet-Soleil compensator.

Tardy Method

Consider a point in the model to be between fringes of the Nth and $(N + 1)$st orders; the following steps are taken when using the Tardy method:

1. With the quarter-wave plates in the plane-polariscope position determine the principal stress directions at the point in question by rotating the crossed polarizer and analyzer filters until an isoclinic fringe intersects the point.
2. Moving only the quarter-wave plates, rotate the plates in proper position to convert to a circular polariscope.
3. Rotate only the analyzer filter until an isochromatic fringe coincides with the point. Determine the angle γ that the analyzer filter rotated.
4. If the $(N + 1)$st order fringe "moves" to the point as the analyzer rotates through the angle γ, the fringe order at the point N_Q is

$$N_Q = (N + 1) - \frac{\gamma}{180} \tag{7-32}$$

Figure 7-36 Light field transmission. Half-order isochromatic fringes ($N = \frac{1}{2}, \frac{3}{2}, \ldots$).

If the Nth order fringe "moves" to the point as the analyzer rotates through the angle γ, the fringe order at the point is

$$N_Q = N + \frac{\gamma}{180} \tag{7-33}$$

When the analyzer is rotated 90°, half-order fringes appear. In this case, the filters can be oriented with respect to the model in any direction. In addition, the light field around the model is no longer dark, and maximum light is being transmitted. This is called *light-field transmission*, as opposed to dark-field transmission, where $\gamma = 0$. An example of light field is shown in Fig. 7-36.

Babinet-Soleil Method

This method requires an additional optical element which has, in effect, an adjustable birefringence. Assume that one has a material which is doubly refracting in its free state, e.g., the material used in wave plates, and that the refraction indices of axes 1 and 2 are n_1 and n_2, respectively; then the phase-shift equation is given by Eq. (7-27), which can be written

$$\Delta = (n_1 - n_2)t \tag{7-34}$$

The phase shift can be adjusted by changing the thickness t. This is accomplished by making two wedges of the same material, as shown in Fig. 7-37a. By moving the wedges as shown, the thickness t_a can be made to vary. For practicality in designing the compensator, a third element is used. A plate of thickness t_b of the same material as the wedges is introduced in the optical path so that the axes are crossed with respect to the wedges, as shown in Fig. 7-37b. When $t_a = t_b$, there is no phase shift through the three-element compensator. If

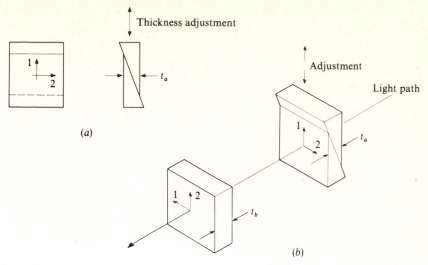

Figure 7-37

the wedges are adjusted so that $t_a > t_b$, there is a phase shift of

$$\Delta = (n_1 - n_2)(t_a - t_b) \tag{7-35}$$

If the compensator is placed in the optical path of the polariscope after the light is transmitted through the model so that the axes of the compensator line up with the principal axes, a shift of the final wavefront entering the analyzer can be accomplished simply by the wedge adjustment. The wedge adjustment can be calibrated to a particular wavelength (usually 22.7 μin) in terms of percentage of full-wave shifts in the compensator. Thus, the optical effect of using the compensator is as if an additional stress field were superposed on the model stress field. The effective stress difference of the compensator will be proportional to the thickness change, and therefore

$$(\sigma_1 - \sigma_2)_c = K_1(t_a - t_b) \tag{7-36}$$

where K_1 is a constant dependent on wavelength and material. The wedge thickness of the compensator can then be increased so that the sum of the model stresses and compensator "stresses" yields equal stresses along axes 1 and 2 (see Fig. 7-38). When $\sigma_1 - \sigma_2 = (\sigma_1)_c - (\sigma_2)_c$, the combined optical effect yields equal stresses σ, where

$$\sigma = (\sigma_1)_c + \sigma_2 = (\sigma_2)_c + \sigma_1$$

Thus, the net effect at the point as seen through the analyzer is a zero-order fringe which is black when white light is used. The compensator is calibrated in terms of fringe orders. The compensator shown in Fig. 7-39 is calibrated to 52

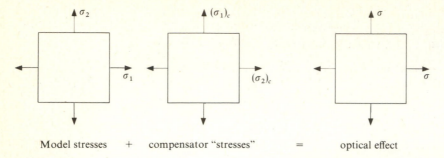

Model stresses   +   compensator "stresses"   =   optical effect

Figure 7-38

Figure 7-39 Babinet-Soleil compensator. *(Courtesy of Photolastic Inc., a subsidiary of Vishay Intertechnology Inc.)*

units per fringe. The adjustment screw is moved until a black ($N = 0$) isochromatic fringe intersects the point in question. If for example, the screw is moved 42 units when the $N = 0$ fringe aligned with some point Q, the fringe order at the point is $N_Q = \frac{42}{52} = 0.81$.

Note if the wedge adjustment causes $(\sigma_1)_c$ to increase relative to $(\sigma_2)_c$, then to obtain a net zero-order fringe with the compensator it must be aligned so that $(\sigma_1)_c$ is aligned in the direction of the principal stress in the model with the lower value σ_2. If lower-order fringes move away from the point in question when the compensator adjustment is increased, the compensator is aligned along the wrong principal axis and must be rotated 90°.

Separation of the Principal Stresses σ_1 and σ_2

For a general state of biaxial stress, three independent pieces of information are necessary. Isochromatic and isoclinic fringes provide full-field visual information which supply two pieces of information at any point. The third is usually in the form of another independent equation in the principal stresses σ_1 and σ_2, since in the general case Eq. (7-30) would contain two unknowns. However, in many two-dimensional problems, the maximum stresses occur on free surfaces, where one of the principal stresses is zero. Thus, in these cases, isochromatic information is sufficient for the determination of the remaining principal stress. For interior points of the model, however, only the stress difference and principal-axis directions are determined from the two sets of fringes. An additional relationship between σ_1 and σ_2 must be found in order to separate σ_1 and σ_2 from Eq. (7-30). Some of the common methods are:

1. Measuring the out-of-plane strain ε_z, where for a thin model undergoing biaxial stress

$$\varepsilon_z = -\frac{\nu}{E}(\sigma_x + \sigma_z) \qquad \text{or} \qquad \varepsilon_z = -\frac{\nu}{E}(\sigma_1 + \sigma_2)$$

   This can be done very accurately using interferometry methods (see Ref. 1, pp. 238–242).
2. Directing the light at an oblique angle to the model, causing new isochromatic data (called the *oblique-incidence method*).
3. Using a technique called the *shear-difference method*, where the finite-difference method is applied to the equilibrium equations [Eqs. (6-17) repeated]

$$\frac{\partial \sigma_x}{\partial x} + \frac{\partial \tau_{xy}}{\partial y} - \bar{F}_x = 0 \tag{7-37a}$$

$$\frac{\partial \tau_{xy}}{\partial x} + \frac{\partial \sigma_y}{\partial y} - \bar{F}_y = 0 \tag{7-37b}$$

   In practice the most common techniques are the oblique-incidence and the shear-difference methods.

   The oblique-incidence method can be performed by rotating the model about the σ_1 axis corresponding to the point in question or by placing a set of prisms aligned along the σ_1 axis, as shown in Fig. 7-40, where σ_1 is perpendicular to the plane of the page.

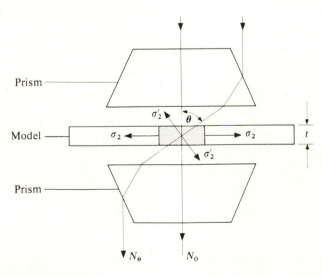

Figure 7-40

The isochromatic fringe order observed at normal incidence N_0 will generally differ from that of oblique incidence N_θ. From Eq. (7-30)

$$\sigma_1 - \sigma_2 = \frac{CN_0}{t}$$

and for the oblique ray

$$\sigma_1 - \sigma_2' = \frac{CN_\theta}{t/(\cos\theta)} \tag{7-38}$$

where $t/(\cos\theta)$ is the distance the oblique ray travels through the model. From Eq. (2-29)

$$\sigma_2' = \sigma_2 \cos^2\theta$$

Combining this with Eqs. (7-30) and (7-38) results in

$$\sigma_1 = \frac{C\cos\theta}{t\sin^2\theta}(N_\theta - N_0\cos\theta) \tag{7-39a}$$

$$\sigma_2 = \frac{C}{t\sin^2\theta}(N_\theta\cos\theta - N_0) \tag{7-39b}$$

The shear-difference method is based on the equilibrium equations. Assuming the absence of body forces, we can write (7-37a) as

$$\frac{\partial\sigma_x}{\partial x} + \frac{\partial\tau_{xy}}{\partial y} = 0$$

For an approximation, this equation can be expressed by finite differences as

$$\left.\frac{\Delta\sigma_x}{\Delta x}\right|_{y=\text{const}} + \left.\frac{\Delta\tau_{xy}}{\Delta y}\right|_{x=\text{const}} = 0$$

or

$$\Delta\sigma_x = -\left.\frac{\Delta\tau_{xy}}{\Delta y}\right|_{x=\text{const}}\left.\Delta x\right|_{y=\text{const}} \tag{7-40}$$

where Δx and Δy are made as small as practical. Starting at a free surface (or any surface where the state of stress is known), construct a grid as shown in Fig. 7-41. Assuming that σ_x at point 0 is known, $(\sigma_x)_0$, we see from Eq. (7-40) that the change in σ_x from point 0 to point 1 is

$$\Delta\sigma_x = (\sigma_x)_1 - (\sigma_x)_0 = -\frac{(\tau_{xy})_a - (\tau_{xy})_b}{\Delta y}\Delta x$$

Thus the value of σ_x at point 1 is

$$(\sigma_x)_1 = (\sigma_x)_0 - \frac{\Delta x}{\Delta y}[(\tau_{xy})_a - (\tau_{xy})_b] \tag{7-41}$$

The shear stress τ_{xy} can be calculated from the photoelastic data. From Mohr's circle of stress it can be seen that

$$\tau_{xy} = \frac{\sigma_1 - \sigma_2}{2}\sin 2\theta \tag{7-42}$$

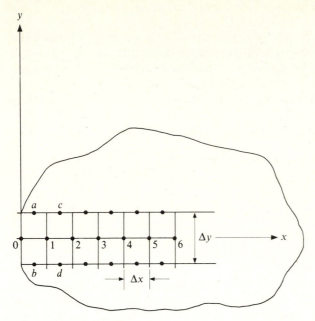

Figure 7-41

where $\sigma_1 - \sigma_2$ can be determined from isochromatic fringe data via Eq. (7-30) and θ is the angle between the x axis and the direction of σ_1, which is determined from the *isoclinic* fringe data at a specific point a. Thus the shear stresses $(\tau_{xy})_a$ and $(\tau_{xy})_b$ can be determined from the photoelastic fringes. Once $(\sigma_x)_1$ is calculated from Eq. (7-41), the principal stresses for point 1 can be determined. From Eq. (2-29) it can be shown that for point i $(i = 1, 2, 3 \ldots)$

$$(\sigma_x)_i = (\sigma_1)_i \cos^2 \theta_i + (\sigma_2)_i \sin^2 \theta_i \qquad (7\text{-}43)$$

Solving Eqs. (7-43) and (7-30) simultaneously yields

$$(\sigma_1)_i = (\sigma_x)_i + \frac{CN}{t} \sin^2 \theta_i \qquad (7\text{-}44a)$$

$$(\sigma_2)_i = (\sigma_x)_i - \frac{CN}{t} \cos^2 \theta_i \qquad (7\text{-}44b)$$

The procedure is then repeated for point 2. That is,

$$(\sigma_x)_2 = (\sigma_x)_1 - \frac{\Delta x}{\Delta y} [(\tau_{xy})_c - (\tau_{xy})_d]$$

etc. The technique can be continued throughout the interior of the model to obtain the stress values at any interior point. Since the method is based on numerical integration, errors are cumulative, and the photoelastic data should be taken carefully.

Reflection Photoelasticity

Reflection techniques can be used on a model of any material. Photoelastic coatings are cemented on areas of investigation, or the entire part can be coated (see Fig. 7-42). In principal, the technique is similar to transmission photoelasticity (see Fig. 7-43).

Since light transmits through the thickness of the coating twice,

$$\sigma_1 - \sigma_2 = \frac{CN}{2t}$$

However, since the coating is used on a specimen which may be of a different material, it is preferable to use the *strain*-optic equation. From Hooke's law,

$$\varepsilon_1 - \varepsilon_2 = \frac{1+\nu}{E}(\sigma_1 - \sigma_2)$$

Thus

$$\varepsilon_1 - \varepsilon_2 = \frac{1+\nu}{E}\frac{CN}{2t} \tag{7-45}$$

Calibration of the coating is generally performed on a beam test specimen using the same coating material and thickness as that used on the prototype. Thus, $[(1 + \nu)/E](C/2t) = f$, where f is a constant for a particular thickness of coating material. Equation (7-45) can then be rewritten as

$$\varepsilon_1 - \varepsilon_2 = fN \tag{7-46}$$

The constant f is the calibration constant with units of strain per fringe.

If the test specimen is thin, the coating itself may carry an appreciable portion of the loading and a correction must be made. The corrective equation is approximated by

$$\varepsilon_1 - \varepsilon_2 = \frac{fN}{C_1} \tag{7-47}$$

where C_1 is given in Fig. 7-44a and b.

Totally coated part

Coating sections

Figure 7-42 Photoelastic coating of an actual machine component. *(Courtesy of Photolastic Inc., a subsidiary of Vishay Intertechnology, Inc.)*

Light source

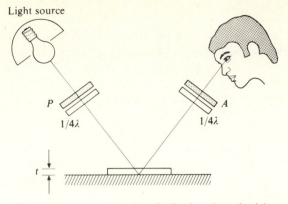

(*a*) Schematic representation of reflection photoelasticity

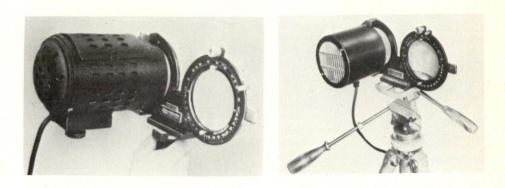

(*b*) Reflection polariscope

Fig. 7-43 (*a, Courtesy of Photolastic Inc., a subsidiary of Vishay Intertechnology, Inc.*)

Example 7-3 The cantilever beam provides a very reliable method for the calibration of a coating. Consider a cantilever beam made of aluminum, $\frac{1}{4}$ in thick and 1 in wide, and coated with a strip of 0.080-in-thick photoelastic plastic bonded to the top surface of the beam, as shown in Fig. 7-45. With a Babinet-Soleil compensator, the fringe value 6 in from the point of loading is found to be $N = 1.54$ after a 20-lb weight is applied to the end of the beam. Determine the fringe constant f for the coating. The material constants for the aluminum are $E_a = 10.3 \times 10^6$ lb/in^2 and $\nu_a = 0.33$.

SOLUTION The stress on the beam (if uncoated) at the measurement point is

$$\sigma_1 = \frac{Mc}{I} = \frac{(6)(20)(0.25/2)}{(\frac{1}{12})(1)(0.25)^3} = 11,500 \text{ lb/in}^2 \qquad \sigma_2 = 0$$

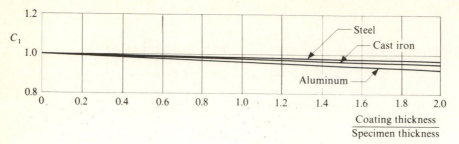

(a) Correction factors for plane-stress problems

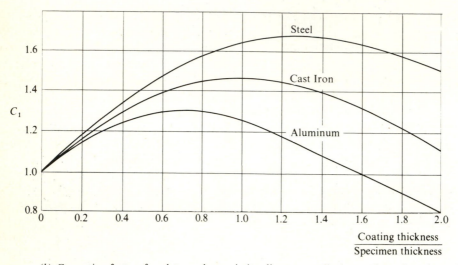

(b) Correction factors for plates or beams in bending, perpendicular to plane of coating

Figure 7-44

Figure 7-45

Since $\varepsilon_1 - \varepsilon_2 = (1 + \nu)(\sigma_1 - \sigma_2)/E$,

$$\varepsilon_1 - \varepsilon_2 = \frac{1 + 0.33}{10.3 \times 10^6}(11{,}500 - 0) = 1490 \ \mu\text{in/in}$$

The thickness ratio is

$$\frac{t_p}{t_s} = \frac{0.08}{0.25} = 0.32$$

Therefore from Fig. 7-44b, $C_1 = 1.2$. Assuming that the strain in the coating is the same as the beam, from Eq. (7-47)

$$f = \frac{C_1}{N}(\varepsilon_1 - \varepsilon_2) = \frac{1.2}{1.54}(1490) = 1160 \; \mu\,\text{in}/(\text{in})(\text{fringe})$$

If the material constants for the coating are $E_c = 340,000 \; \text{lb/in}^2$ and $\nu_c = 0.38$, the C value for the material will be

$$C = \frac{2Etf}{1+\nu} = \frac{(2)(340,000)(0.08)}{1+0.38} \, 1160 \times 10^{-6} = 45.7 \; \text{lb}/(\text{in})(\text{fringe})$$

Stress Freezing in Three-dimensional Photoelasticity

There is a method for locking or freezing the strain field in certain photoelastic materials. Once the strain field is frozen in the model, segments can be cut off the model, and if this is done carefully, the *strain field* is left undisturbed. In this way, three-dimensional investigations can be performed. The technique is called *stress freezing*, but this is a misnomer. The analysis procedure is beyond the scope of this book; detailed description can be found in Ref. 1. In this section, a simple explanation of the mechanics of stress freezing is presented.

Many polymers exhibit diphase behavior; i.e., these materials undergo drastic changes in their mechanical properties at a critical temperature. As an example, the characteristics of PLM-4B epoxy are given in Table 7-4.

Table 7-4 Characteristics of PLM-4B epoxy†

	E, lb/in^2	C, lb/(in)(fringe)	Tensile strength, lb/in^2	ν
Room temperature	450,000	60	9000	0.36
Critical temperature, 240–260°F	2500	2.2	>2000	0.50

† Data from Photolastic Inc.

The diphase behavior is due to the nature of the molecular bonds in the polymer. Above the critical temperature the weak though numerous secondary bonds break down, causing extreme changes in the mechanical properties of the material. To understand how this applies to stress freezing a simplification will suffice. Neglecting the strain induced by the coefficient of expansion of the solid, we have the stress-strain curve shown in Fig. 7-46 for the material at room temperature and at the critical temperature. The slope of the curve for room temperature is $E_{RT} = 450 \times 10^3 \; \text{lb/in}^2$, and at the critical temperature is $E_{CT} = 2.5 \times 10^3 \; \text{lb/in}^2$. Assuming a simple case such as simple tension, the model is first loaded at room temperature to point 1. With the load constant, the temperature is increased gradually to the critical temperature and held there for 2 to 4 h so that the temperature throughout the model is uniform. At this point

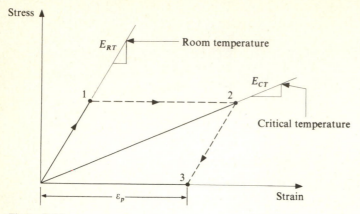

Figure 7-46

the secondary bonds have broken and the strain has increased to a steady value established by point 2. The next step is gradually to decrease the temperature back down to room temperature. During this phase, the secondary bonds re-form but in the deformed state established at point 2. Thus, for all practical purposes, when room temperature is reached, the state of the model is still basically at point 2. At this point, the loading is removed, and since the room-temperature behavior of the material is regained, the slope of the unloading curve is E_{RT} and the final state is given by point 3. The unloaded model can now be viewed through a polariscope, and photoelastic fringe patterns will be present, due to the residual strain field ε_p. Since no residual stresses are present, the three-dimensional model can be sliced and analysis can be performed on two-dimensional slices using a transmission polariscope. For the method of stress freezing, care must be taken for the following:

1. The heating and cooling cycles should be performed slowly so that temperature gradients are minimized, thus reducing the possibility of thermally induced stresses.
2. Attempts should be made to simulate loading with dead weights so that the loading will not change with deflection. An example of a loading arrangement is shown in Fig. 7-47.
3. To avoid machining stresses sharp tools, air cooling, and light cuts at high cutting speeds should be used when slicing the model for analysis.

7-8 BRITTLE COATING†

The brittle-coating method of experimental stress analysis provides a simple and direct approach to many industrial design problems where extreme

† For a comprehensive development of the theory of brittle coating see Ref. 3.

Figure 7-47 Dead weight loading used in stress freezing. *(Courtesy of Photolastic Inc., a subsidiary of Vishay Intertechnology, Inc.)*

accuracy is not required. The coating is applied to as much of the surface of the specimen under investigation as desired. The coating should fail at a low stress level so that the specimen is not overstressed. Since the coating is brittle, it will tend to fail in tension so that fine cracks will form perpendicular to the direction of the maximum principal tensile stress. The main value of the method is that it indicates where maximum stresses occur and the directions of the principal axes. In this way, fewer strain gages are necessary for quantification.

For a general biaxial state of stress it is difficult to achieve good accuracy with this technique. The inaccuracies are due to mismatch of Poisson's ratio of the coating and specimen; the brittle-failure phenomenon; and the variability of the coating due to thickness control, temperature, humidity, and load-time control. In practice, an approximate quantification is based on the assumption of a uniaxial stress field, where the principal stress in the specimen, perpendicular to the coating crack, is computed by

$$(\sigma_1)_s \approx E_s \varepsilon_t \tag{7-48}$$

where $(\sigma_1)_s$ = principal stress in specimen when crack forms

E_s = modulus of elasticity of specimen

ε_t = strain required to crack coating (called the threshold strain)

The threshold strain is determined by coating several beam specimens at the same time as the specimen under investigation.

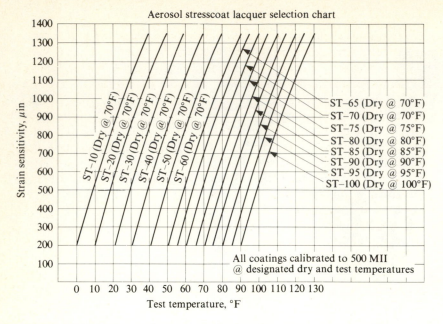

(a)

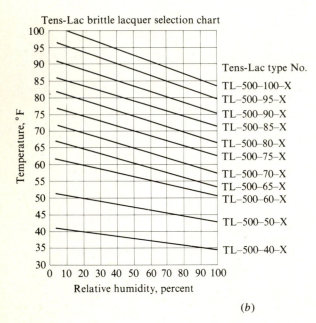

(b)

Figure 7-48 (*a*) Stresscoat brittle coating selection chart. (*Courtesy of Magnaflux Corporation.*) (*b*) Tens-Lac brittle coating selection chart. (*Courtesy of Photolastic Inc., a subsidiary of Vishay Intertechnology, Inc.*)

There are two general classes of brittle coatings. One class has a resin base and is designed for room-temperature use. The second type has a glass base and is used in a moderately elevated temperature range (to 600°F). Magnaflux Corporation manufactures both types under the trade names Stresscoat and Stresscoat All-Temp, respectively. Photolastic Inc. manufactures a room-temperature coating under the name Tens-Lac. The coatings are available in formulations of varying strain sensitivity (threshold strain) dependent on the ambient testing environment (temperature and humidity). Generally, the threshold strain used in applications is 500 μin/in, and the nominal sensitivity of the coating is based on this value. Selection curves for Stresscoat and Tens-Lac are given in Fig. 7-48. Stresscoat and Tens-Lac are both available in aerosol spray cans and in bulk quantities for gun spraying. Figure 7-49 shows the Tens-Lac product line, and Fig. 7-50 shows the Stresscoat materials including spray-gun and accessories.

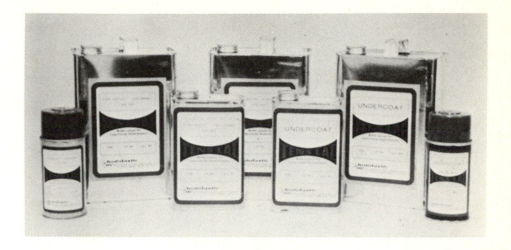

Figure 7-49 Tens-Lac brittle coating product line. (*Courtesy of Photolastic Inc., a subsidiary of Vishay Intertechnology, Inc.*)

Figure 7-50 Stresscoat brittle coating product line. *(Courtesy of Magnaflux Corporation.)*

The manufacturers supply detailed literature covering the methods of surface preparation, reflective undercoating, and the application of the brittle coating. The Stresscoat formulation consists of about one-third zinc resinate dissolved in carbon disulfide with a small amount of plasticizer, dibutyl phthalate. The carbon disulfide makes the material highly toxic and flammable during application. Extreme caution should be exercised; the spray booth should be ventilated, and a mask should be worn. The coating thickness is usually in the range of 0.006 to 0.008 in.

Tens-Lac consists of a resin base with methylene chloride as a solvent; consequently it has low toxicity and is nonflammable. The recommended coating thickness is 0.0015 to 0.005 in.

EXERCISES

7-1 A simply supported steel beam 3 m long of rectangular cross section 75 mm wide and 150 mm deep is to carry a uniform load of 6 kN/m along a section of the span ($E_s = 20 \times 10^{10}$ N/m^2). A photoelastic scale model 500 mm long is to be constructed of a material with a modulus of elasticity of 2×10^{10} N/m^2.

    (*a*) Determine the cross section of the model.

    (*b*) To assure that the stresses in the model do not go beyond the elastic limit of the material, it was decided to use a uniform load of 240 N/m on the model. Determine every scale factor.

    (*c*) The following data were recorded for the model:

$$\text{Maximum bending stress} = 1.4 \times 10^6 \text{ N/m}^2$$

$$\text{Maximum deflection} = 0.050 \text{ mm}$$

Estimate the corresponding values associated with the prototype.

7-2 Figure 7-2 shows a thick-walled disk loaded by diametrically opposing concentrated forces. Consider this to be a test model of a material with $E = 10 \times 10^6$ lb/in^2 and $\nu = 0.34$. The disk is also relatively thin in the axial direction. At point A the tangential strain was found to be 300 μin/in after a load of $P = 500$ lb/in was applied to the disk.

    (*a*) Determine the tangential stress in the model.

    (*b*) If the prototype is steel and 4 times larger than the model and the applied load is 5000 lb/in, determine the tangential stress in the prototype at point A ($E_s = 29 \times 10^6$ lb/in^2, $\nu_s = 0.29$).

    (*c*) The prototype is steel and 4 times larger than the model in the radial direction but extremely long and constrained in the axial direction. If the applied load is 5000 lb/in, determine the tangential stress at point A ($E_s = 29 \times 10^6$ lb/in^2, $\nu_s = 0.29$).

7-3 Figure 7-51 shows a thin plate structure $\frac{1}{4}$ in thick. Throughout the entire structure, show where fewer than three pieces of information are necessary for the quantification of the state of stress at a specific point. For each specific area show the stress element properly oriented and indicate the known and unknown stresses.

7-4 Determine the shear-stress distribution along section A-A of Exercise 3-13.

7-5 A strain pulse is measured with a strain gage, and the output is shown in Fig. 7-52. If the effective gage length is 1.0 in, determine the magnitude and the speed of propagation of the pulse.

7-6 In Exercise 7-5 determine the modulus of elasticity of the specimen if the pulse is an axial strain wave. The weight density of the material is 0.098 lb/in^3.

7-7 A simply supported beam carries a uniform load along its span (see beam D.6 of Appendix D.) The length of the beam is 40 in, the cross section of the beam is rectangular 1.0 in wide and 3 in

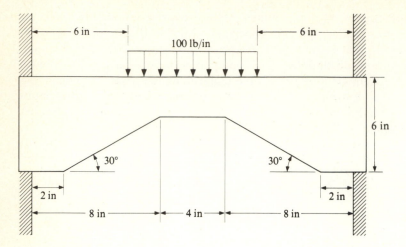

Figure 7-51

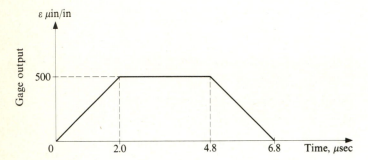

Figure 7-52

deep, and the material is steel, where $E = 29 \times 10^6$ lb/in². A strain gage 4.0 in long is cemented to the bottom of the beam and centered at midspan. The gage is used as a transducer to monitor the magnitude of the uniform load w. If at a given time the strain is measured and found to be 500 μin/in, determine (a) the value of w based on the assumption that the measured strain is the maximum strain at midspan and (b) the exact value of w. Compare this result with part (a).

7-8 A strain-gage rosette is designed as shown in Fig. 7-53. In terms of ε_A, ε_B, and ε_C express the equations for ε_x, ε_y, γ_{xy} and the principal strains. Qualitatively show the locations and relative positions of the strains on Mohr's circle.

7-9 A three-element delta-rosette strain gage is mounted on a steel specimen, where $E_s = 20 \times 10^{10}$ N/m² and $\nu = 0.29$. Under a given load condition the gages read $\varepsilon_A = 546$ μm/m, $\varepsilon_B = -146$ μm/m, and $\varepsilon_C = 0$ μm/m.

    (a) Determine the principal stresses and the location of the principal axes relative to the strain-gage rosette. (In this part of the problem ignore the transverse sensitivity of the gages.)

    (b) Repeat the analysis of part (a) for a transverse gage sensitivity of 5 percent.

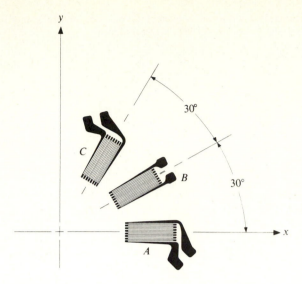

Figure 7-53

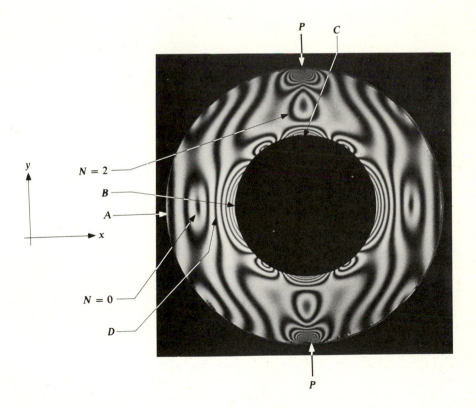

Figure 7-54

7-10 Several sets of three-element rectangular rosettes are applied at various locations on a steel structure. Under a given load condition the following readings are recorded:

Rosette	ε_A, μin/in	ε_B, μin/in	ε_C, μin/in
1	0	500	0
2	546	0	−146
3	692	−400	−692
4	−300	700	700
5	131	594	269

For each case, determine the principal strains, principal stresses, and the orientation of the principal axes relative to the strain-gage rosette.

7-11 Using Eqs. (7-15) and (7-16), derive Eq. (7-17).

7-12 A thick-walled ring made of a photoelastic material is loaded as shown in Fig. 7-54. The photoelastic constant for the material is $C = 60$ lb/(in)(fringe) and the thickness of the material is $\frac{1}{4}$ in. Estimate σ_x, σ_y, and τ_{xy} at points A, B, and C.

7-13 In Exercise 7-12, an oblique incidence adaptor is used with an angle of incidence of 30°. The fringe order at point D is $N_\theta = 1.64$; determine the values of σ_1 and σ_2 at this point.

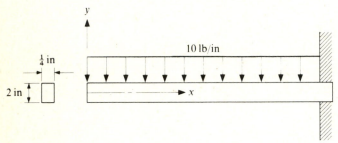

Figure 7-55

7-14 Consider the uniformly loaded rectangular cantilever beam shown in Fig. 7-55. In order to determine how σ_y varies using a photoelastic model, data were taken at $x = 5.0$ in and $x = 5.2$ in, thus making $\Delta x = 0.2$ in. With a Δy of 0.2 in the following data were obtained ($C = 60$ lb/(in)(fringe).

	$x = 5.0$ in		$x = 5.2$ in	
y, in	N	$\theta°$	N	$\theta°$
−0.9	5.63	1	6.09	1
−0.7	4.41	4	4.77	4
−0.5	3.24	8	3.49	8
−0.3	2.15	16	2.31	15
−0.1	1.36	33	1.42	32
0.1	1.43	60	1.50	61
0.3	2.30	75	2.45	76
0.5	3.40	82	3.65	82
0.7	4.58	86	4.93	86
0.9	5.80	89	6.25	89

(*a*) Derive the finite-difference equation corresponding to the equilibrium equation (7-37*b*).

(*b*) Starting at $x = 5.1$ in, $y = -1.0$ in, determine how σ_y varies at $x = 5.1$ in in the range -1 in $\leq y \leq +1$ in.

(*c*) Compare the results of part (*b*) with equation (*e*) of Example 6-5 by plotting a graph of σ_y as a function of *y*.

7-15 A pressure vessel is sprayed with a brittle coating having a threshold strain of 500 μin/in. The vessel is then pressurized, and the first cracks of the coating occur at a specific location at a pressure of 100 lb/in². The pressure is then increased, and similar cracks continue to occur at other locations. At a pressure of 150 lb/in² a second set of cracks forms perpendicular to the first set of cracks and at the same location. At this location it appears obvious that the two principal stresses at the specimen surface are both tensile. Assuming that $\sigma_1 = k_1 p$ and $\sigma_2 = k_2 p$, where k_1 and k_2 are constants and *p* is the pressure, estimate the values of k_1 and k_2. Give a brief explanation of the possible sources of errors in this analysis. For the vessel, $E = 29 \times 10^6$ lb/in² and $\nu = 0.29$.

REFERENCES

1. Dally, J. W., and W. F. Riley: "Experimental Stress Analysis," McGraw-Hill, New York, 1965.
2. Hetényi, M., (ed.): "Handbook of Experimental Stress Analysis," Wiley, New York, 1950.
3. Durelli, A. J., E. A. Phillips, and C. H. Tsao: "Introduction to the Theoretical and Experimental Analysis of Stress and Strain," McGraw-Hill, New York, 1958.

EIGHT

INTRODUCTION TO THE FINITE-ELEMENT METHOD

8-0 INTRODUCTION

The purpose of this chapter is to expose the reader to some of the fundamental concepts of finite elements, and therefore the material is only introductory in nature. The reader is urged to consult Refs. 1 to 5 for further and more advanced detail.

The finite-element method is a numerical technique, well suited to digital computers, in which a structural part is divided into small but finite elements. These elements are then superimposed onto a coordinate grid system, where identifiable points of the elements (called *nodes*) are referenced with respect to the coordinate system. Next, through the use of matrices, the position and elastic properties of the element are defined so that the displacement of each element can be related to the forces on the element. Finally, a composite matrix of the system of every element of the structure is formed which relates the displacements of the nodal points of each element to the external forces on the structure. Once the displacement field is determined, the strains can then be evaluated using the strain-displacement relations, and then finally the stresses are evaluated using the stress-strain relations. Since matrix notation is extensively used, it is recommended that the reader fully understand the treatment of matrices presented in Appendix K before continuing beyond this point.

Many geometric shapes of elements are used in the finite-element method. In this chapter, for simplicity, only two basic elements will be discussed, the one-dimensional axial element and the two-dimensional triangular slab element. When dealing with two-dimensional plane-stress problems (thin, platelike structures), triangular shapes provide a convenient element form.

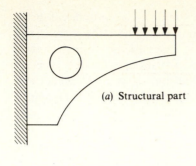

(a) Structural part

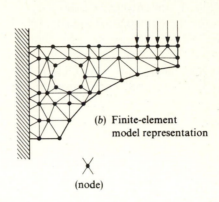

(b) Finite-element
model representation

(node)

Figure 8-1

Figure 8-1 illustrates how a real structure is modeled using triangular elements. Linear elements however, have certain inherent deficiencies:

1. As illustrated in Fig. 8-1b, curved surfaces are difficult to model using the straight-sided triangular element. The solution generally improves if more elements, smaller in size, are used in these areas. However, as will be seen later, this adds to the complexity of the solution, involving additional time and expense. Also, one would think that if the number of elements in these areas were to approach infinity, the solution would converge to the exact one. This is not necessarily true for the triangular element but further discussion of this point is beyond the level of this book.
2. After the structure is loaded, the final shape of the elements remains triangular with straight sides. Although this behavior ensures compatibility of the resulting displacement field, it places a severely unnatural restraint on the displacements, and large errors result unless many elements are used.
3. As will be seen in Sec. 8-3, the triangular slab element is a constant-strain element. This increases the difficulty in modeling even simple states of stress. For example, consider a straight beam in bending; the longitudinal strain varies with respect to the longitudinal and transverse directions. Thus, employing the finite-element method with triangular elements requires using a very large number of elements to arrive at the solution of a very simple class of problem.

The triangular element will be used in this chapter only to demonstrate the mechanics of the finite-element method, as it is a simple element to develop. In practice, however, isoparametric elements are employed, where the same (*iso-*) parameters are used to define the initial geometric shape of the element and to describe the displacements of the element. The advantages of this type of element are that curved surfaces can be used and linear strain elements are possible. The complexity of the mathematical description of these elements makes their treatment beyond the scope of this text.

Creating the composite, or master, matrix of the structure as well as writing the program for each specific problem can be a difficult and time-consuming task. Fortunately, there are many computer programs available today which reduce this difficulty, e.g., NASTRAN. The details of specific programs are left for the student to pursue (some programs are listed at the end of the chapter).

8-1 ONE-DIMENSIONAL MODELS

To begin with, consider the simple spring model shown in Fig. 8-2. The spring constant is k_1 (force/deflection), and the forces applied to the end points (nodes) are defined so that the first subscript denotes the element and the second subscript denotes the node. Since the element in Fig. 8-2 is called element 1 and the nodes are called points 1 and 2, the forces on nodes 1 and 2 are f_{11} and f_{12}, respectively, where f_{ij} denotes an internal force.

When the displacements of points 1 and 2 in the x direction are taken as u_1 and u_2, respectively, the forces necessary for each displacement can be determined. For example,

$$f_{11} = k_1(u_1 - u_2) = k_1 u_1 - k_1 u_2 \qquad (8\text{-}1a)$$

Likewise
$$f_{12} = - k_1(u_1 - u_2) = - k_1 u_1 + k_1 u_2 \qquad (8\text{-}1b)$$

The two equations can be written in matrix form as

$$\begin{Bmatrix} f_{11} \\ f_{12} \end{Bmatrix} = \begin{bmatrix} k_1 & -k_1 \\ -k_1 & k_1 \end{bmatrix} \begin{Bmatrix} u_1 \\ u_2 \end{Bmatrix} \qquad (8\text{-}2)$$

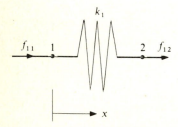

Figure 8-2

In shorthand form, Eq. (8-2) can be written

$$\{f\} = [k]\{u\} \tag{8-3}$$

where

$$\{f\} = \begin{Bmatrix} f_{11} \\ f_{12} \end{Bmatrix}$$

$$[k] = \begin{bmatrix} k_1 & -k_1 \\ -k_1 & k_1 \end{bmatrix}$$

$$\{u\} = \begin{Bmatrix} u_1 \\ u_2 \end{Bmatrix}$$

The matrix $[k]$ is called the element *stiffness matrix* since it describes the forces necessary to produce certain prescribed deflections. Equations (8-1a) and (8-1b) are not independent since $f_{11} = -f_{12}$. This fact explains why the stiffness matrix is singular and the inverse cannot be determined. A solution is obtained quite easily if the displacement of one node and the force on one node are known. For example, if $u_1 = \delta$, where δ is known, and the force f_{12} is known, then from Eq. (8-1b)

$$f_{12} = k_1(-\delta + u_2)$$

from which

$$u_2 = \frac{f_{12}}{k_1} + \delta \tag{8-4}$$

This approach and result are quite simple for a single-element system, but they become impractical for systems with many elements. Since the matrix formulation is well suited to describing large systems of elements, a method of obtaining the correct solution from the stiffness matrix is described.

The stiffness matrix can be transformed into a nonsingular matrix such that an inverse matrix which yields the correct displacements is obtainable. When the stiffness matrix is expressed in standard form, Eq. (8-2) can be rewritten

$$\begin{Bmatrix} f_{11} \\ f_{12} \end{Bmatrix} = \begin{bmatrix} k_{11} & k_{12} \\ k_{21} & k_{22} \end{bmatrix} \begin{Bmatrix} u_1 \\ u_2 \end{Bmatrix} \tag{8-5}$$

where $k_{11} = k_{22} = k_1$ and $k_{12} = k_{21} = -k_1$. As before, assume that the known displacement is u_1. The force at node 1 is f_{11}, and its dependence on u_1 is $k_{11}u_1$. To transform Eq. (8-5) so that the proper displacement relationships are obtained, the constant k_{11} is transformed into an extremely large number by multiplying k_{11} by some number N, where N is extremely large. If $u_1 = \delta$, a known deflection, this will affect the force at node 1 by $Nk_{11}\delta$. Thus Eq. (8-5) becomes

$$\begin{Bmatrix} f_{11} + Nk_{11}\delta \\ f_{12} \end{Bmatrix} = \begin{bmatrix} (1+N)k_{11} & k_{12} \\ k_{21} & k_{22} \end{bmatrix} \begin{Bmatrix} u_1 \\ u_2 \end{Bmatrix} \approx \begin{bmatrix} Nk_{11} & k_{12} \\ k_{21} & k_{22} \end{bmatrix} \begin{Bmatrix} u_1 \\ u_2 \end{Bmatrix} \tag{8-6}$$

The inverse matrix of the transformed stiffness matrix of Eq. (8-6) is

$$\begin{bmatrix} Nk_{11} & k_{12} \\ k_{21} & k_{22} \end{bmatrix}^{-1} = \frac{1}{Nk_{11}k_{22} - k_{12}k_{21}} \begin{bmatrix} k_{22} & -k_{12} \\ -k_{21} & Nk_{11} \end{bmatrix}$$

Since N is very large, $Nk_{11}k_{22} \gg k_{12}k_{21}$ and

$$\begin{bmatrix} Nk_{11} & k_{12} \\ k_{21} & k_{22} \end{bmatrix}^{-1} \approx \frac{1}{Nk_{11}k_{22}} \begin{bmatrix} k_{22} & -k_{12} \\ -k_{21} & Nk_{11} \end{bmatrix}$$

Thus, the inverse for Eq. (8-6) is

$$\begin{Bmatrix} u_1 \\ u_2 \end{Bmatrix} \approx \frac{1}{Nk_{11}k_{22}} \begin{bmatrix} k_{22} & -k_{12} \\ -k_{21} & Nk_{11} \end{bmatrix} \begin{Bmatrix} f_{11} + Nk_{11}\delta \\ f_{12} \end{Bmatrix} \tag{8-7}$$

This matrix equation actually represents two equations,

$$u_1 = \frac{1}{Nk_{11}k_{22}} (f_{11}k_{22} + k_{22}Nk_{11}\delta - k_{12}f_{12}) \tag{8-8a}$$

and

$$u_2 = \frac{1}{Nk_{11}k_{22}} (-k_{21}f_{11} - k_{21}Nk_{11}\delta + Nk_{11}f_{12}) \tag{8-8b}$$

If N is large enough, Eqs (8-8) can be approximated by

$$u_1 \approx \delta \tag{8-9a}$$

and

$$u_2 \approx -\frac{k_{21}}{k_{22}}\delta + \frac{f_{12}}{k_{22}} = \delta + \frac{f_{12}}{k_1} \tag{8-9b}$$

which are the correct displacements.

Although spring problems are rather trivial, they serve to illustrate the basic essentials of the finite-element method. The next step is to build a system of elements. Consider a two-spring system as shown in Fig. 8-3a. First each element is described individually, as shown in Fig. 8-3b. The force-displacement equations are:

Element 1:
$$\begin{Bmatrix} f_{11} \\ f_{12} \end{Bmatrix} = \begin{bmatrix} k_1 & -k_1 \\ -k_1 & k_1 \end{bmatrix} \begin{Bmatrix} u_1 \\ u_2 \end{Bmatrix}$$

Element 2:
$$\begin{Bmatrix} f_{22} \\ f_{23} \end{Bmatrix} = \begin{bmatrix} k_2 & -k_2 \\ -k_2 & k_2 \end{bmatrix} \begin{Bmatrix} u_2 \\ u_3 \end{Bmatrix}$$

The total forces at each node are the external forces, $F_1 = f_{11}$, $F_2 = f_{12} + f_{22}$, and $F_3 = f_{23}$. Thus the two matrices can be combined, resulting in

$$\begin{bmatrix} F_1 \\ F_2 \\ F_3 \end{bmatrix} = \begin{bmatrix} k_1 & -k_1 & 0 \\ -k_1 & k_1 + k_2 & -k_2 \\ 0 & -k_2 & k_2 \end{bmatrix} \begin{Bmatrix} u_1 \\ u_2 \\ u_3 \end{Bmatrix} \tag{8-10}$$

or
$$\{F\} = [k]\{u\}$$

where the force matrix $\{F\}$ represents the external forces at each node. Again

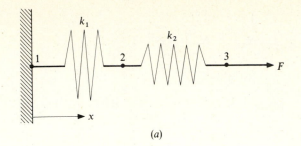

(a)

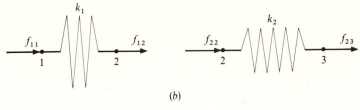

(b)

Figure 8-3

the stiffness matrix $[k]$ is singular since $F_1 + F_2 + F_3 = 0$ and the inverse matrix is indeterminate. As before, the stiffness matrix can be expressed in standard form, and Eq. (8-10) is rewritten as

$$\begin{bmatrix} F_1 \\ F_2 \\ F_3 \end{bmatrix} = \begin{bmatrix} k_{11} & k_{12} & k_{13} \\ k_{21} & k_{22} & k_{23} \\ k_{31} & k_{32} & k_{33} \end{bmatrix} \begin{Bmatrix} u_1 \\ u_2 \\ u_3 \end{Bmatrix}$$

From Fig. 8-3a the displacement at node 1 is known, where $u_1 = \delta = 0$. The stiffness matrix can be transformed as before by replacing k_{11} by Nk_{11}. Since $u_1 = 0$, this does not affect F_1. Thus

$$\begin{bmatrix} F_1 \\ F_2 \\ F_3 \end{bmatrix} = \begin{bmatrix} Nk_{11} & k_{12} & k_{13} \\ k_{21} & k_{22} & k_{23} \\ k_{31} & k_{32} & k_{33} \end{bmatrix} \begin{Bmatrix} u_1 \\ u_2 \\ u_3 \end{Bmatrix}$$

The inverse matrix of the transformed stiffness matrix is

$$\frac{1}{Nk_{11}(k_{22}k_{33} - k_{23}k_{32})} \begin{bmatrix} k_{22}k_{33} - k_{23}k_{32} & k_{13}k_{32} - k_{12}k_{33} & k_{12}k_{23} - k_{13}k_{22} \\ k_{23}k_{31} - k_{21}k_{33} & Nk_{11}k_{33} & -Nk_{11}k_{23} \\ k_{21}k_{32} - k_{31}k_{32} & -Nk_{11}k_{32} & Nk_{11}k_{22} \end{bmatrix}$$

Substituting the original spring constants yields

$$\frac{1}{Nk_1} \begin{bmatrix} 1 & 1 & 1 \\ 1 & N & N \\ 1 & N & N\dfrac{k_1 + k_2}{k_2} \end{bmatrix}$$

Thus
$$
\begin{Bmatrix} u_1 \\ u_2 \\ u_3 \end{Bmatrix} = \frac{1}{Nk_1} \begin{bmatrix} 1 & 1 & 1 \\ 1 & N & N \\ 1 & N & N\dfrac{k_1+k_2}{k_2} \end{bmatrix} \begin{Bmatrix} F_1 \\ F_2 \\ F_3 \end{Bmatrix}
$$

Note that it is unnecessary to know the value of F_1 as this force does not contribute to the solution of u_1, u_2, and u_3. Thus only the applied forces at nodes 2 and 3 need be supplied. Since $F_2 = 0$ and $F_3 = F$, if N is large enough, the above matrix equation reduces to

$$
u_1 \approx 0 \qquad u_2 \approx \frac{F}{k_1} \qquad u_3 \approx \frac{k_1+k_2}{k_1 k_2} F \qquad\qquad (8\text{-}11)
$$

The solution of two springs in series should be familiar to the reader.

Example 8-1 Repeat the previous analysis using three springs (Fig. 8-4).

SOLUTION For each element:

Element 1:
$$
\begin{Bmatrix} f_{11} \\ f_{12} \end{Bmatrix} = \begin{bmatrix} k_1 & -k_1 \\ -k_1 & k_1 \end{bmatrix} \begin{Bmatrix} u_1 \\ u_2 \end{Bmatrix}
$$

Element 2:
$$
\begin{Bmatrix} f_{22} \\ f_{23} \end{Bmatrix} = \begin{bmatrix} k_2 & -k_2 \\ -k_2 & k_2 \end{bmatrix} \begin{Bmatrix} u_2 \\ u_3 \end{Bmatrix}
$$

Element 3:
$$
\begin{Bmatrix} f_{33} \\ f_{34} \end{Bmatrix} = \begin{bmatrix} k_3 & -k_3 \\ -k_3 & k_3 \end{bmatrix} \begin{Bmatrix} u_3 \\ u_4 \end{Bmatrix}
$$

The master stiffness matrix is obtained by adding the internal forces to obtain the external forces, where $F_1 = f_{11}$, $F_2 = f_{12} + f_{22}$, $F_3 = f_{23} + f_{33}$, $F_4 = f_{34}$:

$$
\begin{Bmatrix} F_1 \\ F_2 \\ F_3 \\ F_4 \end{Bmatrix} = \begin{bmatrix} k_1 & -k_1 & 0 & 0 \\ -k_1 & k_1+k_2 & -k_2 & 0 \\ 0 & -k_2 & k_2+k_3 & -k_3 \\ 0 & 0 & -k_3 & k_3 \end{bmatrix} \begin{Bmatrix} u_1 \\ u_2 \\ u_3 \\ u_4 \end{Bmatrix}
$$

Since $u_1 = 0$, transforming the above matrix as before yields

$$
\begin{Bmatrix} F_1 \\ F_2 \\ F_3 \\ F_4 \end{Bmatrix} = \begin{bmatrix} Nk_1 & -k_2 & 0 & 0 \\ -k_1 & k_1+k_2 & -k_2 & 0 \\ 0 & -k_2 & k_2+k_3 & -k_3 \\ 0 & 0 & -k_3 & k_3 \end{bmatrix} \begin{Bmatrix} u_1 \\ u_2 \\ u_3 \\ u_4 \end{Bmatrix}
$$

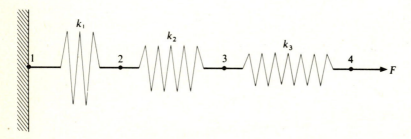

Figure 8-4

Solving for the inverse matrix yields

$$
\begin{Bmatrix} u_1 \\ u_2 \\ u_3 \\ u_4 \end{Bmatrix} = \frac{1}{Nk_1}
\begin{bmatrix}
1 & \dfrac{k_2}{k_1} & \dfrac{k_2}{k_1} & \dfrac{k_2}{k_1} \\[8pt]
1 & N & N & N \\[8pt]
1 & N & N\dfrac{k_1+k_2}{k_2} & N\dfrac{k_1+k_2}{k_2} \\[8pt]
1 & N & N\dfrac{k_1+k_2}{k_2} & N\dfrac{k_1k_2+k_2k_3+k_3k_1}{k_2k_3}
\end{bmatrix}
\begin{Bmatrix} F_1 \\ F_2 \\ F_3 \\ F_4 \end{Bmatrix}
$$

For the example, $F_2 = F_3 = 0$ and $F_4 = F$. Thus, if N is large enough

$$u_1 = 0 \qquad u_2 = \frac{F}{k_1}$$

$$u_3 = \frac{k_1+k_2}{k_1k_2}F \qquad u_4 = \frac{k_1k_2+k_2k_3+k_3k_1}{k_1k_2k_3}F$$

Thus once the stiffness matrix is determined, the displacements can be determined by inverting the transformed stiffness matrix based on boundary conditions. It can be seen that as more elements are used, the stiffness matrix grows and determination of the inverse matrix becomes more difficult. This is where digital computers become very useful.

When solid-body elastic elements are dealt with, the stiffness matrix depends on the elastic properties of the element, which are a function of the cross-sectional dimensions, length, and material properties. To illustrate this, consider an element of constant cross section, as shown in Fig. 8-5.

When it is assumed that the element deforms in a linear manner, the deflection of any point in the element can be described by

$$u = \alpha_1 + \alpha_2 x \tag{8-12}$$

where α_1 and α_2 are constants. Thus

$$u_1 = \alpha_1 + \alpha_2 x_1 \qquad u_2 = \alpha_1 + \alpha_2 x_2$$

Solving the simultaneous equations for α_1 and α_2 yields

$$\alpha_1 = \frac{u_1 x_2 - u_2 x_1}{x_2 - x_1} \qquad \alpha_2 = \frac{u_2 - u_1}{x_2 - x_1}$$

Thus, Eq. (8-12) becomes

$$u = \frac{1}{x_2 - x_1}[(u_1 x_2 - u_2 x_1) + (u_2 - u_1)x]$$

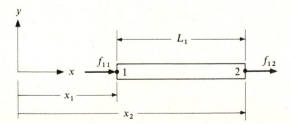

Figure 8-5

Note that the initial length of the element is $L_1 = x_2 - x_1$. Therefore

$$u = \frac{1}{L_1}[(u_1x_2 - u_2x_1) + (u_2 - u_1)x] \tag{8-13}$$

The strain on the element is given by

$$\varepsilon_x = \frac{\partial u}{\partial x}$$

Thus

$$\varepsilon_x = \frac{u_2 - u_1}{L_1} \tag{8-14}$$

In matrix form, Eq. (8-14) can be rewritten as

$$\varepsilon_x = \frac{1}{L_1}\{-1 \quad 1\}\begin{Bmatrix} u_1 \\ u_2 \end{Bmatrix} \tag{8-15}$$

In the next section of this chapter, the matrix $(1/L_1)[-1 \quad 1]$ will be referred to as the $[B]$ matrix; it relates strain to displacement and is basically dependent on the nodal positions of the element.

For a one-dimensional element, the stress is given by $\sigma_x = E_1\varepsilon_x$, where E_1 is the modulus of elasticity of element 1. Thus from Eq. (8-15)

$$\sigma_x = \frac{E_1}{L_1}\{-1 \quad 1\}\begin{Bmatrix} u_1 \\ u_2 \end{Bmatrix} \tag{8-16}$$

Finally, the forces are related to the stress by

$$f_{11} = -\sigma_x A_1 \qquad f_{12} = \sigma_x A_1$$

where A_1 is the area of the element 1 and static equilibrium is assumed. Thus

$$\begin{Bmatrix} f_{11} \\ f_{12} \end{Bmatrix} = \begin{Bmatrix} -1 \\ 1 \end{Bmatrix} A_1\sigma_x \tag{8-17}$$

Substituting Eq. (8-16) for σ_x yields

$$\begin{Bmatrix} f_{11} \\ f_{12} \end{Bmatrix} = \frac{A_1E_1}{L_1}\begin{Bmatrix} -1 \\ 1 \end{Bmatrix}\{-1 \quad 1\}\begin{Bmatrix} u_1 \\ u_2 \end{Bmatrix} \tag{8-18}$$

Multiplying the first two matrices on the right-hand side of Eq. (8-18) yields

$$\begin{Bmatrix} f_{11} \\ f_{12} \end{Bmatrix} = \frac{A_1E_1}{L_1}\begin{bmatrix} 1 & -1 \\ -1 & 1 \end{bmatrix}\begin{Bmatrix} u_1 \\ u_2 \end{Bmatrix} \tag{8-19}$$

Equation (8-19) can be written in a familiar form as

$$\begin{Bmatrix} f_{11} \\ f_{12} \end{Bmatrix} = \begin{bmatrix} k_1 & -k_1 \\ -k_1 & k_1 \end{bmatrix}\begin{Bmatrix} u_1 \\ u_2 \end{Bmatrix} \qquad \text{where } k_1 = \frac{A_1E_1}{L_1} \tag{8-20}$$

A structure with a series of axial elements can be analyzed as before where the axial displacements can be calculated. Once the displacements are known, the strain can be determined using Eq. (8-15) and the stresses using Eq. (8-16).

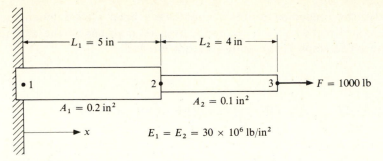

Figure 8-6

Example 8-2 Determine (a) the displacements of points 2 and 3 of the structure shown in Fig. 8-6, (b) the deflections in each element as a function of x, and (c) the strains and stresses in each element.

SOLUTION (a) For element 1,

$$k_1 = \frac{A_1 E_1}{L_1} = \frac{(0.2)(30 \times 10^6)}{5} = 1.20 \times 10^6 \text{ lb/in}$$

For element 2,

$$k_2 = \frac{A_2 E_2}{L_2} = \frac{(0.1)(30 \times 10^6)}{4} = 0.75 \times 10^6 \text{ lb/in}$$

The system of Fig. 8-6a is identical to that of the spring system of Fig. 8-3, where the result is given by Eqs. (8-11). Thus

$$u_1 = 0 \qquad u_2 = \frac{F}{k_1} = \frac{1000}{1.2 \times 10^6} = 8.33 \times 10^{-4} \text{ in}$$

and

$$u_3 = \frac{k_1 + k_2}{k_1 k_2} F = \frac{(1.20 + 0.75) \times 10^6}{(1.2 \times 10^6)(0.75 \times 10^6)}(1000) = 21.67 \times 10^{-4} \text{ in}$$

(b) For element 1 substitution of u_1, u_2, x_1, x_2, and L_1 into Eq. (8-13) yields

$$u = 1.67 \times 10^{-4}x \text{ in}$$

Similarly, for element 2, using u_2, u_3, x_2, x_3 and L_2 yields

$$u = (3.33x - 8.33) \times 10^{-4} \text{ in}$$

(c) From Eq. (8-15) the strain on element 1 is

$$(\varepsilon_x)_1 = \tfrac{1}{5}[-1 \quad 1]\left\{\begin{matrix} 0 \\ 8.33 \times 10^{-4} \end{matrix}\right\} = \tfrac{1}{5}[(-1)(0) + (1)(8.33 \times 10^{-4})] = 167 \times 10^{-6} \text{ in/in}$$

The stress is given by Eq. (8-16) and is

$$(\sigma_x)_1 = \frac{30 \times 10^6}{5}[-1 \quad 1]\left\{\begin{matrix} 0 \\ 8.33 \times 10^{-4} \end{matrix}\right\} = 5000 \text{ lb/in}^2$$

Similarly, for element 2,

$$(\varepsilon_x)_2 = \tfrac{1}{4}[-1 \quad 1]\left\{\begin{matrix} 8.33 \times 10^{-4} \\ 21.67 \times 10^{-4} \end{matrix}\right\}$$

$$= \tfrac{1}{4}[(-1)(8.33 \times 10^{-4}) + (1)(21.67 \times 10^{-4})]$$

$$= 333 \times 10^{-6} \text{ in/in}$$

and

$$(\sigma_x)_2 = \frac{30 \times 10^6}{4}[-1 \quad 1]\left\{\begin{matrix} 8.33 \times 10^{-4} \\ 21.67 \times 10^{-4} \end{matrix}\right\} = 10,000 \text{ lb/in}^2$$

Obviously, the preceding problem could have been solved much more easily by standard methods, but the application of matrices to the example illustrates the basic steps in the finite-element method.

8-2 GENERALIZATION OF THE STIFFNESS MATRIX

The foregoing procedure can be generalized for an elastic element. The analyses in this chapter will be restricted to the two-dimensional plane-stress case and further restricted by neglecting displacement dependence on the third dimension. For thin platelike structures, these restrictions result in negligible errors.

The first step is to characterize the displacement behavior. As before, it will be assumed that the displacements are a linear function of position within the structure. Thus, the displacements within each element in the x and y directions are respectively given by

$$u = \alpha_1 + \alpha_2 x + \alpha_3 y \tag{8-21a}$$

and
$$v = \beta_1 + \beta_2 x + \beta_3 y \tag{8-21b}$$

The six constants α_i and β_i are found as before by substituting the coordinates of each nodal point of the element. Thus it can be seen that the simplest element is triangular with three nodes, each node having two coordinate values. It is important to note that if the displacements are assumed to be linear and the sides of the unstrained element are linear, compatibility will automatically be satisfied.

The next step is to determine the strains using the two-dimensional strain-displacement relations. This results in

$$\varepsilon_x = \frac{\partial u}{\partial x} = \alpha_2 \tag{8-22a}$$

$$\varepsilon_y = \frac{\partial v}{\partial y} = \beta_3 \tag{8-22b}$$

$$\gamma_{xy} = \frac{\partial v}{\partial x} + \frac{\partial u}{\partial y} = \beta_2 + \alpha_3 \tag{8-22c}$$

The constants α_i and β_i will be functions of the geometric size of the element and the positions of each node. In matrix form, this can be expressed as

$$\{\varepsilon\}_n = [B]_n \{\delta\}_n \qquad \text{where } \{\varepsilon\}_n = \begin{Bmatrix} \varepsilon_x \\ \varepsilon_y \\ \gamma_{xy} \end{Bmatrix}_n \tag{8-23}$$

$[B]_n$ is the matrix term due to the geometric size of the nth element predicated by the node locations, and

$$\{\delta\}_n = \begin{Bmatrix} u_1 \\ v_1 \\ u_2 \\ v_2 \\ \vdots \end{Bmatrix}_n$$

is the displacement matrix written in term of the displacements of the nodes of the n th element.

The stress-strain relations for a homogeneous, isotropic plane-stress element are given by Eqs. (1-24) and (1-27):

$$\sigma_x = \frac{E}{1 - \nu^2}(\varepsilon_x + \nu\varepsilon_y) \qquad (8\text{-}24a)$$

$$\sigma_y = \frac{E}{1 - \nu^2}(\nu\varepsilon_x + \varepsilon_y) \qquad (8\text{-}24b)$$

and

$$\tau_{xy} = \frac{E}{2(1 + \nu)}\gamma_{xy} \qquad (8\text{-}24c)$$

This can be written in matrix form for the nth element as

$$\begin{Bmatrix} \sigma_x \\ \sigma_y \\ \tau_{xy} \end{Bmatrix}_n = \frac{E_n}{1 - \nu_n^2} \begin{bmatrix} 1 & \nu & 0 \\ \nu & 1 & 0 \\ 0 & 0 & \frac{1}{2}(1 - \nu) \end{bmatrix}_n \begin{Bmatrix} \varepsilon_x \\ \varepsilon_y \\ \gamma_{xy} \end{Bmatrix}_n \qquad (8\text{-}25)$$

or, in shorthand notation as

$$\{\sigma\}_n = [D]_n\{\varepsilon\}_n \qquad \text{where } \{\sigma\}_n = \begin{Bmatrix} \sigma_x \\ \sigma_y \\ \tau_{xy} \end{Bmatrix}_n \qquad (8\text{-}26)$$

and

$$[D]_n = \frac{E_n}{1 - \nu_n^2} \begin{bmatrix} 1 & \nu & 0 \\ \nu & 1 & 0 \\ 0 & 0 & \frac{1}{2}(1 - \nu) \end{bmatrix}_n \qquad (8\text{-}27)$$

The $[D]_n$ matrix provides a measure of the material-stiffness properties of the nth element. Substituting Eq. (8-23) into (8-26) yields

$$\{\sigma\}_n = [D]_n[B]_n\{\delta\}_n \qquad (8\text{-}28)$$

which provides the relationship between the stress in the element and its nodal displacements.

The final step is to relate the forces on each nodal point to the stress on the element. This is normally done by assuming a general uniform stress distribution on the element and converting this distribution to equivalent forces at each node. This procedure depends on the type of element used in the analysis and will be developed for the triangular element in the next section. To illustrate this, Figure 8-7a shows an element with a pressure p on one face. Figure 8-7b

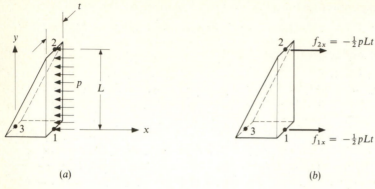

(a)   (b)

Figure 8-7

represents the equivalent nodal forces on the element. In general, the nodal forces can be related to the nth element stresses as

$$\{f\}_n = [A]_n\{\sigma\}_n \qquad \text{where } \{f\}_n = \begin{Bmatrix} f_{1x} \\ f_{1y} \\ f_{2x} \\ f_{2y} \\ \vdots \end{Bmatrix}_n \qquad (8\text{-}29)$$

$\{f\}_n$ are the internal forces in the x and y directions at each node.

The $[A]_n$ matrix is related to the element shape and size, as is the $[B]_n$ matrix, and hence the two matrices are related. This relationship is discussed in the next section. Substituting Eq. (8-28) into Eq. (8-29) yields

$$\{f\}_n = [A]_n[D]_n[B]_n\{\delta\}_n \qquad (8\text{-}30)$$

Equation (8-30) can be rewritten as

$$\{f\}_n = [k]_n\{\delta\}_n \qquad (8\text{-}31)$$

where

$$[k]_n = [A]_n[D]_n[B]_n \qquad (8\text{-}32)$$

The matrix $[k]_n$ is the stiffness matrix of the nth element. Finally, as in the simple one-dimensional problem of the previous section, the stiffness matrices of each element are compiled into a system matrix equation. This can be represented by

$$\{F\}_{\text{sys}} = [K]\{\delta\}_{\text{sys}} \qquad (8\text{-}33)$$

where $\{F\}_{\text{sys}}$ = array of total external forces F_{xi}, F_{yi} at each node
  $[K]$ = master stiffness matrix
  $\{\delta\}_{\text{sys}}$ = array of displacements u_i, v_i of each node in system of elements

Once the master stiffness is determined, the inverse matrix can be evaluated using computer techniques. The final result in shorthand notation is

$$\{\delta\}_{sys} = [K]^{-1}\{F\}_{sys} \tag{8-34}$$

where the inverse matrix $[K]^{-1}$ is evaluated considering boundary conditions as described in Sec. 8-1.

The matrix $\{\delta\}_{sys}$ is then evaluated, which yields the displacements of each node within the system, and finally the strains and stresses are determined by using Eqs. (8-23) and (8-28) for each element.

The material described in this section presents the bare essentials of the finite-element method. More advanced treatments require variational techniques that involve minimizing the total potential energy of the system using energy techniques such as Castigliano's theorem. Obviously, when one is dealing with statically indeterminate systems made up of a large quantity of elements, energy techniques provide a more direct approach.

8-3 TWO-DIMENSIONAL, PLANE-STRESS MODEL

The triangular element provides the simplest two-dimensional element for linear plane-stress problems. Consider the element shown in Fig. 8-8. To determine the $[B]$ matrix, the constants α_i, β_i must first be evaluated. From Eq. (8-21) the displacements of each node are

$$u_1 = \alpha_1 + \alpha_2 x_1 + \alpha_3 y_1 \qquad v_1 = \beta_1 + \beta_2 x_1 + \beta_3 y_1$$

$$u_2 = \alpha_1 + \alpha_2 x_2 + \alpha_3 y_2 \qquad v_2 = \beta_1 + \beta_2 x_2 + \beta_3 y_2$$

$$u_3 = \alpha_1 + \alpha_2 x_3 + \alpha_3 y_3 \qquad v_3 = \beta_1 + \beta_2 x_3 + \beta_3 y_3$$

When the above equations are solved simultaneously, the constants α_i, β_i can

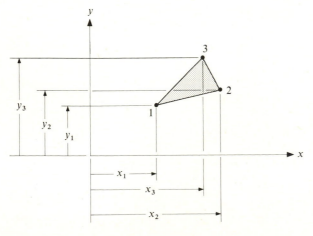

Figure 8-8

be determined in terms of u_i, v_i and x_i, y_i. This results in

$$\alpha_1 = \frac{a_1 u_1 + a_2 u_2 + a_3 u_3}{2a} \qquad (8\text{-}35a)$$

$$\alpha_2 = \frac{b_1 u_1 + b_2 u_2 + b_3 u_3}{2a} \qquad (8\text{-}35b)$$

$$\alpha_3 = \frac{c_1 u_1 + c_2 u_2 + c_3 u_3}{2a} \qquad (8\text{-}35c)$$

$$\beta_1 = \frac{a_1 v_1 + a_2 v_2 + a_3 v_3}{2a} \qquad (8\text{-}35d)$$

$$\beta_2 = \frac{b_1 v_1 + b_2 v_2 + b_3 v_3}{2a} \qquad (8\text{-}35e)$$

$$\beta_3 = \frac{c_1 v_1 + c_2 v_2 + c_3 v_3}{2a} \qquad (8\text{-}35f)$$

where

$$a_1 = x_2 y_3 - x_3 y_2 \qquad b_1 = y_2 - y_3 \qquad c_1 = x_3 - x_2$$

$$a_2 = x_3 y_1 - x_1 y_3 \qquad b_2 = y_3 - y_1 \qquad c_2 = x_1 - x_3$$

$$a_3 = x_1 y_2 - x_2 y_1 \qquad b_3 = y_1 - y_2 \qquad c_3 = x_2 - x_1$$

and

$$2a = a_1 + a_2 + a_3 \qquad (8\text{-}36)$$

It can be shown that a is the area of the element.

Next the strains are determined, recalling that

$$\varepsilon_x = \alpha_2 \qquad \varepsilon_y = \beta_3 \qquad \gamma_{xy} = \alpha_3 + \beta_2$$

Thus

$$\varepsilon_x = \frac{1}{2a}(b_1 u_1 + b_2 u_2 + b_3 u_3) \qquad (8\text{-}37a)$$

$$\varepsilon_y = \frac{1}{2a}(c_1 v_1 + c_2 v_2 + c_3 v_3) \qquad (8\text{-}37b)$$

$$\gamma_{xy} = \frac{1}{2a}(c_1 u_1 + c_2 u_2 + c_3 u_3 + b_1 v_1 + b_2 v_2 + b_3 v_3) \qquad (8\text{-}37c)$$

In matrix form, Eqs. (8-37) are written as

$$\left\{ \begin{array}{c} \varepsilon_x \\ \varepsilon_y \\ \gamma_{xy} \end{array} \right\} = \frac{1}{2a} \begin{bmatrix} b_1 & 0 & b_2 & 0 & b_3 & 0 \\ 0 & c_1 & 0 & c_2 & 0 & c_3 \\ c_1 & b_1 & c_2 & b_2 & c_3 & b_3 \end{bmatrix} \left\{ \begin{array}{c} u_1 \\ v_1 \\ u_2 \\ v_2 \\ u_3 \\ v_3 \end{array} \right\} \qquad (8\text{-}38)$$

or

$$\{\varepsilon\}_n = [B]_n \{\delta\}_n \qquad (8\text{-}39)$$

where the subscript n refers to the nth element and

$$\{\varepsilon\}_n = \begin{Bmatrix} \varepsilon_x \\ \varepsilon_y \\ \gamma_{xy} \end{Bmatrix}_n \qquad \{\delta\}_n = \begin{Bmatrix} u_1 \\ v_1 \\ u_2 \\ v_2 \\ u_3 \\ v_3 \end{Bmatrix}_n$$

and

$$[B]_n = \frac{1}{2a} \begin{bmatrix} b_1 & 0 & b_2 & 0 & b_3 & 0 \\ 0 & c_1 & 0 & c_2 & 0 & c_3 \\ c_1 & b_1 & c_2 & b_2 & c_3 & b_3 \end{bmatrix}_n \qquad (8\text{-}40)$$

The stress-strain relations were given in Sec. 8-2, where the $[D]$ matrix was defined by Eq. (8-27). Thus the next step is to evaluate the $[A]_n$ matrix for the element which relates the nodal forces to the general state of stress. The stresses relative to the element are shown in Fig. 8-9a. Since the stresses are simply the force distributions, they can be converted into equivalent forces at the nodes, as demonstrated in the previous section. First consider the σ_x stress on the left side, as shown in Fig. 8-9b. If thickness of the element is considered to be t, a constant, the total force in the x direction is $-\sigma_x t(y_3 - y_1)$. Recall Eqs. (8-36) $(y_3 - y_1 = b_2)$; then the total force is $-\sigma_x t b_2$. Thus the force on this face is divided equally between nodes 1 and 2. Therefore, due to σ_x on the left face,

$$f_{1x} = f_{3x} = -\frac{t}{2} b_2 \sigma_x \qquad (8\text{-}41a)$$

and

$$f_{2x} = 0 \qquad (8\text{-}41b)$$

On the right face, σ_x produces equivalent forces at all three nodes, as shown in Fig. 8-9c. The total force in the x direction over side 2-3 is

$$\sigma_x t(y_3 - y_2) = -\sigma_x t b_1$$

which is divided equally between nodes 2 and 3. The total force in the x direction over side 1-2 is

$$\sigma_x t(y_2 - y_1) = -\sigma_x t b_3$$

which is divided equally between nodes 1 and 2. Thus the forces on nodes 1, 2, and 3 due to the stress σ_x on the right side of the element are respectively given by

$$f_{1x} = -\frac{t}{2} b_3 \sigma_x \qquad (8\text{-}42a)$$

$$f_{2x} = -\frac{t}{2} (b_1 + b_3)\sigma_x \qquad (8\text{-}42b)$$

$$f_{3x} = -\frac{t}{2} b_1 \sigma_x \qquad (8\text{-}42c)$$

The total forces on nodes 1, 2, and 3 due to σ_x are determined by summing Eqs.

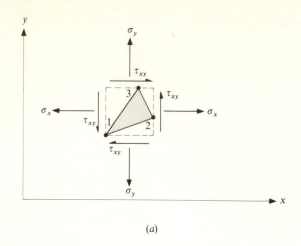

(a)

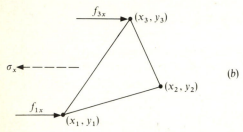

(b)

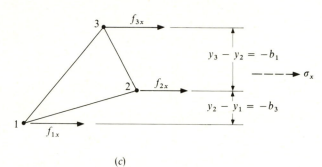

(c)

Figure 8-9

(8-41) and (8-42), or

$$f_{1x} = -\frac{t}{2}(b_2 + b_3)\sigma_x = \frac{t}{2}b_1\sigma_x \qquad (8\text{-}43a)†$$

$$f_{2x} = -\frac{t}{2}(b_1 + b_3)\sigma_x = \frac{t}{2}b_2\sigma_x \qquad (8\text{-}43b)$$

$$f_{3x} = -\frac{t}{2}(b_1 + b_2)\sigma_x = \frac{t}{2}b_3\sigma_x \qquad (8\text{-}43c)$$

†Recall Eqs. (8-36). For example, $b_2 + b_3 = (y_3 - y_1) + (y_1 - y_2) = y_3 - y_2 = -b_1$.

The forces on nodes 1, 2, and 3 due to σ_y are found in a similar manner and are

$$f_{1y} = \frac{t}{2} c_1 \sigma_y \qquad (8\text{-}44a)$$

$$f_{2y} = \frac{t}{2} c_2 \sigma_y \qquad (8\text{-}44b)$$

$$f_{3y} = \frac{t}{2} c_3 \sigma_y \qquad (8\text{-}44c)$$

To account for the shear stress, consider first the shear on the left of the element, as shown in Fig. 8-10. The total force in the y direction on side 1-3 is $-\tau_{xy} t(y_3 - y_1) = -\tau_{xy} t b_2$ and is divided equally between nodes 1 and 3, yielding $-(t/2) b_2 \tau_{xy}$ at each node.

Accounting for the four shear stresses results in the nodal forces

$$f_{1x} = -\frac{t}{2}(c_2 + c_3)\tau_{xy} = \frac{t}{2} c_1 \tau_{xy} \qquad (8\text{-}45a)$$

$$f_{1y} = -\frac{t}{2}(b_2 + b_3)\tau_{xy} = \frac{t}{2} b_1 \tau_{xy} \qquad (8\text{-}45b)$$

$$f_{2x} = -\frac{t}{2}(c_1 + c_3)\tau_{xy} = \frac{t}{2} c_2 \tau_{xy} \qquad (8\text{-}45c)$$

$$f_{2y} = -\frac{t}{2}(b_1 + b_3)\tau_{xy} = \frac{t}{2} b_2 \tau_{xy} \qquad (8\text{-}45d)$$

$$f_{3x} = -\frac{t}{2}(c_1 + c_2)\tau_{xy} = \frac{t}{2} c_3 \tau_{xy} \qquad (8\text{-}45e)$$

$$f_{3y} = -\frac{t}{2}(b_1 + b_2)\tau_{xy} = \frac{t}{2} b_3 \tau_{xy} \qquad (8\text{-}45f)$$

Adding Eqs. (8-43), (8-44), and (8-45) yields the nodal forces due to the complete state of stress

$$f_{1x} = \frac{t}{2}(b_1 \sigma_x + c_1 \tau_{xy}) \qquad (8\text{-}46a)$$

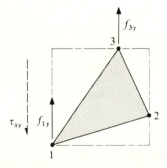

Figure 8-10

$$f_{1y} = \frac{t}{2}(c_1\sigma_y + b_1\tau_{xy}) \tag{8-46b}$$

$$f_{2x} = \frac{t}{2}(b_2\sigma_x + c_2\tau_{xy}) \tag{8-46c}$$

$$f_{2y} = \frac{t}{2}(c_2\sigma_y + b_2\tau_{xy}) \tag{8-46d}$$

$$f_{3x} = \frac{t}{2}(b_3\sigma_x + c_3\tau_{xy}) \tag{8-46e}$$

$$f_{3y} = \frac{t}{2}(c_3\sigma_y + b_3\tau_{xy}) \tag{8-46f}$$

In matrix form, Eqs. (8-46) are written as

$$\begin{Bmatrix} f_{1x} \\ f_{1y} \\ f_{2x} \\ f_{2y} \\ f_{3x} \\ f_{3y} \end{Bmatrix} = \frac{t}{2} \begin{bmatrix} b_1 & 0 & c_1 \\ 0 & c_1 & b_1 \\ b_2 & 0 & c_2 \\ 0 & c_2 & b_2 \\ b_3 & 0 & c_3 \\ 0 & c_3 & b_3 \end{bmatrix}_n \begin{Bmatrix} \sigma_x \\ \sigma_y \\ \tau_{xy} \end{Bmatrix}_n \tag{8-47}$$

or, in shorthand notation,

$$\{f\}_n = [A]_n\{\sigma\}_n \tag{8-48}$$

where

$$[A]_n = \frac{t}{2} \begin{bmatrix} b_1 & 0 & c_1 \\ 0 & c_1 & b_1 \\ b_2 & 0 & c_2 \\ 0 & c_2 & b_2 \\ b_3 & 0 & c_3 \\ 0 & c_3 & b_3 \end{bmatrix}_n \tag{8-49}$$

From observation of the matrix $[B]_n$, Eq. (8-40), we can see that

$$[A]_n = at[B]_n^t \tag{8-50}$$

where $[B]_n^t$ indicates the transpose of matrix $[B]_n$. Recall from Eq. (8-30) that $\{f\}_n = [A]_n[D]_n[B]_n\{\delta\}_n$; then substitution of Eq. (8-50) results in

$$\{f\}_n = at[B]_n^t[D]_n[B]_n\{\delta\}_n \tag{8-51}$$

Since the force is related to the displacement through the stiffness matrix by

$$\{f\}_n = [k]_n\{\delta\}_n \tag{8-52}$$

from Eq. (8-51) the stiffness matrix for the triangular element is given by

$$[k]_n = at[B]_n^t[D]_n[B]_n \tag{8-53}$$

For a triangular element, the stiffness matrix is 6×6. That is,

$$[k]_n = \begin{bmatrix} k_{11} & k_{12} & k_{13} & k_{14} & k_{15} & k_{16} \\ k_{21} & k_{22} & k_{23} & k_{24} & k_{25} & k_{26} \\ k_{31} & k_{32} & k_{33} & k_{34} & k_{35} & k_{36} \\ k_{41} & k_{42} & k_{43} & k_{44} & k_{45} & k_{46} \\ k_{51} & k_{52} & k_{53} & k_{54} & k_{55} & k_{56} \\ k_{61} & k_{62} & k_{63} & k_{64} & k_{65} & k_{66} \end{bmatrix} \tag{8-54}$$

The stiffness matrix is symmetric, where $k_{ij} = k_{ji}$. Evaluating Eqs. (8-53) using Eqs. (8-27) and (8-40) yields

$$k_{11} = \frac{Et}{4a(1-\nu^2)} [b_1^2 + \tfrac{1}{2}(1-\nu)c_1^2] \tag{8-55a}$$

$$k_{12} = k_{21} = \frac{Et}{8a(1-\nu)} b_1 c_1 \tag{8-55b}$$

$$k_{13} = k_{31} = \frac{Et}{4a(1-\nu^2)} [b_1 b_2 + \tfrac{1}{2}(1-\nu)c_1 c_2] \tag{8-55c}$$

$$k_{14} = k_{41} = \frac{Et}{4a(1-\nu^2)} [\nu b_1 c_2 + \tfrac{1}{2}(1-\nu)b_2 c_1] \tag{8-55d}$$

$$k_{15} = k_{51} = \frac{Et}{4a(1-\nu^2)} [b_1 b_3 + \tfrac{1}{2}(1-\nu)c_1 c_3] \tag{8-55e}$$

$$k_{16} = k_{61} = \frac{Et}{4a(1-\nu^2)} [\nu b_1 c_3 + \tfrac{1}{2}(1-\nu)b_3 c_1] \tag{8-55f}$$

$$k_{22} = \frac{Et}{4a(1-\nu^2)} [c_1^2 + \tfrac{1}{2}(1-\nu)b_1^2] \tag{8-55g}$$

$$k_{23} = k_{32} = \frac{Et}{4a(1-\nu^2)} [\nu b_2 c_1 + \tfrac{1}{2}(1-\nu)b_1 c_2] \tag{8-55h}$$

$$k_{24} = k_{42} = \frac{Et}{4a(1-\nu^2)} [c_1 c_2 + \tfrac{1}{2}(1-\nu)b_1 b_2] \tag{8-55i}$$

$$k_{25} = k_{52} = \frac{Et}{4a(1-\nu^2)} [\nu b_3 c_1 + \tfrac{1}{2}(1-\nu)b_1 c_3] \tag{8-55j}$$

$$k_{26} = k_{62} = \frac{Et}{4a(1-\nu^2)} [c_1 c_3 + \tfrac{1}{2}(1-\nu)b_1 b_3] \tag{8-55k}$$

$$k_{33} = \frac{Et}{4a(1-\nu^2)} [b_2^2 + \tfrac{1}{2}(1-\nu)c_2^2] \tag{8-55l}$$

$$k_{34} = k_{43} = \frac{Et}{8a(1-\nu)} b_2 c_2 \tag{8-55m}$$

$$k_{35} = k_{53} = \frac{Et}{4a(1-\nu^2)} [b_2 b_3 + \tfrac{1}{2}(1-\nu)c_2 c_3] \tag{8-55n}$$

$$k_{36} = k_{63} = \frac{Et}{4a(1-\nu^2)}[\nu b_2 c_3 + \tfrac{1}{2}(1-\nu)b_3 c_2] \qquad (8\text{-}55o)$$

$$k_{44} = \frac{Et}{4a(1-\nu^2)}[c_2^2 + \tfrac{1}{2}(1-\nu)b_2^2] \qquad (8\text{-}55p)$$

$$k_{45} = k_{54} = \frac{Et}{4a(1-\nu^2)}[\nu b_3 c_2 + \tfrac{1}{2}(1-\nu)b_2 c_3] \qquad (8\text{-}55q)$$

$$k_{46} = k_{64} = \frac{Et}{4a(1-\nu^2)}[c_2 c_3 + \tfrac{1}{2}(1-\nu)b_2 b_3] \qquad (8\text{-}55r)$$

$$k_{55} = \frac{Et}{4a(1-\nu^2)}[b_3^2 + \tfrac{1}{2}(1-\nu)c_3^2] \qquad (8\text{-}55s)$$

$$k_{56} = k_{65} = \frac{Et}{8a(1-\nu)}b_3 c_3 \qquad (8\text{-}55t)$$

$$k_{66} = \frac{Et}{4a(1-\nu^2)}[c_3^2 + \tfrac{1}{2}(1-\nu)b_3^2] \qquad (8\text{-}55u)$$

Note that the stiffness matrix for the nodal forces of a single triangular element is 6×6. In considering a particular problem, the structure is modeled using a system of triangular elements where normally the solution accuracy increases as more elements are used. This creates an extremely large system stiffness matrix, which in turn makes obtaining the inverse matrix a tremendous task, but the operation can readily be performed on a digital computer.

To obtain the master stiffness matrix, a simple two-element system is presented in the following example.

Example 8-3 Determine the master stiffness matrix for the plate loaded as shown in Fig. 8-11a, using a two-element model as shown in Fig. 8-11b.

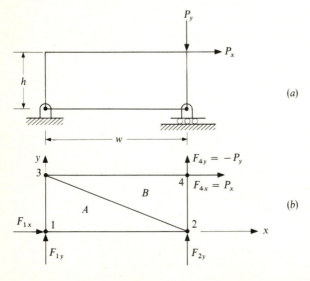

(a)

(b)

Figure 8-11

SOLUTION For element A,

$$
\begin{Bmatrix} f_{1x} \\ f_{1y} \\ f_{2x} \\ f_{2y} \\ f_{3x} \\ f_{3y} \end{Bmatrix} =
\begin{bmatrix}
k_{11} & k_{12} & k_{13} & k_{14} & k_{15} & k_{16} \\
 & k_{22} & k_{23} & k_{24} & k_{25} & k_{26} \\
 & & k_{33} & k_{34} & k_{35} & k_{36} \\
 & & & k_{44} & k_{45} & k_{46} \\
 & \text{sym} & & & k_{55} & k_{56} \\
 & & & & & k_{66}
\end{bmatrix}_A
\begin{Bmatrix} u_1 \\ v_1 \\ u_2 \\ v_2 \\ u_3 \\ v_3 \end{Bmatrix}
$$

The terms in the stiffness matrix for the A element are determined from Eqs. (8-55) using E and ν of the material, t the plate thickness, and Eqs. (8-36), where†

$$a = \tfrac{1}{2}hw \qquad b_1 = y_2 - y_3 = -h \qquad b_2 = y_3 - y_1 = h$$

$$b_3 = y_1 - y_2 = 0 \qquad c_1 = x_3 - x_2 = -w \qquad c_2 = x_1 - x_3 = 0 \qquad c_3 = x_2 - x_1 = w$$

For element B,

$$
\begin{Bmatrix} f_{2x} \\ f_{2y} \\ f_{3x} \\ f_{3y} \\ f_{4x} \\ f_{4y} \end{Bmatrix} =
\begin{bmatrix}
k_{11} & k_{12} & k_{13} & k_{14} & k_{15} & k_{16} \\
 & k_{22} & k_{23} & k_{24} & k_{25} & k_{26} \\
 & & k_{33} & k_{34} & k_{35} & k_{36} \\
 & & & k_{44} & k_{45} & k_{46} \\
 & \text{sym} & & & k_{55} & k_{56} \\
 & & & & & k_{66}
\end{bmatrix}_B
\begin{Bmatrix} u_2 \\ v_2 \\ u_3 \\ v_3 \\ u_4 \\ v_4 \end{Bmatrix}
$$

The terms in the stiffness matrix for the B element are likewise determined from Eqs. (8-55) using E and ν of the material, t the plate thickness, and Eqs. (8-36). However, the nodal points for B are points 2, 4, and 3. Thus, the subscripts for the xy coordinates are raised by 1, and the geometric constants for the B element are

$$a = \tfrac{1}{2}hw \qquad b_1 = y_4 - y_3 = h \qquad b_2 = y_3 - y_2 = h$$

$$b_3 = y_2 - y_4 = -h \qquad c_1 = x_3 - x_4 = -w \qquad c_2 = x_2 - x_3 = w \qquad c_3 = x_4 - x_2 = 0$$

To obtain the master stiffness matrix, the two matrix equations are combined. The sum of the nodal forces equals the total external force at the node. This results in

$$
\begin{Bmatrix} F_{1x} \\ F_{1y} \\ F_{2x} \\ F_{2y} \\ F_{3x} \\ F_{3y} \\ F_{4x} \\ F_{4y} \end{Bmatrix} =
\begin{bmatrix}
(k_{11})^A & (k_{12})^A & (k_{13})^A & (k_{14})^A & (k_{15})^A & (k_{16})^A & 0 & 0 \\
 & (k_{22})^A & (k_{23})^A & (k_{24})^A & (k_{25})^A & (k_{26})^A & 0 & 0 \\
 & & (k_{33})^A + (k_{11})^B & (k_{34})^A + (k_{12})^B & (k_{35})^A + (k_{13})^B & (k_{36})^A + (k_{14})^B & (k_{15})^B & (k_{16})^B \\
 & & & (k_{44})^A + (k_{22})^B & (k_{45})^A + (k_{23})^B & (k_{46})^A + (k_{24})^B & (k_{25})^B & (k_{26})^B \\
 & \text{sym} & & & (k_{55})^A + (k_{33})^B & (k_{56})^A + (k_{34})^B & (k_{35})^B & (k_{36})^B \\
 & & & & & (k_{66})^A + (k_{44})^B & (k_{45})^B & (k_{46})^B \\
 & & & & & & (k_{55})^B & (k_{56})^B \\
 & & & & & & & (k_{66})^B
\end{bmatrix}
\begin{Bmatrix} u_1 \\ v_1 \\ u_2 \\ v_2 \\ u_3 \\ v_3 \\ u_4 \\ v_4 \end{Bmatrix}
$$

Since $u_1 = v_1 = v_2 = 0$, then as in Sec. 8-1, the reaction forces F_{1x}, F_{1y}, and F_{2y} need not be specified. Also, as before, the stiffness coefficients relating those forces and displacements would be modified by multiplying $(k_{11})_A$, $(k_{22})_A$, and $(k_{44})_A + (k_{22})_B$ by N before attempting to find the inverse matrix for the final solution. Recall that N is to be a very large number.

Note that in the previous problem the resulting master stiffness matrix for the two-element model was an 8×8. Further improvement on the solution can be made by increasing the number of elements. Figure 8-12 shows a four-element model where the resulting master stiffness matrix will be a 10×10, since there are five nodes each with two displacement quantities.

† Note that the nodal positions of the triangular element are defined in a counterclockwise order.

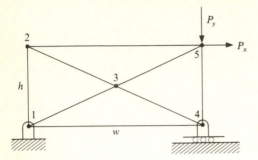

Figure 8-12

8-4 FURTHER DISCUSSION OF THE FINITE-ELEMENT METHOD

Our discussion of the finite-element method has been greatly simplified and rather restrictive, as we are only trying to outline the basis of the method. The method used in obtaining the stiffness matrix employs an equivalent-force approach to determine the nodal forces. This is also called the *direct-stiffness method*. Most of the reference books cited at the end of this chapter use variational methods, such as Castigliano's method, which minimize the total potential energy of the system. However, for the examples presented in this chapter, the variational methods produce the same stiffness matrices. When dealing with elements other than triangular shapes or in problems which are highly statically indeterminate, energy principles are necessary.

The method can also be applied satisfactorily to many three-dimensional problems, and much work is currently being concentrated on various shapes of elements for three-dimensional analyses. When modeling the actual structure, the elements need not all be the same size. The geometry of the mesh often depends on the locations of abrupt changes in loading, structure geometry, and material properties. A node should be employed where load distributions change, as shown in Fig. 8-13. Whenever an abrupt change in the cross section occurs, a steep gradient in stress and strain is expected. This calls for finer-mesh configurations near the boundaries of this change, as shown in Fig.

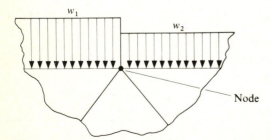

Figure 8-13

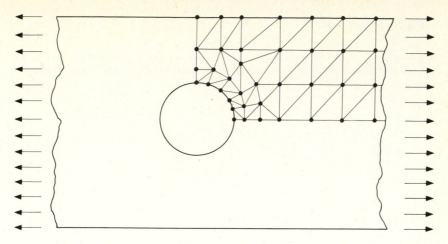

Figure 8-14

8-14. Where changes in material take place, nodal connecting lines should divide the zone of the differing materials, as shown in Fig. 8-15.

The systematic generality of the finite-element method makes the technique a very powerful and versatile approach to a wide range of problems. Many general-purpose computer programs have been and are being constructed. Some examples of programs including a variety of elemental configurations which can be applied to several types of structural problems are NASTRAN (*NA*sa *STR*uctural *AN*alysis, U.S. National Aeronautics and Space Administration), ANSYS (Engineering *A*Nalysis *SYS*tem, Swanson Analysis Systems, Inc.), MAGIC (*M*atrix *A*nalysis via *G*enerative and *I*nterpretive *C*omputation, Air Force Flight Dynamics Laboratory, Wright-Patterson Air Force Base, Ohio), STRUDL (*STRU*ctural *D*esign *L*anguage, Massachusetts Institute of Technology), and SAP (Structural *A*nalysis *P*rogram, E. L. Wilson, University of California, Berkeley).

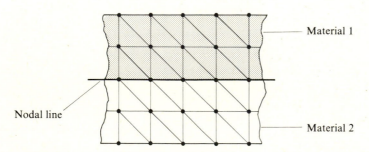

Material 1

Nodal line

Material 2

Figure 8-15

EXERCISES

8-1 Develop the stiffness matrix for a circular rod in torsion and apply the matrix to the rod shown in Fig. 8-16. Obtain the system stiffness matrix and the "transformed" inverse matrix to find the deflection equation. $E = 30 \times 10^6$ lb/in², $\nu = 0.3$.

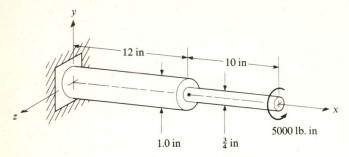

12 in

10 in

5000 lb. in

1.0 in $\frac{3}{4}$ in

Figure 8-16

8-2 A two-dimensional bending element is shown in Fig. 8-17. Prove that the force-displacement equation for the element is

$$
\begin{Bmatrix} V_{11} \\ M_{11} \\ V_{12} \\ M_{12} \end{Bmatrix} = 2\left(\frac{EI_z}{L^3}\right)_1 \begin{bmatrix} 6 & 3L_1 & -6 & 3L_1 \\ 3L_1 & 2L_1^2 & -3L_1 & L_1^2 \\ -6 & -3L_1 & 6 & -3L_1 \\ 3L_1 & L_1^2 & -3L_1 & 2L_1^2 \end{bmatrix} \begin{Bmatrix} v_1 \\ \theta_1 \\ v_2 \\ \theta_2 \end{Bmatrix}
$$

where θ_1 and θ_2 are the slopes of the beam at nodes 1 and 2, respectively. *Hint:* As an example of finding the first row–first column term of the stiffness matrix, consider node 2 to be completely fixed. Then using beams D.1 and D.4 of Appendix D, determine the relationship between V_{11} and M_{11} necessary to give zero slope at node 1. With this, the deflection v_1 at node 1 can be written in terms of V_{11} alone. Using the equilibrium equations, the remaining terms of the first column (which correspond to $\theta_1 = v_2 = \theta_2 = 0$) can also be found from the previous step.

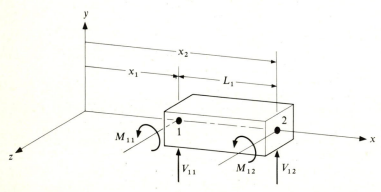

Figure 8-17

8-3 A plate $\frac{1}{8}$ in thick is loaded in tension as shown in Fig. 8-18a. Assume that the equation $\sigma_x = P/A$ is still valid (where A is the area of the plate and a function of x).

(a) Using the conventional strength of materials approach solve for and graph σ_x and ε_x as functions of x. Using direct integration, graph the displacement u as a function of x.

(b) Using the two-element model shown in Fig. 8-18b, evaluate and graph σ_x, ε_x, and u as functions of x and compare the results with part (a).

(c) Repeat part (b) using the three-element model shown in Fig. 8-18c. The material of the plate is steel with $E = 30 \times 10^6$ lb/in² and $\nu = 0.3$.

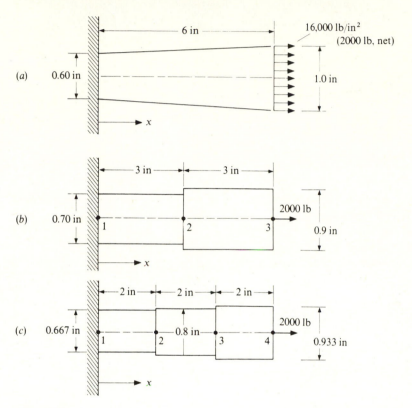

Figure 8-18

8-4 Using the results of Exercise 8-2, determine the slope and deflection at node 3 of the beam shown in Fig. 8-19. The modulus of elasticity of the beam is E and $I_1 = 2I_2$. Compare the deflection obtained with that of Example 4-15.

8-5 Repeat Example 2-11 using the finite-element method. *Hint: Use* the stiffness matrix for two springs in series and impose the condition that $u_3 = 0$ by modifying the stiffness coefficient k_{33}.

8-6 The plate in Exercise 8-3 can be modeled as shown in Fig. 8-20a using triangular slab elements. Thanks to symmetry, the model can be simplified to that of Fig. 8-20b. For elements A, B, and C, determine all displacements, strains, and stresses.

8-7 Determine the system stiffness matrix for the structure shown in Fig. 8-12.

8-8 Example 2-1 pertains to a rod hanging under its own weight. Return to the solution of this example and graph the displacement u as a function of x. Then, using the finite-element method,

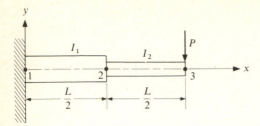

Figure 8-19

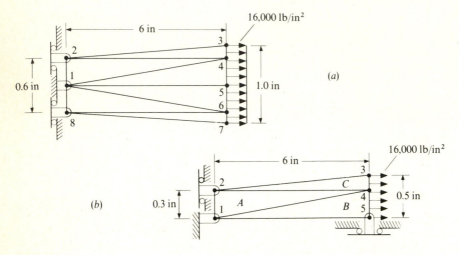

Figure 8-20

graph u as a function of x employing: (a) a one-element model, (b) a two-element model, (c) a three-element model, and (d) a four-element model. Note that the body weight of an element is to be distributed at the nodes of the element. Display the results of Example 2-1 and the results of the finite-element method on the same graph.

8-9 Using three elements solve Exercise 2-2 by the finite-element method. Compare the elastic displacement $u(x)$ of the rotating rod with that obtained by the elastic analysis.

REFERENCES

1. Zienkiewicz, O. C.: "The Finite Element Method in Engineering Science," McGraw-Hill, New York, 1971.
2. Cook, R. D.: "Concepts and Applications of Finite Element Analysis," Wiley, New York, 1974.
3. Gallagher, R. H.: "Finite Element Analysis: Fundamentals," Prentice-Hall, Englewood Cliffs, N.J., 1975.
4. Desai, C. S., and J. F. Abel: "Introduction to the Finite Element Method," Van Nostrand Reinhold, New York, 1972.
5. Paulsen, W. C.: Finite Element Stress Analysis, *Mach. Des.* vol. 43, no. 24, pp. 46–52, September 1971; vol. 43, no. 25, pp. 146–150, October 1971; vol. 43, no. 26, pp. 90–94, October 1971.

USCS AND SI UNITS AND CONVERSIONS

Intermixed throughout this book are problems formulated in USCS and SI†
units. At the time of this writing actual use of SI units in the United States is
still rather limited, but it is only a matter of time when SI units will be used
exclusively in the lay world too. Even though the reader may still be using USCS
units and must continue to do so in his work, eventually it will become necessary
for him to deal with SI units. It is important not only to become familiar with the
nomenclature of SI units but also to be able to think in these units and be
comfortable with the various and differing orders of magnitude. For example,
consider a man weighing 180 lb. If the reader is most familiar with the USCS
gravitational system he has a relatively accurate "feeling" for this quantity.
This is generally learned early in life through comparisons between weights one
becomes familiar with. In converting USCS gravitational units into SI units
there is an additional difficulty in cases such as this example since "weights" of
substances are not generally discussed in the SI system in terms of the force
exerted due to gravitation. Instead, mass units are used. For example, in terms
of SI units the 180-lb man has a mass of 82 kilograms, which on the earth's
surface corresponds to an attraction force of 802 newtons.

The SI units for stress are newtons/meter² (N/m²), or pascals (Pa), which
are directly equivalent. To obtain an approximation when converting stress in
pounds per square inch into newtons per square meter simply multiply by 7000.
This will yield a reasonable approximation. For example, the modulus of
elasticity for steel can be approximated by $E_s \approx 21 \times 10^{10}\,\text{N/m}^2$ or $E_s \approx$

---

† SI is an abbreviation for the International System of Units (from the French, Systéme
International d'Unités).

Table A-1 SI prefixes

Prefix	Symbol	Multiple	Prefix	Symbol	Multiple
giga	G	10^9	centi	c	10^{-2}
mega	M	10^6	milli	m	10^{-3}
kilo	k	10^3	micro	μ	10^{-6}

210 GN/m^2 (G = 10^9, and read "giga"). The most common prefixes used with SI units are given in Table A-1.

Common units in mechanics are described in Table A-2, where the conversions are given.

Table A-2 Conversion factors for USCS units to SI units

To convert from (USCS):	To SI:	Multiply by:
Acceleration:		
ft/s^2	m/s^2	3.048×10^{-1}
in/s^2	m/s^2	2.54×10^{-2}
Area:		
ft^2	m^2	9.29×10^{-2}
in^2	m^2	6.452×10^{-4}
Density:		
slugs/ft^3 (lb·s^2/ft^4)	kg/m^3	5.152×10^2
Energy, work:		
ft·lb	J (N·m)	1.356
Force:		
lb	N	4.448
Length:		
ft	m	3.048×10^{-1}
in	m	2.54×10^{-2}
Mass:		
slugs (lb·s^2/ft)	kg	14.59
Power:		
ft·lb/min	W	2.26×10^{-2}
hp (1 hp = 550 ft·lb/s)	W	7.457×10^2
Pressure, stress:		
lb/ft^2	N/m^2 or Pa	47.88
lb/in^2	N/m^2 or Pa	6.895×10^3
Velocity:		
ft/min	m/s	5.08×10^{-3}
ft/s	m/s	3.048×10^{-1}
Volume:		
ft^3	m^3	2.832×10^{-2}
in^3	m^3	1.639×10^{-5}

RELATIONSHIP BETWEEN THE SHEAR MODULUS G, YOUNG'S MODULUS E, AND POISSON'S RATIOS ν

In Chap. 1 the relationship

$$G = \frac{E}{2(1 + \nu)} \qquad (1\text{-}26)$$

was given. For a homogeneous and isotropic material, this relationship is valid and can be verified by considering a square element in a pure state of shear, as shown in Fig. B-1a. The deformation of the element is illustrated in Fig. B-1b. The normal strain along the diagonal AC is given by

$$\varepsilon_{AC} = \frac{A'C' - AC}{AC}$$

or

$$\varepsilon_{AC} = \frac{A'C'}{AC} - 1 \qquad (\text{B-1})$$

The shear strain γ_{xy} is given by

$$\gamma_{xy} = \frac{\pi}{2} - \beta \qquad (\text{B-2})$$

The length of $A'C'$, is found from the geometry of Fig. B-2b to be

$$A'C' = 2A'B' \cos \tfrac{1}{2}\beta$$

and from Eq. (B-2)

$$A'C' = 2A'B' \cos \frac{1}{2}\left(\frac{\pi}{2} - \gamma_{xy}\right) \qquad (\text{B-3})$$

But

$$\cos \frac{1}{2}\left(\frac{\pi}{2} - \gamma_{xy}\right) = \cos \frac{\pi}{4} \cos \frac{\gamma_{xy}}{2} + \sin \frac{\pi}{4} \sin \frac{\gamma_{xy}}{2}$$

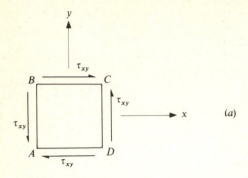

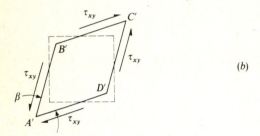

Figure B-1

However, the change in angle γ_{xy} is generally extremely small, so that

$$\cos \frac{\gamma_{xy}}{2} \approx 1 \quad \text{and} \quad \sin \frac{\gamma_{xy}}{2} \approx \frac{\gamma_{xy}}{2}$$

Thus, Eq. (B-3) can be written

$$A'C' \approx \sqrt{2} \; A'B'\left(1 + \frac{\gamma_{xy}}{2}\right) \tag{B-4}$$

If the strains are assumed to be small, then $A'B' \approx AB$ and from Fig. B-1a $AB = (\frac{1}{2}\sqrt{2})AC$. Thus

$$A'C' = AC\left(1 + \frac{\gamma_{xy}}{2}\right)$$

or

$$\frac{A'C'}{AC} = 1 + \frac{\gamma_{xy}}{2} \tag{B-5}$$

Substituting this into Eq. (B-1) results in

$$\varepsilon_{AC} = \frac{\gamma_{xy}}{2}$$

Using the relation $\gamma_{xy} = \tau_{xy}/G$ yields

$$\varepsilon_{AC} = \frac{\tau_{xy}}{2G} \tag{B-6}$$

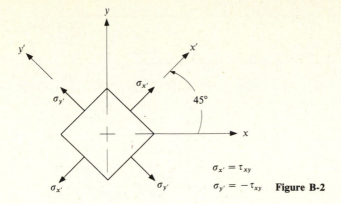

$$\sigma_{x'} = \tau_{xy}$$
$$\sigma_{y'} = -\tau_{xy} \quad \textbf{Figure B-2}$$

From Mohr's circle of stress, the stress state represented in Fig. B-1 is equivalent to that shown in Fig. B-2. Note that

$$\varepsilon_{x'} = \varepsilon_{AC} \tag{B-7}$$

The normal strain in the x' direction is given by

$$\varepsilon_{x'} = \frac{1}{E}(\sigma_{x'} - \nu\sigma_{y'})$$

and since $\sigma_{x'} = \tau_{xy}$ and $\sigma_{y'} = -\tau_{xy}$, we have

$$\varepsilon_{x'} = \frac{1}{E}[\tau_{xy} - \nu(-\tau_{xy})] = \frac{1+\nu}{E}\tau_{xy} \tag{B-8}$$

Equating Eqs. (B-8) and (B-6) yields

$$\frac{\tau_{xy}}{2G} = \frac{1+\nu}{E}\tau_{xy}$$

Solving for G results in

$$G = \frac{E}{2(1+\nu)} \tag{B-9}$$

APPENDIX

C

PROPERTIES OF CROSS SECTIONS

Section	Area	Centroid	Area moments of inertia
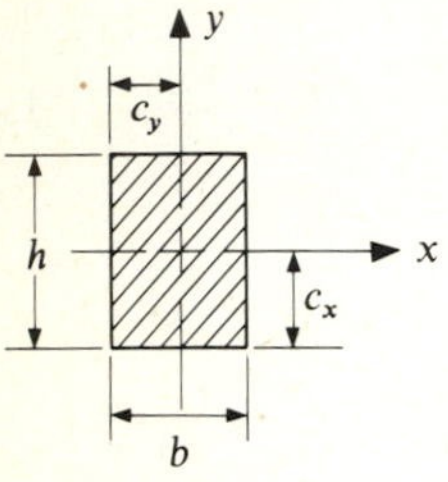 Rectangle	bh	$c_x = \frac{h}{2}$ $c_y = \frac{b}{2}$	$I_x = \frac{bh^3}{12}$ $I_y = \frac{hb^3}{12}$
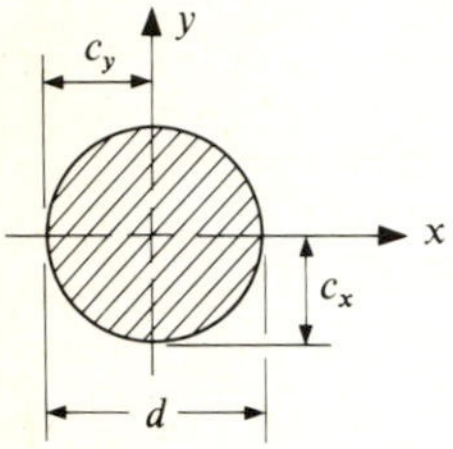 Circle	$\frac{\pi d^2}{4}$	$c_x = \frac{d}{2}$ $c_y = \frac{d}{2}$	$I_x = \frac{\pi d^4}{64}$ $I_y = \frac{\pi d^4}{64}$ $I_z = J = \frac{\pi d^4}{32}$
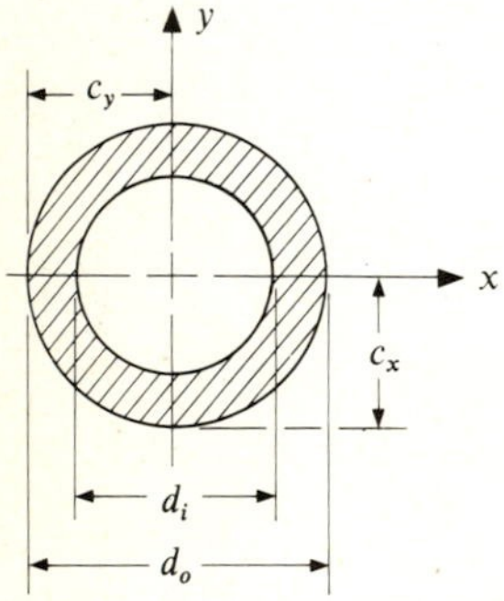 Thick-walled tube	$\frac{\pi}{4}(d_o^2 - d_i^2)$	$c_x = \frac{d_o}{2}$ $c_y = \frac{d_o}{2}$	$I_x = \frac{\pi}{64}(d_o^4 - d_i^4)$ $I_y = \frac{\pi}{64}(d_o^4 - d_i^4)$ $I_z = J = \frac{\pi}{32}(d_o^4 - d_i^4)$

Figure C

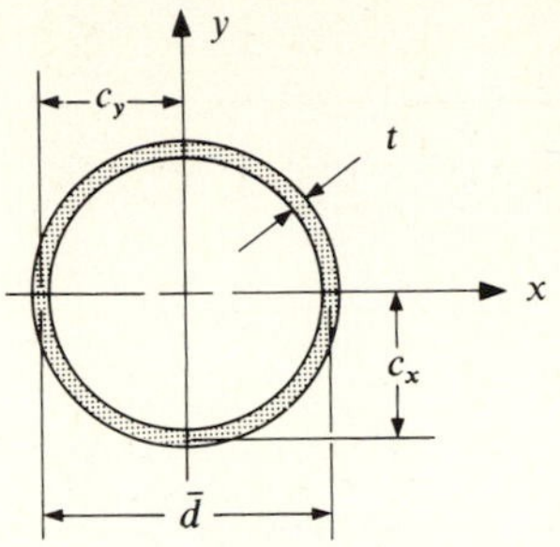 Thin-walled tube ($\bar{d} \gg t$)	$\pi \bar{d} t$	$c_x = \dfrac{\bar{d}}{2}$ $c_y = \dfrac{\bar{d}}{2}$	$I_x = \dfrac{\pi}{8} t \bar{d}^3$ $I_y = \dfrac{\pi}{8} t \bar{d}^3$ $I_z = J = \dfrac{\pi}{4} t \bar{d}^3$
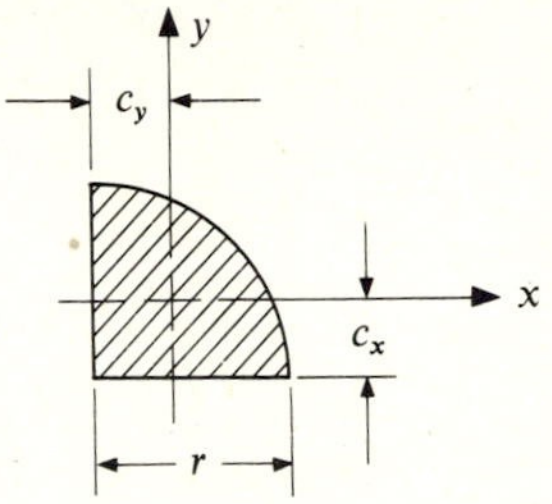 Circular quadrant	$\dfrac{\pi}{4} r^2$	$c_x = \dfrac{4r}{3\pi}$ $c_y = \dfrac{4r}{3\pi}$	$I_x = \left(\dfrac{\pi}{16} - \dfrac{4}{9\pi} \right) r^4$ $I_y = \left(\dfrac{\pi}{16} - \dfrac{4}{9\pi} \right) r^4$ $P_{xy} = \left(\dfrac{1}{8} - \dfrac{4}{9\pi} \right) r^4$
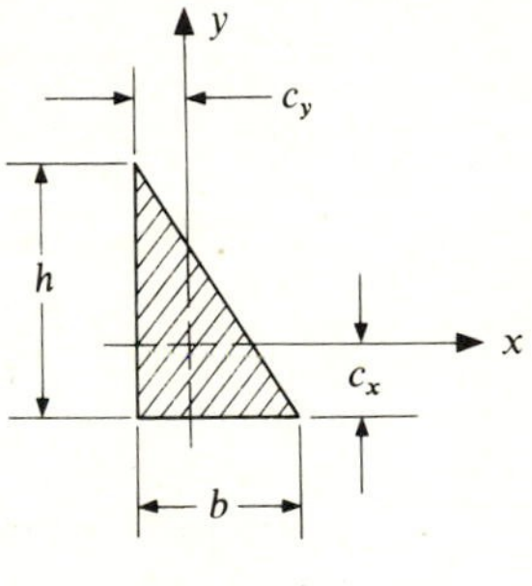 Triangle	$\dfrac{1}{2} bh$	$c_x = \dfrac{h}{3}$ $c_y = \dfrac{b}{3}$	$I_x = \dfrac{bh^3}{36}$ $I_y = \dfrac{hb^3}{36}$ $P_{xy} = -\dfrac{b^2 h^2}{72}$

Figure C (*Continued*)

C-1 COMBINATIONS OF SECTIONS

Example C-1 Find the location of the centroids and the area moments of inertia of the trapezoid section shown in Fig. C-1.

SOLUTION By symmetry, $c_y = a/2$. To determine the properties of the cross section, consider the section as one rectangle and two triangles. The area is

$$A = \frac{1}{2} \frac{a-b}{2} h + bh + \frac{1}{2} \frac{a-b}{2} h = \frac{h}{2} (a+b) \qquad (a)$$

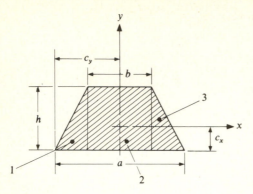

Figure C-1

From the base, the location of the centroid of each triangle is $h/3$ and of the rectangle is $h/2$. Thus

$$Ac_x = A_1c_{1x} + A_2c_{2x} + A_3c_{3x} = 2A_1c_{1x} + A_2c_{2x}$$

or

$$\frac{h}{2}(a+b)c_x = 2\left(\frac{1}{2}\frac{a-b}{2}h\right)\frac{h}{3} + bh\frac{h}{2}$$

Simplifying and solving for c_x yields

$$c_x = \frac{h}{3}\left(\frac{a+2b}{a+b}\right) \tag{b}$$

To find the area moments of inertia, the parallel-axis theorem is used. For the rectangular section, the moment of inertia about its own horizontal centroid is $bh^3/12$. The distance from the centroid of the rectangle to the centroid of the trapezoid section is

$$\frac{h}{2} - \frac{h}{3}\left(\frac{a+2b}{a+b}\right)$$

which simplifies to

$$\frac{h}{6}\left(\frac{a-b}{a+b}\right)$$

Thus, the area moment of inertia of the rectangle about the x axis is

$$(I_x)_2 = \tfrac{1}{12}bh^3 + bh\left(\frac{h}{6}\frac{a-b}{a+b}\right)^2 \tag{c}$$

For each triangular section, the moment of inertia about the horizontal centroid is

$$\frac{1}{36}\frac{a-b}{2}h^3$$

and the distance between the horizontal centroid of the triangular section and the centroid of the trapezoidal section is

$$\frac{h}{3}\frac{a+2b}{a+b} - \frac{h}{3}$$

which simplifies to $hb/3(a+b)$. Thus, the moment of inertia of each triangular section is

$$(I_x)_1 = (I_x)_3 = \frac{1}{36}\frac{a-b}{2}h^3 + \frac{1}{2}\frac{a-b}{2}h\left[\frac{hb}{3(a+b)}\right]^2 \tag{d}$$

From Eqs. (c) and (d) the total moment of inertia of the trapezoidal section about its horizontal centroidal axis is obtained by adding each term. Thus

$$I_x = (I_x)_1 + (I_x)_2 + (I_x)_3$$

and simplifying the final results yields

$$I_x = \frac{h^3}{36}\left(\frac{a^2 + 4ab + b^2}{a+b}\right) \tag{e}$$

Applying the same approach, the reader should verify that the moment of inertia of the trapezoid section about the y axis is

$$I_y = \frac{h}{48}(a+b)(a^2+b^2) \tag{f}$$

BEAMS IN BENDING

D.1 CANTILEVER WITH END LOAD

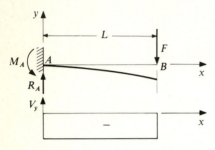

External and internal reactions

$$R_A = -V_y = F$$

$$M_A = FL$$

$$M_z = F(x - L)$$

Deflection

$$y_c = \frac{Fx^2}{6EI}(x - 3L)$$

$$(y_c)_{x=L} = -\frac{FL^3}{3EI}$$

Slope

$$\theta = \frac{dy_c}{dx} = \frac{Fx}{2EI}(x - 2L)$$

D.2 CANTILEVER WITH INTERMEDIATE LOAD

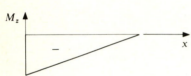

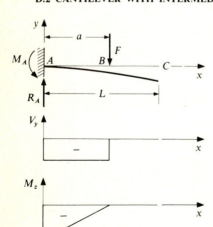

External and internal reactions

$$R_A = -(V_y)_{AB} = F \qquad M_A = Fa$$

$$(M_z)_{AB} = F(x - a) \qquad (M_z)_{BC} = (V_y)_{BC} = 0$$

Deflection

$$(y_c)_{AB} = \frac{Fx^2}{6EI}(x - 3a)$$

$$(y_c)_{BC} = \frac{Fa^2}{6EI}(a - 3x)$$

$$(y_c)_{x=L} = \frac{Fa^2}{6EI}(a - 3L)$$

Slope

$$(\theta)_{AB} = \frac{Fx}{2EI}(x - 2a)$$

$$(\theta)_{BC} = -\frac{Fa^2}{2EI}$$

D.3 CANTILEVER WITH UNIFORM LOAD

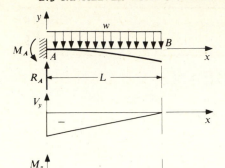

External and internal reactions

$$R_A = wL \qquad\qquad M_A = \frac{wL^2}{2}$$

$$V_y = -w(L-x) \qquad M_z = -\frac{w}{2}(L-x)^2$$

Deflection

$$y_c = -\frac{wx^2}{24EI}(4Lx - x^2 - 6L^2)$$

$$(y_c)_{x=L} = -\frac{wL^4}{8EI}$$

Slope

$$\theta = \frac{wx}{6EI}(3Lx - x^2 - 3L^2)$$

D.4 CANTILEVER WITH MOMENT LOAD

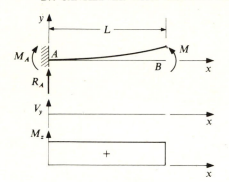

External and internal reactions

$$R_A = V_y = 0 \qquad M_A = M_z = M$$

Deflection

$$y_c = \frac{Mx^2}{2EI} \qquad (y_c)_{x=L} = \frac{ML^2}{2EI}$$

Slope

$$\theta = \frac{Mx}{EI}$$

D.5 SIMPLE SUPPORTS WITH INTERMEDIATE LOAD

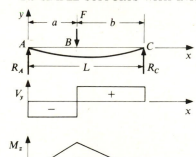

External and internal reactions

$$R_A = \frac{Fb}{L} \qquad\qquad R_C = \frac{Fa}{L}$$

$$(V_y)_{AB} = -R_A \qquad (V_y)_{BC} = R_B$$

$$(M_z)_{AB} = \frac{Fbx}{L} \qquad (M_z)_{BC} = \frac{Fa}{L}(L-x)$$

Deflection

$$(y_c)_{AB} = \frac{Fbx}{6EIL}(x^2 + b^2 - L^2)$$

$$(y_c)_{BC} = \frac{Fa(L-x)}{6EIL}(x^2 + a^2 - 2Lx)$$

Slope

$$\theta_{AB} = \frac{Fb}{6EIL}(3x^2 + b^2 - L^2)$$

$$\theta_{BC} = \frac{Fa}{6EIL}(6Lx - 3x^2 - a^2 - 2L^2)$$

D.6 SIMPLE SUPPORTS WITH UNIFORM LOAD

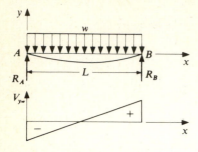

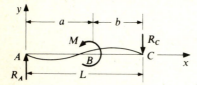

External and internal reactions

$$R_A = R_B = \frac{wL}{2} \qquad V_y = -w\left(\frac{L}{2} - x\right)$$

$$M_z = \frac{wx}{2}(L - x)$$

Deflection

$$y_c = \frac{wx}{24EI}(2Lx^2 - x^3 - L^3)$$

$$(y_c)_{x=\frac{L}{2}} = -\frac{5wL^4}{384EI}$$

Slope

$$\theta = \frac{w}{24EI}(6Lx^2 - 4x^3 - L^3)$$

D.7 SIMPLE SUPPORTS WITH MOMENT LOAD

External and internal reactions

$$R_A = R_C = \frac{M}{L} \qquad V_y = -\frac{M}{L}$$

$$(M_z)_{AB} = \frac{Mx}{L} \qquad (M_z)_{BC} = \frac{M}{L}(x - L)$$

Deflection

$$(y_c)_{AB} = \frac{Mx}{6EIL}(x^2 + 3a^2 - 6aL + 2L^2)$$

$$(y_c)_{BC} = \frac{M}{6EIL}[x^3 - 3Lx^2 + x(2L^2 + 3a^2) - 3a^2L]$$

Slope

$$(\theta)_{AB} = \frac{M}{6EIL}(3x^2 + 3a^2 - 6aL + 2L^2)$$

$$(\theta)_{BC} = \frac{M}{6EIL}(3x^2 - 6Lx + 2L^2 + 3a^2)$$

D.8 SIMPLE SUPPORTS WITH OVERHANGING LOAD

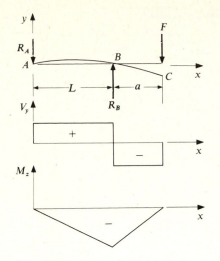

External and internal reactions

$$R_A = \frac{Fa}{L} \qquad\qquad R_B = \frac{F}{L}(L + a)$$

$$(V_y)_{AB} = \frac{Fa}{L} \qquad (V_y)_{BC} = -F$$

$$(M_z)_{AB} = -\frac{Fax}{L} \qquad (M_z)_{BC} = F(x - L - a)$$

Deflection

$$(y_c)_{AB} = \frac{Fax}{6EIL}(L^2 - x^2)$$

$$(y_c)_{BC} = \frac{F(x - L)}{6EI}[(x - L)^2 - a(3x - L)]$$

Slope

$$(\theta)_{AB} = \frac{Fa}{6EIL}(L^2 - 3x^2)$$

$$(\theta)_{BC} = \frac{F}{6EI}[3x^2 - 6x(L + a) + L(3L + 4a)]$$

D.9 CANTILEVER WITH PARTIAL DISTRIBUTED LOAD

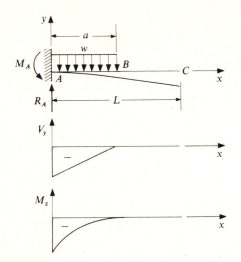

External and internal reactions

$$R_A = wa \quad M_A = \frac{wa^2}{2}$$

$$(V_y)_{AB} = -w(a - x) \quad (V_y)_{BC} = 0$$

$$(M_z)_{AB} = -\frac{w}{2}(a - x)^2 \quad (M_z)_{BC} = 0$$

Deflection

$$(y_c)_{AB} = \frac{wx^2}{24EI}(4ax - x^2 - 6a^2)$$

$$(y_c)_{BC} = \frac{wa^3}{24EI}(a - 4x)$$

Slope

$$(\theta)_{AB} = \frac{wx}{6EI}(3ax - x^2 - 3a^2)$$

$$(\theta)_{BC} = -\frac{wa^3}{6EI}$$

D.10 SIMPLE SUPPORTS WITH PARTIAL DISTRIBUTED LOAD

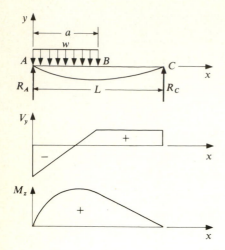

External and internal reactions

$$R_A = \frac{wa}{2L}(2L - a) \qquad R_C = \frac{wa^2}{2L}$$

$$(V_y)_{AB} = \frac{w}{2L}[2L(x - a) + a^2]$$

$$(V_y)_{BC} = \frac{wa^2}{2L}$$

$$(M_z)_{AB} = \frac{wx}{2L}(2aL - a^2 - xL)$$

$$(M_z)_{BC} = \frac{wa^2}{2L}(L - x)$$

Deflection

$$(y_c)_{AB} = \frac{wx}{24EIL}[2ax^2(2L - a) - Lx^3 - a^2(2L - a)^2]$$

$$(y_c)_{BC} = (y_c)_{AB} + \frac{w}{24EI}(x - a)^4$$

Slope

$$(\theta)_{AB} = \frac{w}{24EIL}[6ax^2(2L - a) - 4Lx^3 - a^2(2L - a)^2]$$

$$(\theta)_{BC} = (\theta)_{AB} + \frac{w}{6EI}(x - a)^3$$

D.11 ONE END FIXED AND ONE SIMPLE SUPPORT WITH INTERMEDIATE LOAD

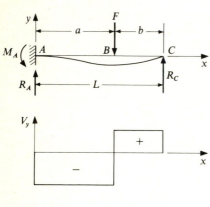

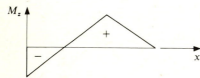

External and internal reactions

$$R_A = \frac{Fb}{2L^3}(3L^2 - b^2) \qquad R_C = \frac{Fa^2}{2L^3}(3L - a)$$

$$M_A = \frac{Fb}{2L^2}(L^2 - b^2)$$

$$(V_y)_{AB} = -R_A \qquad (V_y)_{BC} = R_C$$

$$(M_z)_{AB} = \frac{Fb}{2L^3}[b^2L - L^3 + x(3L^2 - b^2)]$$

$$(M_z)_{BC} = \frac{Fa^2}{2L^3}(3L^2 - 3Lx - aL + ax)$$

Deflection

$$(y_c)_{AB} = \frac{Fbx^2}{12EIL^3}[3L(b^2 - L^2) + x(3L^2 - b^2)]$$

$$(y_c)_{BC} = (y_c)_{AB} - \frac{F(x - a)^3}{6EI}$$

Slope

$$(\theta)_{AB} = \frac{Fbx}{4EIL^3}[2L(b^2 - L^2) + x(3L^2 - b^2)]$$

$$(\theta)_{BC} = (\theta)_{AB} - \frac{F(x - a)^2}{2EI}$$

D.12 ONE END FIXED AND ONE SIMPLE SUPPORT WITH UNIFORM LOAD

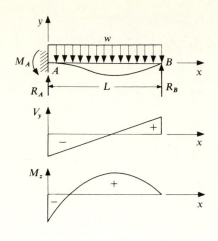

External and internal reactions

$$R_A = \frac{5wL}{8} \qquad R_B = \frac{3wL}{8} \qquad M_A = \frac{wL^2}{8}$$

$$V_y = \frac{w}{8}(8x - 5L)$$

$$M_z = \frac{w}{8}(4x^2 + 5Lx - L^2)$$

Deflection

$$y_c = \frac{wx^2}{48EI}(L - x)(2x - 3L)$$

Slope

$$\theta = \frac{wx}{48EI}(15xL - 8x^2 - 6L^2)$$

D.13 FIXED SUPPORTS WITH INTERMEDIATE LOAD

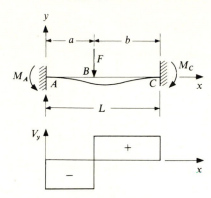

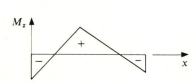

External and internal reactions

$$R_A = \frac{Fb^2}{L^3}(3a + b) \qquad R_C = \frac{Fa^2}{L^3}(3b + a)$$

$$M_A = \frac{Fab^2}{L^2} \qquad M_C = \frac{Fa^2b}{L^2}$$

$$(V_y)_{AB} = -R_A \qquad (V_y)_{BC} = R_C$$

$$(M_z)_{AB} = \frac{Fb^2}{L^3}[x(3a + b) - La]$$

$$(M_z)_{BC} = (M_z)_{AB} - F(x - a)$$

Deflection

$$(y_c)_{AB} = \frac{Fb^2x^2}{6EIL^3}[x(3a + b) - 3aL]$$

$$(y_c)_{BC} = \frac{Fa^2(L - x)^2}{6EIL^3}[(L - x)(3b + a) - 3bL]$$

Slope

$$(\theta)_{AB} = \frac{Fb^2x}{2EIL^3}[x(3a + b) - 2L]$$

$$(\theta)_{BC} = \frac{Fa^2(L - x)}{2EIL^3}[2bL - (L - x)(b + a)]$$

D.14 FIXED SUPPORTS WITH UNIFORM LOAD.

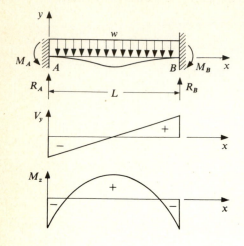

External and internal reactions

$$R_A = R_B = \frac{wL}{2} \qquad M_A = M_B = \frac{wL^2}{12}$$

$$V_y = -\frac{w}{2}(L - 2x)$$

$$M_z = \frac{w}{12}(6Lx - 6x^2 - L^2)$$

Deflection

$$y_c = -\frac{wx^2}{24EI}(L - x)^2$$

Slope

$$\theta = -\frac{wx}{12EI}(L - x)(L - 2x)$$

SINGULARITY FUNCTIONS

E-0 INTRODUCTION

In this appendix, singularity functions are developed for beams in bending, but the functions are applicable as well to problems involving axial or torsional loading. In Chap. 2 the method of section isolation was presented for beams in bending, where it was shown that whenever a discontinuity in the loading occurs, it is necessary to rewrite the shear-force and bending-moment equations. In this way, in each section where the loading is continuous, a separate set of equations exists. For many practical problems, the loading of a beam may be highly discontinuous, and section isolation is cumbersome. In addition, if the deflections of such beams are desired, use of the double-integration method can be a nightmare, as each section must be such that the slopes and deflection of the beam are continuous. It is much simpler to solve such problems by the method of superposition, where each specific type of load, i.e., concentrated forces, distributed forces, concentrated moments, etc., can be analyzed separately and the results combined. The effective use of this technique requires that (1) the analyst have a comprehensive list of solutions for the various forms of loading and support conditions such as in Appendix D;† and (2) the analyst exercise extreme care in calculations, as the superposition of many equations can lead to errors. Singularity functions, if used correctly, provide the most direct solution to a beam loaded in a highly discontinuous manner. Since singularity functions can represent discontinuities, the

---

† Although Appendix D contains many of the common forms of loading and support conditions, the list is far from complete.

load intensity (force per unit length) as a function of axial position can be written in the form of only *one* equation. A further advantage is that this can be done simply by visual inspection of the beam-loading diagram. Direct integration then yields the shear-force equation for the *entire* beam. Integration of the shear-force equation then results in the bending-moment equation for the *entire* beam. Integration of the bending-moment equation provides the equation for the slope of the beam at any point where only one constant of integration results. A final integration results in the equation for deflection of the beam where a second constant of integration arises. Thus, only two constants of integration are necessary corresponding to the support boundary conditions and only *one* set of equations for the load intensity, shear force, bending moment, slope, and deflection results. However, one important thing to keep in mind is that the singularity functions themselves are discontinuous and that the resulting behavior of the beam is still discontinuous. The analyst using singularity functions must have a complete understanding of them and how one performs the necessary integration of them. Before the development of these functions, it is necessary to review the integral relations between load intensity, shear force, bending moment, slope, and deflections.

E-1 INTEGRAL RELATIONS FOR BEAMS IN BENDING

Consider a beam element of infinitesimal length dx as shown in Fig. E-1. V_y and M_z are the shear force and bending moment at position x. The load intensity q (force per unit length) is denoted by $q(x)$ and is positive in the positive y direction. Since the shear force and bending moment, in general, change as functions of x, at $x + dx$ the shear force and bending moment can be described by $V_y + dV_y$ and $M_x + dM_x$.

Summing forces in the y direction results in

$$dV_y + q(x)\, dx = 0$$

or
$$\frac{dV_y}{dx} = -q(x) \tag{E-1}$$

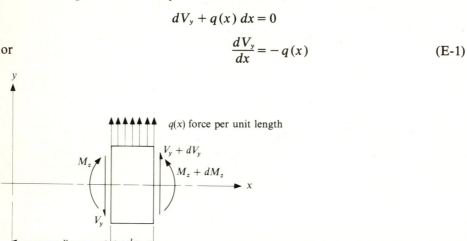

Figure E-1

Therefore the integral relationship is given by

$$V_y = -\int q(x)\, dx \tag{E-2}$$

Summing moments at the center of the element about the z axis yields

$$M_z + dM_z - M_z + V_y \frac{dx}{2} + (V_y + dV_y)\frac{dx}{2} = 0$$

or

$$\frac{dM_z}{dx} + V_y + \tfrac{1}{2}dV_y = 0$$

The first two terms of this equation are finite, whereas dV_y is infinitesimal. Thus

$$\frac{dM_z}{dx} + V_y = 0$$

or

$$\frac{dM_z}{dx} = -V_y \tag{E-3}$$

and the integral relationship is

$$M_z = -\int V_y\, dx \tag{E-4}$$

The integral relations for obtaining the slope and deflection equations were given in Chap. 2 and are

$$\frac{dy_c}{dx} = \int \frac{M_z}{EI}\, dx \tag{E-5}$$

and

$$y_c = \int \left(\frac{dy_c}{dx}\right) dx \tag{E-6}$$

E-2 SINGULARITY FUNCTIONS

Singularity functions, occasionally referred to as *impulse functions* or *Macaulay's functions*, are used in forming a single equation which can describe any discontinuous function. A family of polynomial singularity functions is defined by

$$F_n(x) = \langle x - a \rangle^n \tag{E-7}$$

where n is any integer. The functions $F_n(x)$ have some unique properties and in terms of physical applications will be restricted here to values of n equal to or greater than $n = -2$. The functions $F_n(x)$ behave in the following manner:

$$F_{-2}(x) = \langle x - a \rangle^{-2} = \begin{cases} \pm\infty & \text{when } x = a \\ 0 & \text{when } x \neq a \end{cases} \tag{E-8a}$$

$$F_{-1}(x) = \langle x - a \rangle^{-1} = \begin{cases} \infty & \text{when } x = a \\ 0 & \text{when } x \neq a \end{cases} \tag{E-8b}$$

and for $n \geq 0$,

$$F_n(x) = \langle x - a \rangle^n = \begin{cases} (x-a)^n & \text{when } x > a \\ 0 & \text{when } x \le a \end{cases} \tag{E-8c}$$

For $n < 0$, the singularity function has little physical significance in terms of actual applications. However, after one or two integrations, $F_{-1}(x)$ and $F_{-2}(x)$ will be of the form of Eq. (E-8c), where the function becomes physically meaningful.

Integration of Eqs. (E-8) is performed in the following fashion:

$$\int \langle x - a \rangle^n \, dx = \begin{cases} \langle x - a \rangle^{n+1} + C & \text{for } n < 0 \tag{E-9a} \\[2mm] \dfrac{1}{n+1} \langle x - a \rangle^{n+1} + C & \text{for } n \ge 0 \tag{E-9b} \end{cases}$$

where C is the constant of integration. Thus, for example,

$$\int \langle x - a \rangle^{-2} \, dx = \langle x - a \rangle^{-1} + C \tag{a}$$

and

$$\int \langle x - a \rangle^1 \, dx = \tfrac{1}{2} \langle x - a \rangle^2 + C \tag{b}$$

The singularity functions $F_n(x)$ are shown graphically in Fig. E-2. The function $F_{-2}(x)$ is called a *doublet* and in terms of load intensity (force per unit length) can be used to describe a concentrated moment on a beam. The function $F_{-1}(x)$ is called an *impulse* and can be used to describe a concentrated force. The function $F_0(x)$ is called a *unit step* and can be used to describe the addition of a uniform load initiating at $x = a$. The function $F_1(x)$ is called a unit ramp. Functions of higher order have no specific identification. To understand the application of singularity functions to beams in bending, consider the following example.

Example E-1 For the beam loaded as shown in Fig. E-3a, develop the load-intensity equation using singularity functions.

SOLUTION Figure E-3b shows the free-body diagram for the beam, where R_A and R_E are solvable from the equilibrium equations and all loads are shown in a perspective to show their sign with respect to the positive y direction.

If we use the singularity functions described by Eq. (E-8), the load-intensity equation is described by

$$q(x) = R_A \langle x - 0 \rangle^{-1} - w \langle x - 0 \rangle^0 + w \langle x - a \rangle^0 - M_C \langle x - b \rangle^{-2} - P_D \langle x - c \rangle^{-1} + R_E \langle x - L \rangle^{-1} \tag{a}$$

Terms like $\langle x - 0 \rangle^n$ can be written simply $\langle x \rangle^n$. Thus,

$$q(x) = R_A \langle x \rangle^{-1} - w \langle x \rangle^0 + w \langle x - a \rangle^0 - M_C \langle x - b \rangle^{-2} - P_D \langle x - c \rangle^{-1} + R_E \langle x - L \rangle^{-1} \tag{b}$$

Note that the term $+ w \langle x - a \rangle^0$ is necessary to cancel the uniform load w at $x = a$.

Thus it can be seen that the load-intensity equation for the entire beam can be written by inspection and in the form of only one equation. From Eq. (E-2) the shear-force equation is arrived at by integrating $q(x)$ and changing the sign. To obtain the bending-moment equation, the shear-force equation is integrated, and again the sign is reversed, as indicated by Eq. (E-4).

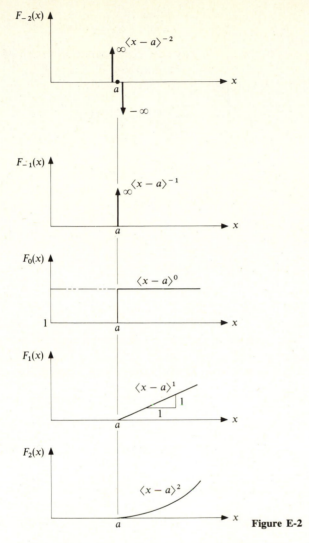

Figure E-2

Example E-2 For Example E-1, express the shear-force and bending-moment equations in terms of singularity functions and graph the results.

SOLUTION Since $V_y = -\int q(x)\,dx$, when Eqs. (E-9) are applied to Eq. (b) of Example E-1, the shear force is

$$V_y = -R_A\langle x\rangle^0 + w\langle x\rangle^1 - w\langle x-a\rangle^1 + M_C\langle x-b\rangle^{-1} + P_D\langle x-c\rangle^0 - R_E\langle x-L\rangle^0 + C_1$$

Since for $x = 0^-$, $V_y = 0$, we have $C_1 = 0$. Thus

$$V_y = -R_A\langle x\rangle^0 + w\langle x\rangle^1 - w\langle x-a\rangle^1 + M_C\langle x-b\rangle^{-1} + P_D\langle x-c\rangle^0 - R_E\langle x-L\rangle^0 \qquad (c)$$

The bending-moment equation results from $M_z = -\int V_y\,dx$; thus

$$M_z = R_A\langle x\rangle^1 - \frac{w}{2}\langle x\rangle^2 + \frac{w}{2}\langle x-a\rangle^2 - M_C\langle x-b\rangle^0 - P_D\langle x-c\rangle^1 + R_E\langle x-L\rangle^1 + C_2$$

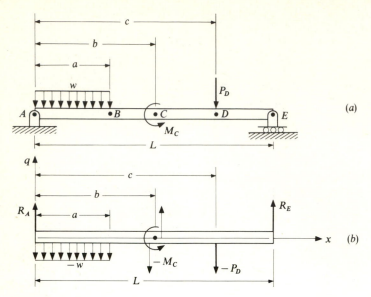

Figure E-3

Since $M_z = 0$ at $x = 0^-$, we have $C_2 = 0$. Therefore

$$M_z = R_A \langle x \rangle^1 - \frac{w}{2} \langle x \rangle^2 + \frac{w}{2} \langle x - a \rangle^2 - M_C \langle x - b \rangle^0 - P_D \langle x - c \rangle^1 + R_E \langle x - L \rangle^1 \qquad (d)$$

In general, the first two integrations of $q(x)$ will not yield any constants of integration, and it is therefore common practice to exclude the constants when obtaining V_y and M_z.

One must be careful when transforming the singularity functions of V_y and M_z into graphical form. Equations (c) and (d) are shown in Fig. E-4.

The impulse $M_C \langle x - b \rangle^{-1}$ in the shear-force diagram has no physical meaning and could be omitted from the V_y diagram.

When integrating M_z / EI_z to obtain the slope dy_c/dx, a constant of integration is necessary to impose boundary conditions. Likewise, integration of dy_c/dx results in y_c, where a second constant of integration is necessary.

Example E-3 For the beam shown in Fig. E-5a, graph the deflection of the centroidal axis y_c as a function of x. For the beam, $E = 10 \times 10^6$ lb/in^2 and $I_z = 0.5$ in^4.

SOLUTION The free-body diagram is shown in Fig. E-5b, where the reader should verify that $R_A = 120$ lb and $R_B = 30$ lb. From Fig. E-5b, the load-intensity equation is

$$q(x) = 120\langle x \rangle^{-1} - \tfrac{20}{15}\langle x \rangle^1 + \tfrac{20}{15}\langle x - 15 \rangle^1 + 20\langle x - 15 \rangle^0 + 30\langle x - 50 \rangle^{-1} \qquad (a)$$

Note that at $x = 15$ in both the slope and the value of 20 lb/in must be eliminated. Integrating (a) and changing signs results in

$$V_y = -120\langle x \rangle^0 + \tfrac{2}{3}\langle x \rangle^2 - \tfrac{2}{3}\langle x - 15 \rangle^2 - 20\langle x - 15 \rangle^1 - 30\langle x - 50 \rangle^0$$

Integrating and changing signs again yields

$$M_z = 120\langle x \rangle^1 - \tfrac{2}{9}\langle x \rangle^3 + \tfrac{2}{9}\langle x - 15 \rangle^3 + 10\langle x - 15 \rangle^2 + 30\langle x - 50 \rangle^1$$

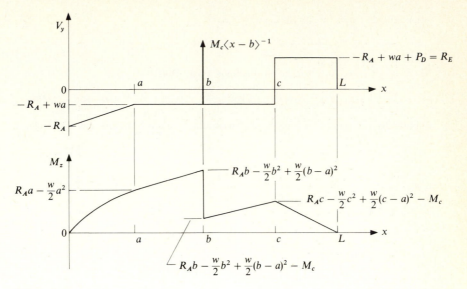

Figure E-4

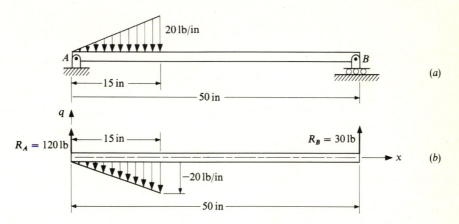

Figure E-5

Since $dy_c/dx = \int (M_z/EI_z)\, dx$, we have

$$\frac{dy_c}{dx} = \frac{1}{EI_z}\left(60\langle x\rangle^2 - \tfrac{1}{18}\langle x\rangle^4 + \tfrac{1}{18}\langle x - 15\rangle^4 + \tfrac{10}{3}\langle x - 15\rangle^3 + 15\langle x - 50\rangle^2 + C_1\right) \qquad (b)$$

Integration of Eq. (b) yields

$$y_c = \frac{1}{EI_z}\left(20\langle x\rangle^3 - \tfrac{1}{90}\langle x\rangle^5 + \tfrac{1}{90}\langle x - 15\rangle^5 + \tfrac{5}{6}\langle x - 15\rangle^4 + 5\langle x - 50\rangle^3 + C_1 x + C_2\right) \qquad (c)$$

The two boundary conditions are $y_c = 0$ at $x = 0$ and 50 in. At $x = 0$, only the first two singularity functions exist. Thus, Eq. (c) is

$$0 = \frac{1}{EI_z}[20(0^3) - \tfrac{1}{90}(0^5) + C_1(0) + C_2]$$

Therefore, $C_2 = 0$. At $x = 50$ in, all functions exist. Thus

$$0 = \frac{1}{EI_z}[(20)(50^3) - (\tfrac{1}{90})(50^5) + (\tfrac{1}{90})(50 - 15)^5 + (\tfrac{5}{6})(50 - 15)^4 + 5(50 - 50)^3 + C_1(50)]$$

Solving for C_1 yields $C_1 = -1.72 \times 10^4$. Substituting C_1 into Eqs. (b) and (c) results in

$$\frac{dy_c}{dx} = \frac{1}{EI_z}(60\langle x \rangle^2 - \tfrac{1}{18}\langle x \rangle^4 + \tfrac{1}{18}\langle x - 15 \rangle^4 + \tfrac{10}{3}\langle x - 15 \rangle^3 + 15\langle x - 50 \rangle^2 - 1.72 \times 10^4) \qquad (d)$$

and
$$y_c = \frac{1}{EI_z}(20\langle x \rangle^3 - \tfrac{1}{90}\langle x \rangle^5 + \tfrac{1}{90}\langle x - 15 \rangle^5 + \tfrac{5}{6}\langle x - 15 \rangle^4 + 5\langle x - 50 \rangle^3 - 1.72 \times 10^4 x) \qquad (e)$$

A graph of Eq. (e) with $E = 10 \times 10^6$ lb/in² and $I_z = 0.5$ in⁴ is presented in Fig. E-6.

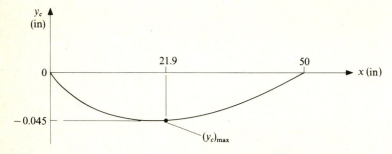

Figure E-6

DETERMINATION OF PRINCIPAL MOMENTS
OF INERTIA AND PRINCIPAL AXES
USING MOHR'S CIRCLE

For beams in bending, one of the conditions that the simplified flexure equation (3-2) is based on is that the coordinate reference axes are principal axes of inertia of the cross section. The principal axes of a cross section are defined as the rectangular axes for which the product of inertia vanishes. For an arbitrary cross section, as shown in Fig. F-1, the yz axes are principal axes provided that

$$P_{yz} = \int_A yz \, dA = 0 \tag{F-1}$$

where P_{yz} is the product of inertia with respect to the yz axes and dA is an infinitesimal area element in the yz plane.

If the product of inertia with respect to the yz axes is not zero, there exists a set of axes (mn) for which the product of inertia disappears. These axes are called the principal axes, and the moments of inertia about these axes (I_m, I_n) are called the principal moments of inertia.

The procedure in locating these axes involves coordinate transformations like those used for stress and strain, and consequently Mohr's circle can be used again. First, the area moments and products of inertia with respect to the yz coordinates are determined, where

$$I_y = \int_A z^2 \, dA \qquad I_z = \int_A y^2 \, dA \qquad P_{yz} = \int_A yz \, dA \tag{F-2}$$

For simple shapes, integration of Eqs. (F-2) may not be necessary, as the

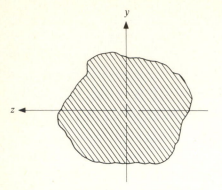

Figure F-1

moments of inertia for common shapes are available in tabular form (see Appendix C). However, it is often necessary to apply the parallel-axis theorem, which in equation form is

$$I_y = \bar{I}_y + \bar{y}^2 A \qquad I_z = \bar{I}_z + \bar{z}^2 A \qquad P_{yz} = \bar{P}_{yz} + \bar{y}\bar{z}A \qquad \text{(F-3)}$$

where I_y, I_z, and P_{yz} are the inertia terms of a particular subsection with respect to an arbitrary yz coordinate system, $\bar{I}_y$, $\bar{I}_z$, and $\bar{P}_{yz}$ are the inertia terms with respect to the centroidal axes of the subsection, and $\bar{y}$, $\bar{z}$ are the distances *from the y, z* axes *to the* subsection centroidal axes, respectively.

Example F-1 Determine I_y, I_z, and P_{yz} for the section shown in Fig. F-2a.

SOLUTION The reader should verify that the given yz axes are the centroidal axes of the total cross section. To determine the moments and product of inertia, the section can be divided into two simple rectangular subsections, as shown in Fig. F-2b. The moment of inertia of

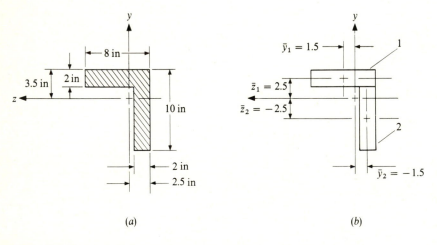

(a)   (b)

Figure F-2

section 1 about the y axis is

$$(I_y)_1 = (\bar{I}_y)_1 + \bar{y}_1^2 A_1 = (\tfrac{1}{12})(2)(8^3) + (1.5)^2(16) = 121.33 \text{ in}^4$$

Likewise for section 2,

$$(I_y)_2 = (\bar{I}_y)_2 + \bar{y}_2^2 A_2 = (\tfrac{1}{12})(8)(2^3) + (-1.5)^2(16) = 41.33 \text{ in}^4$$

thus

$$I_y = (I_y)_1 + (I_y)_2 = 162.7 \text{ in}^4$$

Similarly, for I_z

$$(I_z)_1 = (\bar{I}_z)_1 + \bar{z}_1^2 A_1 = (\tfrac{1}{12})(8)(2^3) + (2.5)^2(16) = 105.33 \text{ in}^4$$

and

$$(I_z)_2 = (\bar{I}_z)_2 + \bar{z}_2^2 A_2 = (\tfrac{1}{12})(2)(8^3) + (-2.5)^2(16) = 185.33 \text{ in}^4$$

Thus

$$I_z = (I_z)_1 + (I_z)_2 = 290.7 \text{ in}^4$$

The product of inertia for section 1 is

$$(P_{yz})_1 = (\bar{P}_{yz})_1 + \bar{y}_1 \bar{z}_1 A_1 = 0 + (1.5)(2.5)(16) = 60 \text{ in}^4$$

and for section 2,

$$(P_{yz})_2 = (\bar{P}_{yz})_2 + \bar{y}_2 \bar{z}_2 A_2 = 0 + (-1.5)(-2.5)(16) = 60 \text{ in}^4$$

Thus

$$P_{yz} = (P_{yz})_1 + (P_{yz})_2 = 120 \text{ in}^4$$

Mohr's circle can be used to determine the location of the principal axes, where the product of inertia is zero.

Example F-2 For the previous example, determine the location of the principal axes (m, n) and the corresponding values of the principal moments of inertia (I_m, I_n).

SOLUTION When we plot P vs. I, as shown in Fig. F-3a, the first point is (I_y, P_{yz}). The second point is $(I_z, -P_{yz})$. This establishes the diameter of the circle, and the circle can be drawn. The radius of the circle is

$$R = \sqrt{\left(\frac{I_y - I_z}{2}\right)^2 + P_{yz}^2} \qquad (F-4)$$

and

$$I_{av} = \frac{I_y + I_z}{2} \qquad (F-5)$$

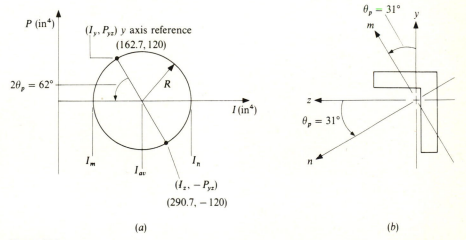

(a)

(b)

Figure F-3

For this example

$$R = \sqrt{\left(\frac{162.7 - 290.7}{2}\right)^2 + 120^2} = 136 \text{ in}^4$$

and

$$I_{av} = \frac{162.7 + 290.7}{2} = 226.7 \text{ in}^4$$

Thus

$$I_m = I_{av} - R = 226.7 - 136 = 90.7 \text{ in}^4$$

and

$$I_n = I_{av} + R = 226.7 + 136 = 362.7 \text{ in}^4$$

and the principal axis m is located θ_p counterclockwise from y, where

$$\tan 2\theta_p = \frac{P_{yz}}{(I_z - I_y)/2} = \frac{120}{(290.7 - 162.7)/2} = 1.875 \tag{F-6}$$

or

$$\theta_p = 31°$$

The principal axes are shown in Fig. F-3b.

STRESS CONCENTRATION FACTORS†

$$\sigma_{max} = K_t \sigma_{nom}$$

G.1

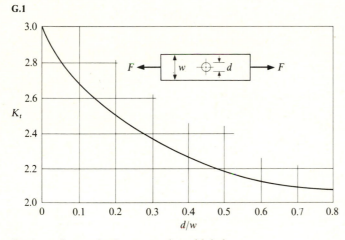

Bar in tension or simple compression with hole;
$\sigma_{nom} = F/A$ where $A = (w - d)t$ and t is the thickness

† R. E. Peterson, Design Factors for Stress Concentration, *Mach. Des.* vol. 23, no. 2, p. 169, February 1951; no. 3, p. 161, March 1951; no. 5, p. 159, May 1951; no. 6, p. 173, June 1951; no. 7, p. 155, July 1951; reproduced with the permission of the author and publisher. For a more comprehensive collection of curves on stress concentrations see R. E. Peterson, "*Stress Concentration Factors,*" Wiley, New York, 1974.

G.2

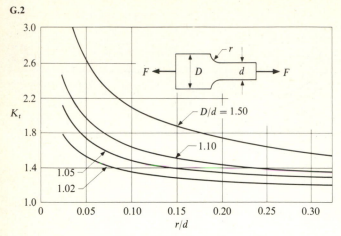

Rectangular filleted bar in tension or simple compression;
$\sigma_{nom} = F/A$ where $A = td$ and t is the thickness

G.3

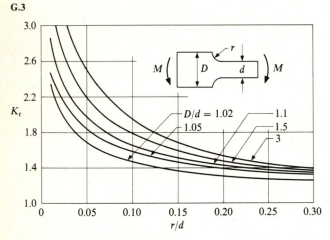

Rectangular filleted bar in bending;
$\sigma_{nom} = Mc/I$ where $c = d/2$ and $I = td^3/12$ where t is the thickness

G.4

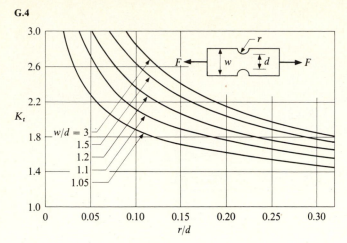

Notched rectangular bar in tension or simple compression;
$\sigma_{\text{nom}} = F/A$ where $A = td$ and t is the thickness

G.5

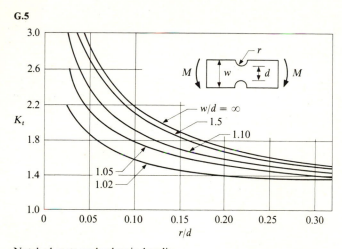

Notched rectangular bar in bending;
$\sigma_{\text{nom}} = Mc/I$ where $c = d/2$ and $I = td^3/12$ where t is the thickness

G.6

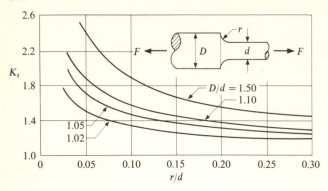

Round shaft with fillet and in tension;
$\sigma_{nom} = F/A$ where $A = \pi d^2/4$

G.7

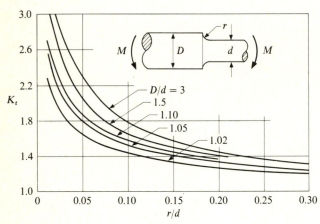

Round shaft with fillet and in bending;
$\sigma_{nom} = Mc/I$ where $c = d/2$ and $I = \pi d^4/64$

G.8

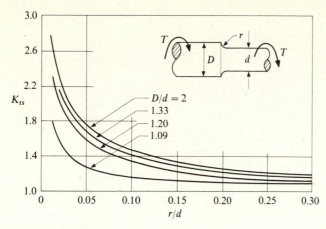

Round shaft with fillet and in torsion;
$\tau_{\text{nom}} = Tc/J$ where $c = d/2$ and $J = \pi d^4/32$

G.9

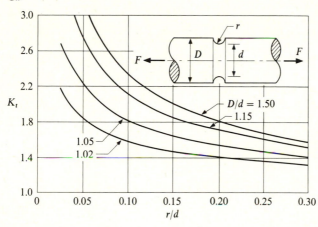

Grooved bar in tension;
$\sigma_{\text{nom}} = F/A$ where $A = \pi d^2/4$

G.10

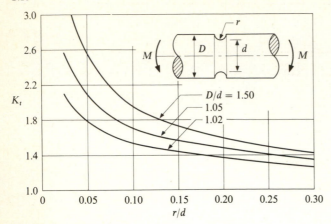

Grooved bar in bending;
$\sigma_{nom} = Mc/I$ where $c = d/2$ and $I = \pi d^4/64$

G.11

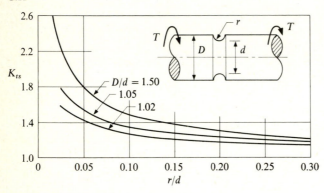

Grooved bar in torsion;
$\tau_{nom} = Tc/J$ where $c = d/2$ and $J = \pi d^4/32$

STRAIN-GAGE INSTALLATION PROCEDURES

The technique given in this appendix may seem lengthy and overcautious, but one should bear in mind the expense for each gage installation. The cost of the gage itself is not as great as the time involved in mounting the gage correctly for accuracy and reliability. *Do it right the first time.* Before an actual application, it is wise to practice the installation procedure on inexpensive gages or out-of-spec gages, which can be obtained from a strain-gage manufacturer.

The following procedure uses Eastman 910† as the adhesive; other appropriate cements are used in a similar manner.

A. *Cementing gage to specimen*
1. Degrease the specimen completely in the area where gage is to be installed.
2. Apply a weak acid (metal conditioner available from strain-gage suppliers) and with a fine wet-dry abrasive paper wet-lap the entire area.
3. Dry the specimen with clean cotton gauze pad.
4. Apply layout lines on specimen to permit accurate location and alignment of gage. The layout lines should be applied by burnishing (not by scribing) using either a fine ball-point pen or hard drafting pencil.
5. Reapply the metal conditioner with a cotton applicator using a fresh surface with each pass until the cotton surface stays clean after each pass.

---

† Selected batches of Eastman 910 are marketed under this name or under the strain-gage manufacturer's name.

6. Dry the specimen with a clean cotton gauze pad.

7. Generously apply an alkaline or neutralizing solution with a cotton applicator.

8. Dry the specimen with a gauze sponge.

9. Select the strain gage to be installed. If a self-temperature-compensating gage is to be used, the gage selection will depend on the specimen material. The characteristics of the gage are completely described by a technical data sheet provided by the manufacturer and packaged with the gage. Each manufacturer uses a numbering code for each gage which designates the grid material, e.g. Advance; carrier or backing, e.g. paper; particular specimen material for which it is temperature-compensated, e.g., stainless steel; and gage configuration, e.g., two-element 90° rosette. For example, the BLH SR-4 code is illustrated in Fig. H-1.

10. Carefully remove the gage from its folder with tweezers, being certain not to touch the foil or wire.

11. Place the gage right side up on a clean glass or plastic working surface. If strain-relief loops are not to be used, disregard instruction 12.

12. Cut a pair of insulated-backed copper terminals and place the terminals on the working surface in the desired position relative to the gage (see Fig. 7-5a).

13. Carefully apply transparent tape over the gage (and terminals if using strain-relief loops). Cut the tape so that over 1 in of tape extends beyond the gage (and terminals) from both ends. Anchoring the tape to the glass on one end, gently wipe the tape down over the gage (and terminal), leaving a short length free at the end for removal.

14. Pull the tape up from the free end at a shallow angle thus removing the gage from the glass.

15. Position the tape over the desired area of the specimen so that the alignment marks on the gage coincide with the layout lines.

16. When the gage is correctly aligned, wipe tape firmly in place and check that the alignment marks coincide with the layout lines.

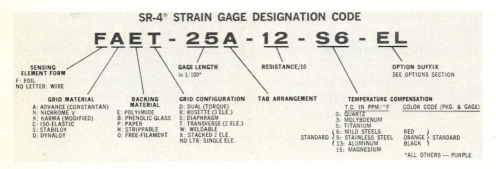

Figure H-1 (Courtesy of Baldwin-Lima-Hamilton Corporation.)

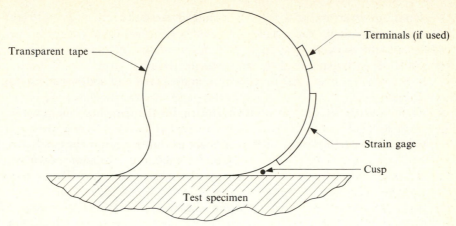

Figure H-2

17. Again, lift the tape at a shallow angle, until the gage (and terminals) are free from the specimen surface. Loop the tape over itself and tack the free end to the specimen so that the underside of the gage faces up (see Fig. H-2).
18. Apply a very small amount of Eastman 910 *catalyst* to the gage (and terminals) and allow to dry for 1 min. See the instructions provided with the Eastman 910.
19. Apply a single drop of Eastman 910 at the cusp formed by the junction of the tape and specimen.
20. Pull the tacked-down free end of the tape from the specimen, and hold the tape lightly taut at a small angle over the specimen. With a piece of Teflon between your free hand and the tape, firmly wipe the tape onto the specimen starting at the anchored end of the tape. This causes the adhesive to spread over the gage area.
21. Apply a firm pressure on the gage using the flat portion of your thumb or a rubber pressure pad and clamp. Apply pressure for over 1 min.
22. Remove tape, peeling it slowly and steadily from surface. The gage is now ready for wiring.

B. *Gage-wiring instructions*

1. Place masking tape over the gage grid, exposing half of the gage terminal tabs (*and* the additional terminals if strain-relief loops are used). This to protect the gage grid during soldering.
2. Tin the gage tabs (and terminals) using rosin-core solder.
3. Strip $\frac{1}{4}$ in of insulation from each wire (electrical wire appropriate to gage installations is marketed by strain-gage manufacturers). If relief loops are not used, strip $\frac{1}{8}$ in of insulation and go to step 10.

4. Separate one strand of wire from each lead, twist the remaining wires of each lead, and tin each lead.
5. Cut off the twisted portions to within $\frac{1}{8}$ in of the insulation.
6. Solder each twisted lead to the separate terminal strip.
7. Bend relief loops so that they touch the grid tabs and solder each loop to gage.
8. Apply a rosin solvent to remove any residual rosin flux and remove masking tape.
9. If possible, anchor wire so that if there is any force on the wire, the connection will not be broken. If the connection is broken, however, the relief-loop wires will break before the gage is damaged. The wiring installation is now complete.
10. If relief loops are not used, twist the $\frac{1}{8}$-in leads separately and solder to the gage tab. This is a risky procedure because if the wire is pulled, the gage can be destroyed quite easily (if relief loops are used, the risk is quite small).

C. *Electrical check*

1. Check the resistance of the gage to determine whether the gage is still intact and not grounded to specimen.

D. *Waterproofing gage*

1. Many waterproofing compounds are supplied by the manufacturers; following their recommendations, coat the finished gage installation.

STRAIN-GAGE ROSETTE EQUATIONS

In Example 7-2, a three-element rectangular rosette was analyzed using Mohr's circle for strain. The principal strains are first determined, then the principal stresses are evaluated using the stress-strain relations. There are basically two forms of three-element rosettes used for general, biaxial stress; the rectangular (45°) rosette, and the delta (120°) rosette (Fig. I-1). The equations for the principal strains and stresses are as follows.

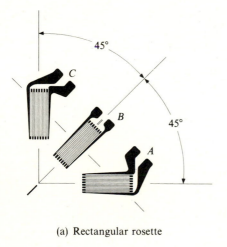

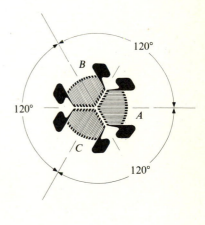

(a) Rectangular rosette   (b) Delta rosette

Figure I-1

I-1 THREE-ELEMENT RECTANGULAR ROSETTE

For the arrangement of Fig. I-1a, the principal strains are

$$\varepsilon_1 = \frac{\varepsilon_A + \varepsilon_C}{2} + \tfrac{1}{2}\sqrt{(\varepsilon_A - \varepsilon_C)^2 + (2\varepsilon_B - \varepsilon_A - \varepsilon_C)^2} \qquad (\text{I-1}a)$$

$$\varepsilon_2 = \frac{\varepsilon_A + \varepsilon_C}{2} - \tfrac{1}{2}\sqrt{(\varepsilon_A - \varepsilon_C)^2 + (2\varepsilon_B - \varepsilon_A - \varepsilon_C)^2} \qquad (\text{I-1}b)$$

and the principal stresses are

$$\sigma_1 = \frac{E}{2}\left[\frac{\varepsilon_A + \varepsilon_C}{1 - \nu} + \frac{1}{1 + \nu}\sqrt{(\varepsilon_A - \varepsilon_C)^2 + (2\varepsilon_B - \varepsilon_A - \varepsilon_C)^2} \right] \qquad (\text{I-2}a)$$

$$\sigma_2 = \frac{E}{2}\left[\frac{\varepsilon_A + \varepsilon_C}{1 - \nu} - \frac{1}{1 + \nu}\sqrt{(\varepsilon_A - \varepsilon_C)^2 + (2\varepsilon_B - \varepsilon_A - \varepsilon_C)^2} \right] \qquad (\text{I-2}b)$$

I-2 THREE-ELEMENT DELTA ROSETTE

For the arrangement of Fig. I-1b, the principal strains are

$$\varepsilon_1 = \frac{\varepsilon_A + \varepsilon_B + \varepsilon_C}{3} + \frac{\sqrt{2}}{3}\sqrt{(\varepsilon_A - \varepsilon_B)^2 + (\varepsilon_B - \varepsilon_C)^2 + (\varepsilon_C - \varepsilon_A)^2} \qquad (\text{I-3}a)$$

$$\varepsilon_2 = \frac{\varepsilon_A + \varepsilon_B + \varepsilon_C}{3} - \frac{\sqrt{2}}{3}\sqrt{(\varepsilon_A - \varepsilon_B)^2 + (\varepsilon_B - \varepsilon_C)^2 + (\varepsilon_C - \varepsilon_A)^2} \qquad (\text{I-3}b)$$

and the principal stresses are

$$\sigma_1 = \frac{E}{3}\left[\frac{\varepsilon_A + \varepsilon_B + \varepsilon_C}{1 - \nu} + \frac{\sqrt{2}}{1 + \nu}\sqrt{(\varepsilon_A - \varepsilon_B)^2 + (\varepsilon_B - \varepsilon_C)^2 + (\varepsilon_C - \varepsilon_A)^2} \right] \qquad (\text{I-4}a)$$

$$\sigma_2 = \frac{E}{3}\left[\frac{\varepsilon_A + \varepsilon_B + \varepsilon_C}{1 - \nu} - \frac{\sqrt{2}}{1 + \nu}\sqrt{(\varepsilon_A - \varepsilon_B)^2 + (\varepsilon_B - \varepsilon_C)^2 + (\varepsilon_C - \varepsilon_A)^2} \right] \qquad (\text{I-4}b)$$

CORRECTIONS FOR CROSS SENSITIVITY OF STRAIN GAGES

The change in resistance of a strain gage does not depend only on the normal strain in the axial direction of the gage. The strain perpendicular to the primary sensing axis of the gage also affects the change in resistance. Normally, however, the error induced by the transverse strain is small. Nevertheless, there are cases when the transverse effects cannot be ignored, and the gage outputs must be corrected. To understand transverse effects, consider a perfectly uniaxial *strain* field. If a linear gage is mounted parallel to the field and the change in resistance is measured for various values of strain, the change in resistance will be proportional to the strain as $\Delta R / R = S_a \varepsilon_a$, where S_a is the axial gage factor and ε_a the axial strain. If the gage is mounted perpendicular to the strain field, the change in resistance is again proportional to the strain, but the relationship is $\Delta R / R = S_t \varepsilon_t$, where S_t is the transverse gage factor and ε_t the transverse strain. Consider now the gage to be mounted in a general biaxial strain field. The change in resistance is thus

$$\frac{\Delta R}{R} = S_a \varepsilon_a + S_t \varepsilon_t$$

or

$$\frac{\Delta R}{R} = S_a (\varepsilon_a + K_t \varepsilon_t) \qquad \text{(J-1)}$$

where $K_t = S_t / S_a$ is the transverse-sensitivity coefficient.

The gage factor S_g, supplied by the manufacturer is generally based on a uniaxial *stress* field, where $\varepsilon_t = -\nu \varepsilon_a$. That is,

$$\frac{\Delta R}{R} = S_g \varepsilon_a \qquad \varepsilon_t = -\nu \varepsilon_a \qquad \text{(J-2)}$$

Substituting $\varepsilon_t = -\nu\varepsilon_a$ into Eq. (J-1) yields

$$\frac{\Delta R}{R} = S_a(1 - \nu K_t)\varepsilon_a \tag{J-3}$$

Thus, it can be seen from Eqs. (J-2) and (J-3) that

$$S_g = S_a(1 - \nu K_t) \tag{J-4}$$

If the gage is not mounted in a uniaxial stress field and Eq. (J-2) is used, an error will develop. To see this, let ε_a and ε_t be the actual strains and $\hat{\varepsilon}_a$ be the strain based on Eq. (J-2). Equating Eqs. (J-2) and (J-1) yields

$$S_g\hat{\varepsilon}_a = S_a(\varepsilon_a + K_t\varepsilon_t) \tag{J-5}$$

Substituting Eq. (J-4) and solving for $\hat{\varepsilon}_a$ gives

$$\hat{\varepsilon}_a = \frac{\varepsilon_a + K_t\varepsilon_t}{1 - \nu K_t} \tag{J-6}$$

The error e using $\hat{\varepsilon}_a$ would be $e = \hat{\varepsilon}_a - \varepsilon_a$. From Eq. (J-6) the error is found to be

$$e = \frac{K_t}{1 - \nu K_t}(\nu\varepsilon_a + \varepsilon_t) \tag{J-7}$$

J-1 TWO-GAGE (90°) ROSETTE

Equation (J-7) provides the measurement error if the exact axial and transverse strains are known, which in general is not the case. However, if the measured values of the axial and transverse strains $\hat{\varepsilon}_a$ and $\hat{\varepsilon}_t$ are known, the actual strains can be determined. The relationship between the measured axial strain and the actual strains is given by Eq. (J-5). In a similar manner the measured transverse strain relationship is

$$S_g\hat{\varepsilon}_t = S_a(\varepsilon_t + K_t\varepsilon_a) \tag{J-8}$$

Solving Eqs. (J-5) and (J-8) simultaneously for ε_a results in

$$\varepsilon_a = \frac{S_g(\hat{\varepsilon}_a - K_t\hat{\varepsilon}_t)}{S_a(1 - K_t^2)}$$

Substituting in Eq. (J-4) yields

$$\varepsilon_a = \frac{1 - \nu K_t}{1 - K_t^2}(\hat{\varepsilon}_a - K_t\hat{\varepsilon}_t) \tag{J-9}$$

Solving for ε_t in a similar manner results in

$$\varepsilon_t = \frac{1 - \nu K_t}{1 - K_t^2}(\hat{\varepsilon}_t - K_t\hat{\varepsilon}_a) \tag{J-10}$$

Thus, if the cross sensitivity is known (usually supplied by the manufacturer) and the axial and transverse strains are measured, the actual values can be determined using Eqs. (J-9) and (J-10).

J-2 THREE-GAGE (45°) RECTANGULAR ROSETTE

For a three-element rectangular rosette as shown in Fig. J-1, gages A and C correspond to axial and transverse gages of a two-element 90° rosette as described previously by Eqs. (J-9) and (J-10). Gage B differs since there is no direct measurement of the strain perpendicular to that grid. The correction equations for the three gages are

$$\varepsilon_A = \frac{1 - \nu K_t}{1 - K_t^2} (\hat{\varepsilon}_A - K_t \hat{\varepsilon}_C) \tag{J-11}$$

$$\varepsilon_B = \frac{1 - \nu K_t}{1 - K_t^2} [\hat{\varepsilon}_B - K_t(\hat{\varepsilon}_A + \hat{\varepsilon}_C - \hat{\varepsilon}_B)] \tag{J-12}$$

$$\varepsilon_C = \frac{1 - \nu K_t}{1 - K_t^2} (\hat{\varepsilon}_C - K_t \hat{\varepsilon}_A) \tag{J-13}$$

where $\hat{\varepsilon}_A$, $\hat{\varepsilon}_B$, $\hat{\varepsilon}_C$ are the measured strains and ε_A, ε_B, ε_C are the actual strains.

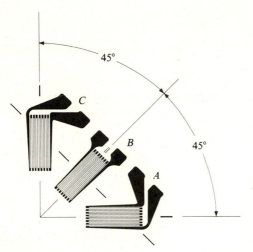

45°

45°

C

B

A

Figure J-1

J-3 THREE-GAGE DELTA (120°) ROSETTE

For the delta rosette shown in Fig. J-2 the corrections are

$$\varepsilon_A = \frac{1 - \nu K_t}{1 - K_t^2} \left[\left(1 + \frac{K_t}{3}\right) \hat{\varepsilon}_A - \tfrac{2}{3} K_t (\hat{\varepsilon}_B + \hat{\varepsilon}_C) \right] \tag{J-14}$$

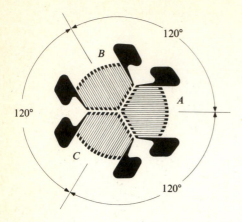

Figure J-2

$$\varepsilon_B = \frac{1 - \nu K_t}{1 - K_t^2}\left[\left(1 + \frac{K_t}{3}\right)\hat{\varepsilon}_B - \tfrac{2}{3}K_t(\hat{\varepsilon}_C + \hat{\varepsilon}_A)\right] \qquad (\text{J-15})$$

$$\varepsilon_C = \frac{1 - \nu K_t}{1 - K_t^2}\left[\left(1 + \frac{K_t}{3}\right)\hat{\varepsilon}_C - \tfrac{2}{3}K_t(\hat{\varepsilon}_A + \hat{\varepsilon}_B)\right] \qquad (\text{J-16})$$

MATRIX ALGEBRA
AND CARTESIAN TENSORS

K-0 INTRODUCTION

A convenient shorthand notation of considerable advantage in dealing with systems of linear equations is the use of matrices. Matrix notation permits a systematic approach to handling large arrays of simultaneous equations and their solutions and provides a convenient form for use in computer programming. In matrix notation, it is generally more convenient to express similar terms using subscript notation. However, since this notation was not used in the main body of this book, it will not be used here. As an example of matrix notation, assume that the displacement field of a body is linearly related to the position within that body. Let u, v, and w be the displacements in the x, y, and z directions respectively. Thus

$$u = a_{11}x + a_{12}y + a_{13}z$$
$$v = a_{21}x + a_{22}y + a_{23}z \tag{K-1}$$
$$w = a_{31}x + a_{32}y + a_{33}z$$

where the a_{ij} terms are constants $(i, j = 1, 2, 3)$. In matrix notation these equations may be represented by

$$\begin{Bmatrix} u \\ v \\ w \end{Bmatrix} = \begin{bmatrix} a_{11} & a_{12} & a_{13} \\ a_{21} & a_{22} & a_{23} \\ a_{31} & a_{32} & a_{33} \end{bmatrix} \begin{Bmatrix} x \\ y \\ z \end{Bmatrix} \tag{K-2}$$

Equation (K-2) can be further abbreviated by using the notation

$$\{\delta\} = [a]\{s\} \tag{K-3}$$

where

$$\{\delta\} = \begin{Bmatrix} u \\ v \\ w \end{Bmatrix} \qquad [a] = \begin{bmatrix} a_{11} & a_{12} & a_{13} \\ a_{21} & a_{22} & a_{23} \\ a_{31} & a_{32} & a_{33} \end{bmatrix} \quad \text{and} \quad \{s\} = \begin{Bmatrix} x \\ y \\ z \end{Bmatrix}$$

This formulation should become clearer when matrix multiplication is discussed. Note that each bracketed array is a matrix but matrices with only one column or one row are denoted by braces whereas matrices with more than one column are denoted by square brackets. For the present example, the matrix with the a_{ij} terms, is written so that the i term represents the row and the j term the column where the constant exists. Thus the term a_{23} exists in row 2 and column 3 of the matrix a_{ij}. In addition, the a_{ij} matrix is a *square matrix* since the number of rows equals the number of columns. In terms of notation the column matrices $\{\delta\}$ and $\{s\}$ are 3×1 matrices since they contain three rows and one column, where the a_{ij} matrix is 3×3 since it contains three rows and three columns.

A square matrix which contains only the diagonal terms a_{11}, a_{22}, and a_{33}, all other constants being zero, is called a *diagonal matrix*

$$\begin{bmatrix} a_{11} & 0 & 0 \\ 0 & a_{22} & 0 \\ 0 & 0 & a_{33} \end{bmatrix}$$

A square matrix for which $a_{ij} = a_{ji}$ is called a *symmetric matrix*

$$\begin{bmatrix} a_{11} & b & c \\ b & a_{22} & d \\ c & d & a_{33} \end{bmatrix}$$

A shorthand form commonly used for the symmetric matrix is

$$\begin{bmatrix} a_{11} & b & c \\ & a_{22} & d \\ \text{sym} & & a_{33} \end{bmatrix}$$

K-1 MATRIX ALGEBRA

Addition

The sum of two matrices can be made only if the two matrices have equal rows and columns; it is performed simply by adding corresponding elements, i.e.,

$$[a_{ij}] + [b_{ij}] = [c_{ij}] \tag{K-4}$$

where $c_{ij} = a_{ij} + b_{ij}$. For example,

$$\begin{bmatrix} 2 & 0 & -1 \\ 6 & -3 & 2 \end{bmatrix} + \begin{bmatrix} 0 & -2 & 1 \\ 1 & 0 & -1 \end{bmatrix} = \begin{bmatrix} 2 & -2 & 0 \\ 7 & -3 & 1 \end{bmatrix}$$

Scalar Multiplication

The product of a scalar and a matrix is the matrix formed by multiplying each element of the matrix by the scalar. Thus

$$s[a_{ij}] = [sa_{ij}] \tag{K-5}$$

For example,

$$3 \begin{bmatrix} 1 & 3 & 0 \\ 2 & -1 & 1 \\ 0 & 2 & -2 \end{bmatrix} = \begin{bmatrix} 3 & 9 & 0 \\ 6 & -3 & 3 \\ 0 & 6 & -6 \end{bmatrix}$$

Matrix Multiplication

The product of two matrices can be defined only when the number of columns in the first matrix equals the number of rows of the second matrix. In this case, the order of multiplication is important. For example, a 3×1 matrix cannot be premultiplied by a 1×2 matrix. However, a 1×2 matrix can be premultiplied by a 3×1 matrix. The formal definition of the product of two matrices is

$$[c] = [a][b] \tag{K-6}$$

where

$$c_{ij} = \sum_{k=1}^{n} a_{ik} b_{kj} \tag{K-7}$$

and n is the number of columns in the $[a]$ matrix, which equals the number of rows of the $[b]$ matrix. For example, let

$$[a] = \begin{bmatrix} 2 & 1 \\ -3 & 2 \end{bmatrix} \quad \text{and} \quad [b] = \begin{bmatrix} 1 & 3 & 0 \\ -2 & 0 & 1 \end{bmatrix}$$

Note that $[a]$ has two columns and $[b]$ has two rows. Thus the product of $[a][b]$ can be found. The product of $[b][a]$ cannot be found since the number of columns of $[b]$ does not equal the number of rows of $[a]$. Using Eq. (K-7) each term of the product matrix $[c]$, where $[c] = [a][b]$, can be determined:

$$c_{11} = a_{11}b_{11} + a_{12}b_{21} = (2)(1) + (1)(-2) = 0$$

$$c_{12} = a_{11}b_{12} + a_{12}b_{22} = (2)(3) + (1)(0) = 6$$

$$c_{13} = a_{11}b_{13} + a_{12}b_{23} = (2)(0) + (1)(1) = 1$$

$$c_{21} = a_{21}b_{11} + a_{22}b_{21} = (-3)(1) + (2)(-2) = -7$$

$$c_{22} = a_{21}b_{12} + a_{22}b_{22} = (-3)(3) + (2)(0) = -9$$

$$c_{23} = a_{21}b_{13} + a_{22}b_{23} = (-3)(0) + (2)(1) = 2$$

Thus
$$\begin{bmatrix} 2 & 1 \\ -3 & 2 \end{bmatrix}\begin{bmatrix} 1 & 3 & 0 \\ -2 & 0 & 1 \end{bmatrix} = \begin{bmatrix} 0 & 6 & 1 \\ -7 & -9 & 2 \end{bmatrix}$$

The product matrix will have as many rows as the first matrix and as many columns as the second matrix. To perform the multiplication quickly, note that c_{ij} is found by multiplying each term of the i row of the $[a]$ matrix separately and in order by each term of the j column of the $[b]$ matrix and then adding the results. For example, to obtain c_{23}, isolate row 2 of $[a]$ and column 3 of $[b]$:

$$\begin{bmatrix} 2 & 1 \\ -3 & 2 \end{bmatrix} \begin{bmatrix} 1 & 3 & 0 \\ -2 & 0 & 1 \end{bmatrix} = \begin{bmatrix} & & \\ \hline & & 2 \end{bmatrix} \begin{array}{l} \text{row 2} \\ c_{23} \end{array}$$

$\uparrow$ $\nearrow$ $\uparrow$

row 2   column 3   column 3

Multiplying each term in row 2 of $[a]$ by each term in column 3 of $[b]$ in order results in $(-3)(0) = 0$ and $(2)(1) = 2$. Adding the results yields $c_{23} = 0 + 2 = 2$.

If the $[a]$ and $[b]$ matrices are both square matrices, the products of $[a][b]$ and $[b][a]$ can both be performed, but in general

$$[a][b] \neq [b][a]$$

Thus, the order of multiplication is still important. The associative law for matrix multiplication does hold when multiplying three matrices; i.e.,

$$[a]([b][c]) = ([a][b])[c]$$

Inverting Matrices

The process of division can be thought of as multiplication involving a reciprocal. For example, consider scalar division; if

$$y = Qx$$

and the solution for x is required, both sides of the equation are divided by Q. In other words, each side is multiplied by the reciprocal of Q:

$$Q^{-1}y = Q^{-1}Q^1x$$

However, since $Q^{-1}Q^1 = 1$, we have

$$x = Q^{-1}y$$

Let us reconsider the matrix equations

$$\{\delta\} = [a]\{s\} \tag{K-3}$$

If an inverse matrix $[a]^{-1}$ exists such that

$$[a]^{-1}[a] = [1] \qquad \text{where} \quad [1] = \begin{bmatrix} 1 & 0 & 0 \\ 0 & 1 & 0 \\ 0 & 0 & 1 \end{bmatrix}$$

then multiplying both sides of Eq. (K-3) yields

$$[a]^{-1}\{\delta\} = [a]^{-1}[a]\{s\} = [1]\{s\} = \{s\}$$

or $$\{s\} = [a]^{-1}\{\delta\} \tag{K-8}$$

This, in reality, is the same procedure one takes to solve a set of simultaneous equations. For example, return to the long form of Eq. (K-3), given by

$$u = a_{11}x + a_{12}y + a_{13}z$$

$$v = a_{21}x + a_{22}y + a_{23}z \tag{K-1}$$

$$w = a_{31}x + a_{32}y + a_{33}z$$

if u, v, w, and the coefficients a_{ij} are considered known, Eqs. (K-1) can be solved simultaneously for x, y, and z provided that certain conditions are met. This can be done with Cramer's rule using determinants. For example, solving for x results in

$$x = \frac{\begin{vmatrix} u & a_{12} & a_{13} \\ v & a_{22} & a_{23} \\ w & a_{32} & a_{33} \end{vmatrix}}{\begin{vmatrix} a_{11} & a_{12} & a_{13} \\ a_{21} & a_{22} & a_{23} \\ a_{31} & a_{32} & a_{33} \end{vmatrix}}$$

where vertical lines denote a determinant. Expanding the determinant in the numerator yields

$$\begin{vmatrix} u & a_{12} & a_{13} \\ v & a_{22} & a_{23} \\ w & a_{32} & a_{33} \end{vmatrix} = (a_{22}a_{33} - a_{23}a_{32})u$$

$$+ (a_{13}a_{32} - a_{12}a_{33})v$$

$$+ (a_{12}a_{23} - a_{13}a_{22})w$$

Thus it can be seen that x is linearly related to u, v, and w provided that the determinant $|a|$ is nonzero. In a similar manner it can be shown that y and z are also linearly related to u, v, and w. The final results can be written

$$\begin{Bmatrix} x \\ y \\ z \end{Bmatrix} = \frac{\begin{bmatrix} b_{11} & b_{12} & b_{13} \\ b_{21} & b_{22} & b_{23} \\ b_{31} & b_{32} & b_{33} \end{bmatrix} \begin{Bmatrix} u \\ v \\ w \end{Bmatrix}}{|a|} \tag{K-9}$$

where it has already been shown that

$$b_{11} = a_{22}a_{33} - a_{23}a_{32}$$

$$b_{12} = a_{13}a_{32} - a_{12}a_{33}$$

$$b_{13} = a_{12}a_{23} - a_{13}a_{22}$$

Reapplying Cramer's rule to find y and z results in

$$b_{21} = a_{23}a_{31} - a_{21}a_{33}$$

$$b_{22} = a_{11}a_{33} - a_{13}a_{31}$$

$$b_{23} = a_{13}a_{21} - a_{11}a_{23}$$

$$b_{31} = a_{21}a_{32} - a_{22}a_{31}$$

$$b_{32} = a_{12}a_{31} - a_{11}a_{32}$$

$$b_{33} = a_{11}a_{22} - a_{12}a_{21}$$

The $[b]$ matrix can be generalized; it is called the *transpose* of the *cofactor* of the matrix $[a]$ and written as

$$[b] = [A]^t$$

The matrix $[A]^t$ is formed by first finding the cofactor matrix of $[a]$. The cofactor matrix $[A]$ is found in the following manner. The matrix $[a]$ is

$$\begin{bmatrix} a_{11} & a_{12} & a_{13} \\ a_{21} & a_{22} & a_{23} \\ a_{31} & a_{32} & a_{33} \end{bmatrix}$$

The row i, column j term in the cofactor is found by striking out row i, column j of the $[a]$ matrix and evaluating the determinant of the remaining terms. The sign of the result is established by adding the row and column numbers; if it is even, the sign is plus; if it is odd, the sign is minus. For example, the row 1, column 1, cofactor is

$$\begin{bmatrix} \cancel{a_{11}} & \cancel{a_{12}} & \cancel{a_{13}} \\ \cancel{a_{21}} & a_{22} & a_{23} \\ \cancel{a_{31}} & a_{32} & a_{33} \end{bmatrix}$$

or

$$A_{11} = \begin{vmatrix} a_{22} & a_{23} \\ a_{32} & a_{33} \end{vmatrix} = a_{22}a_{33} - a_{23}a_{32}$$

Since $i = 1$, $j = 1$, we have $i + j = 2$, which is even. Thus the sign is positive. For row 1, column 2, the cofactor is

$$\begin{bmatrix} \cancel{a_{11}} & \cancel{a_{12}} & \cancel{a_{13}} \\ a_{21} & \cancel{a_{22}} & a_{23} \\ a_{31} & \cancel{a_{32}} & a_{33} \end{bmatrix}$$

or

$$A_{12} = -\begin{vmatrix} a_{21} & a_{23} \\ a_{31} & a_{33} \end{vmatrix} = -(a_{21}a_{33} - a_{23}a_{31})$$

The sign is negative since $i = 1$, $j = 2$; $i + j = 3$ (odd). Thus the cofactor matrix $[A]$ can be found:

$$\begin{bmatrix} a_{22}a_{33} - a_{23}a_{32} & a_{23}a_{31} - a_{21}a_{33} & a_{21}a_{32} - a_{22}a_{31} \\ a_{13}a_{32} - a_{12}a_{33} & a_{11}a_{33} - a_{13}a_{31} & a_{12}a_{31} - a_{11}a_{32} \\ a_{12}a_{23} - a_{13}a_{22} & a_{13}a_{21} - a_{11}a_{23} & a_{11}a_{22} - a_{12}a_{21} \end{bmatrix}$$

The final step is to find the transpose of this matrix, which is done simply by interchanging the rows and columns of the cofactor matrix. Thus $[A]^t$ is

$$\begin{bmatrix} a_{22}a_{33} - a_{23}a_{32} & a_{13}a_{32} - a_{12}a_{33} & a_{12}a_{23} - a_{13}a_{22} \\ a_{23}a_{31} - a_{21}a_{33} & a_{11}a_{33} - a_{13}a_{31} & a_{13}a_{21} - a_{11}a_{23} \\ a_{21}a_{32} - a_{22}a_{31} & a_{12}a_{31} - a_{11}a_{32} & a_{11}a_{22} - a_{12}a_{21} \end{bmatrix}$$

which is identical to the $[b]$ matrix of Eq. (K-9). The inverse of a square matrix $[a]$ is defined as

$$[a]^{-1} = \frac{[A]^t}{|a|} \tag{K-10}$$

where $[A]^t$ is the transpose of the cofactor matrix of $[a]$ and $|a|$ is the determinant of $[a]$.

The condition necessary to find the inverse is that the determinant of the matrix $[a]$ be nonzero. If $|a| = 0$, the determinant is said to be *singular.* As an example, find the inverse of the matrix

$$[a] = \begin{bmatrix} 3 & 0 & 1 \\ -1 & 2 & 0 \\ 1 & -2 & 1 \end{bmatrix}$$

The inverse $[a]^{-1}$ is

$$[a]^{-1} = \frac{[A]^t}{|a|}$$

The cofactor of $[a]$ is

$$[A] = \begin{bmatrix} 2 & 1 & 0 \\ -2 & 2 & 6 \\ -2 & -1 & 6 \end{bmatrix}$$

and the transpose of the cofactor is

$$[A]^t = \begin{bmatrix} 2 & -2 & -2 \\ 1 & 2 & -1 \\ 0 & 6 & 6 \end{bmatrix}$$

The determinant of $[a]$ is

$$|a| = (3)[(2)(1) - (0)(-2)] - (0)[(-1)(1) - (0)(1)] + (1)[(-1)(-2) - (2)(1)] = 6$$

Thus, the inverse of $[a]$ is

$$[a]^{-1} = \frac{1}{6} \begin{bmatrix} 2 & -2 & -2 \\ 1 & 2 & -1 \\ 0 & 6 & 6 \end{bmatrix}$$

K-2 TENSORS

A tensor is a matrix that generally describes a property of state of which is invariant with respect to the coordinate system. For example, a vector is a tensor since regardless of the coordinate system, the vector remains un-

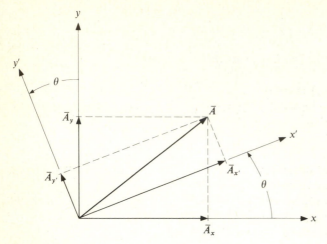

Figure K-1

changed. The vector shown in Fig. K-1 can be described either with respect to the xy or the $x'y'$ coordinate systems. Thus,

$$\bar{A} = \bar{A}_x + \bar{A}_y \qquad \text{or} \qquad \bar{A} = \bar{A}_{x'} + \bar{A}_{y'}$$

If the relationship between the xy and $x'y'$ coordinates is known, and if tensors are known in one set of coordinates, the tensor can be determined (or transformed) through matrix multiplication with respect to the other coordinates. This is done by determining an appropriate transformation matrix. For example, returning to Fig. K-1, assume that the angle θ is known and that the vector $\bar{A}$ is described by its x and y components, A_x and A_y, respectively.

Both A_x and A_y have components in the x' direction, whose sum yields $A_{x'}$. Thus

$$A_{x'} = A_x \cos \theta + A_y \sin \theta \qquad \text{(K-11a)}$$

Likewise

$$A_{y'} = - A_x \sin \theta + A_y \cos \theta \qquad \text{(K-11b)}$$

Equations (K-11) can be written using matrix notation:

$$\begin{Bmatrix} A_{x'} \\ A_{y'} \end{Bmatrix} = \begin{bmatrix} \cos \theta & \sin \theta \\ -\sin \theta & \cos \theta \end{bmatrix} \begin{Bmatrix} A_x \\ A_y \end{Bmatrix} \qquad \text{(K-12)}$$

The square matrix in Eq. (K-12) is called the transformation matrix which transforms the vector components in the xy coordinate system into the vector components in the $x'y'$ coordinate system.

The previous example is considered a two-dimensional transformation, but in general the same procedure holds true for three-dimensional transformations. Thus

$$\begin{Bmatrix} A_{x'} \\ A_{y'} \\ A_{z'} \end{Bmatrix} = \begin{bmatrix} a_{11} & a_{12} & a_{13} \\ a_{21} & a_{22} & a_{23} \\ a_{31} & a_{32} & a_{33} \end{bmatrix} \begin{Bmatrix} A_x \\ A_y \\ A_z \end{Bmatrix} \qquad \text{(K-13)}$$

or, further simplifying,

$$\{A\}_{x'y'z'} = [a]\{A\}_{xyz} \tag{K-14}$$

The transformation matrix $[a]$ has some unique properties:

1. The determinant of $[a]$ is unity.
2. The inverse of $[a]$ is simply the transpose of $[a]$. Thus the inverse of Eq. (K-14) can be written as

$$\{A\}_{xyz} = [a]^{-1}\{A\}_{x'y'z'} \qquad \text{where} \quad [a]^{-1} = [a]^t$$

3. The a_{ij} terms are the *directional cosines* of the $x'y'z'$ coordinates with respect to the xyz coordinates. For example, in Fig. K-1 the directional cosines of the x' coordinate relative to the xy coordinates are $\cos\theta$ and $\cos(90-\theta)$, respectively. This constitutes the first row of the transformation matrix since $\cos(90-\theta) = \sin\theta$.

   Likewise, the directional cosines of the y' coordinate relative to the xy coordinates are $\cos(90+\theta)$ and $\cos\theta$, respectively. This constitutes the second row of the transformation matrix since $\cos(90+\theta) = -\sin\theta$.

A tensor of order n has 3^n components. Thus a vector is a tensor of order 1 since 3 components are necessary to describe a vector. A scalar is a tensor of order 0 since only one component is needed to describe a scalar. Stress and strain are second-order tensors, since in general, 9 terms are necessary to describe the state of stress or strain at a point.

A second-order tensor $[A]$ is a matrix which transforms as

$$[A]_{x'y'z'} = [a][A]_{xyz}[a]^t \tag{K-15}$$

where the $[a]$ matrix is the same as discussed with vectors. For example, consider a biaxial state of stress where the stress tensor reduces to a 2×2 matrix. Assume that the stresses are known with respect to the xy coordinate system (as shown in Fig. K-1). That is,

$$[\sigma]_{x,y} = \begin{bmatrix} \sigma_x & \tau_{xy} \\ \tau_{xy} & \sigma_y \end{bmatrix} \tag{K-16}$$

Using the transformation matrix of Eq. (K-12), we see that the stress tensor in the $x'y'$ coordinate system is given by Eq. (K-15), where

$$[\sigma]_{x'y'} = [a][\sigma]_{xy}[a]^t$$

$$\begin{bmatrix} \sigma_{x'} & \tau_{x'y'} \\ \tau_{x'y'} & \sigma_{y'} \end{bmatrix} = \begin{bmatrix} \cos\theta & \sin\theta \\ -\sin\theta & \cos\theta \end{bmatrix} \begin{bmatrix} \sigma_x & \tau_{xy} \\ \tau_{xy} & \sigma_y \end{bmatrix} \begin{bmatrix} \cos\theta & -\sin\theta \\ \sin\theta & \cos\theta \end{bmatrix}$$

Multiplication of the second and third matrices on the right-hand side of this equation yields

$$\begin{bmatrix} \sigma_x & \tau_{xy} \\ \tau_{xy} & \sigma_y \end{bmatrix} \begin{bmatrix} \cos\theta & -\sin\theta \\ \sin\theta & \cos\theta \end{bmatrix} = \begin{bmatrix} \sigma_x\cos\theta + \tau_{xy}\sin\theta & -\sigma_x\sin\theta + \tau_{xy}\cos\theta \\ \tau_{xy}\cos\theta + \sigma_y\sin\theta & -\tau_{xy}\sin\theta + \sigma_y\cos\theta \end{bmatrix}$$

Finally, premultiplying this by $\begin{bmatrix} \cos\theta & \sin\theta \\ -\sin\theta & \cos\theta \end{bmatrix}$ yields $\begin{bmatrix} \sigma_{x'} & \tau_{x'y'} \\ \tau_{x'y'} & \sigma_{y'} \end{bmatrix}$,

where

$$\sigma_{x'} = \sigma_x \cos^2\theta + \sigma_y \sin^2\theta + 2\tau_{xy}\sin\theta\cos\theta$$

$$\sigma_{y'} = \sigma_x \sin^2\theta + \sigma_y \cos^2\theta - 2\tau_{xy}\sin\theta\cos\theta \qquad (K\text{-}17)$$

$$\tau_{x'y'} = (-\sigma_x + \sigma_y)\sin\theta\cos\theta + \tau_{xy}(\cos^2\theta - \sin^2\theta)$$

Note that the equations for $\sigma_{x'}$, $\tau_{x'y'}$, and $\sigma_{y'}$ agree with Eqs. (2-29), (2-30), and (2-33), respectively.

The advantage of tensor notation is in its compactness and mathematical simplicity. Once the transformation matrices are known, any tensor can be transformed in any direction with respect to coordinate systems with little difficulty and without resorting to the actual properties of the tensor. For example, Eqs. (2-29) and (2-30) are obtained from free-body diagrams whereby the stresses are first converted into net forces acting over the appropriate surfaces and the equilibrium equations are then necessary. With tensor transformation this process is not used.†

---

† When dealing with strain, the matrix $\begin{bmatrix} \varepsilon_x & \gamma_{xy} \\ \gamma_{xy} & \varepsilon_y \end{bmatrix}$ is *not* a tensor, as it will not transform. However, the matrix $\begin{bmatrix} \varepsilon_x & \frac{1}{2}\gamma_{xy} \\ \frac{1}{2}\gamma_{xy} & \varepsilon_y \end{bmatrix}$ is a tensor.

ANSWERS TO SELECTED PROBLEMS

CHAPTER 1

1-1 $R_B = 8000$ lb, $R_D = 1000$ lb

1-3 $(R_A)_x = 100$ lb, $(M_A)_y = -600$ lb·in, $(M_A)_z = -300$ lb·in

1-5 $F_x = 1500$ lb, $M_z = -4500$ lb·in

1-7 $\sigma_x = 9340$ lb/in², $\sigma_y = -2200$ lb/in², $\epsilon_z = -214 \ \mu$in/in

1-11 $\epsilon_x = -175 \ \mu$m/m, $\epsilon_y = -45 \ \mu$m/m, $\epsilon_z = 280 \ \mu$m/m, $\gamma_{xy} = 130 \ \mu$m/m, $\gamma_{yz} = 520 \ \mu$m/m, $\gamma_{zx} = -390 \ \mu$m/m

CHAPTER 2

2-1 (a) $(\sigma_x)_1 = \dfrac{2F}{at} \dfrac{E_1}{E_1 + E_2}$ $\qquad (\sigma_x)_2 = \dfrac{2F}{at} \dfrac{E_2}{E_1 + E_2}$

(b) $e = \dfrac{a}{4} \dfrac{3E_1 + E_2}{E_1 + E_2}$

2-3 $F = 24{,}000$ lb, $e = 4.44$ in

2-5 (a) AB: $\tau_{max} = 20{,}370$ lb/in², BC: $\tau_{max} = 2550$ lb/in², CD: $\tau_{max} = 2600$ lb/in²

(b) $\theta_D = 1.3°$

2-6 (a) $\sigma_{max} = 6620$ lb/in², $\sigma_{min} = -3970$ lb/in²

(b) 460 lb/in²

2-12 $(\sigma_\theta)_{max} = 2080$ lb/in₂

2-13 $(y_c)_A = -0.071$ in, $\theta_A = 0.093°$

2-15 (a) $T_A = 760$ lb·in, $T_D = 7240$ lb·in

(b) $\tau_{max} = 37{,}600$ lb/in²

(c) $\theta_{max} = 1.84°$

2-19 $\sigma = 6930$ lb/in², $\tau = 2000$ lb/in², $\tau' = 270$ lb/in²

2-23 $(\sigma_x)_{max} = 4120$ lb/in², $\tau = 2264$ lb/in²

CHAPTER 3

3-1 $\sigma_{max} = 21.6$ MN/m^2, $\sigma_{min} = -21.3$ MN/m^2

3-4 $\sigma_{max} = 15,460$ lb/in^2, $\sigma_{min} = -15,460$ lb/in^2, $T_{wall} \approx 73$ lb·in

3-5 1.36 in left of center of vertical wall

3-7 $M_{max} = 23,570$ lb·ft

3-10 12.6 kN

3-14 $K_i = 1.097$, $K_o = 0.903$

3-17 (a) $\delta_{max} = 0.00205$ in

3-20 $\sigma_r = -33,300$ lb/in^2, $\sigma_\theta = 41,700$ lb/in^2, $\tau_{r\theta} = 400$ lb/in^2

3-25 $M_p = 59.3$ kN·m, SF = 1.52

3-26 $h_p = 0.225c$, $x_p = \dfrac{L}{6}$

CHAPTER 4

4-1 $U = 10.62$ in·lb

4-2 $u = 2.02$ N·m/m^3

4-5 $(\delta_C)_v = (8k_1 + k_2)P^3/16$

4-7 $\delta = 0.0167$ in

4-11 $\theta_3 = 5.41P/AE$ clockwise

4-14 $\delta_y = -40.6$ mm

4-15 Straight-beam approximation: $\delta = 0.0409$ in, Eq. (4-44): $\delta = 0.0437$ in

4-17 Vertical reaction: $w_o L/2$

Horizontal reaction: $\dfrac{w_o L^3 h}{4I_1(2h^3/I_2 + 3L/A_1 + 3Lh^2/I_1)}$

4-20 $R_A = R_B = P/2$, $M_A = M_B = Ph/2$

$T_A = T_B = Pl^2/4[(1 + \nu)h + l]$

4-25 $(\delta_C)_v = 1.833PL/AE + 0.289\alpha_T L\,\Delta T$

4-27 $\delta_{max} = 0.3023 M_e L^2/EI_z$, $\delta_{res} = 0.0023 M_e L^2/EI_z$

4-28 (a) $\delta = 0.119wL^4/EI_z$, (b) $\delta = 0.125wL^4/EI_z$

4-31 Differential equation:

$y_c(x) = -\dfrac{P}{Fk}[\sin kx + (\tan kL)(1 - \cos kx) - kx]$, where $k = \sqrt{\dfrac{F}{EI_z}}$

$(y_c)_{max} = 23.15$ mm

Energy technique: $(y_c)_{max} = 23.15$ mm

CHAPTER 5

5-1 (a) $\sigma_1 = 10,000$ lb/in^2, $\sigma_2 = \sigma_3 = 0$, $\tau_{max} = 5000$ lb/in^2, $\sigma_{vM} = 10,000$ lb/in^2

(b) $\sigma_1 = 5000$ lb/in^2, $\sigma_2 = 0$, $\sigma_3 = -5000$ lb/in^2, $\tau_{max} = 5000$ lb/in^2, $\sigma_{vM} = 8660$ lb/in^2

(c) $\sigma_1 = 10,000$ lb/in^2, $\sigma_2 = 5000$ lb/in^2, $\sigma_3 = 0$, $\tau_{max} = 5000$ lb/in^2, $\sigma_{vM} = 8660$ lb/in^2

(d) $\sigma_1 = \sigma_2 = 5000$ lb/in^2, $\sigma_3 = 0$, $\tau_{max} = 2500$ lb/in^2, $\sigma_{vM} = 5000$ lb/in^2

(e) $\sigma_1 = 9000$ lb/in^2, $\sigma_2 = 0$, $\sigma_3 = -1000$ lb/in^2, $\tau_{max} = 5000$ lb/in^2, $\sigma_{vM} = 9540$ lb/in^2

(f) $\sigma_1 = 14,000$ lb/in^2, $\sigma_2 = 0$, $\sigma_3 = -6000$ lb/in^2, $\tau_{max} = 10,000$ lb/in^2, $\sigma_{vM} = 17,780$ lb/in^2

5-4 $\tau_{max} = 56.9$ MN/m^2, $\sigma_{vM} = 101$ MN/m^2

5-5 (a) $\delta = 0.000968$ in, $p = 9200$ lb/in^2

(b) Inner member: $(\sigma_r)_{max} = 0$, $(\sigma_r)_{min} = -9200$ lb/in^2, $(\sigma_\theta)_{max} = -23,900$ lb/in^2, $(\sigma_\theta)_{min} = -33,200$ lb/in^2

Outer member: $(\sigma_r)_{max} = 0$, $(\sigma_r)_{min} = -9200$ lb/in^2, $(\sigma_\theta)_{max} = 32,900$ lb/in^2, $(\sigma_\theta)_{min} = 23,700$ lb/in^2

(c) Inner member:

$(\sigma_r)_{max} = -5000$ lb/in², $(\sigma_r)_{min} = -10{,}500$ lb/in², $(\sigma_\theta)_{max} = -19{,}300$ lb/in², $(\sigma_\theta)_{min} = -24{,}800$ lb/in²

Outer member:

$(\sigma_r)_{max} = 0$, $(\sigma_r)_{min} = -10{,}500$ lb/in², $(\sigma_\theta)_{max} = 37{,}500$ lb/in², $(\sigma_\theta)_{min} = 27{,}000$ lb/in²

5-6 Maximum-shear theory: 42.5 kN, maximum-distortion theory: 44.4 kN

5-9 3500 r/min

5-10

Part	Maximum-shear-stress theory	Distortion theory
(a)	$n = 2$	$n = 2$
(b)	$n = 1.31$	$n = 1.31$
(c)	$n = 1$	$n = 1.15$
(d)	$n = 2$	$n = 2.25$

5-12 (a) 0.996, 0.969, 0.909, 1.01, (b) $n = 1.15$

5-14 7810 lb/in²

5-16 (a) 1.093 in, (b) 1.030 in

5-17 $w_L = 11.66 M_p/L^2$

5-19 (a) 4210 lb, (b) 6624 lb

5-21 60.3 kN

5-22 (a) $n = 4.05$, (b) $n = 1.29$

CHAPTER 6

6-1

$$\sigma' = \begin{bmatrix} 3 + \sqrt{3}/2 & -1 & -1 + \sqrt{3} \\ -1 & \frac{1}{4}(9 - \sqrt{3}/2) & \frac{1}{4}(\frac{5}{2} + \sqrt{3}) \\ -1 + \sqrt{3} & \frac{1}{4}(\frac{5}{2} + \sqrt{3}) & -\frac{1}{4}(5 + \frac{3}{2}\sqrt{3}) \end{bmatrix} \times 1000 \text{ lb/in}^2$$

6-3 (a) $\sigma = 492$ MN/m², $\tau = 259$ MN/m²

(b) $\sigma = 372$ MN/m², $\tau = 357$ MN/m²

(c) $\sigma = 529$ MN/m², $\tau = 81$ MN/m²

6-5 $\sigma_1 = 4317$ lb/in², $\sigma_2 = 2000$ lb/in², $\sigma_3 = -2317$ lb/in², $n_x(1) = 0.976$, $n_y(1) = 0.154$, $n_z(1) = -0.155$, $n_x(2) = 0$, $n_y(2) = 0.707$, $n_z(2) = 0.707$, $n_x(3) = 0.218$, $n_y(3) = -0.690$, $n_z(3) = 0.690$

6-8 (a) $\epsilon_x = 900\ \mu$in/in, (b) $\epsilon_{x'} = 2300\ \mu$in/in

6-18 Stress-function solution:

$$\sigma_r = \frac{4M}{bK}\left[\left(\frac{r_i r_o}{r}\right)^2 \ln\frac{r_o}{r_i} - r_o^2 \ln\frac{r_o}{r} - r_i^2 \ln\frac{r}{r_i}\right]$$

$$\sigma_\theta = \frac{4M}{bK}\left[r_o^2 - r_i^2 - \left(\frac{r_i r_o}{r}\right)^2 \ln\frac{r_o}{r_i} - r_o^2 \ln\frac{r_o}{r_i} - r_i^2 \ln\frac{r}{r_i}\right]$$

$$\tau_{r\theta} = 0 \qquad \text{where } K = (r_o^2 - r_i^2)^2 - 4\left(r_i r_o \ln\frac{r_o}{r_i}\right)^2$$

Solution via Eqs. (3-15) and (3-17):

$$\sigma_r = \frac{2M}{bC}\left(\frac{r_o}{r}\ln\frac{r}{r_i} + \frac{r_i}{r}\ln\frac{r_o}{r} - \ln\frac{r_o}{r_i}\right)$$

$$\sigma_\theta = \frac{2M}{bC}\left[\ln\frac{r_o}{r_i} - \frac{1}{r}(r_o - r_i)\right]$$

where $C = (r_o^2 - r_i^2)\ln\dfrac{r_o}{r_i} - 2(r_o - r_i)^2$

6-22 $u = \dfrac{P}{6EI_z}y[3x(2L-x) - 3(1+\nu)c^2 + (2+\nu)y^2]$

$v = -\dfrac{P}{6EI_z}[x^2(3L-x) + 3(1+\nu)c^2x + 3\nu y^2(L-x)]$

6-28 $(\tau_x)_{max} = 92.1$ MN/m^2 at $y = \pm12$ mm, $z = 0$, $\theta' = 3.57°/m$
6-30 $\tau_{max} \approx 20{,}170$ lb/in^2

CHAPTER 7

7-1 (a) 12.5 by 25 mm
   (b) $L = 6$, $F = 150$, $\sigma = P = \frac{25}{6}$, $E = 10$, $M = 900$, $\epsilon = \frac{5}{12}$, $\delta = \frac{5}{2}$
   (c) $\sigma_p = 5.83$ MN/m^2, $\delta_p = 0.125$ mm
7-2 (a) 3000 lb/in^2, (b) 7500 lb/in^2, (c) 8200 lb/in^2
7-5 1200 μin/in, 2.083×10^5 in/s
7-8 $\epsilon_1 = (\epsilon_A - \epsilon_B + \epsilon_C) + \sqrt{(\epsilon_B - \epsilon_C)^2 + \frac{1}{3}(3\epsilon_B - 2\epsilon_A - \epsilon_C)^2}$

$\epsilon_2 = (\epsilon_A - \epsilon_B + \epsilon_C) - \sqrt{(\epsilon_B - \epsilon_C)^2 + \frac{1}{3}(3\epsilon_B - 2\epsilon_A - \epsilon_C)^2}$

7-13 $\sigma_1 = -77$ lb/in^2, $\sigma_2 = -557$ lb/in^2

CHAPTER 8

8-3 (a) $\sigma_x = \dfrac{24\times10^4}{x+9}$ lb/in^2, $\epsilon_x = \dfrac{8\times10^{-3}}{x+9}$ in/in, $u = (8\times10^{-3})\ln\left(1+\dfrac{x}{9}\right)$ in

   (b) $(\sigma_x)_1 = 22{,}860$ lb/in^2, $(\sigma_x)_2 = 17{,}770$ lb/in^2, $u_1 = 0$, $u_2 = 2.29\times10^{-3}$ in, $u_3 = 4.06\times10^{-3}$ in
   (c) $(\sigma_x)_1 = 24{,}000$ lb/in^2, $(\sigma_x)_2 = 19{,}950$ lb/in^2, $(\sigma_x)_3 = 17{,}190$ lb/in^2
   $u_1 = 0$, $u_2 = 1.60\times10^{-3}$ in, $u_3 = 2.93\times10^{-3}$ in, $u_4 = 4.08\times10^{-3}$ in

8-4 $v_3 = -\dfrac{3}{16}\dfrac{PL^3}{EI_2}$, $\theta_3 = -\dfrac{5}{16}\dfrac{PL^2}{EI_2}$

8-8 $u(x) = \dfrac{W}{2AE}x\left(2 - \dfrac{x}{L}\right)$

   (a) One element: $u_{x=0} = 0$, $u_{x=L} = \dfrac{1}{2}\dfrac{WL}{AE}$

   (b) Two elements: $u_{x=0} = 0$, $u_{x=L/2} = \dfrac{3}{8}\dfrac{WL}{AE}$, $u_{x=L} = \dfrac{1}{2}\dfrac{WL}{AE}$

   (c) Three elements: $u_{x=0} = 0$, $u_{x=L/3} = \dfrac{5}{18}\dfrac{WL}{AE}$, $u_{x=2L/3} = \dfrac{4}{9}\dfrac{WL}{AE}$, $u_{x=L} = \dfrac{1}{2}\dfrac{WL}{AE}$

   (d) Four elements: $u_{x=0} = 0$, $u_{x=L/4} = \dfrac{7}{32}\dfrac{WL}{AE}$, $u_{x=L/2} = \dfrac{3}{8}\dfrac{WL}{AE}$, $u_{x=3L/4} = \dfrac{15}{32}\dfrac{WL}{AE}$, $u_{x=L} = \dfrac{1}{2}\dfrac{WL}{AE}$